Lecture Notes in Computer Science 7851

Commenced Publication in 1973
Founding and Former Series Editors:
Gerhard Goos, Juris Hartmanis, and Jan van Leeuwen

Michel Daydé   Osni Marques
Kengo Nakajima (Eds.)

High Performance Computing for Computational Science – VECPAR 2012

10th International Conference
Kobe, Japan, July 17-20, 2012
Revised Selected Papers

 Springer

Volume Editors

Michel Daydé
University of Toulouse
INPT (ENSEEIHT) - IRIT
31062 Toulouse Cedex 9, France
E-mail: michel.dayde@irit.fr

Osni Marques
Lawrence Berkeley National Laboratory
Berkeley, CA 94720-8139, USA
E-mail: oamarques@lbl.gov

Kengo Nakajima
The University of Tokyo
Information Technology Center
Tokyo 113-8658, Japan
E-mail: nakajima@cc.u-tokyo.ac.jp

ISSN 0302-9743   e-ISSN 1611-3349
ISBN 978-3-642-38717-3   e-ISBN 978-3-642-38718-0
DOI 10.1007/978-3-642-38718-0
Springer Heidelberg Dordrecht London New York

Library of Congress Control Number: 2013939789

CR Subject Classification (1998): F.2, C.4, B.2, B.8, G.1, D.3, D.4, C.2, J.2-3, I.3, I.4, I.6

LNCS Sublibrary: SL 1 – Theoretical Computer Science and General Issues

Typesetting: Camera-ready by author, data conversion by Scientific Publishing Services, Chennai, India

Printed on acid-free paper

Springer is part of Springer Science+Business Media (www.springer.com)

Preface

VECPAR is an international conference series dedicated to the promotion and advancement of all aspects of high-performance computing for computational science, as an academic discipline and a technique for real-world applications, extending the frontier of the state of the art and the state of practice. The audience and participants of VECPAR are researchers in academia, laboratories and industry. More information on the conference is available at http://vecpar.fe.up.pt.

The 10th edition of the conference, VECPAR 2012, was held in Kobe, Japan, during July 17–20, 2012. It was the fifth time the conference was held outside Porto (Portugal), where it started, succeeding Valencia (Spain) in 2004, Rio de Janeiro (Brazil) in 2006, Toulouse (France) in 2008, and Berkeley (USA) in 2010.

The conference program consisted of seven invited talks, 28 papers and 15 posters. The invited talks were presented by Horst Simon, *Barriers to Exascale Computing*; Richard Vuduc, *A Theory for Co-Designing Algorithms and Architectures Under Power and Die-Area Constraints*; Takashi Furumura, *Peta-scale FDM Simulation of Strong Ground Motion and Tsunami: Toward Disaster Prediction and Mitigation;* Ryutaro Himeno, *Grand Challenge in Life Science on the K Computer;* Yoshinobu Kuramashi, *Lattice QCD - From Quarks to Nuclei;* Kenji Ono, *HPC/PF — High-Performance Computing Platform: An Environment That Accelerates Large-Scale Simulations*; and Atsushi Oshiyama, *Materials Design Through Computing: Nanostructures of Silicon and Carbon*. The major themes of the conference (thus the accepted papers and posters) were:

- Large-scale simulations in CS&E
- Parallel and distributed computing
- Numerical algorithms for CS&E
- Multiscale and multiphysics problems
- Data-intensive computing
- Imaging and graphics
- Performance analysis

The most significant contributions of VECPAR 2012 are available in the present book, edited after the conference, and after a second review of all orally presented papers. The first round of reviews was based on an eight-page extended abstract. Each paper was reviewed by three reviewers; in some cases a fourth reviewer helped in the final decision. Out of 59 submissions, 30 were accepted for presentation. For the second round of reviews, authors were given a larger page budget, so they could better address reviewers' comments and suggestions.

In addition, two related events were organized on the day before the conference:

- The 7th International Workshop on Automatic Performance Tuning (iWAPT 2012), whose contributions are also available in the present book

- Tutorial on High-Performance Numerical Tools for the Development and Scalability of High-End Computer Applications.

VECPAR 2012 took place at the RIKEN Advanced Institute for Computational Science (AICS) and the Integrated Research Center of Kobe University. Participants had the opportunity to visit the K Computer, at the time the second system in the top500 list. Paper submissions were managed with the EasyChair conference system; the conference website and registration process were managed by the University of Tokyo.

The success of VECPAR and the long life of the series result from the work of many people. As on all previous occasions, a large number of collaborators were involved in the organization and promotion of the conference. Here, we mention just a few but through them we would like to express our gratitude to those who dedicated their time to the success of VECPAR 2012: Kimihiko Hirao, Akinori Yonezawa, Yoshio Oyanagi, Atsushi Hori, Nobuyuki Kaya, Sameer Shende, Tony Drummond, and the iWAPT organizers, in particular Rich Vuduc, Takeshi Iwashita, and Hiroyuki Takizawa.

We thank all authors that contributed to this book, for adhering to the deadlines and responding to the reviewers' comments and suggestions, and all members of the Scientific Committee, who greatly helped us with the paper selection process.

January 2013

Michel Daydé
Osni Marques
Kengo Nakajima

Organization

Organizing Committee

Kimihiko Hirao	RIKEN, Japan, Co-chair
Akinori Yonezawa	RIKEN, Japan, Co-chair
Atsushi Hori	RIKEN, Japan
Takeshi Iwashita	Kyoto University, Japan
Akira Kageyama	Kobe University, Japan
Takahiro Katagiri	University of Tokyo, Japan
Nobuyuki Kaya	Kobe University, Japan
Kengo Nakajima	University of Tokyo, Japan
Hiroshi Nakashima	Kyoto University, Japan
Yoshio Oyanagi	Kobe University, Japan
Tetsuya Sakurai	University of Tsukuba, Japan
Mitsuhisa Sato	University of Tsukuba/RIKEN, Japan
Hideyuki Usui	Kobe University, Japan
Yusaku Yamamoto	Kobe University, Japan

Steering Committee

Michel Daydé	University of Toulouse/INPT (ENSEEIHT) - IRIT, France - Chair
Álvaro Coutinho	COPPE/UFRJ, Brazil
Jack Dongarra	University of Tennessee, USA
Inês Dutra	University of Porto, Portugal
José Fortes	University of Florida, USA
Vicente Hernandez	Technical University of Valencia, Spain
Osni Marques	Lawrence Berkeley National Laboratory, USA
Ken Miura	National Institute of Informatics, Japan
Kengo Nakajima	University of Tokyo, Japan
José Palma	University of Porto, Portugal

Scientific Committee

Osni Marques	Lawrence Berkeley National Laboratory, USA - Chair
Xing Cai	Simula Research Laboratory, Norway - Vice-Chair
Edmond Chow	Georgia Institute of Technology, USA - Vice-Chair

Hiroshi Nakashima	Kyoto University, Japan
Satoshi Ohshima	University of Tokyo, Japan
Hiroshi Okuda	University of Tokyo, Japan
Kenji Ono	RIKEN, Japan
Serge Petiton	Université des Sciences et Technologies de Lille, France
Thierry Priol	INRIA Rennes, France
Heather Ruskin	Dublin City University, Ireland
Damian Rouson	Sandia National Laboratories, USA
Tetsuya Sakurai	University of Tsukuba, Japan
Mitsuhisa Sato	University of Tsukuba, Japan
Sameer Shende	University of Oregon, USA
A. Augusto Sousa	University of Porto, Portugal
Frederic Suter	IN2P3 ENS Lyon, France
Domenico Talia	University of Calabria, Italy
Osamu Tatebe	University of Tsukuba, Japan
Francisco Tirado	Complutense University of Madrid, Spain
Stan Tomov	University of Tennessee Knoxville, USA
Miroslav Tuma	Academy of Sciences of the Czech Republic, Czech Republic
Paulo Vasconcelos	University of Porto, Portugal
Xavier Vasseur	CERFACS, France
Richard Vuduc	Georgia Institute of Technology, USA
Roland Wismüller	Universität Siegen, Germany

Invited Speakers

Takashi Furumura	The University of Tokyo, Japan
Ryutaro Himeno	RIKEN, Japan
Yoshinobu Kuramashi	University of Tsukuba/RIKEN, Japan
Kenji Ono	RIKEN Advanced Institute for Computational Science AICS, Japan
Atsushi Oshiyama	The University of Tokyo, Japan
Horst Simon	Lawrence Berkeley National Laboratory, USA
Richard Vuduc	Georgia Institute of Technology, USA

Sponsoring Organizations

The Organizing Committee is very grateful to the following organizations for their kind support of VECPAR 2012:

Cray Japan, Inc.
Cybernet Systems Co., Ltd.
Fujitsu Limited
Hitachi, Ltd.
IBM Japan, Ltd.
NEC Corporation
SGI Japan Ltd.

Posters

Large-Scale Magnetostatic Domain Decomposition Analysis Using the Minimal Residual Method, Hiroshi Kanayama (Kyushu University), Masao Ogino (Nagoya University), Shin-Ichiro Sugimoto (The University of Tokyo) and Seigo Terada (Kyushu University)

Parallelized Adaptive Mesh Refinement PIC Scheme with Dynamic Domain Decomposition, Yohei Yagi (Kobe University), Masaharu Matsumoto (Kobe University/JST-CREST), Masanori Nunami (NIFS) and Hideyuki Usui (Kobe Univerisity/JST-CREST)

Development of a Scalable PIC Simulator for Spacecraft–Plasma Interaction Problems, Yohei Miyake (Kobe University), Hiroshi Nakashima (Kyoto University) and Hideyuki Usui (Kobe University)

An Architecture Concept for the Scalable Simulation of Dendritic Growth, Andreas Schafer (University Erlangen-Nuremberg) and Dietmar Fey (University Erlangen-Nuremberg)

Parallel Numerical Simulation of Navier-Stokes and Transport Equations on GPUs, Wesley Menenguci (Universidade Federal do Espirito Santo), Lucia Catabriga (Universidade Federal do Espirito Santo), Alberto De Souza (Universidade Federal do Espirito Santo) and Andrea Valli (Universidade Federal do Espirito Santo)

An Implementation of Development Support Middleware for Finite Element Method Applications, Takeshi Kitayama (The University of Tokyo), Takeshi Takeda (The University of Tokyo) and Hiroshi Okuda (The University of Tokyo)

MGCUDA: An Easy Programming Model for CUDA-Based Multiple GPUs Platform, Cheng Luo (The University of Tokyo) and Reiji Suda (The University of Tokyo)

Construction of Approximated Invariant Subspace for a Real Symmetric Definite Generalized Eigenproblem Using a Linear Combination of Resolvents as the Filter, Hiroshi Murakami (Tokyo Metropolitan University)

OpenMP/MPI Implementation of Tile QR Factorization Algorithm on Multi-Core Cluster, Tomohiro Suzuki (University of Yamanashi), Hideki Miyashita (Software Laboratory Inc.) and Hidetomo Nabeshima (University of Yamanashi)

Parallel Block Gram-Schmidt Orthogonalization with Optimal Block-size, Yoichi Matsuo (Keio University) and Takashi Nodera (Keio University)

Implementation of ppOpen-AT into OpenFOAM, Satoshi Ito (The University of Tokyo/JST-CREST), Satoshi Ohshima (The University of Tokyo) and Takahiro Katagiri (The University of Tokyo)

File Composition Technique for Improving Access Performance of a Number of Small Files, Yoshiyuki Ohno (RIKEN AICS), Atsushi Hori (RIKEN AICS) and Yutaka Ishikawa (The University of Tokyo/RIKEN AICS)

Message from the Chairs of iWAPT 2012

The International Workshop on Automatic Performance Tuning (iWAPT) brings together researchers studying how to automatically adapt algorithms and software for high performance on modern machines. The workshop convened for the seventh consecutive year on July 17, 2012, at the RIKEN Advanced Institute for Computational Science in Kobe, Japan.

If one were to identify a theme for this year's program, it might arguably be *model-driven domain-specific optimization.* The two invited keynote speakers — Paolo Bientinesi (RWTH Aachen), who spoke on "A Domain-Specific Compiler for Linear Algebra Operations," and Jakub Kurzak (University of Tennessee, Knoxville), who spoke on "Autotuning BLAS for GPUs" — showcased the state of the art on this theme.

The remaining presentations reinforced various aspects of this theme, including new algorithmic techniques, new programming model support and methods tailored to massively parallel GPU architectures. There were ten such technical presentations, chosen after a review of 18 submitted manuscripts (of 15 pages each). This number of submissions was the highest since the first meeting in 2006. The manuscripts of the final ten papers appear in this volume.

Many people and organizations helped to make this workshop a success. We are grateful to the VECPAR Organizing Committee, especially Osni Marques and Kengo Nakajima, for their logistical and intellectual support; the iWAPT Steering Committee, especially Reiji Suda and Takahiro Katagiri, for their guidance; and the Program Committee for volunteering their time to help assemble an outstanding program. Furthermore, the workshop would not be possible without the generous financial support of the Japan Science and Technology Agency, whose contributions have made Japan a leading international player in autotuning research. Lastly, we wish of course to thank the invited speakers, authors, and meeting participants for their insights and thoughtful debate throughout the workshop.

January 2013

Hiroyuki Takizawa
Richard Vuduc
Takeshi Iwashita

Organization

iWAPT 2012 Organizing Committees

General Chair

Hiroyuki Takizawa   Tohoku University, Japan

Finance Chair

Toshiyuki Imamura   The University of Electro-communication,
                           Japan

Steering Committee Liaison

Reiji Suda   The University of Tokyo, Japan

Web Chair

Hisayasu Kuroda   Ehime University, Japan

Publicity Chair

Akihiro Fujii   Kogakuin University, Japan

Local Arrangements Chair

Takaharu Yaguchi   Kobe University, Japan

Steering Committee

Jonathan T. Carter	Lawrence Berkeley National Laboratory, USA
John Cavazos	University of Delaware, USA
Victor Eijkhout	Texas Advanced Computing Center, University of Texas, USA
Toshiyuki Imamura	The University of Electro-Communication, Japan
Domingo J. Canovas	University of Murcia, Spain
Takahiro Katagiri	The University of Tokyo, Japan
Ken Naono	Hitachi Ltd., Japan
Osni Marques	Lawrence Berkeley National Laboratory, USA

Markus Pueschel	ETH Zurich, Switzerland
Reiji Suda	The University of Tokyo, Japan
Richard Vuduc	Georgia Institute of Technology, USA
R. Clint Whaley	The University of Texas at San Antonio, USA
Yusaku Yamamoto	Kobe University, Japan

Program Committee

Richard Vuduc	*Program Chair*, Georgia Institute of Technology, USA
Takeshi Iwashita	*Program Vice-Chair*, Kyoto University, Japan
Toshiyuki Imamura	The University of Electro-communication, Japan
Akira Naruse	Fujitsu Laboratories Ltd., Japan
Franz Franchetti	Carnegie Mellon University, USA
Takahiro Katagiri	The University of Tokyo, Japan
Osni Marques	Lawrence Berkeley National Laboratory, USA
David Padua	University of Illinois at Urbana Champaign, USA
Markus Pueschel	ETH Zurich, Switzerland
Takao Sakurai	Hitachi Ltd., Japan
Hiroyuki Takizawa	Tohoku University, Japan
Keita Teranishi	Cray Inc., USA
Yusaku Yamamoto	Kobe University, Japan
Masahiro Yasugi	Kyushu Institute of Technology, Japan

Table of Contents

Applications

Finite Element Method from Various Viewpoints

Cloud and Visualization

Parallel Iterative Solvers on Multicore Architectures

The Seventh International Workshop on Automatic Performance Tuning

Barriers to Exascale Computing

Horst D. Simon

Lawrence Berkeley National Laboratory, Mail Stop 50A-4133,
Berkeley, CA 94720, USA
hdsimon@lbl.gov

Abstract. The development of an exascale computing capability with machines capable of executing $O(10^{18})$ operations per second by the end of the decade will be characterized by significant and dramatic changes in computing hardware architecture from current (2012) petascale high-performance computers. From the perspective of computational science, this will be at least as disruptive as the transition from vector supercomputing to parallel supercomputing that occurred in the 1990s. This was one of the findings of a 2010 workshop on crosscutting technologies for exascale computing. The impact of these architectural changes on future applications development for the computational sciences community can now be anticipated in very general terms. While the community has been investigating the road to exascale worldwide in the last several years, there are still several barriers that need to be overcome to obtain general purpose exascale performance. This short paper will summarize the major challenges to exascale, and how much progress has been made in the last five years.

Keywords: exascale computing, energy efficient computing, resilience, massive parallelism, heterogeneous computing, technology trends, TOP500.

1   Introduction

It may come as surprise to many who are currently deeply engaged in research and development activities that could lead us to exascale computing, that it has been already exactly five years, since the first set of community town hall meetings were convened in the U.S. to discuss the challenges for the next level of computing in science. It was in April and May 2007, when three meetings were held in Berkeley, Argonne and Oak Ridge that formed the basis for the first comprehensive look at exascale [1].

What is even more surprising is that in spite of numerous national and international initiatives that have been created in the last five years, the community has not made any significant progress towards reaching the goal of an Exaflops system. If one reflects and looks back at early projections, for example in 2010, it seemed to be possible to build at least a prototype of an exascale computer by 2020. This view was expressed in documents such as [2], [3]. I believe that the lack of progress in the intervening years has made it all but impossible to see a working exaflops system by

M. Daydé, O. Marques, and K. Nakajima (Eds.): VECPAR 2012, LNCS 7851, pp. 1–3, 2013.

2020. Specifically, I do not expect a working Exaflops system to appear on the #1 spot of the TOP500 list with a RMAX performance exceeding 1 Exaflop/s by November 2019. More recent revisions of the earlier plans have taken this view, and the most recent DOE Exascale Strategy in the U.S. expects delivery of the First Prototype Exascale Cabinet in 2020, and a prototype system to be available in 2021.

There are four major technology challenges that need to be addressed in order to build an exascale system. Today these challenges are effective barriers to reaching one exaflop/s level performance by 2020.

Energy Challenge: reduce the power consumption of all elements of the system so that the operational cost is within reasonable power budgets.

Parallelism Challenge: develop a programming model and system software that allows a software developer to use effectively unprecedented parallelism, while also managing data locality and energy-efficiency.

Resilience Challenge: achieve resilience to faults so that they have no impact on development and operations of a system.

Memory and Storage Challenge: develop energy efficient technologies and architectures that can provide 100s of Petabytes memory and high storage capacity, low power requirements, and ability to move large amounts of data.

These challenges were well indentified already in the early workshops and elaborated in great detail in late 2009. What accounts for the lack of progress? Initial estimates for carrying out a successful research program that would engage an incentivize systems and software vendors, and technology companies was estimated to cost about $300 -$400M per year over the next decade in addition to the already existing investments in high performance computing. This price tag of about $3-4B additional funding for an Exascale initiative is not affordable under the current budget realities in the U.S.

The original thinking envisioned a model of close collaboration between industry, labs, and academia that would create a tight feedback loop and funding for industry, so that the unique technology challenges for exascale would be addressed by industry. This model has fallen by the wayside. While the U.S. can fortunately count on several strong technology partners in industry that are willing to be engaged in Exaflops computing, the model has however fundamentally changed. We are "back to the future", in the sense that the next five to ten years will see a replay of the successful "leverage COTS" model that ASCI pioneered in the mid 1990s. In short, there will be no *specific* exascale technology development in industry. Instead just like the Path Forward program of 15 years ago, there will a few targeted investments that will make commercially developed technology more useful in the HPC context. The recent acquisition of the Cray interconnect network technology by Intel can be interpreted in this context.

In my lecture I will discuss how the above challenges are actual barriers that probably cannot be overcome by the "back to the future COTS model" that the US community is deploying. We will get to Exaflops eventually, but not by 2020.

References

1. Simon, H., Zacharia, T., Stevens, R.: Modeling and Simulation at the Exascale for Energy and Environment, Berkeley, Oak Ridge, Argonne (2007),
   http://science.energy.gov/ascr/
   news-and-resources/program-documents/
2. Stevens, R., White, A.: Crosscutting Technologies for Computing at Exaflops, San Diego (2009),
   http://science.energy.gov/ascr/news-and-resources/
   workshops-and-conferences/grand-challenges/
3. Shalf, J., Dosanjh, S., Morrison, J.: Exascale Computing Technology Challenges. In: Palma, J.M.L.M., Daydé, M., Marques, O., Lopes, J.C. (eds.) VECPAR 2010. LNCS, vol. 6449, pp. 1–25. Springer, Heidelberg (2011)

Toward a Theory of Algorithm-Architecture Co-design

Richard Vuduc and Kenneth Czechowski

Georgia Institute of Technology, Atlanta, GA 30332-0765, USA

We are carrying out a research program that asks whether there is a useful mathematical framework for reasoning at a high-level about the behavior of an algorithm on a supercomputer with respect to the physical constraints of energy, power, and die area. By "high-level," we mean that we wish to explicitly relate characteristics of an algorithm, such as its inherent parallelism or memory and communication behavior, with parameters of an architecture, such as the number of cores, structure of the memory hierarchy, or network topology. Our ultimate goal is to say, in broad but also quantitative terms, how macroscopic changes to an architecture might affect the execution time, scalability, accuracy, and power-efficiency of a computation; and, conversely, identify what classes of computation might best match a given architecture. The approach we shall outline marries abstract algorithmic complexity analysis with caps on power and die area, which are arguably the central first-order constraints on the extreme-scale systems of 2018 and beyond [1, 16, 21, 29, 41]. We refer to our approach as one of *algorithm-architecture co-design*.

We emphasize the term, "algorithm-architecture," rather than "hardware-software" or other equivalent expression. The former evokes a high-level mathematical process that precedes and complements traditional methods based on cycle-accurate architecture simulation of code artifacts and detailed traces [3, 8, 9, 15, 19, 23, 25, 37, 39, 42]. Our approach draws from prior work on high-level analytical performance analysis and modeling [4, 5, 20, 20, 22–24, 26–28, 32, 34–36, 47, 49], as well as more classical work on models for circuits and very large-scale integration [30, 31, 38, 40, 43]. However, our methods return to higher-level I/O-centric complexity analysis [2, 6, 7, 11, 12, 17, 18, 44–46, 48], pushing it further by trying to resolve constants [10, 13], which is necessary to connect algorithmic analysis with the hard physical constraints imposed by power and die area. Our aim is not to achieve the level of cycle-accuracy possible through detailed simulation. Instead, our belief is that freedom from the artifacts of current hardware and software implementations, while nevertheless incorporating costs that reflect the reality of physical machines, may lead to new insights and research directions for achieving the next level in performance and scalability.

Abstractly, the formal co-design problem might look as follows. Let a be an algorithm from a set A of algorithms that all perform the same computation within the same desired level of accuracy. The set A might contain different algorithms, such as "A = {fast Fourier transform, F-cycle multigrid}," for the Poisson equation model problem [14, 33]. Or, A may be a set of tuning parameters for one algorithm, such as the set of all possible tile sizes for one-level tiled

M. Daydé, O. Marques, and K. Nakajima (Eds.): VECPAR 2012, LNCS 7851, pp. 4–8, 2013.

matrix multiply. Next, let μ be a machine architecture from a set M, and suppose that each processor of μ has a die area of $\chi(\mu)$. Lastly, suppose $T(n; a, \mu)$ is the time to execute a on μ for a problem of size n, while using a maximum instantaneous power of $\Phi(\mu)$. Then, our goal is to determine the algorithm a and architecture μ that minimize time subject to constraints on total power and die area. That is, in principle we wish to solve the mathematical optimization problem,

$$(a^*, \mu^*) = \operatorname*{argmin}_{(a \in A,\ \mu \in M)} T(n; a, \mu) \tag{1}$$

$$\text{subject to:} \quad \Phi(\mu) = \Phi_{\max} \tag{2}$$

$$\chi(\mu) = \chi_{\max} \tag{3}$$

where $\Phi_{\max}$ and $\chi_{\max}$ are caps on total system power and die area per chip, respectively.

Such an analysis framework explicitly binds characteristics of algorithms and architectures, Equation (1), with physical hardware constraints, Equations (2)–(3). The fundamental research problem is to find the right forms of $T(n; a, \mu)$, $\Phi(\mu)$, and $\chi(\mu)$, and then see what algorithms and architectures emerge as solutions to the optimization problem. In our talk, we shall outline how one might instantiate such a framework and shows the kinds of insights that emerge.

References

[1] The Potential Impact of High-End Capability Computing on Four Illustrative Fields of Science and Engineering. The National Academies Press, Washington, DC (2008)

[2] Arge, L., Goodrich, M.T., Nelson, M., Sitchinava, N.: Fundamental parallel algorithms for private-cache chip multiprocessors. In: Proceedings of the Twentieth Annual Symposium on Parallelism in Algorithms and Architectures, SPAA 2008, p. 197. ACM Press, New York (2008)

[3] Badia, R.M., Rodriguez, G., Labarta, J.: Deriving analytical models from a limited number of runs. In: Proceedings of Parallel Computing, ParCo, Minisymposium on Performance Analysis, pp. 1–6 (2003)

[4] Barker, K., Benner, A., Hoare, R., Hoisie, A., Jones, A., Kerbyson, D., Li, D., Melhem, R., Rajamony, R., Schenfeld, E., Shao, S., Stunkel, C., Walker, P.: On the Feasibility of Optical Circuit Switching for High Performance Computing Systems. In: ACM/IEEE SC 2005 Conference, SC 2005. IEEE (2005),
http://ieeexplore.ieee.org/xpls/abs_all.jsp?arnumber=1559968&tag=1

[5] Barker, K.J., Hoisie, A., Kerbyson, D.J.: An early performance analysis of POWER7-IH HPC systems. In: Proceedings of 2011 International Conference for High Performance Computing, Networking, Storage and Analysis on SC 2011, p. 1. ACM Press, New York (2011)

[6] Blelloch, G.E.: Programming parallel algorithms. Communications of the ACM 39(3), 85–97 (1996)

[7] Blelloch, G.E., Gibbons, P.B., Simhadri, H.V.: Low depth cache-oblivious algorithms. In: Proc. ACM Symp. Parallel Algorithms and Architectures, SPAA, Santorini, Greece (June 2010)

[8] Carrington, L., Snavely, A., Wolter, N.: A performance prediction framework for scientific applications. Future Generation Computer Systems 22(3), 336–346 (2006)

[9] Casas, M., Badia, R.M., Labarta, J.: Prediction of behavior of MPI applications. In: 2008 IEEE International Conference on Cluster Computing, pp. 242–251. IEEE (September 2008)

[10] Chandramowlishwaran, A., Choi, J.W., Madduri, K., Vuduc, R.: Towards a communication optimal fast multipole method and its implications for exascale. In: Proc. ACM Symp. Parallel Algorithms and Architectures, pp. 182–184. ACM, New York (2012),
http://dl.acm.org/citation.cfm?id=2312039

[11] Chowdhury, R.A., Silvestri, F., Blakeley, B., Ramachandran, V.: Oblivious algorithms for multicores and network of processors. In: 2010 IEEE International Symposium on Parallel & Distributed Processing, IPDPS, pp. 1–12. IEEE (2010)

[12] Culler, D., Karp, R., Patterson, D., Sahay, A., Schauser, K.E., Santos, E., Subramonian, R., von Eicken, T.: LogP: Towards a realistic model of parallel computation. ACM SIGPLAN Notices 28(7), 1–12 (1993)

[13] Czechowski, K., McClanahan, C., Battaglino, C., Iyer, K., Yeung, P.-K., Vuduc, R.: On the communication complexity of 3D FFTs and its implications for exascale. In: Proc. ACM Int'l. Conf. Supercomputing, ICS, San Servolo Island, Venice, Italy (June 2012) (to appear)

[14] Demmel, J.W.: Applied Numerical Linear Algebra. SIAM (1997)

[15] Desprez, F., Markomanolis, G.S., Quinson, M., Suter, F.: Assessing the Performance of MPI Applications through Time-Independent Trace Replay. In: 2011 40th International Conference on Parallel Processing Workshops, pp. 467–476. IEEE (September 2011)

[16] Dongarra, J., Beckman, P., Aerts, P., Cappello, F., Lippert, T., Matsuoka, S., Messina, P., Moore, T., Stevens, R., Trefethen, A., Valero, M.: The International Exascale Software Project: A call to cooperative action by the global high performance community. In: Int'l. J. High-Performance Computing Applications, IJHPCA, vol. 23(4), pp. 309–322 (2009),
http://hpc.sagepub.com/content/23/4/309

[17] Frigo, M., Leiserson, C.E., Prokop, H., Ramachandran, S.: Cache-oblivious algorithms. In: Proc. Symp. Foundations of Computer Science, FOCS, New York, NY, USA, pp. 285–297 (October 1999)

[18] Ghosh, S., Martonosi, M., Malik, S.: Cache miss equations: A compiler framework for analyzing and tuning memory behavior. ACM Trans. Programming Languages and Systems (TOPLAS) 21(4), 703–746 (1999)

[19] Gonzalez, J., Gimenez, J., Casas, M., Moreto, M., Ramirez, A., Labarta, J., Valero, M.: Simulating Whole Supercomputer Applications. IEEE Micro 31(3), 32–45 (2011)

[20] Guz, Z., Bolotin, E., Keidar, I., Kolodny, A., Mendelson, A., Weiser, U.: Many-Core vs. Many-Thread Machines: Stay Away From the Valley. IEEE Computer Architecture Letters 8(1), 25–28 (2009)

[21] Hemmert, K.S., Vetter, J.S., Bergman, K., Das, C., Emami, A., Janssen, C., Panda, D.K., Stunkel, C., Underwood, K., Yalamanchili, S.: IAA Interconnection Networks Workshop 2008. Technical Report FTGTR-2009-03, Future Technologies Group, Oak Ridge National Laboratory (April 2009),
http://ft.ornl.gov/pubs-archive/iaa-ic-2008-workshop-report-final.pdf

[22] Hill, M.D., Marty, M.R.: Amdahl's Law in the multicore era. IEEE Computer 41(7), 33–38 (2008)

[23] Hoefler, T., Schneider, T., Lumsdaine, A.: LogGOPSim: Simulating large-scale applications in the LoGOPS model. In: Proceedings of the 19th ACM International Symposium on High Performance Distributed Computing, HPDC 2010, p. 597. ACM Press, New York (2010)

[24] Hoisie, A., Johnson, G., Kerbyson, D.J., Lang, M., Pakin, S.: A performance comparison through benchmarking and modeling of three leading supercomputers: Blue Gene/L, Red Storm, and Purple. In: Proc. ACM/IEEE Conf. Supercomputing, SC, number 74, Tampa, FL, USA (November 2006)

[25] Jagode, H., Knupfer, A., Dongarra, J., Jurenz, M., Mueller, M.S., Nagel, W.E.: Trace-based performance analysis for the petascale simulation code FLASH. International Journal of High Performance Computing Applications (December 2010)

[26] Kerbyson, D.J., Alme, H.J., Hoisie, A., Petrini, F., Wasserman, H.J., Gittings, M.: Predictive performance and scalability modeling of a large-scale application. In: Proceedings of the 2001 ACM/IEEE Conference on Supercomputing (CDROM) - Supercomputing 2001, p. 37. ACM Press, New York (2001)

[27] Kerbyson, D.J., Hoisie, A., Wasserman, H.: Modelling the performance of large-scale systems. In: IEE Proceedings–Software, vol. 150, pp. 214–221 (August 2003)

[28] Kerbyson, D.J., Jones, P.W.: A Performance Model of the Parallel Ocean Program. International Journal of High Performance Computing Applications 19(3), 261–276 (2005)

[29] Kogge, P., Bergman, K., Borkar, S., Campbell, D., Carlson, W., Dally, W., Denneau, M., Franzon, P., Harrod, W., Hill, K., Hiller, J., Karp, S., Keckler, S., Klein, D., Lucas, R., Richards, M., Scarpelli, A., Scott, S., Snavely, A., Sterling, T., Williams, R.S., Yelick, K.: Exascale Computing Study: Technology challenges in acheiving exascale systems (September 2008),
http://users.ece.gatech.edu/~mrichard/
ExascaleComputingStudyReports/ECS_reports.htm

[30] Kung, H.: Let's design algorithms for VLSI systems. In: Proceedings of the Caltech Conference on VLSI: Architecture, Design, and Fabrication, pp. 65–90 (1979)

[31] Lengauer, T.: VLSI theory. In: Handbook of Theoretical Computer Science, ch. 16, pp. 837–865. Elsevier Science Publishers G.V. (1990)

[32] Lively, C.W., Taylor, V.E., Alam, S.R., Vetter, J.S.: A methodology for developing high fidelity communication models for large-scale applications targeted on multi-core systems. In: Proc. Int'l. Symp. Computer Architecture and High Performance Computing, SBAC-PAD, Mato Grosso do Sul, Brazil, pp. 55–62 (October 2008)

[33] Mandel, J., Parter, S.V.: On the multigrid F-cycle. Applied Mathematics and Computation 37(1), 19–36 (1990)

[34] Numrich, R.W.: Computational force: A unifying concept for scalability analysis. In: Advances in Parallel Computing, vol. 15. IOS Press (2008)

[35] Numrich, R.W.: A metric space for computer programs and the Principle of Computational Least Action. J. Supercomputing 43(3), 281–298 (2008)

[36] Numrich, R.W., Heroux, M.A.: Self-similarity of parallel machines. Parallel Computing 37(2), 69–84 (2011)

[37] Rodrigues, A.F., et al.: The structural simulation toolkit. ACM SIGMETRICS Performance Evaluation Review 38(4), 37 (2011)

[38] Rosenberg, A.L.: Three-Dimensional VLSI: a case study. Journal of the ACM 30(3), 397–416 (1983)

[39] Rosenfeld, P., Cooper-Balis, E., Jacob, B.: DRAMSim2: A Cycle Accurate Memory System Simulator. IEEE Computer Architecture Letters 10(1), 16–19 (2011)

[40] Savage, J.E.: Models of Computation: Exploring the power of computing. CC-3.0, BY-NC-ND, electronic edition (2008)

[41] Simon, H., Zacharia, T., Stevens, R.: Modeling and simulation at the exascale for energy and the environment. Technical report, Office of Science, U.S. Dept. of Energy (May 2008),
http://www.sc.doe.gov/ascr/ProgramDocuments/Docs/TownHall.pdf

[42] Snavely, A., Wolter, N., Carrington, L.: Modeling application performance by convolving machine signatures with application profiles. In: Proceedings of the Fourth Annual IEEE International Workshop on Workload Characterization, WWC-4 (Cat. No.01EX538), pp. 149–156. IEEE

[43] Thompson, C.D.: Area-time complexity for VLSI. In: Proceedings of the Eleventh Annual ACM Symposium on Theory of Computing, STOC 1979, pp. 81–88. ACM Press, New York (1979)

[44] Toledo, S.: Locality of reference in LU decomposition with partial pivoting. SIAM J. Matrix Anal. Appl. 18(4), 1065–1081 (1997)

[45] Valiant, L.G.: A bridging model for parallel computation. Communications of the ACM 33(8), 103–111 (1990)

[46] Valiant, L.G.: A bridging model for multi-core computing. In: Halperin, D., Mehlhorn, K. (eds.) ESA 2008. LNCS, vol. 5193, pp. 13–28. Springer, Heidelberg (2008)

[47] van Gemund, A.J.: Symbolic performance modeling of parallel systems. IEEE Transactions on Parallel and Distributed Systems 54(7), 922–927 (2005)

[48] Wickremesinghe, R., Arge, L., Chase, J.S., Vitter, J.S.: Efficient sorting using registers and caches. J. Experimental Algorithmics (JEA) 7, 9 (2002)

[49] Woo, D.H., Lee, H.-H.S.: Extending Amdahl's Law for energy-efficient computing in the many-core era. IEEE Computer 41(12), 24–31 (2008)

Visualization of Strong Ground Motion from the 2011 Off Tohoku, Japan (Mw=9.0) Earthquake Obtained from Dense Nation-Wide Seismic Network and Large-Scale Parallel FDM Simulation

Takashi Furumura[1,2]

[1] Center for Integrated Disaster Information Research,
Interfaculty Initiative in Information Studies, The University of Tokyo
[2] The Earthquake Research Institute, The University of Tokyo
1-1-1 Yayoi, Bunkyo-ku, Tokyo 113-0032, Japan
furumura@eri.u-tokyo.ac.jp

Abstract. Strong ground motion from the 2011 off Tohoku, Japan (Mw=9.0) earthquake is demonstrated by high-resolution 3D parallel FDM simulation of seismic wave propagation using the Earth Simulator supercomputer. Complicated wavefield accompanying the earthquake in connection with the radiation of seismic wave from complex source rupture process and strong amplification of ground motion in complicated subsurface structure beneath populated cities are demonstrated by the comparison of visualized seismic wavefield derived by the computer simulation and observation from dense nation-wide seismic network. Good correspondence between simulation results and observed actual seismic wavefield promising us the effectiveness of the present simulation model for ground motion simulation which is applicable not only for reproducing strong ground motions for the past events but also for mitigating earthquake disasters associated with future earthquakes.

Keywords: FDM simulation, 2011 off Tohoku, Japan, earthquake, Earth Simulator, parallel computing.

1 Introduction

On March 11, 2011, a destructive, Mw 9.0 earthquake occurred off the coast of Japan in the Pacific Ocean causing extreme disasters in northeastern Japan with estimated toll of dead and missing persons more than 18,000 due to high tsunami waves and strong ground motions.

This earthquake starts from off Miyagi, where large earthquake of Mw=7.5-8.0 had been repeatedly occurred with a recurrent period of about 40 years. Therefore, it was anticipated that the next earthquake should occur within 30 years with a probability of 99 %. However, the occurred earthquake was a much larger, mega-thrust event where fault rupture spreads entirely over the area of 500 km by 200 km covering off Miyagi, off Fukushima, and off Ibaraki earthquake nucleation zones.

M. Daydé, O. Marques, and K. Nakajima (Eds.): VECPAR 2012, LNCS 7851, pp. 9–16, 2013.

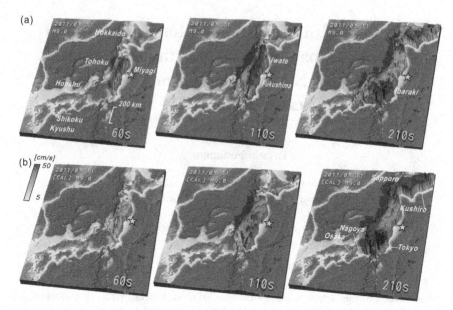

Fig. 1. Snapshots of seismic wave propagation following the 2011 off Tohoku, Japan (Mw=9.0) earthquake at time 60, 110 and 210 s after the earthquake starts, derived by (a) observation from dense K-NET and KiK-net strong motion network and (b) results of FDM simulation of seismic wave propagation. Star indicates the hypocenter of this earthquake.

Strong ground motions from such destructive earthquake were well recorded by the nation-wide K-NET and KiK-net seismic network in Japan. The strong motion network consists of over 1,800 seismic stations with an almost uniform station interval of about 20-25 km across Japan, and the observed waveform data is opened by the NIED web page [1] for public immediately after the earthquake occurs. We could explore the source rupture process and wave propagation properties of this event by making full use of dense observational data to study the cause of strong ground motions disasters during the earthquake.

Also, we will conducted a computer simulation of seismic wave propagation for this earthquake by using a detail source-rupture model over the plate boundary and a high-resolution subsurface structural model in order to understand the seismic wave propagation process in detail with compliment the observations. For the large scale simulation of seismic wave propagation we will employ the Earth Simulator supercomputer (ES) with a suitable parallel finite-difference method (FDM) code for solving equation of motions in 3D heterogeneous structure.

In this study, we will compare visualized seismic wavefield derived by dense seismic observation and high-resolution computer simulation.

2   Visualization of Wave Propagation by Dense Strong Motion Network

The visualized seismic wavefield derived by the 1,189 K-NET and KiK-net strong motion stations during the 2011 off Tohoku earthquake is illustrated in Figure 1 following the visualization procedure of the seismic waves [2][3]. The observed ground motion at each station of three-component ground accelerations are first integrated to construct a velocity record after applying a band-pass filter with a pass-band frequency of 0.05 - 10 Hz to remove instrument noise. The ground motion as gridded data is then obtained by interpolation of the ground velocity record using a conventional gridding algorithm. Since the intensity of ground motion and strong motion damages manifests on a logarithmic scale of horizontal ground velocity motion, a scalar value of the strength of the ground motion is calculated from the root mean square of the observed two horizontal-component velocity motions and the record of vertical ground motion is not used in this study.

The resulting scalar represents the energy of the seismic wave at each point of the regular mesh is then used to render the wavefront of the seismic wave using the "height_field" function of the POV-Ray rendering software [4]. For visualizing seismic wavefield more naturally we adopt a simple color which assign an intense (red) color for stronger wavefield and gentle (yellow) color for weaker wavefield rather than commonly using garish color tables such as red-blue and rainbow etc. In order to highlight the wavefront of intense ground motions and eliminate weak scattering wavefield, a proper opacity function proportional to the logarithmic amplitude of the ground motion is applied for the rendering of the seismic wavefield. The figure is protracted onto a surface topography image or a satellite photograph for realistic representation of seismic waves propagating on the Earth surface.

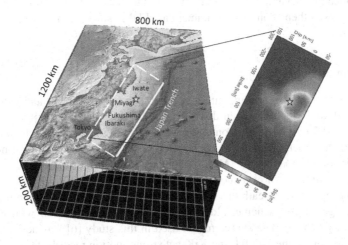

Fig. 2. Structural model of northern Japan used in the 3D simulation of seismic wave propagation, showing the configuration of crust and upper-mantle structure and subducting Pacific Plate. The source-slip model [6] for the 2011 off Tohoku earthquake used in this study is shown in right.

The resulting combined image offers direct means to study propagation of the seismic waves in heterogeneous structure and development of strong ground motions due to the amplification of ground motions in populated areas with soft and thick sediments beneath. In the first frame of the snapshot at 60 s after source initiation, we see that large ground motions are built up from the radiation produced by a bilaterally rupturing fault from a hypocenter at off Miyagi (marked by star) from north and to south, illustrating the extent of a rectangular rupture area with raised ground motions. In the second (110 s) frame of the snapshot, a second large shock, almost as large as the first, spreads again over northern Japan, producing intense and long-term shaking of ground motions over northern Japan. As the strong ground motions propagate to Ibaraki, about 200 km southwest of the hypocenter, a third shock illuminates the surface area around Ibaraki. Then, the overlap of these strong ground motions extends the large, prolonged shaking area from Ibaraki to Tokyo (110 and 210 s). We see amplified and prolonged ground shaking in populated cities, such as Tokyo, Nagoya, and Osaka due to the resonance of long-period ground motions within large and thick sedimentary basins. Large ground motions in the basin continued for several minutes.

3   Parallel FDM Simulation of Seismic Wave Propagation from the 2011 Off Tohoku Earthquake

To compliment the observation and to seek further insights into the understanding of the complicated seismic wavefield during the destructive 2011 off Tohoku earthquake, we then conduct a numerical FDM simulation of seismic wave propagation.

3.1   Simulation Model

The simulation model represents an area of 800 km by 1200 km and extends to a depth of 200 km, which has been discretized into grid point with uniform resolution of 0.5 km by 0.5 km in the horizontal direction and 0.25 km in the vertical direction. The subsurface structural model is constructed based on J-SHIS sediment model [5], lateral variation of crust/mantle (Moho) boundary and depth of the subducting Paficic Plate. We assumed a lower-most shear-wave velocity of Vs=1.0 km/s in the sedimentary layer just beneath the surface, and thus, the FDM simulation using a 16-th-order staggered-grid scheme allow wave propagation simulation with maximum frequency of f=0.5 Hz. The source model used in this study [6] was derived from an joint inversion using the K-NET and KiK-net strong motion records, teleseismic body waveforms, geodetic (GPS) data, and tsunami waveform data. The inferred subsurface structural model of surface topography and subducting Pacific-Plate and source-slip distributions over the subducting Plate are shown in Figure 2. The source model represents a very large (>50 m) slip over shallow (<10 km) part of the subducting plate near the Japan trench. In this simulation the source rupture over the fault source plane is represented by a number of point sources arranged over the subducting Pacific Plate at interval of 0.5 km.

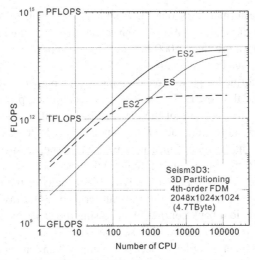

Fig. 3. Parallel performance of the 3D FDM simulation code for seismic wave propagation derived by strong scaling test using former ES and new ES (ES2) after performance tuning. Dashed line denote parallel performance for ES2 before code tuning.

The seismic wave propagation at each grid point of the subdomain is calculated by solving the equation of motion in 3D. For the parallel FDM simulation the 3D simulation model is partitioned vertically into many subdomains of equal grid numbers, and each subdomain is assigned into each node of the ES. The ES is constructed by 160 nodes of NEC SX-9 supercomputers each consists of eight vector processors with a large SMP memory, and are connected by a large fat-tree inter-node computer network. We adopted a hybrid parallel scheme for the parallel FDM simulation [7] in which a thread (automatic) parallelism is adopted for in-node parallel computing and a MPI is used for inter-node parallel computing.

3.2   Performance Tuning of the FDM Code Suitable for New Earth Simulator

The operation of the ES start in 2002 and then it was updated by the new model (hereafter we denote ES2) in 2009. Since the former ES had a high memory bandwidths relative to the CPU operation speed (4 Byte/FLOPS) and an effective single-stage crossbar switches (128 GB/s) for intra-node communications, it was able to pull out very high actual performance over 50 % of its peak performance (8 GFLOPS/processor) and also an effective parallel performance more than 99.99 % even though no special tuning was performed to our parallel FDM code. However, the memory bandwidth of the ES2 dropped to 2.5 Byte/FLOPS and the network system was changed into a conventional fat-tree structure. As a result, the actual computational performance of our FDM simulation dropped to 22.7 %, and a parallelization rate is also dropped greatly to 99.9472 %. Although the theoretical performance of the new ES2 (102.8 GFLOPS/processor) is 12.8 times faster than the old model, actual performance of our FDM simulation remained 5.8 times in fact. Furthermore, the theoretical speed-up of our parallel FDM simulation using large number of processors more than 1,000 became slower than before due to the fall of parallel performance (see broken line in Figure 3).

We therefore performed a performance tuning of our parallel FDM code suitable for the architecture of the new ES2. The log of the computing profile obtained from the new ES2 showed that the rate of the memory bank conflict increased 3 times as large as before, and moreover, the idling time of the MPI communication was increasing 1.5 times than before. Since new ES2 has a special memory distribution structure to share 256 memories among eight SMP processors via 16 set of memory router, a bank conflict occurs very frequently and its penalty at the time of occurring is very large. Until now, we had set the first subscript of 3D array variables as odd number in order to avoid a bank conflict by accessing the same memory bank during loop of calculation. In addition, we also set the second subscript of the 3D array variables into odd number in order to access a memory bank at random during loop calculation. Furthermore, we changed an order of the FDM calculation, so that required calculation for a given variable might be performed at once after loading the variable, and reduced useless memory loading and storing procedures. In order to use effectively a small capacity (256 KB) cache memory (ADB; Assignable Data Buffer) with which the new model ES2 was firstly equipped, we wrote a directive of the ADB for notifying reusable variables storing into the cache. Also we applied a loop unrolling technique to make a cache friendly code.

In the former parallel FDM program on ES we used a conventional function such as MPI_TYPE_CREATE_SUBARRAY() which is equipped in the MPI2 in order to pack and unpack data between variables at a large interval and MPI communication buffers before exchanging data between neighbour processors. However, the performance of the MPI communication improved drastically when this procedure was described explicitly by ourselves in the program. Moreover, MPI communication data for the same destination can be merged into one large data to enlarge communication speed and to reduce total number of communications. The MPI communication buffer can be allocating onto a global memory of the ES2 to reduce the time of the data copy between memory during the MPI.

Finally we attained 1.36 times as large performance ratios on the ES2 than before as a result of such memory and MPI tuning mentioned above. Effective performance rate of our present FDM simulation has improved dramatically from 22.7% to 31.3%, and the parallel performance rate has improved significantly from 99.9472 % to 99.9959 %. However, the influence of dramatic drop of memory bandwidth from ES to ES2 is mortal for our FDM simulation even carrying out extensive performance tuning, and it is not expectable to extract large sustain performance as before.

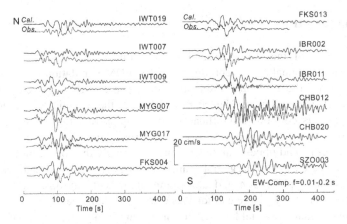

Fig. 4. Comparison of waveform of east-west component ground motions derived by computer FDM simulation (Cal; thick lines) and observation at 12 stations (Obs; thin lines). Waveforms are aligned from north to south and the K-NET station names are shown in right top.

3.3   Simulation Results and Comparison between Observations

Figure 1b shows a set of snapshot of the seismic wavefield derived by the 3D FDM simulation and using the visualization procedure of seismic wave propagating on the surface. Ground motion on sea is masked for direct comparison of observed seismic wavefield shown in Figure 1a. The computation of seismic wave propagation in 360 sec took CPU (wall-clock) time of 2 hours using 32 node (256 processors) of the ES2.

It is confirmed a good correspondence between the observed and simulated wavefield, except for a slight overestimation of the amplitude of the ground motion for the computer simulation. As we saw in the observed seismic wavefield the effect of thick sedimentary basin in major population centres such as Tokto, Nagoya, and Osaka can also be clearly seen as a region of anomalous localized amplification and prolonged ground motions due to development of long-period ground motions.

Band-pass (f=0.01-0.2 Hz) filtered waveforms of east-west component ground velocity motions derived from the simulation are compared with 12 K-NET strong motion record, showing a good match with the observation is confirmed in terms of the arrival time of the P and S waves, waveshape, and duration of seismic waves, though the amplitude of simulated ground motion are slightly overestimated. This is probably due to the mismatch between the 3D structural model used in the present FDM simulation and 1D layered structural model used for source inversion study [6], where localized structure such as sedimentary basins are not represented well, although most of the K-NET stations are placed in sedimentary basins.

Anyhow, fairly good match between simulatted and observed wavefield in terms of the snapshots of seismic wave propagation and synthetic seismograms demonstrated the effectiveness of the present simulation model including detail subsurface structure of crust and upper-mantle structure and near-surface sedimentary layers and large scale parallel FDM simulation of seismic wave propagation using supercomputers.

4   Conclusion

The development of the strong ground motions due to the destructive 2011 off Tohoku, Japan, earthquake was reproduced by 3-D FDM simulation of seismic wave propagation using the ES2 and observation of dense K-NET and KiK-net strong motion network across Japan. The characteristics of the seismic wavefield during large earthquake is controlled significantly by complex radiation properties of the seismic wave from earthquake source and propagation of seismic wave in heterogeneous crust and upper-mantle structure. Especially the soft sedimentary layer beneath populated cities with large impedance contrast between rigid bedrock beneath cause significant amplification and elongation of ground motions which leads in significant disasters during large earthquakes even for distant events.

Such complicated wavefield observed in Japan during the 2011 off Tohoku earthquake should probably be a common characteristics for most sedimentary basins in the world to caused the similar damage in the past and in future. Applying the FDM simulation of seismic wave using detail 3D structural model implemented on powerful supercomputers is expecting to mitigate future earthquake disasters instead of reappearance of the damage during the past events.

Acknowledgement. We acknowledge the National Institute for Earth Science and Disaster Prevention Research (NIED), Japan for providing the K-NET and the KiK-net strong motion data. We also acknowledge the Earth Simulator Center, the Japan Marine Science and Technology Center (JAMSTEC) for providing CPU time of the Earth Simulator.

References

1. K-NET, KiK-net Strong Motion Network of the National Research Institute for Earth Science and Disaster Prevention, http://www.kyoshin.bosai.go.jp/kyoshin/
2. Furumura, T., Takemura, S., Noguchi, S., Takemoto, T., Maeda, T., Iwai, K., Padhy, S.: Strong Ground Motions from the 2011 Off- the Pacific- Coast- of- Tohoku, Japan (Mw=9.0) Earthquake Obtained from a Dense Nation-wide Seismic Network. Landslides 8(3), 333–338 (2011)
3. Furumura, T., Kennett, B.L.N., Koketsu, K.: Visualization of 3D wave propagation from the 2000 Tottori-ken Seibu, Japan, earthquake: observation and numerical simulation. Bull. Seism. Soc. Am. 93, 870–881 (2003)
4. POV-Ray- The Persistence of Vision Raytracer, http://www.povray.org/
5. Japan Seismic Hazard Information Station, http://www.j-shis.bosai.go.jp/
6. Lee, S.-J., Huang, B.-S., Ando, M., Chiu, H.-C., Wang, J.-H.: Evidence of large scale repeating slip during the 2011 Tohoku-Oki earthquake, Geophys. Res. Lett. 38, L19306 (2011), doi:10.1029/2011GL049580
7. Furumura, T., Chen, L.: Parallel simulation of strong ground motions during recent and historical damaging earthquakes in Tokyo, Japan. Parallel Computing 31, 149–165 (2005)
8. Earth Simulator, http://www.jamstec.go.jp/es/en/

Grand Challenge in Life Science on K Computer

Ryutaro Himeno

RIKEN Advanced Center for Computing and Comunication,
2-1, Wako, Saitama, 351-0198 Japan
himeno@riken.jp

Abstract. In 2006, we started a grand challenge project called ISLiM for K computer to demonstrate its performance. The ISLiM stands for Integrated Simulation of Living Matter to reproduce life phenomena on a supercomputer for understanding them and developing new medicine or new treatments. We have 6 research teams: Molecular scale team, cell scale team, organ and body scale team, data analysis fusion team, brain and neural system team and HPC team. We are developing a high performance software package for life science for K computer which contains 31 application software. Currently two third of them are running on K computer and several ones shows more than 30 percent of theoretical peak performance of K computer.

Keywords: HPC, Life Science Application, Molecular Dynamics, Blood Flow, Heart Simulation.

1 Introduction

In 2006, we started our grand challenge project in Life Science to show how K computer is effective using real application. This project is called as ISLiM which stands for Integrated Simulation of Living Matter. The ISLiM is included in the Next Generation Supercomputer Research and Development which is developing K computer. In the early 2000, there were a few codes in life science to show good scalability on thousand processors. To accelerate supercomputer usage in life science, we decided to develop a HPC software package for life science researchers which contains variety of allocations from molecule and cell scale to organ and whole body scale. In addition, we are developing data processing application to get information from experimental data. We also develop codes for Brain and Neural simulation. 34 application codes are included in the package.

2 Basic Concept and Teams of the ISLiM

The ISLiM is developing a software package which can be utilized for reproducing life phenomena on a supercomputer for understanding them and developing new medicine or new treatments. We have 6 research teams: Molecular scale team, cell scale team, organ and body scale team, data analysis fusion team, brain and neural

M. Daydé, O. Marques, and K. Nakajima (Eds.): VECPAR 2012, LNCS 7851, pp. 17–22, 2013.

system team and HPC team(figure 1).   We are developing a high performance software package for life science for K computer which contains 34 application software (see table 1).

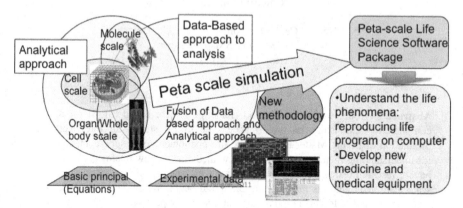

Fig. 1. Basic concept of the ISLiM[1]

Table 1. ISLiM Software Package[2]

	Applications	No. of application Software
Molecule scale	MD, Quantum Chemistry, Coarse Grain MD	9
Cell scale	Voxel-based multi-compartment transport-diffusion simulator	1
Organ/whole body scale	Heart sim., Lung sim., HIFU sim., Fluid-structure sim. for blood flow,	6
Brain & neural system	Neural sim., whole brain sim., cortical micro circuit sim.,	5
Data based analysis	Whole genome association study, data assimilation, prediction of protein-protein interaction, genome sequencer data processing, Haplotype whole genome association study	9
HPC	Parallel middleware, parallel visualization software, high-speed software core library	4

3   Current Status of the Software

31 out of 34 application codes are targeted to be utilized on K computer and the current developing status are shown in figure 2. We have promised to tune a few

codes exceeding effective 1 Peta FLOPS once we can use full system of K computer. Three codes, ZZ-EFSI, UT-Heart and cppmd have already achieved more than 1 peta FLOPS effective performance. Figure 3 shows number of nodes we tested using each application codes. The total number of nodes of K computer is about 83,000. 11/31 application codes show good scalability more than 10,000 nodes.

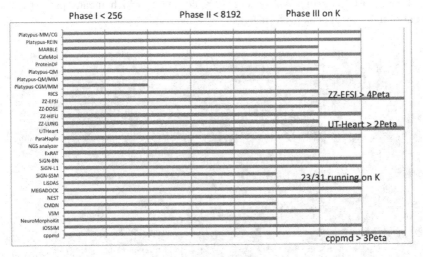

Fig. 2. Current status of the code development

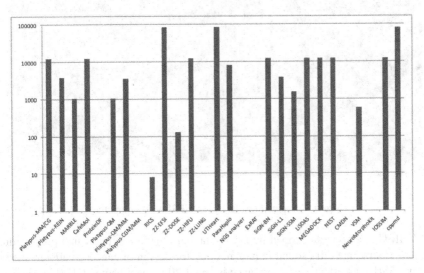

Fig. 3. Tested maximum number of nodes using each application code

4   Codes Which Show Effective Performance

Four codes show more than 20 percent of effective performance using 10,000 nodes of K computer. Those are ZZ-EFSI, ZZ-HIFU, UT-Heart and cppmd. ZZ-EFSI is a code to solve fluid-solid coupled problem for analyzing blood flow with elastic cells shown in figure 4[3]. This code developed by Dr. Sugiyama and others showed more than 4 Peta FLOPS using full system of K comouter which means more than 40 percent of theoretical peak performance.

Fig. 4. Computed results by ZZ-EFSI on K computer

ZZ-HIFU is a code developed by Dr. Okita and others to solve sound propagation equation to simulate ultra sound in human body for design of cancer treatment shown in figure 5[4]. Its effective performance is about 20 percent using 10,000 nodes of K computer.

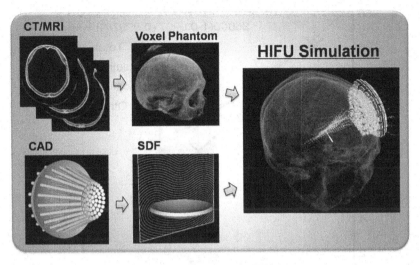

Fig. 5. Target of ZZ-HIFU

UT-Heart is a heart simulation software developed by Prof. Hisada and his group in many years in The University of Tokyo[5]. It is based on multi-scale simulation model coupled with coronary artery circulation with capillary shown in figure 6. UT-Heart achieved more than 2 Peta FLOPS on the full system of K computer.

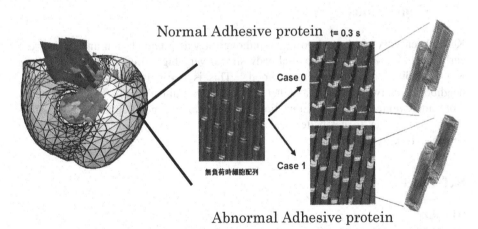

Fig. 6. UT-Heart simulation model

Cppmd is a molecular dynamics code developed by HPC team of the ISLiM[6]. This code achieved more than 3 Peta FLOPS on the full system of K computer. Its weal scaling performance is shown in figure 7 up to 16,384 nodes on K computer.

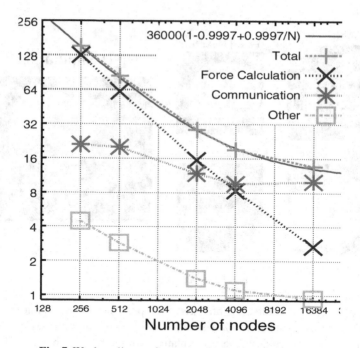

Fig. 7. Weak scaling performance of cppmd on K computer

5   Conclusions

K has been very stable and powerful although it was a new design and at the very early stage. Several codes have already shown very high effective performance on K. Scalability on K shows very good. This is pertly because of the effective neighboring network TOFU and ICC. Currently, Fujitsu's c/c++ compiler needs more improvement (SIMD optimization). Computation time of application is getting longer and longer to get better scalability on peta-scale system. This may make design of Exa-scale system more difficult.

References

[1] http://www.csrp.riken.jp/index_e.html
[2] http://www.csrp.riken.jp/application_e.html
[3] http://www.csrp.riken.jp/application_o_e.html#O1
[4] http://www.csrp.riken.jp/application_o_e.html#O4
[5] http://www.csrp.riken.jp/application_o_e.html#O6
[6] http://www.csrp.riken.jp/application_h_e.html#H1

HPC/PF - High Performance Computing Platform: An Environment That Accelerates Large-Scale Simulations

Kenji Ono[1], Tomohiro Kawanabe[2], and Toshio Hatada[3]

[1] Institute of Industrial Science, The University of Tokyo
4-6-1, Komaba, Meguro-ku, Tokyo 153-8505, Japan
Advanced Institute for Computational Science, RIKEN,
7-1-26, Minatojima-minami-machi, Kobe 650-0047, Japan
keno@riken.jp
[2] Institute of Industrial Science, The University of Tokyo
tkawanab@iis.u-tokyo.ac.jp
[3] Institute of Industrial Science, The University of Tokyo
hatada@iis.u-tokyo.ac.jp

Abstract. Peta-scale supercomputers enable us to tackle very large-scale problems of which results provide useful information to understand physical phenomena or to improve performance of a product design. The large-scale simulation is becoming dawn to earth due to parallel computing techniques, however, inherent barriers on the distributed parallel environment are still remains in simulation process, i.e., grid generation, visualization and data analysis. In this paper, the authors would like to clarify the issues to be resolved for productive support environment of the large-scale simulation, and to propose a foundational framework to enhance the utilization of huge computational resources.

1 Introduction

Recent progress of hardware and software development yields the ability of PFLOPS computation and allows us to tackle very large-scale problems. The K computer[1] is the first machine achieved over 10 PFLOPS in LINPACK benchmark and will be started to operate at the end of 2012. The K computer is planned to use variety of fields including science and engineering, especially, is anticipated from industry field. Skilled manufacturing field is one of the important application fields of the computer simulation. The utilization of state-of-the-art HPC system, however, requires some degree of skills for users, therefore this becomes a barrier in terms of the promotion of utilization. For example, the number of grid points increases so that the solution with higher order is obtained. In addition, the number of process and associated files generated reach the order of 10^5. Increase in the number of files handled by the scale of the problem will affects the performance of input and output and the file system not only greatly increase the cost of management. Since the increase of the number makes

M. Daydé, O. Marques, and K. Nakajima (Eds.): VECPAR 2012, LNCS 7851, pp. 23–27, 2013.

the simulation process complicated, work efficiency will be reduced. Engineers utilize the HPC to yield excellent results in the field of their product design. In order to increase outcome by simulation on HPC, it is necessary to resolve these issues unique to large-scale problem.

This paper describes a construction of the framework that can be efficiently utilized in the large-scale parallel simulations.

2   Concept of HPC/PF

HPC/PF is a support environment for executing large-scale parallel simulations efficiently. This system also has integrated aspects of the system in addition to so-called pre-post function, management of simulation execution, automated processing by the workflow, database management of results. The proposed system is designed and implemented as a collection of subsystems, so that various use-case are able to be supported by provided many functions as shown in Fig.1. Each subsystem is loosely connected by script language. Since the system is composed of relatively highly independent software components, it is easy to introduce open source software as a component and/or to replace by the component that has same features and higher performance.

In the heart of this HPC/PF system, various physical simulators will be incorporated. The core physical simulators, we introduce a simulation program has been developed in various fields in national projects in Japan so far. For example, FFB[2], FFR, FISTR, Adventure, UPACS[3], VCAD[4], and so on.

2.1   Assumed Use-Case

Assumed typical use-cases for the HPC/PF are follows.

Use of Database

Documents and Case Studies. In order to provide basic information of each core simulator, the document, e.g., manual, user guide, tutorial, installation guide, will be registered for the end-users. Case study examples are also registered including a model file, input parameters, computed results, visualized and analyzed data, and associated experimented data. Accumulated validation data tells us the guideline of the utilization of the simulator and increase the reliability of simulators. This verification and validation processes are important for not only end-users but also application developers. Especially, when the version control system of software and associated examples are operated, it is useful to develop application efficiently.

Reproduction of Examples. The database system enables us to provide registered examples. User can install delivered software application onto the user's environment, then can be reproduced the examples provided by the database. This step is a practice of using the simulation, and confirmation process. This experience helps the user to get better understand.

Customization Based on Provided Templates. Users to deepen understanding of the simulator are now working on their own problems. They would be helpful to download the project files from the database example, then they can customize to their concerned problems. This mechanism will help greatly to improve productivity.

Parameter Study. Optimization is one of the powerful strategies to obtain better shape and/or performance of the products, and requires many runs corresponding to different parameters. Even a simple comparison of the computed results for different parameters, we can derive useful information from the comparison result.

3 Key Components in HPC/PF

In this section, key components in HPC/PF are described.

3.1 Project Management

Both parallel computation and parameter study uses and generates many files, e.g., grid files, parameter files, result files, analyzed files, and so on. Since many files are found in a directory of a file system, we need to organize the files essential to the simulations. In our approach, all files needed to a specific simulation are managed by an asset list, which describes necessary files. The asset list contains all file names required to the smallest unit of simulation, which is defined as a case. We often need to manage a number of cases in the parametric study that is performed in practical simulation. Multiple cases often form a project defined as one group that has parameters associated each other.

3.2 File Handling

In parallel computation, each core reads and writes a file at the same time, which degrades file I/O performance. Therefore, the most important and critical issue is a mechanism of file handling in terms of performance of the proposed system. MPI-I/O is one of the candidates to improve the performance by its parallel file accessing. In addition to incorporate the parallel I/O libraries, we also introduce a file index system that provides a mechanism to treat a lot of files as it looks one file. The index contains the information related to subdomains in domain decomposition method, e.g., the coordinate value of bounding box, a rank number of MPI process, index of computational space, and file names. This index plays a role of managing raw files and supplying the information in other subsystems. For example, in data processing phase, the application access the index file at first, then the application is able to know the file name to be processed.

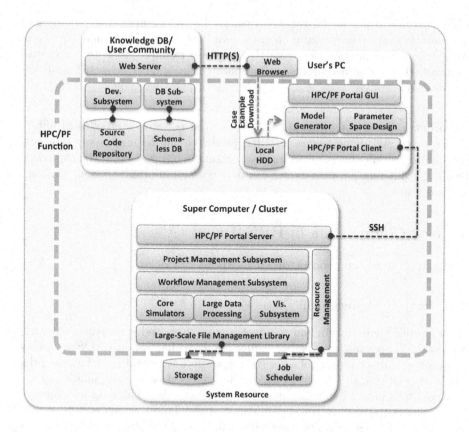

Fig. 1. A schematic of logical concept of HPC/PF. Main functions are implemented on a server, and applications that need user-interface are operated on user's computer. A database subsystem is hosted on a web server and can be distributed. An ssh protocol is required for the communication between different platforms.

3.3   Workflow

In practical simulation cases, the engineer repeats predetermined tasks to obtain the information to be needed in the design or for the optimization. Through entire process of above mentioned parameter study, we need to manage parameter space and its execution case. To do so, it is required to incorporate the functions of the design of parameter space, definition of workflow and automatic execution including batch job queuing, and file resource management. Xcrypt[5] provides functions to perform sequential/parallel runs of a program with different parameters. Since Xcrypt is written by Perl script language, the portability is very high even on the most supercomputers.

3.4 Visualization and Data Analysis

Computed results on the parallel computer are written in a large number of files and the files cannot be moved any more because the limitation of disk space and the operation time of the data operation. Thus, it is required that the visualization system has the ability to handle many files at the same time and to visualize the data in distributed parallel environment. In visualization and analysis process, users have their own way to analyze the data, and the visualization environment is depending on users platform. We have planned to introduce the well-established visualization platform VisIt[6] into the proposed HPC/PF. For data analysis, the map-reduce framework[7], which can be effectively work as the data supplier from the raw dataset, is combined to another filter programs. The derived metadata will be registered to the database and utilized.

4 Concluding Remarks

We have been outlined the issues on large-scale parallel simulation and a basic concept of an infrastructure to provide useful functions. The HPC/PF system was organized by independent program stacks including open source software so that the system enables users to support their simulation scenario efficiently and to enhance productivity.

Acknowledgement. A portion of this research was supported by the grant for Strategic Program on HPCI Field No. 4: Industrial Innovations from the Ministry of Education, Culture, Sports, Science, and Technology (MEXT)s Development and Use of Advanced, High-Performance, General-Purpose Supercomputers Project, and carried out in partnership with the University of Tokyo. We express our thanks to all parties involved.

References

1. http://www.aics.riken.jp/en/
2. http://www.ciss.iis.u-tokyo.ac.jp/rss21/en/
3. Takaki, R., Yamamoto, K., Yamane, T., Enomoto, S., Mukai, J.: The Development of the UPACS CFD Environment. In: Veidenbaum, A., Joe, K., Amano, H., Aiso, H. (eds.) ISHPC 2003. LNCS, vol. 2858, pp. 307–319. Springer, Heidelberg (2003)
4. http://vcad-hpsv.riken.jp/en/
5. Hiraishi, T., Abe, T., Iwashita, T., Nakashima, H.: Xcrypt: a Perl Extension for Job Level Parallel Programming. In: Second International Workshop on High-performance Infrastructure for Scalable Tools WHIST 2012 (Held as part of ICS 2012), Venice, Italy (2012)
6. https://wci.llnl.gov/codes/visit/
7. Pike, R., Dorward, S., Griesemer, R., Quinlan, S.: Interpreting the data: Parallel analysis with Sawzall. Scientific Programming Journal 13(4), 227–298 (2005)

Programming the LU Factorization
for a Multicore System with Accelerators

Jakub Kurzak[1], Piotr Luszczek[1], Mathieu Faverge[1], and Jack Dongarra[1,2,3]

[1] University of Tennessee, Knoxville TN 37919, USA
[2] Oak Ridge National Laboratory, Oak Ridge TN 37831, USA
[3] University of Manchester, Manchester M13 9PL, UK
{kurzak,luszczek,faverge,dongarra}@eecs.utk.edu

Abstract. LU factorization with partial pivoting is a canonical numerical procedure and the main component of the High Performance LINPACK benchmark. This article presents an implementation of the algorithm for a hybrid, shared memory, system with standard CPU cores and GPU accelerators. Performance in excess of one TeraFLOPS is achieved using four AMD Magny Cours CPUs and four NVIDIA Fermi GPUs.

1 Introduction

This paper presents an implementation of the canonical formulation of the LU factorization, which relies on partial (row) pivoting for numerical stability. It is equivalent to the `DGETRF` function in the LAPACK numerical library. Since the algorithm is coded in double precision, it can serve as the basis for an implementation of the *High Performance LINPACK* benchmark (HPL) [2]. The target platform is a hybrid, multi-CPU, multi-GPU shared memory system.

2 Background

The LAPACK block LU factorization is the main point of reference here, and LAPACK naming convention is followed. The LU factorization of a matrix M has the form $M = PLU$, where L is a unit lower triangular matrix, U is an upper triangular matrix and P is a permutation matrix. The LAPACK algorithm proceeds in the following steps: Initially, a set of nb columns (*the panel*) is factored and a pivoting pattern is produced (`DGETF2`). Then the elementary transformations, resulting from the panel factorization, are applied to the remaining part of the matrix (*the trailing submatrix*). This involves swapping of up to nb rows of the trailing submatrix (`DLASWP`), according to the pivoting pattern, application of a triangular solve with multiple right-hand-sides to the top nb rows of the trailing submatrix (`DTRSM`), and finally, application of matrix multiplication of the form $C = C - A \times B$ (`DGEMM`), where A is the panel without the top nb rows, B is the top nb rows of the trailing submatrix, and C is the trailing submatrix without the top nb rows. Then the procedure is applied repeatedly, descending down the diagonal of the matrix.

M. Daydé, O. Marques, and K. Nakajima (Eds.): VECPAR 2012, LNCS 7851, pp. 28–35, 2013.

3   The Solution

The main hybridization idea is captured on Figure 1 and relies on representing the work as a *Directed Acyclic Graph* (DAG) and dynamic task scheduling, with CPU cores handling the complex fine-grained tasks on the *critical path* (the longest path through the DAG), and GPUs handling the coarse-grained data-parallel tasks outside of the critical path. Some number of columns (*lookahead*) are assigned to the CPUs, and the rest of the matrix is assigned to the GPUs in a 1D block-cyclic fashion. In each step of the factorization, the CPUs factor a panel and update their portion of the trailing submatrix, while the GPUs update their portions of the trailing submatrix. After each step, one column of tiles shifts from the GPUs to the CPUs.

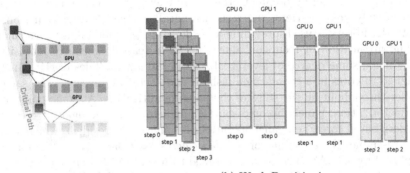

(a) Task Graph Scheduling   (b) Work Partitioning

Fig. 1. Scheduling the task graph of the LU factorization, with fine-grained tasks on the critical path being dispatched to individual CPU cores and coarse-grained tasks outside of the critical path being dispatched to GPU devices

The implementation relies on a number of state-of-the-art solutions such as: tile data layout, block-cyclic data distribution, parallel recursive panel factorization, GPU kernel autotuning, the technique of *lookahead*, the use of superscalar scheduling and communication-computation overlapping.

3.1   Tile Data Layout

The matrix is laid out in square tiles on the CPU side (*host memory*), where each tile occupies a continuous region of memory. Tiles are stored in column-major and elements within tiles are stored in column-major. This layout, referred to as *Column-Column Rectangular Block* (CCRB) [4] is the native layout of the PLASMA library[7]. Tiles are transposed on the GPU side (*device memory*), i.e. the layout is translated to *Column-Row Rectangular Block* (CRRB), which is critical to the performance of the row swap (DLASWP) operation. This tile-wise transposition is trivial to code and fast to execute.

3.2 CPU Kernels

CPUs are responsible for the panel factorization and a portion of the update of the trailing submatrix. The update is relatively straightforward and requires three operations: row swap (DLASWP), triangular solve (DTRSM) and matrix multiplication (DGEMM). In the case of DLASWP, one core is responsible for swaps in one column of tiles. The LAPACK DLASWP function cannot be used, because of the use of tile layout, so DLASWP is hand-coded. In the case of DTRSM and DGEMM one core is responsible for one tile. Calls to Intel *Math Kernel Library* (MKL) are used, with layout set to column-major and the *leading dimension* set to tile size (nb).

The LAPACK panel factorization (DGETF2) is sequential and memory bound, and can deliver performance of roughly 2.0 Gflop/s, which is completely inadequate for a hybrid LU implementation. Running at such speed, panel factorizations would completely dominate the entire execution time. A fast alternative is absolutely critical. Here, the recursive-parallel panel factorization from the PLASMA library is used, providing an order of magnitude higher performance.

The application of recursion allows for a decrease in memory intensity by introducing some degree of level 3 BLAS operations [3]. Tiles of the panel are assigned to cores in a *round-robin* fashion and each core preserves the same set of tiles throughout all the steps of the panel factorization. At some point in the LU factorization, panels become short enough to fit in the aggregate cache of the designated cores, i.e., panel operations become cache-resident, which at some level resembles the technique of *Parallel Cache Assignment* (PCA) [1] currently employed by ATLAS. The cores are forced to work in lock-step, but can benefit from a high level of cache reuse. The ultra-fine granularity of operations requires very light-weight synchronization. Synchronization is implemented using *busy-waiting* on volatile variables and works at the speed of hardware cache-coherency.

3.3 GPU Kernels

The update of the trailing submatrix on the GPUs requires kernels for three operations: row swap (DLASWP), triangular solve (DTRSM) and matrix multiplication (DGEMM). Also, a tile-wise transposition is required to convert between the CCRB layout in the host memory and the CRRB layout in the device memory. This transposition follows the transfer of each panel from the host memory to the device memory and precedes the transfer of each column returning from the device memory to the host memory.

DLASWP is implemented by creating nb (tile size) threads per multiprocessor and assigning one column to each thread. DTRSM (an in-place operation) is replaced by an inversion of the diagonal block (application of the L factor to identity) on a CPU, followed by a DGEMM on the GPUs (out-of-place). The transposition is implemented by spanning the column being transposed with a block-grid / thread-grid, such that each individual thread transposes one element (no loops in the kernel). These straightforward implementations are sufficient to make the impact of the operations negligible in comparison to the DGEMM.

The DGEMM kernels are produced using the *Automatic Stencil TuneR for Accelerators* (ASTRA) system [5], which follows the principles of *Automated Empirical Optimization of Software* (AEOS), popularized by the *Automatically Tuned Linear Algebra Software* (ATLAS) [9]. The same process is currently used to produce DGEMM kernels for the MAGMA project [6].

The kernel is expressed through a parametrized *stencil*, creating a large search space of possible implementations. The search space is aggressively pruned, using mostly constraints related to the usage of hardware resources. On NVIDIA GPUs, one of the main selection criteria is *occupancy*, i.e. the capability of the kernel to launch a big number of *Single Instruction Multiple Threads* (SIMT) threads. The pruning process identifies a few tens of kernels for each tile size. The final step of autotuning is benchmarking these kernels to find the best performing ones.

There are two differences between the kernels used here and the MAGMA kernels. MAGMA kernels operate on matrices in canonical FORTRAN 77 column-major layout, compliant with the *Basic Linear Algebra Subroutines* (BLAS) standard. The kernels used here operate on matrices in CRRB tile layout [4]. Also, MAGMA kernels are tuned for the case where all three input matrices are square, while the kernels used here are tuned for the *block outer product* operation in the LU factorization, i.e., $C = C - A \times B$, where the width of A and the height of B are equal to the matrix tile size nb.

DGEMM kernels achieve the best performance when texture reads are used for read-only data (A and B input matrices). The complete LU factorization applies matrix multiplications exceeding this limit by splitting them into a sequence of multiple DGEMM calls (two or three). Here the tuning is done for the largest case where texture mapping can be used without such splitting ($\sim$12K$\times$12K). Table 1 lists the performance of the autotuned kernels along with their most important tuning parameters (the blocking factors, i.e., the size of DGEMM performed by each multiprocessor in the outermost loop).

Table 1. Autotuned block outer product GPU DGEMM kernels

TILE SIZE	32	64	96	128	160	192	224	256	288
BLOCKING	32×32×8	64×64×16	32×32×6	64×64×16	32×32×8	64×64×16	32×32×8	64×64×16	32×32×6
GFLOPS	208	250	255	272	265	278	269	277	274

3.4   Superscalar Scheduling

Manually multithreading the hybrid LU factorization would be tedious, given the three different levels of granularity involved: single tile, one column, a large block (submatrix). Here the scheduling infrastructure of the PLASMA library is used, namely the QUARK superscalar scheduler [8]. The LU factorization code is expressed with the canonical serial loop nest, where calls to CPU and GPU kernels are augmented with information about sizes of affected memory regions and directionality of arguments (IN, OUT, INOUT). QUARK schedules the work by resolving data hazards (RaW, WaR, WaW) at runtime. Two important extensions are critical to the implementation of the hybrid LU factorization: variable-length list of dependencies and support for nested parallelism.

CPU tasks, such as panel factorizations and row swaps, affect columns of the matrix of variable height. For such tasks, the list of dependencies is created incrementally, by looping over the tiles involved in the operation. It is a similar situation for the GPU tasks, which involve large blocks of the matrix (large arrays of tiles). The only difference is that here transitive (redundant) dependencies are manually removed, to decrease scheduling overheads, while preserving correctness.

The second crucial extension of QUARK is support for nested parallelism, i.e., superscalar scheduling of tasks, which are internally multithreaded. The hybrid LU factorization requires parallel panel factorization for the CPUs to be able to keep pace with the GPUs. At the same time, the ultra-fine granularity of the panel operations prevents the use of QUARK inside the panel. Instead, the panel is manually multithreaded using cache coherency for synchronization, and scheduled by QUARK as a single task, entered at the same time by multiple threads.

3.5 Communication

Each panel factorization is followed by a broadcast of the panel to all the GPUs. After each update, the GPU in possession of the leading leftmost column sends that column back to the CPUs (host memory). These communications are expressed as QUARK tasks with proper dependencies linking them to the computational tasks. Because of the use of lookahead, the panel factorizations can proceed ahead of the trailing submatrix updates and so can transfers, which allows for perfect overlapping of communication and computation, as further discussed in the following section.

4   Results

The system used for this work couples one CPU board with four sockets and one GPU board with four sockets. The CPU board is a H8QG6 Supermicro system with 4 AMD Magny Cours chips, 12 cores each, clocked at 2.1 GHz. The GPU board is an NVIDIA Tesla S2050 system with 4 Fermi chips, 14 multiprocessors each, clocked at 1.147 GHz.

The theoretical peak of a single CPU socket amounts to $2.1\ GHz \times 12\ cores \times 4\ ops\ per\ cycle \simeq 101\ Gflop/s$, making it ~403 Gflop/s for all four CPU sockets. The theoretical peak of a single GPU amounts to $1.147\ GHz \times 14\ cores \times 32\ ops\ per\ cycle \simeq 514\ Gflop/s$, making it ~2055 Gflop/s for all four GPUs. The combined CPU-GPU peak is ~2459 Gflop/s.

The system runs Linux kernel version 2.6.35.7 (Red Hat distribution 4.1.2-48). The CPU part of the code is built using GCC 4.4.4. Intel MKL version 2011.2.137 is used for BLAS calls on the CPUs. The GPU part of the code is built using CUDA 4.0.

Figure 2a shows the overall performance of the hybrid LU factorization, and Table 2 lists the exact performance number for each point along with values of tuning parameters. Tuning is done by exhaustive search across all parameters. Matrix size goes up to 34,560. Beyond that point the size of memory on all GPUs is exceeded. Each GPU can provide 2.6 GB of *Error Correcting Code* (ECC) protected memory.

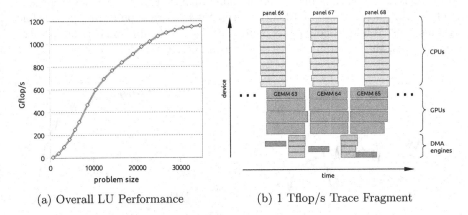

(a) Overall LU Performance   (b) 1 Tflop/s Trace Fragment

Fig. 2. (a) Overall performance of the LU factorization. (b) Trace fragment of the run which exceeded execution rate of 1 Tflop/s.

Figure 2b shows a small fragment in the middle of a 23,040 run (the smallest size exceeding 1 Tflop/s performance). In the CPU part, only the panel factorizations are shown. The steps shown on the figure correspond to factoring submatrices of size ~12,000. Due to the deep lookahead, panel factorizations on the CPUs run a few steps ahead of trailing submatrix updates on the GPUs. This allows for perfect overlapping of CPU work and GPU work. It also allows for perfect overlapping of communication between the CPUs and the GPUs, i.e., between the host memory and the device memories. Each panel factorization is followed by a broadcast of the panel to the GPUs (light gray DMA). Each trailing submatrix update is followed by returning one column to the CPUs (dark gray DMA).

Table 2. LU performance and values of tuning parameters

MATRIX SIZE [K]	0.6	1.9	3.2	4.5	5.8	6.7	8.6	10.6	12.5	14.4	16.6	19.2	21.1	23.0	25.0	26.9	28.8	30.7	32.6	34.6
TILE SIZE			64					96			128					192				
LOOKAHEAD			1					2		3	5			12			13		14	
PANEL CORES										12										
GFLOPS	6	39	95	163	249	315	465	598	690	768	838	912	976	1022	1068	1098	1121	1142	1150	1160

Figure 3a shows the performance of the panel factorization throughout the largest run (34,560), using different numbers of cores, for panels of width 192. The jagged shape of the lines reflects the fact that the panel cores have to compete for main memory with the other cores, applying updates at the same time. Generally, more cores provide higher performance, due to more computing power and larger capacity of their combined caches. However, 24 cores (two sockets) provide only a small performance improvement over 12 cores (single socket) due to the higher cost of inter-socket communication over communication within the same socket. In actual LU runs, the use of 12 cores turns out to always be optimal, even for large matrices. While 12-core panel factorizations are capable of keeping up with GPU updates, the remaining cores can be committed to CPU updates.

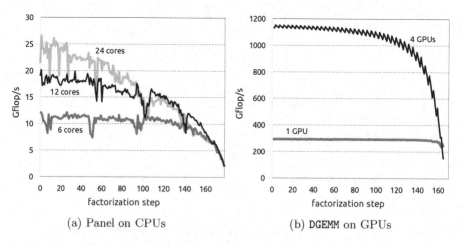

(a) Panel on CPUs   (b) DGEMM on GPUs

Fig. 3. (a) Performance of the panel factorization on CPUs at each step of the LU factorization. Panel width = tile size = 192. (b) Performance of the 4-GPU DGEMM task and performance of a single-GPU portion of that task.

Figure 3b shows the performance of the GPU DGEMM kernel throughout the entire factorization. The gray line shows the DGEMM kernel performance on a single GPU. The black line shows the performance of the 4-GPU DGEMM task. The jagged shape of the line is due to the load imbalance among the GPUs. The high peaks correspond to the calls where the load is perfectly balanced, i.e., the number of columns updated by the GPUs is divisible by 4. When this is not the case, the number of columns assigned to different GPUs can differ by one. The load imbalance can be completely eliminated by scheduling the GPUs independently, although potential performance benefits are on the order of a few percent.

5   Conclusions

The results reveal the challenges of programming a hybrid multicore system with accelerators. There is a disparity in the performance of the CPUs and the GPUs to start with. It turns into a massive disproportion when the CPUs are given the difficult (synchronization-rich and memory-bound) task of panel factorization, and the GPUs are given the easy (data-parallel and compute-bound) task of matrix multiplication. While the performance of panel factorization on the CPUs is roughly at the level of 20 Gflop/s, the performance of matrix multiplication on the GPUs is almost at the level of 1,200 Gflop/s (two orders of magnitude). The same disproportion applies to the computational power of the GPUs versus the communication bandwidth between the CPU memory and the GPU memory (host to device). The key to achieving good performance under such adverse conditions is overlapping of CPU processing and GPU processing and overlapping of communication. The work also reveals that the PLASMA framework can easily adopt GPU acceleration, perhaps showing a path for the eventual merge of the PLASMA and MAGMA projects into a single cohesive multicore/manycore software package.

References

1. Castaldo, A.M., Whaley, R.C.: Scaling LAPACK panel operations using parallel cache assignment. In: ACM SIGPLAN Symposium on Principles and Practice of Parallel Programming, PPoPP 2010. ACM, Bangalore (2010), doi:10.1145/1693453.1693484 (accepted to ACM TOMS)
2. Dongarra, J.J., Luszczek, P., Petitet, A.: The LINPACK benchmark: Past, present and future. Concurrency Computat.: Pract. Exper. 15(9), 803–820 (2003), doi:10.1002/cpe.728
3. Gustavson, F.G.: Recursion leads to automatic variable blocking for dense linear-algebra algorithms. IBM J. Res. Dev. 41(6), 737–756 (1997), doi:10.1147/rd.416.0737
4. Gustavson, F.G., Karlsson, L., Kågström, B.: Parallel and cache-efficient in-place matrix storage format conversion. Tech. Rep. UMINF 10.05, Department of Computer Science, Umeå University (2010), http://www8.cs.umu.se/research/uminf/reports/2010/005/part1.pdf (accepted to ACM TOMS)
5. Kurzak, J., Tomov, S., Dongarra, J.: Autotuning GEMMs for Fermi. Tech. Rep. UT-CS-11-671, Electrical Engineering and Computer Science Department, University of Tennessee (2011), http://www.netlib.org/lapack/lawnspdf/lawn245.pdf (accepted to IEEE TPDS)
6. MAGMA, http://icl.eecs.utk.edu/magma/
7. PLASMA, http://icl.eecs.utk.edu/plasma/
8. QUARK, http://icl.eecs.utk.edu/quark/
9. Whaley, R.C., Petitet, A., Dongarra, J.: Automated empirical optimizations of software and the ATLAS project. Parallel Comput. Syst. Appl. 27(1-2), 3–35 (2001), doi:10.1016/S0167-8191(00)00087-9

Efficient Two-Level Preconditioned Conjugate Gradient Method on the GPU

Rohit Gupta, Martin B. van Gijzen, and Cornelis Kees Vuik

Delft University of Technology, Mekelweg 4, 2628CD, Delft, The Netherlands
{rohit.gupta,m.b.vangijzen,c.vuik}@tudelft.nl

Abstract. We present an implementation of a Two-Level Preconditioned Conjugate Gradient Method for the GPU. We investigate a Truncated Neumann Series based preconditioner in combination with deflation. This combination exhibits fine-grain parallelism and hence we gain considerably in execution time when compared with a similar implementation on the CPU. Its numerical performance is comparable to the Block Incomplete Cholesky approach. Our method provides a speedup of up to 16 for a system of one million unknowns when compared to an optimized implementation on one core of the CPU.

1 Introduction

Our work is motivated by the Mass-Conserving Level Set approach [6] to solve the Navier Stokes equations for multi-phase flow. The most time consuming step in this approach is the solution of the (discretized) pressure-correction equation, which is a Poisson equation with discontinuous coefficients. The discretized equation takes the form of a linear system,

$$Ax = b, \ A \in \mathbb{R}^{N \times N}, \ N \in \mathbb{N} \tag{1}$$

where N is the number of degrees of freedom. A is symmetric positive definite (SPD). Due to the large contrast in the densities of the fluids involved, the matrix A has a large condition number κ, which results in slow convergence when the system (1) is solved using the iterative Conjugate Gradients (CG) method.

1.1 Focus of This Research

To overcome the slow convergence it is imperative to use preconditioning. The resulting system then looks like, $M^{-1}Ax = M^{-1}b$, where the matrix M is symmetric and positive definite. The choice of M is such that the operation $M^{-1}y$, for some vector y, is computationally cheap and M can also be stored efficiently. This research aims to find preconditioning schemes that can exploit the computing power of the GPU. To this end the preconditioning schemes should offer fine-grain parallelism. At the same time they should prove effective in bringing down the condition number of $M^{-1}A$. We use a two-level preconditioner. The

M. Daydé, O. Marques, and K. Nakajima (Eds.): VECPAR 2012, LNCS 7851, pp. 36–49, 2013.

first level preconditioner is based on the Truncated Neumann series of the triangular factors of the coefficient matrix A. After this, we apply Deflation to treat the remaining small eigenvalues in the spectrum of the preconditioned matrix. We compare our schemes with Block-Incomplete Cholesky (Block-IC) Preconditioners, as a benchmark to check their quality. The numerical performance of the preconditioners we introduce in this paper comes close to their Block-IC counterparts for our model problem and they also offer fine-grain parallelism making them very suitable for the GPU.

1.2 Related Work

Preconditioning has been studied previously for GPU implementations of the Conjugate Gradient method. The preconditioner in [2] offers as much parallelism as the number of degrees of freedom, N (or the number of unknowns). However, our experiments [1] show its use is limited for two-phase (high condition number, (κ)) flow problems. An extension to [2] is provided in [8] wherein a relaxation factor is utilized. In [9] an incomplete LU decomposition based preconditioner with fill-in is used combined with reordering using multi-coloring.

One of the first works [4] using GPU computing used Multigrid with CG. More recently in [3] also multigrid has been investigated for solving Poisson type problems. In [11] a comparative study is presented between deflation and multigrid. It shows that the former is a competitive technique in comparison with the latter.

This paper is organized as follows: in the next section we present the test problem. A brief overview of the preconditioning schemes and their features can be found in Section 3. We discuss the approach of two level preconditioning in Section 4. In Section 5 we introduce the Conjugate Gradient Algorithm with Preconditioning and Deflation. Furthermore we comment on two different implementations for this method in Section 5.1. In Section 6 we present our results and we end with a discussion in Section 7.

2 Problem Definition

We define a test problem in order to test our preconditioning schemes. We define a unit square as our computational domain in 2D (Figure 2). It has two fluids with a density contrast ($\rho_1 = 1000$, $\rho_2 = 1$). It has an interface layer (at $y = 0.5$), where there is a jump in coefficient values due to contrast in densities of the two fluids. This jump is also visible in the eigenvalue spectrum as shown in Figure 1. Boundary conditions are applied to this domain as indicated in Figure 2. The resulting discretization matrix A is sparse and SPD. It has a pentadiagonal structure due to the 5-point stencil discretization. For a grid of dimensions $(n + 1) \times n$ the matrix A is of size $N = n \times n$. Stopping criteria are defined for convergence as $\| r_i \|_2 \leq \| b \|_2 \, \epsilon$, where r_i is the residual at the i-th step, b is the right-hand side and ϵ is the tolerance. For our experiments we have kept ϵ at 10^{-6}. The initial guess (x_0) is a random vector to avoid artificially fast convergence due to a smooth initial error.

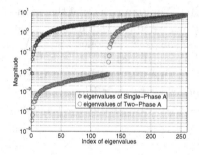

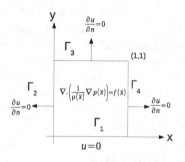

Fig. 1. 2D grid (16 × 16) with 256 unknowns. Jump at the interface due to density contrast.

Fig. 2. unit square with boundary conditions

Through this test case we can ascertain the effectiveness of deflation for such problems on the GPU. The final goal however remains to be able to make a solver capable of handling the linear systems arising in bubbly flow problems.

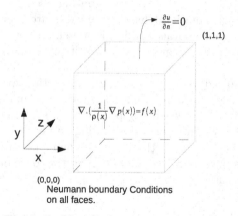

Fig. 3. Problem Definition. Unit cube in 3-D.

To this end we also define a test case with a unit cube with bubbles. This 3D formulation poses additional challenges and is a harder problem to solve due to many more small eigenvalues corresponding to the number of bubbles in the system. In Figure 4 we present two cases where there is a single bubble and 8 bubbles in the domain presented in Figure 3. The contrast between the densities of the bubble and the surrounding medium is of the same order as in the 2D problem.

In the 3D case we apply Neumann boundary conditions on all faces. The matrix is SPD and has a septadiagonal structure. The problem size is $N = n \times n \times n$. We maintain the same stopping criteria, tolerance and initial conditions as the 2D problem. The bubbles are placed symmetrically in the test cases (depicted in Figure 4) whose results we present in Section 6.2.

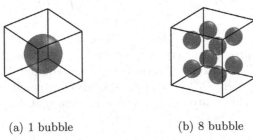

(a) 1 bubble   (b) 8 bubble

Fig. 4. 3 Geometries for Realistic Problems

3   Preconditioning Schemes

Preconditioning operation $y_i = M^{-1}r_i$ involves the preconditioning matrix M and the residual vector, r_i at the i-th iteration. The preconditioner matrix M, for our problem, is sparse.

We compare our results to the standard Block Incomplete Cholesky preconditioner (for which $M = LL^T$). We apply the block structure to A and generate L as suggested in [10]. The Block Incomplete Cholesky preconditioners in our results are suffixed with a number like $2n$, $4n$ etc. which denotes the block-size. So for example in a Block-IC preconditioner with blocksize $8n$ where the matrix A has $N = n \times n$ unknowns the preconditioner will be named like $M^{-1}_{Blk-IC(8n)}$. Since the data parallelism in Block Incomplete schemes is limited by the block-size (refer [1] for details) we turn our attention to preconditioners that have more inherent parallelism.

3.1   Neumann Series Based Preconditioning

We define the preconditioning matrix, $M = (I + LD^{-1})D(I + (LD^{-1})^T)$, where L is the strictly lower triangular part and D is the diagonal of A, the coefficient matrix. We apply the truncated Neumann Series for approximation of M^{-1}. Specifically for $(I + LD^{-1})$ (and similarly for $(I + (LD^{-1})^T)$) the series can be defined as

$$(I + LD^{-1})^{-1} = I - LD^{-1} + (LD^{-1})^2 - (LD^{-1})^3 + \cdots \text{ if } \| LD^{-1} \|_\infty < 1. \tag{2}$$

In our problem $\| LD^{-1} \|_\infty < 1$, hence the Neumann Series is a valid choice for approximating the inverse of $(I + LD^{-1})$. So we can redefine M^{-1} as

$$M^{-1} = (I - D^{-1}L^T + \cdots)D^{-1}(I - LD^{-1} + \cdots). \tag{3}$$

For making our preconditioners (computationally) feasible we truncate the series (2) after 1 or 2 terms. We refer to these as the Neu1 and Neu2 Preconditioners. Note that

$$M^{-1}_{Neu1} = (I - D^{-1}L^T)D^{-1}(I - LD^{-1}) \tag{4}$$

$$M^{-1}{}_{Neu2} = (I - D^{-1}L^T + (D^{-1}L^T)^2)D^{-1}(I - LD^{-1} + (LD^{-1})^2). \quad (5)$$

We define $K = (I - LD^{-1})$ for M^{-1}_{Neu1} and $K = (I - LD^{-1} + (LD^{-1})^2)$ for $M^{-1}{}_{Neu2}$. For the preconditioners as given by (4) and (5) we only store LD^{-1} and calculate $K^T D^{-1} K x$ term-by-term every time required. Every term in the expansion of $M^{-1}x = K^T D^{-1} K x$ can be (roughly) computed at the cost of one $LD^{-1}x$ operation. This is around $2N$ multiplications and N additions. This is only true for the stencil we discuss in this paper.

4   Deflation

To improve the convergence of our method further we also use a second level of preconditioning. Deflation aims to remove the remaining bad eigenvalues from the preconditioned matrix, $M^{-1}A$. This operation increases the convergence rate of the Preconditioned Conjugate Gradient (PCG) method. We define the matrices $P = I - AQ, Q = ZE^{-1}Z^T, E = Z^T AZ$, where $E \in \mathbb{R}^{d \times d}$ is the invertible Galerkin matrix, $Q \in \mathbb{R}^{N \times N}$ is the correction matrix, and $P \in \mathbb{R}^{N \times N}$ is the deflation operator. $Z \in \mathbb{R}^{N \times d}$ is the so-called 'deflation-subspace matrix' whose d columns are called 'deflation' or 'projection' vectors. The deflated system is now

$$PA\hat{x} = Pb. \quad (6)$$

The vector $\hat{x}$ is not necessarily a solution of the original linear system, since x might contain components in the null space of PA, $\mathcal{N}(PA)$. Therefore this 'deflated' solution is denoted as $\hat{x}$ rather than x. The final solution has to be calculated using the expression $x = Qb + P^T\hat{x}$. The deflated system (6) can be solved using a symmetric positive definite (SPD) preconditioner, M^{-1}. We therefore seek a solution of $M^{-1}PA\hat{x} = M^{-1}Pb$. The resulting method is called the Deflated Preconditioned Conjugate Gradient (DPCG) method as listed in

---

Algorithm 1. Deflated Preconditioned Conjugate Gradient Algorithm

1: Select x_0. Compute $r_0 := b - Ax_0$ and $\hat{r}_0 = Pr_0$, Solve $My_0 = \hat{r}_0$ and set $p_0 := y_0$.

2: **for** i:=0,..., until convergence **do**
3:     $\hat{w}_i := PAp_i$
4:     $\alpha_i := \frac{(\hat{r}_i, y_i)}{(p_i, \hat{w}_i)}$
5:     $\hat{x}_{i+1} := \hat{x}_i + \alpha_i p_i$
6:     $\hat{r}_{i+1} := \hat{r}_i - \alpha_i \hat{w}_i$
7:     Solve $My_{i+1} = \hat{r}_{i+1}$
8:     $\beta_i := \frac{(\hat{r}_{i+1}, y_{i+1})}{(\hat{r}_i, y_i)}$
9:     $p_{i+1} := y_{i+1} + \beta_i p_i$
10: **end for**
11: $x_{it} := Qb + P^T x_{i+1}$

Algorithm 1. We choose Sub-domain Deflation and use piecewise constant deflation vectors.We make stripe-wise deflation vectors (see Figure 7) unlike the block deflation vectors suggested in [7]. These deflation vectors lead to a regular structure for AZ and, therefore, an efficient storage of AZ.

In order to implement deflation on the GPU we have to break it down into a series of operations,

$$a_1 = Z^T r, \tag{7a}$$

$$a_2 = E^{-1} a_1, \tag{7b}$$

$$a_3 = AZ a_2, \tag{7c}$$

$$s = r - a_3. \tag{7d}$$

(7b) shows the solution of the inner system that results during the implementation of deflation.

5 Two Level Preconditioned Conjugate Gradient Implementation

The implementation of the Deflated Preconditioned Conjugate Gradient(DPCG) method follows Algorithm 1. The deflation operation requires solving the system $Ea_2 = a_1$ in every iteration. Also a matrix-vector product, AZa_2 is required in every iteration. The first operation can be performed in two different ways as we will see in Section 5.1. To optimize the second operation we store AZ in such a format such that we get the same number of operations, memory access pattern and (approximately) performance as the sparse matrix vector product Ax.

5.1 GPU Implementation of Deflation

We store the matrix A in the Diagonal (DIA) format and follow the implementation as detailed in [5]. For deflation, every iteration we have to solve the system $Ea_2 = a_1$. This can be done in two ways.

1. Calculating E^{-1} explicitly so that the $E^{-1}a_1$ becomes a dense matrix-vector product which can be calculated using the *gemv* routine from MAGMA BLAS library for the GPU.
2. Using triangular solve routines from the MAGMA BLAS library. Specifically we use the *dpotrs* and *dpotrf* functions ([12]).

The parallelism available in the second method drops for larger systems compared to the first method which is embarrassingly parallel on the GPU. On the other hand, in the first method calculation of E^{-1} (which is only done once in the setup phase) becomes expensive as the number of deflation vectors increases. In case of our test problem the setup times for the second method are one-third when compared to the first method (details in [1]). However, this one time calculation can make the operation $a_2 = E^{-1}a_1$ very quick on the GPU. So a selection of high-quality deflation vectors (such that $d << N$), which lead to a smaller E matrix and hence computationally cheaper inversion provides an advantage for a GPU implementation.

5.2   Storage of the Matrix AZ

The structure of the matrix AZ stored as an $N \times d$ matrix, where d is the number of domains/deflation vectors, can be seen in Figure 5. In Figures 5 to 7 it must be noted that $d = 2n$ here and $N = n \times n = 64, n = 8$. The AZ matrix is formed by multiplying the Z matrix (a part of which is shown in the adjoining figure of matrix AZ in Figure 5) with the coefficient matrix, A. The colored boxes indicate non-zero elements in AZ. They have been color coded to provide reference for how they are stored in the compact form. The red elements are in the same space as the deflation vector. The green elements result from the horizontal fill-in and the blue elements result from the vertical fill-in. The arrangement of the deflation vectors (on the grid) is shown in Figure 7. Each ellipse corresponds to the non-zero part of the corresponding deflation vector in matrix Z. The trick to store AZ in an efficient way (for the GPU) is to make sure that memory accesses are coalesced. For this we need to have a look at how the operation $a_3 = AZa_2$ works, where a_2 is a $d \times 1$ vector. For each element of the resulting vector a_3 we need an element from at most 5 different columns of the AZ matrix. Now it must be recalled that in case of A times x we have 5 elements of A in a single row multiplied with 5 elements of x as detailed in [5]. So we start looking at the different colored elements and group them so that the access pattern to calculate each element of a_3 is similar to the Sparse-Matrix

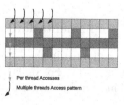

Fig. 6. AZ matrix after compression

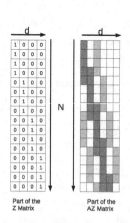

Part of the Z Matrix   Part of the AZ Matrix

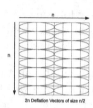

2n Deflation Vectors of size n/2

Fig. 5. Parts of Z and AZ matrix. number of deflation vectors $=2n$.

Fig. 7. Deflation vectors for the 8×8 grid

Vector Product operation. Wherever there is no element in AZ we can store a zero. Thus in the compacted form the $N \times d$ matrix AZ can be stored in $5N$ elements as illustrated in Figure 6. The golden arrows in Figure 6 show how each thread on the GPU can compute one element when the operation AZa_2 is performed where a_2 is a $d \times 1$ vector. The black arrows show the accesses done by multiple threads. This is similar to the DIA format of storage and calculating Sparse Matrix Vector Product as suggested in [5].

5.3   Extension to Real (Bubble) Problems and 3D

This storage format can be extended to include bubbles in the domain. In this case, only the values of coefficients change but the structure of the matrix remains the same. For a 3D problem, deflation vectors that correspond to planes or stripes can lead to an AZ matrix that is similar in structure compared to the matrix A and hence can be stored using the ideas presented in the previous section.

In Figure 8 we provide an example for a 3D scenario in order to explain what planar and stripe-wise vectors look like. One can notice that stripe wise vectors are piecewise constant vectors. We briefly talk about stripe-wise vectors. Every vector has length N. Each vector has ones for the row on which it is defined and zeroes for the rest of the column. Planar vectors are an extension of stripe-wise vectors and are defined on n^2 cells (have n^2 ones and rest of the column has zeroes). It must be noted that for a 3D problem the number of unknowns or problem size is $N = n^3$ where n is the size of the grid in any one dimension.

For our experiments in Section 6.2 we use n^2 stripe-wise and n planar vectors.

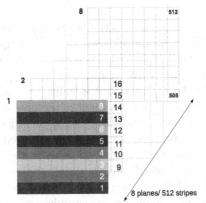

Fig. 8. Planes and stripes for a 8^3 uniform cubic mesh

6   Numerical Results

We performed our experiments on the hardware available with the Delft Institute of Applied Mathematics.

- For the CPU version of the code we used a single core of Q9550 @ 2.83 Ghz with 12MB L2 cache and 8 GB main memory.

- For the GPU version we used a NVIDIA Tesla(Fermi) C2070 with 6GB memory.

We use optimized BLAS libraries (MAGMA and ATLAS) on both GPU and CPU for daxpys, dot products and calculation of norms.

All times reported in this section are measured in seconds. The time we report for our implementations is the time taken (this excludes the setup time, specifically the steps 2 to 10 in Algorithm 1) for iterations required for convergence. In our results, speedup is measured as a ratio of this iteration time on the CPU versus the GPU. The effect of setup time vis-a-vis the iteration time is reported in detail in [1] (Figure 11 and 12 in Appendix A for quick reference). The setup phase includes the assigning and initializing the memory and the operations required to be done before entering the iteration loop, namely,

1. Assigning space to variables required for temporary storage during the iterations.
2. Making matrix AZ.
3. Making matrix E.
4. Populating x, b.
5. Doing the operations as specified in the first line of Algorithm 1 in Section 4.

It also involves the setup for the operation $Ea_2 = a_1$ using either of the two approaches mentioned in Section 5.1.

6.1   Stripe-Wise Deflation Vectors - Experiments with 2D Test Problem

For the 2D problem we have used $2n$ deflation vectors unless otherwise mentioned. For the DPCG implementation which uses Block Incomplete Cholesky as the first-level preconditioner, the difference in speedup between the two different implementations to compute coarse grid solution ($Ea_1 = a_2$) as mentioned in the previous section is negligible (Figure 9 in Appendix A). This is due to the fact that in this case the majority of the time is spent in the preconditioning step and it dominates the iteration time, so the effect of the deflation operation is overshadowed. However, for the Truncated Neumann Series based preconditioners the difference between GPU and CPU execution times is significant (Figure 10 in Appendix A) since preconditioner is highly parallelizable. Consequently the choice of inner solve in the deflation step becomes decisive in the length of execution time. The speedup attainable for the complete solver with explicit inverse (E^{-1}) based calculation of a_2 is four times that of the triangular solve strategy (Figure 10 in Appendix A). A comparison of how the wall-clock times for the different preconditioning algorithms vary for the DPCG method is presented in Table 1. Grid Size is $N = 1024 \times 1024$, $n = 1024$ and $2n$ deflation vectors have been used. These times and number of iterations shown in Table 2 are presented for the deflation implementation with explicit E^{-1} calculation.

In Table 2 we present the number of iterations required for convergence of different preconditioning schemes. The number of iterations is not affected by

Table 1. Wall Clock Times for DPCG on a 2D problem with $N = 1024 \times 1024$

Preconditioning Variant	CPU	GPU
$M^{-1}{}_{Blk-IC(2n)}$	28.4	9.8
$M^{-1}{}_{Blk-IC(4n)}$	25.48	10.15
$M^{-1}{}_{Blk-IC(8n)}$	22.8	11.28
$M^{-1}{}_{Neu1}$	20.15	1.29
$M^{-1}{}_{Neu2}$	25.99	1.47

Table 2. Iterations required for Convergence of 2D problem using DPCG with $2n$ deflation vectors

	Grid Sizes			
Preconditioning Variant	128^2	256^2	512^2	1024^2
$M^{-1}{}_{Blk-IC(2n)}$	76	118	118	203
$M^{-1}{}_{Blk-IC(4n)}$	61	98	98	178
$M^{-1}{}_{Blk-IC(8n)}$	56	86	91	156
$M^{-1}{}_{Neu1}$	76	117	129	224
$M^{-1}{}_{Neu2}$	61	92	101	175

the choice of implementations for the Deflation Method discussed in Section 5.1. It can be noticed that the results for the second type (Neu2) of Neumann Series based Preconditioner (with $K = (I - LD^{-1} + (LD^{-1})^2)$) lie between the Block-IC scheme with block sizes $4n$ and $8n$.

6.2 Stripe and Plane-Wise Deflation Vectors - Experiments with 3D Problems

It is possible to use stripes for 3D problems and problems involving bubbles as well. However, stripe-wise deflation vectors are not the best choice one can make for the deflation subspace. For 3D experiments we measure our results against an optimized CPU implementation that utilizes Sub-domain deflation vectors (block-shaped vectors). Block vectors do not suit the storage pattern that we have utilised for this study but they can also give good results. In Table 3 and 4 we see the results for a case when we have 3D geometries. For the first set of results presented in Table 3 the geometry is that of slabs of different material. It must be noted now that $N = n^3$ and <u>not</u> n^2. The computational domain is now a unit cube. We present the results with n plane and n^2 stripe-wise deflation vectors. There are three slabs in the unit cube. The middle slab is 0.5 units thick. Its density is 10^{-3} times the density of the surrounding slabs.

As we can see in the results of Table 3 the speedup drops. This is a consequence of the fact that the inner system takes a lot of time to solve now and the data structure and the associated kernels for the operation AZa_2 do not perform very well for very large number of deflation vectors. Moreover, if more (n^2) vectors are used the setup times become prohibitive and there is no speedup

Table 3. $3D$ Problem (128^3 points in the grid) with 3 layers. Middle layer 0.5 units thick. Tolerance set at 10^{-6}. Density contrast 10^{-3}. Comparison of CPU and GPU implementations.

	CPU[1]	GPU[2]	
	8 block vectors	128 plane vectors	16384 stripe vectors
	DICCG(0)	DPCG(neu2)	
Number of Iterations	206	324	259
Setup Time	0.3	0.36	148.5
Iteration Time	35.18	7.66	112
Speedup	-	4.59	–

at all. The iteration times are high since we use the triangular solve method for inner system. In Table 4 we continue to have a unit cube but instead of slabs of different material we now consider bubbles in the system. In particular, we have a single bubble with its center coinciding with the center of the cube and another case when we have eight bubbles, 2 in each dimension and equally spaced (refer Figure 4). It can be noticed from the results that the speedup becomes worse for the problem with more bubbles and that can be explained by the fact that stripe-wise vectors cut the bubbles and are poor approximations of the eigenvectors of the preconditioned matrix.

Table 4. $3D$ Problem (128^3 points in the grid) with 1 and 8 bubbles. Tolerance set at 10^{-6}. Density contrast 10^{-3}. Comparison of CPU and GPU implementations.

1 bubble		
	CPU[1]	GPU[2]
	8 block vectors	128 plane vectors
	DICCG(0)	DPCG(neu2)
Number of Iterations	237	287
Setup Time	0.31	0.64
Iteration Time	40.44	6.79
Speedup	-	5.95
8 bubble		
Number of Iterations	142	402
Setup Time	0.3	0.36
Iteration Time	24.4	9.51
Speedup	-	2.56

In Tables 3 and 4 the GPU version uses triangular solves for the inner system since with explicit solve and stripe-wise vectors the round-off errors in the solution of the inner system (due to explicit inverse calculation) grow very quickly

---

[1] CPU version uses CG for inner system solve.

[2] GPU version uses triangular factorization based inner solve.

and convergence is never achieved. We only show the results with n vectors in Table 4 since with n^2 vectors there is no speedup.

7 Conclusions and Future Work

We have shown how two level preconditioning can be adapted to the GPU for computational efficiency. In order to achieve this we have investigated preconditioners that are suited to the GPU. At the same time we have made new data structures in order to optimize deflation operations.

Through our results we demonstrate that the combination of Truncated Neumann based preconditioning and deflation proves to be computationally efficient on the GPU. At the same time its numerical performance is also comparable to the established method of Block-Incomplete Cholesky Preconditioning.

The approach of using stripe-wise vectors is applicable to 3D problems and problems with bubbles in the domain. However, these deflation vectors, though simple to implement are not the most effective choice for the deflation of more ill-conditioned problems.

Through this study we have learnt that the choice made in the implementation of deflation method is crucial for the overall run-time of the method. We are now continuing to extend our work on 3D problems with bubbles. We believe that the approach of calculating inverse of the matrix E explicitly can be very effective for the GPU. In order to overcome the possibly large setup time of this scheme and to avoid delayed convergence we are now working on better deflation vectors based on Level-Set Sub-domain deflation. A small number of these vectors can capture the small eigenvalues and result in an effective deflation step (this is discussed in [7]). This directly translates to a low setup time and overall gain in this approach of implementing deflation.

A Detailed Results

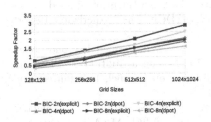

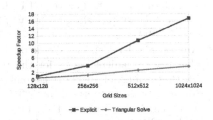

Fig. 9. Comparison of Explicit versus triangular solve strategy for DPCG. Block-IC Preconditioning with $2n$, $4n$ and $8n$ block sizes.

Fig. 10. Comparison of Explicit versus triangular solve strategy for DPCG. Neumann Series based Preconditioners $M^{-1} = K^T D^{-1} K$, where $K = (I - LD^{-1} + (LD^{-1})^2)$

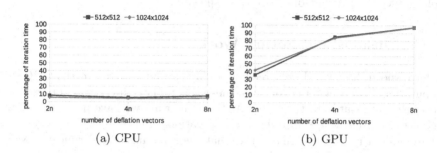

(a) CPU   (b) GPU

Fig. 11. Setup Time as percentage of the total (iteration+setup) time for triangular solve approach across different sizes of deflation vectors for DPCG

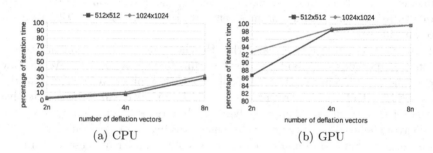

(a) CPU   (b) GPU

Fig. 12. Setup Time as percentage of the total (iteration+setup) time for explicit E^{-1} approach across different sizes of deflation vectors for DPCG

References

1. Gupta, R., van Gijzen, M.B., Vuik, C.: Efficient Two-Level Preconditioned Conjugate Gradient Method on the GPU, Reports of the Department of Applied Mathematical Analysis, Delft University of Technology, Delft, The Netherlands. Report 11–15 (2011)
2. Ament, M., Knittel, G., Weiskopf, D., Strasser, W.: A Parallel Preconditioned Conjugate Gradient Solver for the Poisson Problem on a Multi-GPU Platform. In: Proceedings of the 18th Euromicro Conference on Parallel, Distributed, and Network-based Processing, pp. 583–592 (2010)
3. Jacobsen, D.A., Senocak, I.: A Full-Depth Amalgamated Parallel 3D Geometric Multigrid Solver for GPU Clusters. In: Proceedings of 49th AIAA Aerospace Sciences Meeting including the New Horizons Forum and Aerospace Exposition, AIAA 2011-946, pp. 1–17 (January 2011)
4. Bolz, J., Farmer, I., Grinspun, E., Schröder, P.: Sparse matrix solvers on the GPU: conjugate gradients and multigrid. ACM Trans.Graph. 22, 917–924 (2003)
5. Bell, N., Garland, M.: Efficient Sparse Matrix-Vector Multiplication on CUDA, NVR-2008-04, NVIDIA Corporation (2008)

6. Van der Pijl, S.P., Segal, A., Vuik, C., Wesseling, P.: A mass conserving Level-Set Method for modeling of multi-phase flows. International Journal for Numerical Methods in Fluids 47, 339–361 (2005)
7. Tang, J.M., Vuik, C.: New Variants of Deflation Techniques for Pressure Correction in Bubbly Flow Problems. Journal of Numerical Analysis, Industrial and Applied Mathematics 2, 227–249 (2007)
8. Helfenstein, R., Koko, J.: Parallel preconditioned conjugate gradient algorithm on GPU. Journal of Computational and Applied Mathematics 236, 3584–3590 (2011)
9. Heuveline, V., Lukarski, D., Subramanian, C., Weiss, J.P.: Parallel Preconditioning and Modular Finite Element Solvers on Hybrid CPU-GPU Systems. In: Proceedings of the Second International Conference on Parallel, Distributed, Grid and Cloud Computing for Engineering, paper 36 (2011)
10. Meijerink, J.A., van der Vorst, H.A.: An Iterative Solution Method for Linear Systems of Which the Coefficient Matrix is a Symmetric M-Matrix. Mathematics of Computation 31, 148–162 (1977)
11. Jönsthövel, T.B., van Gijzen, M., MacLachlan, S., Vuik, C., Scarpas, A.: Comparison of the deflated preconditioned conjugate gradient method and algebraic multigrid for composite materials. Computational Mechanics 50, 321–333 (2012)
12. MAGMABLAS documentation, http://icl.cs.utk.edu/magma/docs/

Parallelization of the QR Decomposition with Column Pivoting Using Column Cyclic Distribution on Multicore and GPU Processors

Andrés Tomás[1], Zhaojun Bai[1], and Vicente Hernández[2]

[1] Department of Computer Science,
University of California, Davis, CA 95616, USA
{andres,bai}@cs.ucdavis.edu
[2] Dept. Sistemas Informáticos y Computaciòn,
Universitat Politècnica de València E-46022 Valencia, Spain
vhernand@dsic.upv.es

Abstract. The QR decomposition with column pivoting (QRP) of a matrix is widely used for rank revealing. The performance of LAPACK implementation (DGEQP3) of the Householder QRP algorithm is limited by Level 2 BLAS operations required for updating the column norms. In this paper, we propose an implementation of the QRP algorithm using a distribution of the matrix columns in a round-robin fashion for better data locality and parallel memory bus utilization on multicore architectures. Our performance results show a 60% improvement over the routine DGEQP3 of Intel MKL (version 10.3) on a 12 core Intel Xeon X5670 machine. In addition, we show that the same data distribution is also suitable for general purpose GPU processors, where our implementation obtains up to 90 GFlops on a NVIDIA GeForce GTX480. This is about 2 times faster than the QRP implementation of MAGMA (version 1.2.1).

Topics. Parallel and Distributed Computing.

1 QR Decomposition with Column Pivoting

The QR decomposition with column pivoting (QRP) is proposed for computing a rank revealing QR factorization (RRQR) [7]. Although QRP may fail to reveal the numerical rank correctly, it is still a popular and economical method in many applications. The QRP is also used as the first step to more robust RRQR methods [2,8] and for accelerating a Jacobi method for computing the singular value decomposition [5,6].

The QRP decomposition of $A \in \mathbb{R}^{m \times n}$ is defined by an orthonormal Q and a upper triangular matrix R such that

$$AP = QR,$$

where P is a permutation matrix chosen so that

$$|r_{11}| \geq |r_{22}| \geq \cdots \geq |r_{nn}|$$

M. Daydé, O. Marques, and K. Nakajima (Eds.): VECPAR 2012, LNCS 7851, pp. 50–58, 2013.
© Springer-Verlag Berlin Heidelberg 2013

and moreover, for each i,

$$|R_{ii}| \geq \|R_{(k:j,j)}\|_2 \quad \text{for } j = i+1, \ldots, n.$$

An outline of the Householder QRP algorithm is shown in Algorithm 1. Note that the formula for the column norm updating is simplified here. The current LAPACK implementation uses a more robust approach [4]. The omitted detail is not relevant to the parallelization discussed in this paper.

Algorithm 1. QR decomposition with column pivoting

```
1   p_{1:n} = 1 : n
2   c_{1:n} = ||Ae_{1:n}||²₂
3   for j = 1 : n
4       Choose i such that c_i = max(c_{j:n})
5       if i ≠ j
6           swap(p_i, p_j); swap(A_i, A_j); swap(c_i, c_j)
7       end
8       Determine a Householder matrix H_j such that
9           H_j A_{j:m,j} = ±||A_{j:m,j}||₂ e₁
10      A_{j:m,j+1:n} = H_j A_{j:m,j+1:n}
11      c_{j+1:n} = c_{j+1:n} - A_{j,j+1:n} · A_{j,j+1:n}
12  end
```

The main difference among the various implementations of Algorithm 1 is how the Householder reflectors are applied. Since a Householder reflector H is a rank-one modification of the identity,

$$H = I - \tau v v^T,$$

its application requires the computation

$$HA = A - \tau v v^T A,$$

which can be implemented using three levels of BLAS. The current LAPACK implementation (xGEQP3) is block based. It groups several rank-one updates for exploiting the Level 3 BLAS operations [9].

Figure 1 shows the performance of Intel MKL[1] routine DGEQP3 on a 12 core (6 cores per socket) Intel Xeon X5670 machine. Here the execution is set up to use one thread per core. The poor performance of DGEQP3 compared to DGEQRF is because of the extensive use of the Level 2 BLAS operation DGEMV for column norm updates, which is limited by the memory bandwidth and do not scale with the number of cores. Figure 2 shows the amount of time spent by DGEQP3 on calls to DGEMV and DGEMM relative to the total execution time on the same platform. The reported amount of time for DGEMV does not include the unblocked part of DGEQP3 which processes the last block of the matrix.

---

[1] Intel MKL version 10.3 http://software.intel.com/en-us/intel-mkl

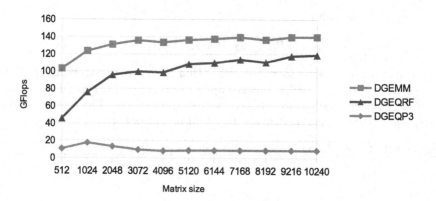

Fig. 1. Performance of Intel MKL routines DGEMM, DGEQRF, and DGEQP3 on a 12 core (6 cores per socket) Intel Xeon X5670 machine

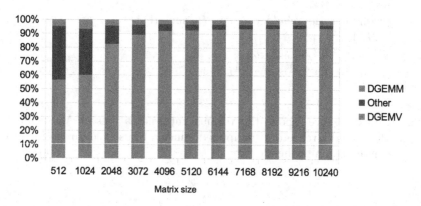

Fig. 2. Execution time in percentage of Intel MKL routines DGEMV and DGEMM in DGEQP3 on a 12 core (6 cores per socket) Intel Xeon X5670 machine

Although DGEQP3 uses the YTY^T representation [10] like DGEQRF, it is not fully blocked because of the column norm updating. The column norm updating requires to compute the row vector $v^T A$ for each Householder reflection applied to A. DGEQP3 updates the columns of A every k (block size) times. However, it still has to fetch the whole trailing matrix for the matrix-vector product. As the matrix size gets too large to fit entirely in cache memory, the performance of DGEQP3 decreases to the level of DGEMV.

2   Parallel QRP for Multicore Processors

The design of LAPACK DGEQP3 routine is to enclose parallelism inside BLAS routines. In a typical multicore implementation this means that each BLAS routine contains at least one OpenMP parallel section. Therefore, for each call to BLAS, a whole set of threads is started and stopped. This thread management overhead is negligible for Level 3 BLAS operations, but it could be very significant for Level 1 and 2 BLAS operations due to the low computational intensity, namely, the low average number of floating point operations per memory access.

In contrast, we propose the following Algorithm 2 to use only one OpenMP parallel section. The parallelism here is not inside BLAS operations, but among the vector computations required for all columns. The critical parts of the algorithm are implemented with synchronization primitives, which is more efficient than starting and stopping threads.

Algorithm 2 is a block algorithm, where the block size is denoted as b. We assume without loss of generality that the matrix size is an exact multiple of b. The loop from lines 6 to 29 performs the panel factorization, that is, the QR factorization of the first b columns. The only sequential part of this loop is the pivot selection and computation of the Householder transform (lines 7 to 13). The rest of the loop updates the matrix F used to accumulate part of the Householder matrices application,

$$H_1 H_2 \cdots H_k A = A - YTY^T A = A - YF^T.$$

The last loop in Algorithm 2 applies the Householder matrices to the rest of the matrix (lines 30 to 35). Recent work on using a parallel cache assignment approach to speed up the panel factorization can be found in [3].

Algorithm 2 processes the columns of the matrix in their natural order from left to right. On a parallel machine, it is natural to group the processors into a logical ring and deal columns in a round-robin fashion. This technique staggers the computation across the processors and guarantees a load balanced computation. This distribution was first proposed in the context of the parallel implementation of a QR decomposition with local pivoting [1]. The selection of which columns are processed by each thread is not left to the OpenMP runtime, but explicitly controlled in lines 8, 15, and 31.

With the column cyclic distribution, each thread is ensured to work with the same subset of matrix columns during all the processes. If the OpenMP runtime guarantees processor affinity, this will provide good memory locality at the lower levels of memory hierarchy. This is important in modern multicore processors where each core has typically its own L1 cache. As each thread works only with a subset of the columns, there is a good probability of accessing a column already stored in the L1 cache inside the core associated with this thread.

Algorithm 2. OpenMP parallel QRP using column cyclic distribution

```
1   p_{1:n} = 1 : n
2   #pragma omp parallel
3       i = omp_get_thread_num(); t = omp_get_num_threads()
4       c_{1:n} = ||Ae_{1:n}||_2^2
5       for r = 1 : b : n − 1
6           for k = 1 : b
7               #pragma omp barrier
8               if r + k − 1 mod t = i
9                   Choose u such that c_u = max(c_{j:n})
10                  if u ≠ j then swap(p_u, p_j); swap(A_u, A_j); swap(c_u, c_j)
11                  Apply previous transformations to A_j ← A_j + YF^T
12                  Determine Householder matrix H_j
13              end
14              #pragma omp barrier
15              for j = r : r + k − 1
16                  if j mod t = i then T_{j−r+1,k} = −τ_{r+k}Y_{r+k−1}Y_j
17              end
18              #pragma omp barrier
19              for j = r : n
20                  if j mod t = i
21                      F_{k,j} = F_{:,j}T_{:,k}
22                      if j > r + k − 1
23                          F_{k,j} ← F_{k,j} − τ_{r+k}Y_{r+k−1}A_j
24                          A_{r+k−1,j} ← Y_{r+k−1,:}F_{:,j}^T
25                          c_r = c_r − A_{j,r}^2
26                      end
27                  end
28              end
29          end
30          #pragma omp barrier
31          for j = r + b : n
32              if j mod t = i
33                  A_{j:m,j} ← A_{j:m,j} + F_{:,j}^T Y_{j,:}
34              end
35          end
36      end
```

As the number of cores increases on modern multicore processors, the architecture is gearing towards a non-uniform memory access (NUMA) model. On these architectures, each core has direct access to a part of the memory, but the rest of the memory must be accessed via some communication network to other core. This network is implemented by the cache hardware and is transparent to the user. Therefore, these processors can be still programmed using the same shared memory model as previous multicore processors. However, the memory

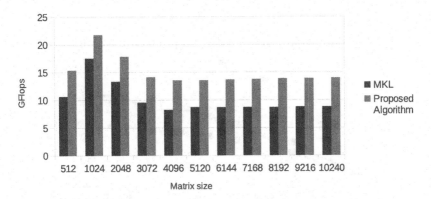

Fig. 3. Performance comparison of the proposed algorithm and Intel MKL routine DGEQP3 on a 12 core (6 cores per socket) Intel Xeon X5670 machine

access latency could have large variations depending on which part of the memory is accessed. In order to get good performance on these processors, techniques from the distributed memory programming paradigm can be used to reduce the communication among cores.

By the column distribution of Algorithm 2, a straightforward memory distribution can be easily derived. The memory physically close to each core should contain the columns updated by the thread associated to this core. This can be implemented in current operating systems by allocating and filling this memory from the thread itself. This technique is known in the literature as *first touch* policy. As Algorithm 2 creates all threads at the start, this initialization can be efficiently performed at the start of the process.

To guarantee that each consecutive column is stored in a different physical memory page, the matrix must padded so the column length is a multiple of the page size. This memory overhead could be important for small matrices, because the typical page size on current platforms is 4096 bytes.

3 Performance Results

Figure 3 shows the results from an OpenMP implementation of Algorithm 2 on a 12 core (6 cores per socket) Intel Xeon X5670 machine. Here the execution of the proposed algorithm is set up to use one thread for each available core. The block size for the proposed algorithm is set to 48, which it has been determined empirically as the optimal for the platform. Moreover, the performance of the proposed algorithm includes the cost of initializing the data distribution from the standard Fortran matrix storage.

The proposed algorithm shows about 60% improvement over the optimized DGEQP3 in Intel MKL for the matrix sizes tested. This improvement comes from the data distribution, which allows faster memory access as a result of better data locality and parallel memory bus utilization. Moreover, the proposed algorithm

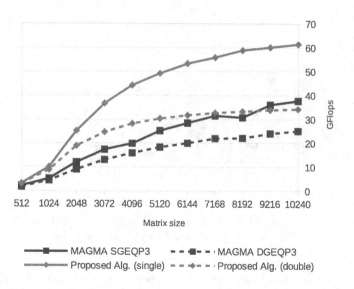

Matrix size

MAGMA SGEQP3 ▪▪▪▪▪▪ MAGMA DGEQP3
Proposed Alg. (single) ▬ ▬ ◆ ▬ ▬ Proposed Alg. (double)

Fig. 4. Performance of the proposed GPU implementation and MAGMA implementation of xGEQP3 on NVIDIA Tesla C2050

has less thread management overhead than the implementation which keeps parallelism inside BLAS operations.

4   Parallel QRP for GPU Processors

In order to achieve good performance on a general purpose GPU processor, the computation must be divided into independent parallel subtasks. Moreover, each subtask must be also suitable to efficient parallelization by a certain number of processors (typically a multiple of 32). The parallel distribution of Algorithm 2 can be easily adapted to the GPU parallel model, assigning each column to a block of threads. If the matrix is sufficiently large, there is enough work to keep all processors in a block busy, and enough blocks to keep the whole GPU busy. The main difference with the multicore version is that the memory distribution is not required, because the memory access on current GPU processors is uniform.

Figures 4 and 5 compare the GPU performance of Algorithm 2 and MAGMA's xGEQP3 routines[2] on two NVIDIA Fermi platforms. In single precision, the GPU implementation of Algorithm 2 obtains about 60 GFlops on Tesla C2050 and 90 GFlops on GeForce GTX480. This is about two times faster than the MAGMA implementation using the same hardware. The improvement is because our implementation runs entirely on the GPU, with no memory transfers from the CPU. In contrast, the panel factorization of xGEQP3 in MAGMA is performed on the CPU and the trailing matrix update on the GPU. This approach works quite well for the LU decomposition with partial pivoting and the QR without

---

[2] MAGMA version 1.2.1 released on June 29, 2012. http://icl.cs.utk.edu/magma/

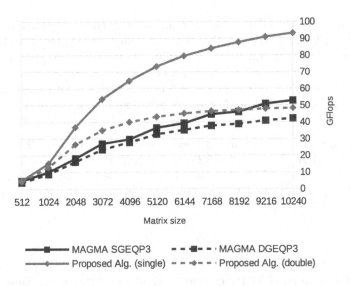

Matrix size

———■——— MAGMA SGEQP3   – – ■ – – ·MAGMA DGEQP3
———◆——— Proposed Alg. (single) – – ◆ – – ·Proposed Alg. (double)

Fig. 5. Performance of the proposed GPU implementation and MAGMA implementation of xGEQP3 on NVIDIA GeForce GTX480

pivoting, because the panel factorization can be computed in parallel while the GPU is still updating the trailing matrix with the previous block. However, the overlap of computation and communication is not possible for the QR decomposition with pivoting. The pivoting criteria requires that the trailing matrix update must be completed before starting the panel factorization.

The performance results confirm that the QRP decomposition is limited by memory speed. Therefore a GPU platform is more adequate for computing the QRP than a traditional CPU because of raw memory bandwidth. Another interesting observation is that the low-end GeForce GTX480 obtains better performance than the high-end Tesla C2050 even in double precision. This is because the C2050 has less memory bandwidth (in part due to ECC checking, which could not be disabled in our experiments).

5   Conclusions

We proposed a parallel algorithm for computing the QRP decomposition on multicores. This algorithm uses a column cyclic memory distribution and only one parallel OpenMP section. With the column cyclic distribution each processor works with a subset of the columns, improving memory access bandwidth and data locality. Moreover, it has lower thread management overhead than the implementations which only use parallelism inside BLAS operations. The proposed algorithm is about 60% faster than Intel MKL routine DGEQP3 on a 12 core Intel Xeon X5670 machine.

Although the column cyclic data distribution is not required on a GPU, the same strategy is employed to allocate work among the processors. Our CUDA implementation of the QRP is about 2 times faster than the MAGMA version of xGEQP3 on NVIDIA Fermi GPUs. Our implementation runs entirely on the GPU, while MAGMA's implementation splits the work between CPU and GPU, which requires expensive data transfers. In other decompositions, such as LU or QR, these transfers can be overlapped with computations, but this optimization cannot be applied to QRP because of the pivoting selection criteria.

Acknowledgment. Tomás and Bai were supported in part by the U.S. DOE SciDAC grant DOE-DE-FC0206ER25793 and NSF grant PHY1005502. This research used resources of the National Energy Research Scientific Computing Center, which is supported by the Office of Science of the U.S. DOE under Contract No. DE-AC02-05CH11231.

References

1. Bischof, C.H.: A parallel QR factorization algorithm with controlled local pivoting. SIAM J. Sci. Stat. Comput. 12, 36–57 (1991)
2. Chandrasekaran, S., Ipsen, I.C.F.: On rank-revealing factorisations. SIAM J. Matrix Anal. Appl. 15, 592–622 (1994)
3. Castaldo, A.M., Whaley, R.C.: Scaling LAPACK panel operations using parallel cache assignment. In: 15th ACM SIGPLAN Annual Symposium on Principles and Practice of Parallel Programming, pp. 223–231 (2010)
4. Drmač, Z., Bujanović, Z.: On the failure of rank-revealing QR factorization software – a case study. ACM Trans. Math. Softw. 35, 12:1–12:28 (2008)
5. Drmač, Z., Veselić, K.: New fast and accurate Jacobi SVD algorithm I. SIAM J. Matrix Anal. Appl. 29, 1322–1342 (2008)
6. Drmač, Z., Veselić, K.: New fast and accurate Jacobi SVD algorithm II. SIAM J. Matrix Anal. Appl. 29, 1343–1362 (2008)
7. Golub, G.H.: Numerical methods for solving linear least squares problems. Numer. Math. 7, 206–216 (1965)
8. Gu, M., Eisenstat, S.: Efficient algorithms for computing a strong rank-revealing QR factorization. SIAM J. Sci. Comput. 17, 848–869 (1996)
9. Quintana-Orti, G., Sun, X., Bischof, C.H.: A BLAS-3 version of the QR factorization with column pivoting. SIAM J. Sci. Comput. 19, 1486–1494 (1998)
10. Schreiber, R., van Loan, C.: A storage-efficient WY representation for products of Householder transformations. SIAM J. Sci. Stat. Comput. 10, 53–57 (1989)

A High Performance SYMV Kernel
on a Fermi-core GPU

Toshiyuki Imamura[1,2,4], Susumu Yamada[3,4], and Masahiko Machida[3,4]

[1] University of Electro-Communications, Chofu-shi, Tokyo 182-8585, Japan
[2] RIKEN Advanced Institute for Computational Science, Kobe-shi,
Hyogo 650-0047, Japan
[3] CCSE, Japan Atomic Energy Agency, Kashiwa-shi, Chiba 277-8587, Japan
[4] CREST, Japan Science and Technology Agency

Abstract. A high-performance SYMV kernel is implemented on Fermi-core GPUs using an atomic-operation based algorithm. The algorithm is effective for the memory bandwidth and reduced memory usage. On a Tesla C2050, sustained double-precision and single-precision performances of approximately 43 GFLOPS and 78 GFLOPS, respectively, were achieved. The proposed SYMV kernel also performs on a GeForce GTX580 with 72 GFLOPS and 128 GFLOPS in the double-precision and single-precision modes, respectively. The proposed SYMV kernel outperforms major CUDA BLAS kernels, CUBLAS, MAGMABLAS, and CULA-BLAS. This performance improvement has a significant impact when the SYMV kernel is plugged into user codes.

1 Introduction

In the development of our eigensolver [1], it was very difficult to speed up Householder tridiagonalization. Detailed cost analysis revealed the cost of the kernel SYMV to be extremely high, and cost reduction is a very important problem. With emerging GPGPU, costly kernels such as SYMV and SYR2K in Householder tridiagonalization can be accelerated. Nath et al. [2] reported the optimization of the SYMV kernel in their MAGMABLAS library, which had faster performance than the CUBLAS library. Consequently, their Householder tridiagonalization routine, magma_dsytrd, performs very well. In the present paper, we present a new implementation of the SYMV kernel in order to realize a new eigenvalue solver, and we demonstrate the performance of this kernel on a single Fermi core GPU, such as an NVIDIA Tesla C2050 or an NVIDIA GeForce GTX580.

2 SYMV Kernel

The SYMV is a kernel function for a symmetric-matrix vector product and is categorized in Level 2 BLAS. Since the cost of operation and the amount of data in SYMV are each $O(n^2)$, performance bounds are based on memory bandwidth.

M. Daydé, O. Marques, and K. Nakajima (Eds.): VECPAR 2012, LNCS 7851, pp. 59–71, 2013.
© Springer-Verlag Berlin Heidelberg 2013

It is difficult to achieve sufficient performance on a general CPU, on which the memory bandwidth is limited to at most 30 or 40 [GB/s]. Therefore, better performance is expected to contribute to a wider global memory bandwidth of the GPU (140 [GB/s] on a GTX280, for example). Compared with other Level 2 kernels (GEMV, for example), the symmetry of the matrix helps to reduce data accesses between memory and processor. In other words, improving the algorithm is expected to provide higher performance.

2.1 SYMV Algorithms

Taking symmetry into account, the Fortran code can be written as follows:

```
w(1:n)=0
do i=1,n
   y0=0
   do j=1,i-1
      y0 =y0 +a(j,i)*x(j)
      w(j)=w(j)+a(j,i)*x(i)
   enddo
   y(i)=y0+a(i,i)*x(i)
enddo
y(1:n)=y(1:n)+w(1:n)
```

The vector w, which can be replaced by y, is introduced in order to clarify the CUDA algorithm. Based on the source lines shown above, the framework of the SYMV algorithm specified with the 'U' option for a CUDA environment is presented in Figure 1. The algorithm is divided primarily into three kernels: pre-processing (corresponding to w(1:n)=0)), post-processing (corresponding to y(1:n)+=w(1:n)), and the main process. The main kernel consists of three parts, calculation on non-diagonal blocks, calculation on a diagonal block, and sumup of the registers on each thread. In Figure 1, the outer-most loop represented by counter i is expanded UX times, and the thread-block is organized into a one-dimensional array of threads, where BLOCK_SIZE is the number of threads issued.

Since Figure 2 shows the data access pattern for a specific thread-block represented in block.id, data updating of vector w is performed by multiple thread-blocks. In order to secure updating of vector w, we need an exclusive control mechanism. There are several variations of implementations in which vector w is multiplexed and exclusive (or mutex) control is fully obligated. In the remainder of this section, we would like to explain three algorithms with regard to an exclusive control mechanism on the SYMV kernel.

2.2 Atomic Algorithm

The algorithm shown in Figure 3 (top) uses atomic operations or a mutex mechanism, and so is referred to as the *Atomic algorithm*. Since CUDA 4.x does not

```
<01> // variable n refers to the dimension of the matrix A.
<02> // variables y_{*} and w_{0} refer to registers.
<03> // array s is on shared memory.
<04> // assume blockDim>=UX.
<05> // sumup adds up the value of the specified register in a block.
<06> kernel kernel_preprocess
<07>     set j := thread.id + block.id * block.Dim.
<08>     if j < n then
<09>         w[j] := 0.
<10>     endif
<11> endkernel
<12> kernel kernel_main
<13>     define j ≡ j̄ + thread.id.
<14>     foreach (thread, block) do
<15>         set i:=UX*block.id.
<16>         y_{0} := ... := y_{UX−1} := 0.
<17>         // part one / sweep along the column block
<18>         CORE of either the Algorithm Atomic, Blocked, or Ticket in Fig. 3
is called here.
<19>         // part two / calculation on a diagonal part
<20>         for j̄:=thread.id to UX-1 do
<21>             s(thread.id, j) := s(j, thread.id) := a(i+thread.id, i + j).
<22>         endfor
<23>         sync threads in a block
<24>         if thread.id<UX then
<25>             y_{k} += s(thread.id, k) * x[i + k]  for k ∈ [0, UX).
<26>         endif
<27>         // part three
<28>         s(k, thread.id) := sumup(y_{k})  for k ∈ [0, UX).
<29>         if thread.id<UX then
<30>             y[i + thread.id] += s(thread.id, thread.id).
<31>         endif
<32>     endfor
<33> endkernel
<34> kernel kernel_postprocess
<35>     set j := thread.id + block.id * block.Dim.
<36>     if j < n then
<37>         y[j] += w[j], or y[j] += ∑ w(j, :).
<38>     endif
<39> endkernel
```

Fig. 1. Framework of the SYMV Algorithm

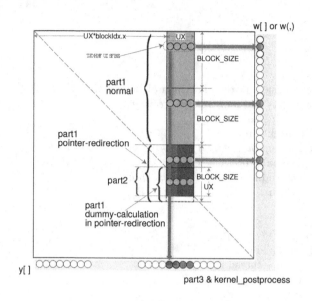

Fig. 2. Schematic diagram of the data access pattern on the SYMV kernel

support atomicAdd for double precision, critical section controls (mutex lock and unlock) are implemented using `atomicCAS` and `atomicExch` functions.

In our implementation, the atomic functions are issued on the master threads to avoid incurring the serialization of all working threads. Thus, loss of thread serialization can be minimized. Following the CUDA thread model, it is reasonable to secure the update of vector w, even if the operations to be in the Atomic algorithm are relatively small. Consequently, we expect that the total cost on exclusive controls is reduced. This lock-and-unlock mechanism is a general model and can be applied to other BLAS kernel implementations, such as GEMV and spMV, to reduce memory usage on working buffers.

2.3 Blocked Algorithm

Figure 3 (middle) shows the variation for exclusive control of the Atomic algorithm. This algorithm adds a second index to the variable w and requires no exclusive control. Thus, data access to w is implicitly blocked, and we refer to this algorithm as the *Blocked algorithm*. This is also known as the 'scatter and gather technique'. From the viewpoint of the multiplicity of vector w, this is similar to the algorithm adopted in MAGMABLAS when UX=1.

2.4 Ticket Algorithm

On the other hand, we can generate another intermediate algorithm between the Atomic algorithm and the Blocked algorithm. In Figure 3 (bottom), vector w

```
// CORE of the Atomic Algorithm
    for j̄:=0 to i − 1 − thread.id step BLOCK_SIZE do
        y_{k} += a(j, i + k) * x[j]  for k ∈ [0, UX).
        w_{0} := ∑_{k∈[0,UX)} a(j, i + k) * x[i + k].
        // equivalent to atomic (w[j] += w_{0}) in the CUDA semantics.
        mutex_lock @ the masterthread and syncthreads.
        w[j] += w_{0}.
        syncthreads and mutex_unlock @ the masterthread.
    endfor
```

```
// CORE of the Blocked Algorithm
    for j̄:=0 to i − 1 − thread.id step BLOCK_SIZE do
        y_{k} += a(j, i + k) * x[j]  for k ∈ [0, UX).
        w_{0} := ∑_{k∈[0,UX)} a(j, i + k) * x[i + k].
        w(j, block.id) += w_{0}.
    endfor
```

```
// CORE of the Ticket Algorithm
    ticket_id = get_ticket_id( ).
    for j̄:=0 to i − 1 − thread.id step BLOCK_SIZE do
        y_{k} += a(j, i + k) * x[j]  for k ∈ [0, UX).
        w_{0} := ∑_{k∈[0,UX)} a(j, i + k) * x[i + k].
        w(j, ticket_id) += w_{0}.
    endfor
// 'int Ticket_id_master' is initialized at the preprocessing step.
__device__ int function get_ticket_id( )
    __shared__ int shred_buff;
    if is_master_thread then
        shared_buff := atomicInc( &Tickect_id_master, gridDim.x );
    endif
    syncthreads; block_id := mod(shared_buff, M); syncthreads
    return  block_id
end function
```

Fig. 3. Core Algorithms (Top: Atomic, Middle: Blocked, Bottom: Ticket)

is multiplexed M times (M should be a multiple of the number of SM's). This limits mutex control among activated thread-blocks and consequently reduces the number of atomic operations. In this case, since exclusive control is open not only for a single thread-block but also for multiple thread-blocks, this algorithm is similar to a seat reservation model, such as for reserving train tickets. Therefore, this algorithm is referred to as the *Ticket algorithm*.

2.5   Another Kernel Implementation (L+U Algorithm)

There is another implementation of SYMV. A symmetric matrix A is represented by the sum of upper and lower triangle matrices by taking into account the symmetry of the matrix

$$A = L + D + U = \tilde{L}(= L + D) + U(= L^t).$$

Using the decomposition, SYMV can be calculated by $Ax = \tilde{L}x + Ux$. We refer to the above expression as the *L+U algorithm*, which can be easily implemented using GEMV kernels modified for upper and lower triangle matrices. In the present paper, the GEMV kernel codes presented in [3] are modified and used for the SYMV kernel. Since this algorithm requires two kernels, the upper triangular part (Ux) and the lower triangular part ($\tilde{L}x$), the overhead to start up GPU kernels is quite small. Therefore, the L+U algorithm offers better performance when the matrix dimension is small.

2.6   Pointer Redirection Optimization

Pointer redirection [2] is an optimization technique for CUDA programming that produces coherent running threads on a specific loop. When all of the threads in a thread-block proceed with the loop and a number of these threads access out of bound on array accesses, invalid pointers are modified in order to access proper addresses. In the present case, invalid accesses to matrix a and vector x are adjusted to access their top element. In the case of vector w, an invalid pointer is redirected to a dummy variable on global memory in order to protect w from an invalid overwrite.

3   Experiment Results

The primary experiments were conducted on an NVIDIA Tesla C2050 GPU and an NVIDIA GeForce GTX580, and we use an NVIDIA GeForce GTX280 for a preliminary test. Their specifications and software environment are summarized in Table 1. All of the performance tests include only kernel execution without host-device data transfer. In other words, CUDA BLAS kernels operate in the non-thunking mode. The performance parameters (BLOCK_SIZE, UX) are chosen to be (256, 16) and (128, 26) for the DP and SP modes, respectively.

Table 1. Hardware and software specifications of GPU's used in the present study
(*Theoretical peak performance is referred from [5])

	Tesla C2050	GTX580	GTX280
The core architecture	Fermi	Fermi	GT200
The number of CUDA cores	448	512	240
Processor core clock [GHz]	1.15	1.544	1.296
Peak performance [DP/SP GFLOPS]*	515/1030	393/1573	78/933
Memory capacity [GB]	3	1.5	1
Memory bandwidth [GB/s]	144	192.4	141.7
Host CPU	Core i7-860	Core i7-2600K	Phenom 9750
Frequency [GHz]	2.8	3.4	2.4
CUDA Compute Capability	2.0	2.0	1.3
CUDA version	4.0	4.1	4.1
NVIDIA Linux driver	275.09.07	295.71	295.71
GNU gcc version	4.4.5	4.5.3	4.5.3

3.1 Preliminary Performance Prediction

We first theoretically discuss the performance bound for the SYMV. As a result of symmetry, memory access to an element of matrix A corresponds to two multiply-add operations. Thus, the memory requirement per DP operation is computed by 'BF' := sizeof(element)[Byte]/4[flop] = 2 [Byte/flop]. Based on the benchmark reports, e.g., [4], the sustained memory bandwidth of a C2050 with ECC switched on is calculated to be approximately 99 [GB/s]. The optimal SYMV performance is 99/BF [GFLOPS]. Therefore, the SYMV kernel is bounded by 49.5 and 99 [GFLOPS] in cases of the DP and SP modes, respectively.

3.2 Comparison of Four Algorithms

The Atomic, Ticket, and Blocked algorithms have no difference in the core computation part, except for exclusive operation and sumup of vector w. Table 2 summarizes the differences in parts 1 and 3 of these algorithms. Table 2 indicates that the Atomic algorithm has a significant advantage with respect to the cost of w, the computational cost, and memory usage. In contrast, the Atomic algorithm requires far more atomic operations than other algorithms.

Here, we should have two scenarios according to the cost of the atomic operations. If the cost of atomic operations is too high, we expect *Atomic > Ticket < Blocked*. Thus, the Ticket algorithm has an advantage in such cases. On the other hand, if the cost of atomic operations is small, we expect *Atomic < Ticket < Blocked*, which leads us to select the Atomic algorithm.

Table 3 shows the elapsed time and the overhead cost for the above three algorithms as well as the L+U algorithm on an NVIDIA Tesla C2050. Tables 2 and 3 suggest that the overhead cost is proportional to the memory usage of w. This tendency was reported by Nath et al. [2], and the results of the present

Table 2. Complexity analysis for three algorithms ($L \equiv \lceil N/\text{BLOCK\_SIZE}\rceil$)

	words of extra data	cost of sumup w	max(atomic ops. per thread-block)
Atomic	$N + L$	N	$2L$
Ticket	$NM + M$	NM	2
Blocked	$N\lceil N/\text{UX}\rceil$	$N\lceil N/\text{UX}\rceil$	0

Table 3. Analysis of SYMV calculation for four algorithms in the DP mode on a Tesla C2050 (top: total time [s], bottom: overhead time [s])

			matrix dimension		
	1,000	2,000	4,000	6,000	8,000
Atomic	.252E-3	.395E-3	.968E-3	.189E-2	.319E-2
	.118E-3	.126E-3	.127E-3	.119E-3	.124E-3
Ticket	.429E-3	.611E-3	.127E-2	.234E-2	.380E-2
	.306E-3	.343E-3	.383E-3	.407E-3	.450E-3
Blocked	.426E-3	.632E-3	.160E-2	.294E-2	.492E-2
	.266E-3	.376E-3	.694E-3	.103E-2	.159E-2
L+U	.139E-3	.406E-3	.147E-2	.321E-2	.566E-2

Table 4. Analysis of SYMV calculation for four algorithms in the DP mode on a GeForce GTX280 (top: total time [s], bottom: overhead time [s])

			matrix dimension		
	1,000	2,000	4,000	6,000	8,000
Atomic	.275E-2	.514E-2	.914E-2	.110E-1	.149E-1
	.387E-4	.364E-4	.399E-4	.405E-4	.400E-4
Ticket	.398E-3	.579E-3	.119E-2	.211E-2	.347E-2
	.236E-3	.264E-3	.305E-3	.360E-3	.415E-3
Blocked	.346E-3	.556E-3	.142E-2	.270E-2	.454E-2
	.236E-3	.289E-3	.503E-3	.855E-2	.139E-2
L+U	.167E-3	.470E-3	.163E-2	.361E-2	.624E-2

study are similar. Furthermore, the cost of atomic operations is not significant compared to the total computational time. Table 4 also shows the elapsed time and the overhead cost for an NVIDIA GeForce GTX280, which is prior to the Fermi GPU core architecture. Since the costs of the Ticket algorithm and the Blocked algorithm on a GeForce GTX280 are almost equivalent to those on a Tesla C2050, the large performance difference between a Tesla C2050 and a GeForce GTX280 stems from the cost of the atomic operations. A technical paper by NVIDIA reported that a Fermi core adopts fast and greatly improved atomic memory operations, compared to GT200 cores [6], and that the cost of atomic operations on a GTX280 is quite high. The Ticket algorithm is a better algorithm in such a case. Thus, the Atomic algorithm has better overall performance than the Ticket and Blocked algorithms in the case of the current Fermi core architecture. However, if the cost imbalance between calculation and atomic operations is severe in a future architecture, the selected algorithm will differ.

Furthermore, Tables 3 and 4 suggest that the L+U algorithm performs with a tiny overhead. In fact, the L+U algorithm has better performance when the matrix dimension is smaller ($N = 1,000$). Therefore, on a Tesla C2050, we switch the algorithm in the DP mode, using the L+U algorithm when $N < 2,020$ and the Atomic algorithm when $N \geq 2,020$. In the case of the SP mode, the border between the L+U algorithm and the Atomic algorithm is $N = 2,990$.

3.3 Performance Test

In the present paper, we measured the performance of SYMV kernels for the present implementation (hybrid of the Atomic and L+U algorithms) and three major BLAS implementations: CUBLAS [7], MAGMABLAS [8] (run in the L-mode), and BLAS in CULA [9]. Here, the performance is calculated by $2N^2/$ 'elapsed time'.

Tesla C2050. The performance on a single NVIDIA Tesla C2050, which is a high-end Fermi-core GPU card, is presented in Figure 4. The release versions of the CUDA BLAS's are as follows: CUBLAS R4.0, MAGMABLAS R1.2.1 (run in the L-mode), and CULABLAS R12. In [2], the performance on the SP mode for MAGMABLAS on a C2050 exceeds 80 GFLOPS; however, we could not obtain that performance in the present C2050 environment. In order to compare the throughput from the FLOPS rates fairly, the results on a C2050 are based on the performance measurement in the present paper.

In the DP and SP modes, the performance of the proposed SYMV kernel reaches 43 and 78 GFLOPS, respectively, when the matrix dimension is 18,000. These values represent speed-ups of approximately 2.5 times and 4.0 times, respectively, when CUBLAS is set as a baseline. Furthermore, performances of 86% and 78% of the upper bound, respectively, are achieved.

GeForce GTX580. The performance on a single NVIDIA GeForce GTX580, which is also a Fermi-core GPU, is presented in Figure 5. The release versions of the CUDA BLAS's are as follows: CUBLAS R4.1, MAGMABLAS R1.2.1 (run in the L-mode), and CULABLAS R14. In the DP and SP modes, the performance of the proposed SYMV kernel achieves 71 and 128 GFLOPS, respectively, when the matrix dimension is 12,000. These values represent speed-ups of approximately 2.8 times and 3.3 times, respectively, in comparison to CUBLAS.

3.4 Discussion and Related Research

Compared with other implementations of BLAS, small fluctuations in the SYMV appear upon varying the matrix dimension, whereas the performance of the

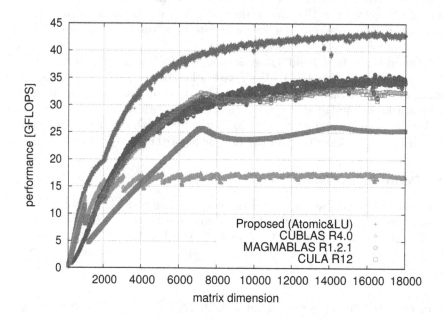

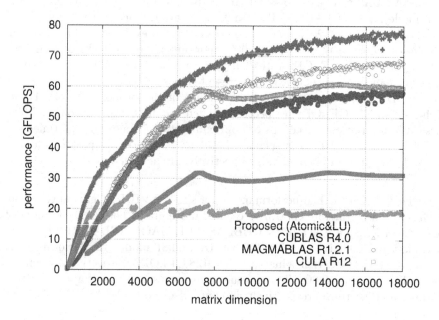

Fig. 4. Performance results on a Tesla C2050 in non-thunking mode (Top: DSYMV = DP mode, bottom: SSYMV = SP mode)

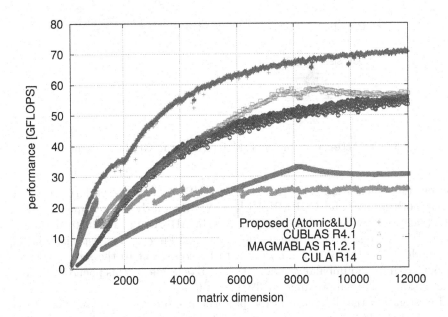

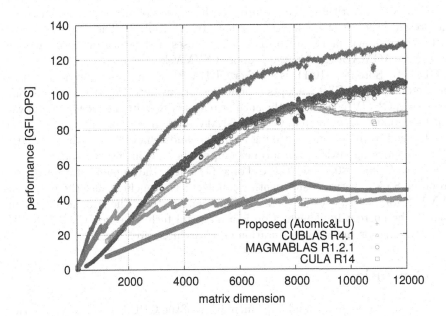

Fig. 5. Performance results on a GeForce GTX580 in non-thunking mode (Top: DSYMV = DP mode, bottom: SSYMV = SP mode)

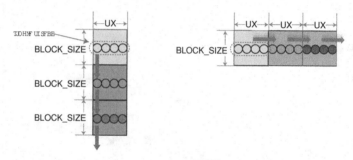

Fig. 6. Access patterns in GEMV-T (left) and GEMV-N (right)

present implementation is the most stable. CUBLAS exhibits performance with a sawtooth profile, and CULABLAS behaves irregularly for multiples of 32 or 64 dimensions, which is also true for MAGMABLAS on a C2050 in the SP mode. The proposed SYMV kernel also has a small fluctuation, especially in the SP mode. Since the period is observed to be equivalent to the size of UX, we recognize that load imbalance among thread-blocks affects this phenomena. The fluctuation reaches approximately 3% and is negligible.

Investigating the difference in the performance profile by reading the source code of MAGMABLAS reveals that MAGMABLAS is realized by the GEMV-N-based algorithm (access through the same row direction). On the other hand, the GEMV-T algorithm (access through the same column direction) used in the present study is the enhanced version presented in [3]. Their typical access patters are shown in Figure 6. This also causes a difference in performance between MAGMABLAS and the proposed SYMV kernel.

GLAS by Sørensen [10,11] is another CUDA BLAS implementation that uses atomic operations. GLAS adopts atomic operations not in a SYMV kernel but rather in a GEMV-N kernel at the current release, wherein `atomicAdd` or equivalent functionality is emulated. GLAS's GEMV implementation on a Tesla C2050 is optimized by an automatic tuning technique and achieved approximately 90% of the memory bandwidth (performance upper bound) in the SP mode.

For the Tesla C2050 and the GeForce GTX580, the implementation of the proposed SYMV kernel is sufficiently optimized. We conclude that the proposed SYMV algorithm is, overall, the most stable and fastest. The Atomic algorithm used in the current implementation is a powerful technique in CUDA GPGPU programming for the Fermi generation GPU architecture.

4   Conclusion

We have presented an optimal implementation of the GPU kernel for symmetric-matrix vector multiplication, referred to as the SYMV kernel. The proposed SYMV kernel uses the Atomic algorithm, which requires very little extra working memory. In the DP and SP modes, the proposed SYMV kernel performs at 43 and 78 GFLOPS on a Tesla C2050, respectively. The implementation of the SYMV kernel is herein demonstrated to provide remarkable memory consumption and performance. Since a generic processor performs at from 2 to 5 GFLOPS on DSYMV, the impact of the proposed SYMV kernel is enormous.

In the future, we would like to examine the proposed SYMV kernel on Kepler, which is a new GPU core architecture. Furthermore, we would like to apply the SYMV kernel to a previously proposed eigensolver [1].

Acknowledgements. The present study was supported in part by the Ministry of Education, Science, Sports and Culture through a Grant-in-Aid for Scientific Research (B), 21300013, and through a Grant-in-Aid for Scientific Research on Innovative Areas, 22104003.

References

1. Imamura, T., Yamada, S., Machida, M.: Development of a High Performance Eigensolver on the Peta-Scale Next Generation Supercomputer System, the Atomic Energy Society of Japan. Progress in Nuclear Science and Technology 2, 643–650 (2011)
2. Nath, R., Tomov, S., et al.: Optimizing symmetric dense matrix-vector multiplication on GPUs. In: Proc. of the Intl. Conf. High Performance Computing, Networking, Storage and Analysis, SC 2011 (2011)
3. Imamura, T.: Performance-stabilization and automatic performance tuning for DGEMV routines on a CUDA environment. IPSJ Journal, Transaction of Advanced Computing Systems, ACS 4(4), 158–168 (2011) (in Japanese)
4. Schäfer, A., Fey, D.: High Performance Stencil Code Algorithm for GPGPUs. In: Proc. of ICCS 2011, Procedia Computer Science, vol. 4, pp. 2077–2036 (2011)
5. Hwu, W.W. (ed.): GPU Computing Gems Jade Edition (Applications of GPU Computing Series). Morgan Kaufmann (2011)
6. NVIDIA: whitepaper NVIDIA's Next Generation CUDA Compute Architecture: Fermi,
   http://www.nvidia.com/content/PDF/
   fermi_white_papers/NVIDIAFermiComputeArchitectureWhitepaper.pdf
7. NVIDIA: CUDA CUBLAS Library, http://developer.download.nvidia.com
8. Agullo, E., Demmel, J., et al.: Numerical linear algebra on emerging architectures: The PLASMA and MAGMA projects. J. of Physics: Conference Series 180 (2009)
9. Humphrey, J.R., Price, D.K., et al.: CULA: Hybrid GPU Accelerated Linear Algebra Routines. In: SPIE Defense and Security Symposium (DSS) (2010)
10. Sørensen, H.H.B.: Auto-tuning Dense Vector and Matrix-Vector Operations for Fermi GPUs. In: Wyrzykowski, R., Dongarra, J., Karczewski, K., Waśniewski, J. (eds.) PPAM 2011, Part I. LNCS, vol. 7203, pp. 619–629. Springer, Heidelberg (2012)
11. GPUlab: GLAS library version 0.0.2,
    http://gpulab.imm.dtu.dk/docs/glas_v0.0.2_C2050_cuda_4.0_linux.tar.gz

Optimizing Memory-Bound SYMV Kernel on GPU Hardware Accelerators

Ahmad Abdelfattah[1], Jack Dongarra[2], David Keyes[1], and Hatem Ltaief[3]

[1] KAUST Division of Mathematical and Computer Sciences and Engineering,
Thuwal, Saudi Arabia
[2] Innovative Computing Laboratory, University of Tennessee, Knoxville TN USA
[3] KAUST Supercomputing Laboratory, Thuwal, Saudi Arabia

Abstract. Hardware accelerators are becoming ubiquitous high performance scientific computing. They are capable of delivering an unprecedented level of concurrent execution contexts. High-level programming language extensions (e.g., CUDA), profiling tools (e.g., PAPI-CUDA, CUDA Profiler) are paramount to improve productivity, while effectively exploiting the underlying hardware. We present an optimized numerical kernel for computing the symmetric matrix-vector product on nVidia Fermi GPUs. Due to its inherent memory-bound nature, this kernel is very critical in the tridiagonalization of a symmetric dense matrix, which is a preprocessing step to calculate the eigenpairs. Using a novel design to address the irregular memory accesses by hiding latency and increasing bandwidth, our preliminary asymptotic results show 3.5x and 2.5x fold speedups over the similar CUBLAS 4.0 kernel, and 7-8% and 30% fold improvement over the Matrix Algebra on GPU and Multicore Architectures (MAGMA) library in single and double precision arithmetics, respectively.

1 Introduction

GPUs have been, for a long time, dedicated for graphics processing. However, their increasing level of parallelism and computing capability have drawn attention in the HPC community, as low cost, low power, and high Gflop/s processing units. The latest architecture released by nVidia, codenamed Fermi, has a theoretical peak of 1 Tflop/s for single precision (SP), and about 500 Gflop/s for double precision (DP). Fermi has been highlighted as the first complete GPU computing architecture [5], with a complete memory hierarchy, ECC support, IEEE 754-2008 compliant floating point performance, and many novel features. Due to the drastic change from the previous GPU architecture, further tuning of existing numerical kernels is required to efficiently exploit new features in the Fermi architecture, in order to boost the performance.

One of the critical numerical kernels in dense linear algebra is the symmetric matrix-vector multiplication (SYMV). The kernel is, by nature, memory-bandwidth (BW) bound. It is a core step in computing the eigenpairs of a dense symmetric matrix. Having irregular memory access pattern due to the symmetric

M. Daydé, O. Marques, and K. Nakajima (Eds.): VECPAR 2012, LNCS 7851, pp. 72–79, 2013.

property of the matrix, the kernel design on GPUs is challenging. We present a novel design of the SYMV kernel. We try to exploit the new features introduced in Fermi. Most of the techniques used in this design target hiding memory latency and increasing memory bandwidth. When it comes to GPU programming for high performance, there are a lot of knobs to tune a kernel design. However, investigating all these knobs is daunting and time consuming. Therefore, we rely on performance counters to profile existing SYMV kernels in order to detect and identify weak points, where possible improvements can be made. PAPI CUDA Component [3] and the nVidia Compute Profiler [2] were the main performance counter tools used during the design process. The new kernel design is tested against two open-source SYMV kernels: the nVidia's CUBLAS 4.0 implementation and the Matrix Algebra on GPU and Multicore Architectures (MAGMA) 1.0.0-rc5 [1] implementation. MAGMA SYMV kernel [9] was tuned for Fermi. Our preliminary design is 3.5x better than CUBLAS 4.0 and 7-8% better than MAGMA in SP, while the speedup is about 2.5x over CUBLAS 4.0 and 1.3x over MAGMA in DP.

The rest of the paper is organized as follows. Section 2 discusses some previous work. Section 3 describes our proposed design in the SYMV kernel. Sections 4 and 5 present experimental and profiler results, respectively. Section 6 shows the impact of the new design on the overall symmetric eigenvalue problem. We summarize and propose some future work in Section 7.

2   Related Work

Accelerator-based hardware are employed in many HPC software libraries and applications, where they often outperform homogeneous x86 architecture in performance, power consumption, and cost-effectiveness. The STI Cell processor and GPUs have already been used in accelerating dense linear algebra ([7], [11] and [10]) as well as stencil computations [4].

An up-to-date highly tuned SYMV kernel was recently presented in [9]. The basic idea is to divide the matrix A into square blocks. Each Streaming

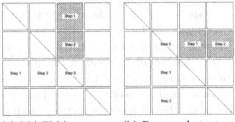

(a) MAGMA strategy (b) Proposed strategy

Fig. 1. Proposed computation strategy against MAGMA strategy. The vertical movement of thread blocks in (b) is more suitable for column major formats.

Multiprocessor (SM) is responsible for one or more blocks. The kernel launches as many thread blocks as the number of diagonal matrix blocks. Each thread block is responsible for exactly one block-row. Figure 1(a) shows an example thread block movement. Each non-diagonal block is computed in two fashions: transposed and non-transposed. Partial results from transposed computations are written to global memory so that the correct thread blocks can consume them. The MAGMA implementation is, therefore, divided into two kernel calls. The first one does the computation. The second kernel is a final reduction step through global memory. Recursive blocking [9] was used to save shared memory usage in GPUs. In addition, pointer redirecting was adopted to handle matrix dimensions that are not multiples of the block dimension. The next section describes the design outlines of our proposed kernel and how it differs from the MAGMA kernel strategy.

3   Kernel Description

GPU kernels are conceptually designed following two main strategies. The first one (the *block-level strategy*) is how thread blocks travel throughout the matrix blocks. The second one (the *thread-level strategy*) is how a single matrix block is processed by one thread block. The first strategy has to optimize memory accesses through global memory and L2 cache, while the second strategy goes deeper into the memory hierarchy i.e., registers and L1 cache/shared memory, to optimize processing block elements through efficient use of single SM's limited resources.

The new design has similar block-level strategy to the MAGMA kernel, with the exception it organizes memory accesses more efficiently. Moreover, as opposed to MAGMA, there are three successive kernel calls in the proposed design. The first kernel is a computation kernel for diagonal blocks only. The second one is a computation kernel for the non-diagonal blocks. The third kernel is a final reduction step done through global memory, which is very similar to the MAGMA kernel. The reason for separating the computation into three kernels will be shortly apparent. The proposed design divides the matrix into 64×64 blocks. This is an auto-tuning result obtained from MAGMA's internal parameters. In the first kernel, we launch as many thread blocks as the number of the diagonal blocks. When a thread block finishes computation, the partial result (64 element-vector representing the block row) is written into global memory.

The second kernel has the same number of threads as the first kernel. Each thread block travels vertically through the matrix (Figure 1(b)). This is a more memory-friendly scheme compared to MAGMA, since blocks are fetched in compliance with the data layout (column-major format). This scheme achieves thus better profiling in terms of number of load instructions from global memory and L2 cache than MAGMA (see Section 5).

Going at a lower level in the kernel design (the *thread-level strategy*), each diagonal block computation produces a partial result, a 64-element vector. A non-diagonal block computation produces two 64-element vectors. We enumerate the new contributions in this strategy.

Separating Different Computation Patterns. Diagonal blocks have different processing strategy than non-diagonal blocks. Therefore, they require different resources in terms of registers and shared memory. Since one SM can host multiple thread blocks, separating different computation strategies can allow multiple thread blocks/SM for kernels that are not resource-consuming. This is the main reason why the diagonal block computation has been separated from non-diagonal block computation.

Data Prefetching. Data prefetching [6] arises almost everywhere in our design. Each block is divided into smaller pieces, which we refer to as *chunks*. A software pipeline is implemented to hide the memory latency by prefetching the next chunk of data, while a current chunk is being processed. This is a burden on the GPU memory resources, so organizing the work between threads has to be within the physical resource limit allowed per thread as well as per SM. Figures 2(a) and 2(b) describe how data prefetching is applied to diagonal and non-diagonal blocks, respectively. In the non-diagonal case, prefetching spans blocks; while processing the second chunk of a given block, the first chunk of the next matrix block is being prefetched.

Using More Registers. A very important feature of our kernel is that it completely avoids computing partial products in shared memory. Shared memory is used only in a final reduction step before a partial result of an entire block is written into global memory. This feature avoids paying a penalty in terms of potential shared memory bank conflicts. It also reduces the occurrences of synchronization points. Using registers pays off very well, especially when register spilling to local memory is avoided. This is guaranteed on Fermi as long as each thread uses 63 registers or less.

4   Experimental Results

All experiments were executed on a single Fermi C2070 GPU, with 448 cores and 6 GB of DRAM, connected to a machine with dual socket quad core Intel Xeon processor, running at 2.67GHz, and with 24 MB of main memory. The kernel is implemented using CUDA C v4.0 and originally designed for matrices of dimensions that are multiples of 64. For other irregular dimensions, the matrix is padded with zeros inside the SM shared memory and registers. No padding is done in global memory.

Figures 3(a) and 3(b) show the performance results (in Gflop/s) for SP and DP, respectively. The proposed design is far better than the CUBLAS 4.0 kernel. There are some dips in the SP performance, which we are trying to resolve. Overall, there is a 7-8% improvement over MAGMA in SP. The performance gap widens in DP and reaches more than 30%. Although our kernel is mainly tuned for DP, the smaller improvement seen for SP against MAGMA is explained below along with the memory performance analysis.

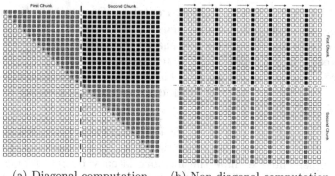

(a) Diagonal computation.   (b) Non-diagonal computation.

Fig. 2. Computation strategy inside a block. In (a), diagonal blocks are processed as two chunks. Hashed elements are loaded from DRAM then overwritten in a mirroring step. Black elements are not loaded at all from memory. Their values are loaded from shared memory during the mirroring step. In (b), non-diagonal blocks are also divided into two chunks. Threads are originated at the black elements. As threads move from left to right in the upper chunk, they prefetch hashed elements from the lower chunk in their registers.

Since the kernel is memory bound, the reported performance numbers are far below the theoretical floating point peak performance. However, we can get intuition about the quality of the kernel design by translating Gflop/s into GB/s to see how close we are from the Fermi peak memory bandwidth. Fermi C2070 GPUs have theoretical peak memory bandwidth of 144 GB/s (with ECC turned on). However, the actual (sustainable) peak memory bandwidth is about 103 GB/s (when ECC is on). This information is obtained by running a CUDA implementation of the STREAM benchmark [8]. The memory bandwidth is calculated by dividing the amount of useful data loaded/stored from/into global memory by the total runtime of the kernel. For the SYMV kernel, and a matrix of dimension N, the total amount of useful data is from A, X, and Y, that is, $\frac{1}{2}N(N+1) + 2N$ elements, where each element consumes 4 bytes in SP and 8 bytes in DP.

Figures 3(c) and 3(d) show the memory bandwidths of the SP and DP kernel versions. Our kernel scores about 70% (SP) and 80% (DP) of the actual peak memory bandwidth. This is 7-8% (SP) and 30% (DP) better than MAGMA, and 250% (SP) and 140% (DP) better than CUBLAS 4.0. It is interesting to see how the improvements in memory bandwidth matches those of performance. As previously mentioned, memory bandwidth improvement in SP is less than in DP. Running the same DP kernel for SP means saving more registers per thread, and loading less data each time. We thought that doubling the block size as well as the number of threads would result in memory bandwidth similar to the DP case. However, we were not able to double the number of threads in an SM because we are already using the maximum possible number on Fermi.

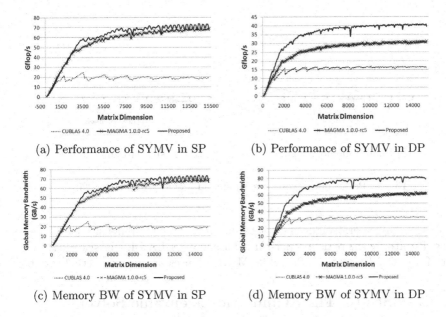

(a) Performance of SYMV in SP

(b) Performance of SYMV in DP

(c) Memory BW of SYMV in SP

(d) Memory BW of SYMV in DP

Fig. 3. Performance of the SYMV kernel in SP and DP on Fermi C2070

5 Performance Analysis

In this section, we analyze the performance of the new kernel, by studying the performance counters obtained from the nVidia and PAPI-CUDA [3] profilers. All three kernels were tested for matrix dimensions up to 10000. We selected the most relevant performance counters to the proposed kernel study. All results in this section are for the DP kernel. The first performance counter is the number of 64-bit load instructions made to the global memory. In general, going to global memory is a penalty, so the less we refer to global memory the better. Our experiments shows that the proposed design achieves 17% less load instructions than CUBLAS, and 13% less load instructions than MAGMA. Although the improvement is not significant, it could potentially have strong impact on performance, due to the huge penalty of going to global memory.

In addition, shared memory has higher latency than registers. Since we minimize the usage on shared memory, Figures 4(a) and 4(b) show that we refer less to shared memory and thus, pay much less penalty in terms of bank conflicts. The burden is rather put on registers, which are faster to read and compute, and do not have restrictions of the load pattern. It is noteworthy to mention that CUBLAS does not encounter any bank conflicts, though being the slowest kernel.

Two final performance counters are SM activity and registers-per-thread usage. Surprisingly, CUBLAS 4.0 took the lead for occupancy at 98.36%, followed by our design at 94.54%, and MAGMA at 80.90%. This result again shows it is indeed critical to consider all performance counters, when judging the kernel

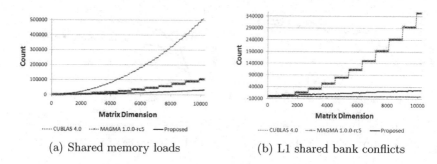

(a) Shared memory loads   (b) L1 shared bank conflicts

Fig. 4. Performance counters for shared memory on Fermi C2070

quality. A single performance metric cannot reflect a comprehensive performance view. Regarding the registers-per-thread usage, CUBLAS 4.0 uses the least number of registers/thread i.e., 29, while MAGMA uses 51 and our kernel uses 63.

6   Case Study: The Symmetric Eigenvalue Solver

The proposed DSYMV (in DP) was integrated into MAGMA, and a test was made for the tridiagonalization routine (DSYTRD) and the overall symmetric eigensolver (DSYEVD). We repeated the tests for MAGMA, and for CUBLAS. Results are shown in Figures 5(a) and 5(b). The new DSYTRD improves asymptotically by 88% with CUBLAS SYMV and by 20% with MAGMA SYMV. Looking at the overall symmetric eigensolver, the new DSYEVD is about 66% better with CUBLAS SYMV and about 17% better with MAGMA SYMV.

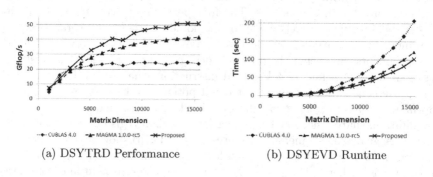

(a) DSYTRD Performance   (b) DSYEVD Runtime

Fig. 5. Impact of tuned SYMV kernels on DSYTRD and DSYEVD

7   Summary and Future Work

This paper introduces an optimized kernel for computing the symmetric matrix-vector product on Fermi GPUs. The kernel achieves 3.5x (SP) and 2.5x (DP) fold speedups over CUBLAS 4.0, and 7-8% (SP) and 30% (DP) improvement over MAGMA, similarly to the memory bandwidth. One possible extension to the work presented in this paper is to consider the load imbalance in the block-level strategy. The vertical movement encounters different loads for thread blocks. We intend to apply a 1D block cyclic distribution of non-diagonal blocks. Non-diagonal blocks are to be mapped in a periodic manner over the available number of SMs (14 on Fermi C2070), as done in [7]. Although this scheme might not be friendly with respect to the column-major data layout, we expect that the load balance can compensate for this penalty, especially if a tile data layout within each block is considered.

Acknowledgements. We would like to thank Timothy Lanfear (nVidia), for providing the STREAM benchmark for CUDA, and Rajib Nath (UCSD) for his help in understanding MAGMA design outlines of the SYMV kernel.

References

1. Matrix Algebra on GPU and Multicore Architectures. Innovative Computing Laboratory, University of Tennessee, http://icl.cs.utk.edu/magma/
2. Nvidia visual profiler, http://developer.nvidia.com/nvidia-visual-profiler
3. Performance Application Programming Interface (PAPI). Innovative Computing Laboratory, University of Tennessee, http://icl.cs.utk.edu/papi/
4. Datta, K., Williams, S., Volkov, V., Carter, J., Oliker, L., Shalf, J., Yelick, K.: Auto-tuning the 27-Point Stencil for Multicore. In: Proc. iWAPT 2009: The Fourth International Workshop on Automatic Performance Tuning (2009)
5. Glaskowsky, P.N.: nVidia's Fermi: The first complete gpu computing architecture. Technical report (2009)
6. Kirk, D., Mei Hwu, W.: Programming Massively Parallel Processors, A Hands-on Approach. Morgan Kaufmann (2010)
7. Kurzak, J., Buttari, A., Dongarra, J.J.: Solving systems of linear equations on the CELL processor using Cholesky factorization. IEEE Transactions on Parallel and Distributed Systems 19(9), 1–11 (2008)
8. McCalpin, J.: Stream: Sustainable memory bandwidth in high performance computers, http://www.cs.virginia.edu/stream/
9. Nath, R., Tomov, S., Dong, T., Dongarra, J.: Optimizing symmetric dense matrix-vector multiplication on gpus. In: Proceedings of 2011 International Conference for High Performance Computing, Networking, Storage and Analysis, SC 2011, pp. 6:1–6:10. ACM, New York (2011)
10. Nath, R., Tomov, S., Dongarra, J.: Accelerating GPU Kernels for Dense Linear Algebra. In: Palma, J.M.L.M., Daydé, M., Marques, O., Lopes, J.C. (eds.) VECPAR 2010. LNCS, vol. 6449, pp. 83–92. Springer, Heidelberg (2011)
11. Volkov, V., Demmel, J.W.: Benchmarking GPUs to Tune Dense Linear Algebra. In: Proceedings of the 2008 ACM/IEEE Conference on Supercomputing, SC 2008, pp. 31:1–31:11. IEEE Press, Piscataway (2008)

Numerical Simulation of Long-Term Fate of CO_2 Stored in Deep Reservoir Rocks on Massively Parallel Vector Supercomputer

Hajime Yamamoto[1], Shinichi Nanai[1], Keni Zhang[2], Pascal Audigane[3], Christophe Chiaberge[3], Ryusei Ogata[4], Noriaki Nishikawa[5], Yuichi Hirokawa[5], Satoru Shingu[5], and Kengo Nakajima[6]

[1] Taisei Corporation, Yokohama, Japan
[2] Beijing Normal University, Beijing, China
[3] Bureau de Recherches Geologiques et Minieres, Orléans, France
[4] NEC Corporation, Tokyo, Japan
[5] Japan Agency for Marine-Earth Science and Technology, Yokohama, Japan
[6] The University of Tokyo, Tokyo, Japan
hajime.yamamoto@sakura.taisei.co.jp

Abstract. As one of the promising approaches for reducing greenhouse-gas content in the atmosphere, CCS (carbon dioxide capture and storage) has been recognized worldwide. CO_2 is captured from large emission sources and injected and stored in deep reservoir rocks, including saline aquifers, depleted oil and gas field. Under typical pressure and temperature conditions at deep reservoirs (depths > 800m), CO_2 will be stored in supercritical state, subsequently dissolving in groundwater, and eventually forming carbonate minerals through geochemical reactions in a long-term (e.g., thousands of years). To ensure the safety and permanence of the storage, numerical simulation is considered as the most powerful approach for predicting the long-term fate of CO_2 in reservoirs. A parallelized general-purpose hydrodynamics code TOUGH2-MP has been used on scalar architectures where it exhibits excellent performance and scalability. However, on the Earth Simulator (ES2), which is a massively parallel vector computer, extensive tune-ups were required for increasing the vector operation ratio. In this paper, the performance of the modified TOUGH2-MP code on ES2 is presented with some illustrative numerical simulations of long-term fate of CO_2 stored in reservoirs.

Keywords: CCS, Hydrodynamics, Vector processors, The Earth Simulator.

1 Introduction

CCS (carbon dioxide capture and storage) is an emerging and promising technology for reducing greenhouse-gas content in the atmosphere, through capturing CO_2 from large emission sources and injecting and storing it into deep reservoir rocks, including saline aquifers, depleted oil and gas field [1]. Under typical deep reservoir conditions (depths > 800m), CO_2 will be stored in supercritical state, subsequently dissolving in

M. Daydé, O. Marques, and K. Nakajima (Eds.): VECPAR 2012, LNCS 7851, pp. 80–92, 2013.

groundwater, and eventually forming carbonate minerals through geochemical reactions in a long-term (e.g., hundreds to thousands of years). Numerical simulation is regarded as the most powerful approach for predicting the long-term fate of CO_2 in reservoirs, to ensure the safety and permanence of the storage. The simulations are generally conducted by using numerical simulators of multi-component, multi-phase fluid flow in porous media, but can often be computationally demanding for large-scale, high-resolution models because of complex non-linear processes involved.

In this study, we implemented a hydrodynamics code TOUGH2-MP on a massively parallelized vector computer, the Earth Simulator (ES2) in Japan. The code was extensively modified for the vector computer including the replacement of the original matrix solver to another suitable for vector processors. This paper presents the performance of the improved code on ES2 for high-resolution simulations of CO_2 behavior in deep reservoirs on the following two topics 1) CO_2 migration in highly heterogeneous geologic formations; 2) DDC (dissolution-diffusion convection) process in CO_2 – brine system, which are both important for predicting long-term fate of CO_2 stored in reservoirs.

2   General-Purpose Flow Simulator TOUGH2-MP

TOUGH2 [2] is a general-purpose numerical simulator for multi-dimensional fluid and heat flows of multiphase, multicomponent fluid mixtures in porous and fractured media. TOUGH2 solves mass and energy balance equations that describe fluid and heat flow in multiphase, multicomponent systems.

$$\frac{d}{dt}\int_{V_n} M^\kappa \, dV_n = \int_{\Gamma_n} \mathbf{F}^\kappa \cdot \mathbf{n} d\Gamma_n + \int_{V_n} q^\kappa \, dV_n \tag{1}$$

where, M^κ: energy or mass of component κ (e.g., water, CO_2, NaCl) per volume, $\mathbf{F}^\kappa$: mass or heat flux, q^κ: sink and sources, $\mathbf{n}$: normal vector on the surface element $d\Gamma_n$ pointing inward into V_n. The mass accumulation term in the left hand side is,

$$\text{Mass:} \quad M^\kappa = \phi \sum_\beta S_\beta \rho_\beta X_\beta^\kappa \tag{2}$$

$$\text{Heat:} \quad M^h = (1-\phi)\rho_R C_R T + \phi \sum_\beta S_\beta \rho_\beta U_\beta \tag{3}$$

where, ϕ: porosity, S_β: the saturation of phase β, ρ_β: the density of phase β, X_β^κ: the mass fraction of component κ present in phase β, ρ_β: grain density, T: temperature (°C), C_R: specific heat of the rock, U_β: specific internal energy of phase β.

Fluid advection is described with a multiphase extension of Darcy's law.

$$\mathbf{F}_\beta = \rho_\beta \mathbf{u}_\beta = -k \frac{k_{r\beta} \rho_\beta}{\mu_\beta} \left(\nabla P_\beta - \rho_\beta \mathbf{g} \right) \tag{4}$$

Here $\mathbf{u}_\beta$ is the Darcy velocity (volume flux) in phase β, k is absolute permeability, $k_{r\beta}$ is relative permeability to phase β, μ_β is viscosity, and P_β is the fluid pressure in phase

β (=P+$P_{c\beta}$, $P_{c\beta}$: the capillary pressure). The $k_{r\beta}$ and $P_{c\beta}$ are normally given as a non-linear function of saturation S_β, which is subject to change at each time step of transient simulations.

Heat flux includes conductive and convective components

$$\mathbf{F}^h = -\lambda \nabla T + \sum_\beta h_\beta \mathbf{F}_\beta \tag{5}$$

where, λ is thermal conductivity, and h_β is specific enthalpy in phase β.

Space discretization is made directly from the integral form of the basic conservation equations, without converting them into partial differential equations (IFDM, integral finite difference method). Time is discretized fully implicitly as a first-order backward finite difference.

$$R_n^\kappa(\mathbf{x}^{t+1}) = M_n^\kappa(\mathbf{x}^{t+1}) - M_n^\kappa(\mathbf{x}^t)$$
$$- \frac{\Delta t}{V_n} \left\{ \sum_m A_{nm} F_{nm}^\kappa(\mathbf{x}^{t+1}) + V_n q_n^{\kappa,t+1} \right\} = 0 \tag{6}$$

where, $\mathbf{x}^t$: the independent primary variables (i.e., pressure, temperature, saturation …) at time step t, R_n^κ: the residuals, M_n^κ: mass or heat accumulation term averaged over the element (gridblock) n with volume V_n, Δt: time step length, F_{nm} is the average value of the (inward) normal component of flux $\mathbf{F}$ over the surface segment A_{nm} between volume elements V_n and V_m. The equations (6) can be iteratively solved by Newton/Raphson method.

ECO2N [3] is a fluid property (EOS, equation of state) module designed for applications to geologic CO_2 storage, including a comprehensive description of the thermodynamics and thermophysical properties of H_2O-NaCl-CO_2 mixtures, modeling single and/or two-phase isothermal or nonisothermal flow processes, two-phase mixtures, fluid phases appearing or disappearing, as well as salt precipitation or dissolution. The nonlinear processes include interactions of immiscible multi-phase fluids in porous media; thermo-physical properties of supercritical CO_2 fluid that can changes rapidly against the pressure and temperature conditions at reservoir depths.

In this study, we use a parallel simulator TOUGH2-MP [4] with ECO2N fluid property module, which is a three-dimensional, fully implicit model that solves large, sparse linear systems arising from discretization of the partial differential equations for mass and energy balance. The original TOUGH2-MP uses MPI for parallel implementation, the METIS software package [5] for simulation domain partitioning, and the iterative parallel linear solver package Aztec [6] for solving linear equations by multiple processors. On scalar architecture machines, it exhibits excellent performance and scalability. In fact, "super-linear speedup" meaning speedup higher than expected linear speed up (a speedup of more than p when using p processors) has been reported on multi-core PCs [4].

3   Code Implementation and Modification on ES2

The Earth Simulator (ES) is a massively parallel vector supercomputer operated by JAMSTEC, originally developed for, and extremely used in, global climate change simulations. The ES had been the most powerful supercomputer in the world from 2002 to 2004, and recently it was upgraded to new ES2 in March 2009. ES2 is an NEC SX-9/E system and consists of 160 nodes with eight vector processors (the peak performance of each processors is 102.4Gflop/s) and 128 GB of computer memory at each node. For a total of 1280 processors and 20 TB of main memory, the total peak performance is 131 Tflop/s.

TOUGH2-MP was ported to the Earth Simulator, but a special tune-up to increase its vector operation ratio (VOR) was needed for the efficient use of the ES vector processors. The original source code of TOUGH2-MP over 40,000 lines was originally written assuming the use on scalar computers. Thus it contains many obstacles for increasing the vector operation ratio, such as frequent conditional branches and short loop lengths. Especially deadly short loop lengths in the matrix solver were found to be the key issue of the improvement, because it limits upper bound of the average vector length and thus decreases the vector operation ratio that should be 95% or more to get reasonable performance on the vector architecture computer. In the original code, an iterative parallel linear solver package Aztec [6] was employed. The Aztec solver uses a distributed variable block row (DVBR) format (a generalization of the VBR format) as a matrix storage format, which is highly memory-efficient; however the innermost loop is relatively short, order of number of off-diagonal components for matrix-vector operations.

In order to achieve efficient parallel/vector computation for applications with unstructured grids, the following three issues are critical [7]: (1) local operation and no global dependency, (2) continuous memory access, and (3) sufficiently long innermost loops for vectorization. Nakajima [7] suggested that DJDS (descending-order jagged diagonal storage) reordering is suitable for efficient length of innermost loops, producing 1D arrays of coefficient with continuous memory access and sufficient length of innermost loops. Based on the considerations, we replaced the Aztec solver to another matrix solver that employs DJDS format. The solver was found among GeoFEM [8], which has been implemented and optimized for various types of parallel computers, from PC cluster to the Earth Simulator. In addition, we performed loop-unrolling and inline expansion wherever possible and effective, and rewrote bottleneck computations (loops) in the fluid property (EOS) module.

On ES2, the modified code is about 60 times faster than the original code with Aztec solver. The speed of the new solver is 10-14 GFlops/PE (10-14% of peak performance; VOR > 99.5%; Figure 1a), while that of the original Aztec solver on ES2 is 0.15 GFlops/PE with VOR=80%. As expected, the speed up was achieved largely due to the change of the matrix storage format that greatly helps the speed-up of matrix-vector product calculations in the sparse matrix solver. Additionally, exclusions or modifications of many conditional branches equipped for the general-purpose code also contribute to the speedup considerably.

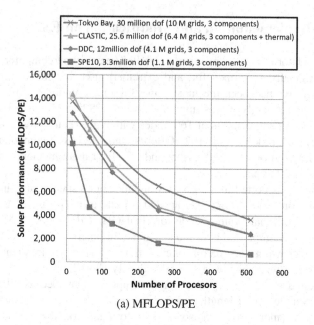

(a) MFLOPS/PE

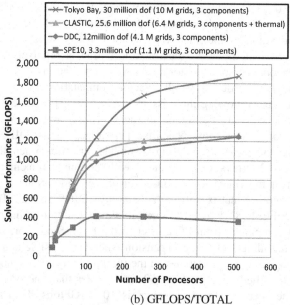

(b) GFLOPS/TOTAL

Fig. 1. Computation performance of the new solver of TOUGH2-MP on the Earth Simulator 2. In addition to the two problems in this paper (SPE10 in section 4.1 and DDC in section 4.2), two larger problems (Tokyo Bay [13] and QLASTIC [14]) are also included.

Figure1 shows the scalability of the new solver of TOUGH2-MP on the ES2. In addition to the two problems shown below in this paper (SPE10 in section 4.1 and DDC in section 4.2), two larger problems (Tokyo Bay [13] and QLASTIC [14]) are also included. The solver performance is considerably reduced with increasing number of PEs probably because of the load increase of communication among PEs, which is pronounced for smaller models (i.e., SPE10).

4   Numerical Simulation of CO_2 Behaviors in Deep Reservoirs

This study intends to demonstrate potential benefits from high performance computing for simulating CO_2 behavior in deep reservoirs for important scientific and engineering topics. Here, we investigate uncertainties due to grid resolution effects on the two topics: 1) CO_2 behaviors in highly heterogeneous reservoir formations; 2) the diffusion-dissolution-convection process that may cause gravity instability that greatly enhances the convective mixing of dissolved CO_2 in reservoirs in long-term.

4.1 CO_2 Behaviors in Highly Heterogeneous Reservoirs

Reservoir formations for storing CO_2 are usually regarded as heterogeneous porous media, often consisting of alternating sand and mud layers having quite different permeability and porosity. Obviously, because CO_2 preferentially migrates into higher permeable portion of the reservoir, the heterogeneity of reservoirs should be properly considered and represented in simulation models. However, it is well known that transient simulation of multi-phase flow in heterogeneous media generally requires very long computational time, because the heterogeneity of hydraulic properties strongly limits the length of time steps, resulting in a huge number of time steps to be solved in the simulation. As a practice in such as oil and gas industries, for performing simulations in a practically reasonable time, the heterogeneity is spatially averaged with reducing number of grid cells of the computation model (i.e., up-scaling). In this study, with the help of high-performance computing, we directly solved a highly heterogeneous reservoir model without such simplifications.

Figure 2 shows the heterogeneous model known as SPE-10 model [9], representing irregular nature of sand/shale distribution in a reservoir with 1.122×10^6 ($60 \times 220 \times 85$) grid cells in the dimension of 6400m $\times$ 8800m $\times$ 170m. In this simulation, CO_2 in supercritical state is injected at the injector with the rate of 390 k tons / year, with producing brine groundwater with the rate of 580 k tons / year [10].

The distribution of CO_2 after 20-years injection is shown in Figure 3. For a comparison, a simulation result obtained from a homogeneous model with a unique average permeability and porosity is also shown. Because the density of supercritical CO_2 is smaller than that of groundwater, injected CO_2 tend to overrides on denser groundwater. In the homogeneous model, the override effect is prominent and the lower part of the reservoir volume cannot effectively be used for storing CO_2. In addition, the CO_2 plume spread widely on the top of the reservoir, suggesting higher risks of CO_2 leakage through undetected high-permeable features such as faults. On the other hand, in the heterogeneous model, CO_2 tends to migrate in sand portions with higher permeability, suppressing the gravity override. The more tortuous flow paths of CO_2 in the heterogeneous model results in larger contact area of CO_2 and groundwater, and thus enhances the dissolution of CO_2 in groundwater, which is

Fig. 2. SPE-10 model, a highly heterogeneous reservoir model. The colors indicate porosity of the sand and mud in the reservoir. The CO_2 injector and the water producer used in the simulation are shown in the figure [10].

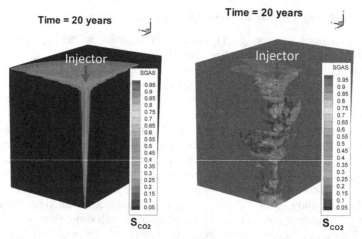

Fig. 3. CO2 plume spreading from the injector in the homogeneous (left) and heterogeneous model (right). S_{CO2}: saturation of gaseous CO_2.

deemed as more stable form of storage than buoyant supercritical CO_2 plume. Figure 4 includes the change of CO_2 status in the reservoir over time. The dissolution of CO_2 in groundwater is enhanced nearly double in the heterogeneous model than that in the homogeneous model.

As seen in the above, although the heterogeneity of reservoir is an important key in predicting the efficiency and risks of CO_2 storage, two-phase flow simulations in heterogeneous porous media computationally demanding in general. The 20-years simulation of the homogeneous model was finished only in about 3 node-hours (about 650 time steps), while it took more than 900 node-hours (about 40,000 time-steps) for the heterogeneous model. The simulation was performed by using 2 or 4 nodes (16 to 32 PE) of ES2, taking into account the limited scalability for this SPE10 model as shown in Figure 1.

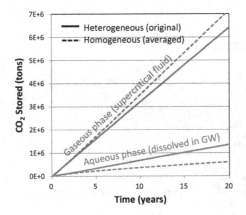

Fig. 4. Time evolution of the CO_2 form stored in the reservoir

4.2   Diffusion-Dissolution-Convection Process of CO_2-Brine System

As mentioned above, injected supercritical CO_2 generally tends to override over native groundwater in the reservoir (Figure 5a). The supercritical CO_2 on top gradually dissolves in surrounding groundwater, and increases its density. This results in a situation that denser fluid laid on lighter fluid as schematically shown in the red-box in Figure 5a. In a certain time, the Rayleigh-Taylor instability invokes convective mixing of the groundwater [11]. The mixing would significantly enhance the CO_2 dissolution into groundwater, and reduce the amount of CO_2 in buoyant supercritical state, and eventually attain more stable storage.

The simulation of the Rayleigh-Taylor instability is sensitive to numerical errors. The simulation grids should be fine enough to resolve incubation times for onset of convection, and spatial evolution of convective fingers or tongues [11, 12]. Figure 5b shows a simulation result for a local-scale model of a small region (1m×1m×4m) with 1cm spacing, resulting in about 4 million gridblocks [12]. Starting from the initial condition that supercritical CO_2 is stagnate on the top of groundwater, CO_2 gradually dissolves in groundwater, developing a thickness of a CO_2 diffuse layer. When the thickness of the diffusive layer reaches a critical thickness, the convection with developing downward fingers of high-CO_2 water occurs.

Massively parallel computation would make it possible to simulate not only the local process shown above, but also the whole system in reservoir scales (i.e., more than hundreds to thousands of meters) on the resolution of centimeter-scale grids. Figure 6 shows a preliminary simulation result performed for a homogeneous and horizontal reservoir model of 10km radius and 40m thickness. The model is finely discretized with about 10cm grid spacing. We successfully simulated the evolution of the convectional mixing process in entire reservoir, showing that the fingering gradually developed from centimeter to tens of meters scale enhancing the dissolution of CO_2 in groundwater, and the CO_2 in supercritical state stay on the top of the reservoir was eventually lost.

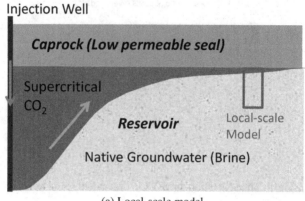

(a) Local-scale model

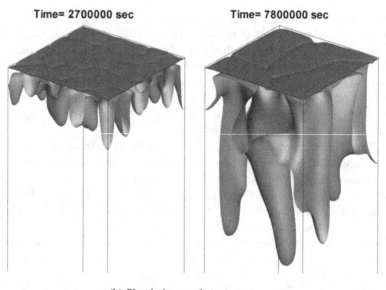

(b) Simulation results

Fig. 5. (a) The concept of the local-scale model and (b) the simulated results of CO_2 concentration dissolved in groundwater. The model size is 1m×1m×4m. The iso-surfaces in the right figures are colored for CO_2 mass fractions (brown: 0.025, green: 0.008).

Injection Well

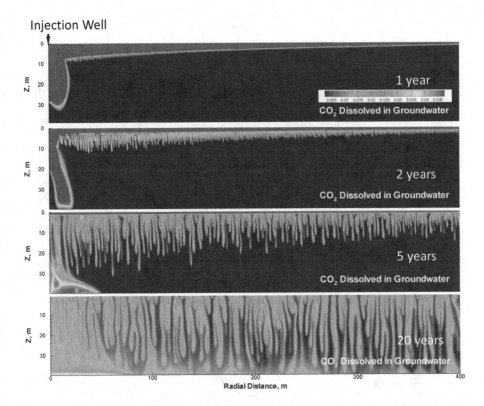

Fig. 6. A preliminary simulation result of the diffusion-dissolution-convection process in a radially symmetric homogeneous model at a reservoir-scale (10km radius and 40m thickness). CO_2 is injected in supercritical state with the rate of 100k tons/year for one year. The contours show the time evolution of CO_2 mass fraction in the aqueous phase (groundwater) in the post-injection period. Due to the gravity convection, the supercritical CO_2 which has overridden over groundwater during the injection period is dissolved promptly, and eventually disappears (completely dissolved in the groundwater). The permeability and porosity are 1darcy and 20% respectively, employing the Corey equation for relative permeability and neglecting capillarity.

Figure 7 compares the result obtained from two models with different grid-spacings. In the 'coarse' model, the grid-spacings in vertical and lateral direction are both increased 10 times from the original 'fine' model shown above. In Figure 7a, the supercritical CO_2 laid along the reservoir top is recognized as the red colored portion of high CO_2 mass fraction in aqueous phase. It was seen that the coarseness of the grid artificially stabilizes the layer of CO_2-saturated groundwater on the top [11]. Figure 7b shows the time evolution of the amount of CO_2 dissolved in groundwater. It is seen that the rough model underestimates the dissolution of CO_2 by almost half, and thus underestimates the long-term stability of CO_2 stored in the underground reservoir. Usually 3D field scale simulations employ grid model with the size of tens

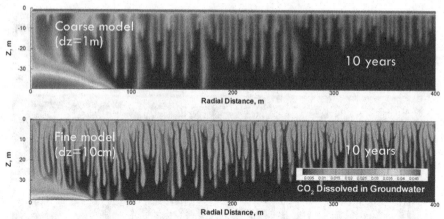

(a) CO_2 mass fraction in the aqueous phase at 10 years

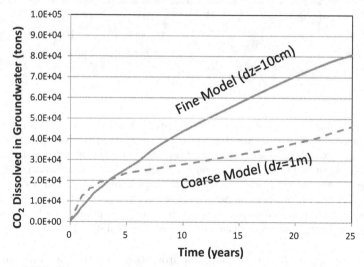

(b) The time evolution of the amount of CO_2 dissolved in groundwater

Fig. 7. Impact of grid spacing on the prediction of CO_2 dissolution in groundwater

to hundreds of meters, but finer grid model will be required to predict the long-term fate of CO_2 appropriately.

These above simulations were both performed by using 2 to 16 nodes (8 to 128 PE) of ES2. Vector efficiency was observed in the same level as in other measurements.

5   Conclusion

A general purpose hydrodynamics code TOUGH2-MP was successfully implemented on the Earth Simulator (ES2). The performance of the TOUGH2-MP code on ES2 was considerably improved by the efforts including (1) the replacement of matrix solver and (2) restructuring and rewriting of the EOS routines. So far, the computational performance of 10 to 14 GFlops/PE (approximately 10-14% of peak performance of a vector processor on ES2 with a vector operation ratio > 99.5%) has been achieved. The achieved performance is satisfactory for the general purpose code, which was originally developed for scalar architectures, but not for vector architectures.

Using the code, uncertainties due to grid resolution effects on the two topics were investigated: 1) CO_2 behaviors in highly heterogeneous reservoir formations; 2) the diffusion-dissolution-convection process that may cause gravity instability greatly enhancing convective mixing of dissolved CO_2 in reservoirs in long-term. These simulations illustrate the practical necessity of fine grid-resolution in numerical reservoir models, and thus importance of high-performance computing for predicting the long-term fate of CO_2 even in actual storage projects in the future.

Acknowledgments. The use of the Earth Simulator was supported by the "Open Advanced Facilities Initiative for Innovation (Strategic Use by Industry)" funded by the Ministry of Education, Culture, Sports, Science, and Technology of Japan.

References

1. Bachu, S., Gunter, W.D., Perkins, E.H.: Energy Convers. Manage 35(4), 269–279 (1994)
2. Pruess, K., Oldenburg, C., Moridis, G.: TOUGH2 User's Guide, Version 2.0, Rep. LBNL-43134. Lawrence Berkeley National Lab. (1999)
3. Pruess, K.: ECO2N : A TOUGH2 Fluid Property Module for Mixtures of Water, NaCl, and CO_2, Rep. LBNL-57952 (2005)
4. Zhang, K., Wu, Y.S., Pruess, K.: User's Guide for TOUGH2-MP, Rep. LBNL-315E. Lawrence Berkeley National Lab. (2008)
5. Karypsis, G., Kumar, V.: METIS V4.0, Technical Report, University of Minnesota (1998)
6. Tuminaro, R.S., Heroux, M., Hutchinson, S.A., Shadid, J.N.: Official Aztec user's guide, Ver 2.1, Sandia National Laboratories (1999)
7. Nakajima, K.: Applied Numerical Mathematics 5, 4237–4255 (2005)
8. Nakajima, K.: Parallel Iterative Solvers of GeoFEM with Selective Blocking Preconditioning for Nonlinear Contact Problems on the Earth Simulator. In: ACM/IEEE Proceedings of SC 2003. Phoenix, AZ (2003)
9. Christie, Blunt: Tenth SPE comparative solution project: a comparison of upscaling techniques, SPE 66599, SPE Reservoir Simulation Symposium, Houston, Texas (2001)
10. Qi, R., LaForce, T.C., Blunt, M.J.: Design of carbon dioxide storage in aquifers. International Journal of Greenhouse Gas Control 3, 195–205 (2009)
11. Ennis-King, J. P., Paterson, L.: Role of convective mixing in the long-term storage of carbon dioxide in deep saline formations. SPE J. 10(3), 349–356. SPE-84344-PA (2005)

12. Pruess, K., Zhang, K.: Numerical modeling studies of the dissolution–diffusion–convection process during CO_2 storage in saline aquifers. Technical Report LBNL-1243E. Lawrence Berkeley National Laboratory, California (2008)
13. Yamamoto, H., Zhang, K., Karasaki, K., Marui, A., Uehara, H., Nishikawa, N.: Int. J. Greenhouse Gas Control 3, 586–599 (2009)
14. Audigane, P.D., Michel, A., Trenty, L., Yamamoto, H., Gabalda1, S., Sedrakian, A., Chiaberge, C.: CO2 injection modeling in large scale heterogeneous aquifers. Eos Trans. AGU 92(51) (2011); Fall Meet. Suppl., Abstract H51H-1302 (2011)

High Performance Simulation of Complicated Fluid Flow in 3D Fractured Porous Media with Permeable Material Matrix Using LBM

Jinfang Gao and H.L. Xing

Earth Systems Science Computational Centre, School of Earth Sciences,
The University of Queensland, Brisbane, QLD 4072, Australia
{j.gao1,h.xing}@uq.edu.au

Abstract. To analyze and depict complicated fluid behaviours in fractured porous media with various permeable material matrix across different scales, an Enhanced Heterogeneous Porous Media Computational Model is proposed based on Lattice Boltzmann method (LBM). LBM is widely employed to model basic fluid dynamics within disordered structures due to its powerful applicability to mesoscopic fluid mechanics and its potential performance of parallel computing. This paper combines with the force model, statistical material physics and the parallel algorithm to effectively describe the feature changes while fluid passes through the fractured porous media with diverse permeable material matrix of high resolution by using supercomputers. As an application example, a 3D sandstone sample is reconstructed with 36 million grids using the scanned CT images and characterized with different feature values at each lattice grid to distinguish pores, impermeable solids and permeable material matrix by stating its local physical property. The calculation and comparison results with the conventional LBM are discussed to demonstrate the advantages of our method in modeling complicated flow phenomena in fractured porous media with variable permeable material matrix across different scales, and its sound computing performance that keeps the parallel speedup linearly with the number of processors.

Keywords: fluid dynamics, fractured porous media, diverse permeable material matrix, interfacial dynamics, Lattice Boltzmann, high performance computing.

1 Introduction

The imperative of complex fluid flow and transport in porous media becomes increasingly apparent in both scientific research and industrial applications [1-2] (i.e., CO_2 geosequestration, mining and material engineering). Among conventional approaches, flow dynamics are usually described from the perspective of macroscale view by solving macroscopic continuum equations (i.e., mass, momentum and energy) with finite differential method, finite volume method and finite element method [3-4]. However, heterogeneous materials typically possess a multitude of mechanically significant scales and each of these scales requires appropriate

M. Daydé, O. Marques, and K. Nakajima (Eds.): VECPAR 2012, LNCS 7851, pp. 93–104, 2013.
© Springer-Verlag Berlin Heidelberg 2013

modeling. Heterogeneous structures on the microscale (i.e., material grains, pore space and micro-cracks) are often close in size to a characteristic length for the macroscopic pattern for such as subsurface flow and transport. Hence these structures strongly influence macroscale processes. While conducting simulation in the microscale or nanoscale, these kinds of conventional methods show disadvantages in simulating the complicated fluid dynamics in dissimilar porous media, as well as its complex boundary and surface conditions.

Derived from particulate natures, LBM has enormous strengths to model basic fluid dynamics compared with other conventional numerical techniques, which is advanced for studying the complicated fluid flow behaviours such as in porous media, together with the complexity of structural geometry and pore scale flow uncertainties. Most of applications in LBM simulating fluid flow in porous media have focused on pore scale with simplified impermeable solids [5][8]. However for some cases, such as the fractured porous media with pores/fractures across the different scales inside, the matrix cannot be simply treated as impermeable solids, while under certain resolutions, some material matrix even has diverse permeability.

To get detailed information about the microstructure of heterogeneous porous media, digital images are nowadays widely applied to describing complicated structures with multiple colors such as in mining, medicine and material sciences [14]. A digital image is made up of a rectangular array of equal-sized picture elements. Such elements are usually referred to as "voxels". Therefore, it is possible to describe and measure complex heterogeneous structures and material properties in the format of images (i.e., pores and different kinds of materials). Based on the scanned CT/MRI images, accurate reconstruction of the 2D/3D numerical domain of heterogeneous porous media can be conducted [7], and such kinds of digital models can be easily converted into LBM grids across different resolutions [13].

To consider nonlinear pore scale processes and their related influence on the heterogeneous structures and material properties variation, this paper will focus on modeling of fluid dynamics in permeable matrix through extending the force model algorithm [9], together with the statistical calculation of diverse permeable material parameters at each LBM grid on the advanced parallel computers (in the section 2), and the comparison of its calculation results and parallelization with the conventional impermeable solids case (in the section 3).

2   Methodology and Modeling

2.1   Lattice Boltzmann Method

Instead of solving the conventional Navier-Stokes equations or its simplified form with macroscopic properties, the discrete Boltzmann equation is solved here to simulate the fluid flow with collision models such as Bhatnagar-Gross-Krook (BGK) [10-11]. LBM models the fluid consisting of fictitious particles, and such particles collide with one another and propagate over a discrete lattice mesh. By simulating streaming and collision processes across a limited number of particles, the intrinsic particle interactions evince a microcosm of viscous flow behaviour applicable to the greater mass [5-6].

The generalized BGK Lattice Boltzmann equation (LBM BGK) is described as below [9][12]:

$$f_i(\mathbf{x}+\mathbf{e}_i\Delta t, t+\Delta t)-f_i(\mathbf{x},t)=-\frac{1}{\tau}\left[f_i(\mathbf{x},t)-f_i^{eq}(\mathbf{x},t)\right] \tag{1}$$

The left term of the equation is the propagation part, while the right side is the collision part. Especially, in the LBM BGK model, τ is a constant relaxation number. Parameter $\mathbf{e}_i$ means the direction of related neighbours of one certain resting particle. $f_i(\mathbf{x},t)$ and $f_i(\mathbf{x}+\mathbf{e}_i\Delta t,t+\Delta t)$ are the fluid population distribution in the $\mathbf{e}_i$ direction on lattice node $\mathbf{x}$ at time step t and $t+\Delta t$, respectively. In the classical LBM structure $DmQn$ (where m is the domain dimension, and n is the number of the resting particle and its related neighbours), $\mathbf{e}_i$ has n directions. For example, the classical model $D3Q19$ is applied for 3D calculation in this paper with one resting particle having 18 neighbours (Fig.1). $f_i^{eq}(\mathbf{x},t)$ is the equilibrium distribution of fluid particle population in the $\mathbf{e}_i$ direction at time step t on the same lattice node. While simulating incompressible or slightly compressible fluid flow in porous media using the LBM BGK model, $f_i^{eq}(\mathbf{x},t)$ is defined as:

$$f_i^{eq}(\mathbf{x},t)=\omega_i\rho\left[1+\frac{\mathbf{e}_i\cdot\mathbf{u}}{c_s^2}+\frac{(\mathbf{e}_i\cdot\mathbf{u})}{2\phi c_s^4}-\frac{\mathbf{u}^2}{2\phi c_s^2}\right] \tag{2}$$

where c_s is 1/3 speed of sound running over the lattice mesh, and ϕ is the average porosity of porous media. ρ is fluid density related to pressure p $(p=c_s^2\rho/\phi)$. $\mathbf{u}$ is the macroscopic velocity of fluid flow. ω_i is the coefficient weight of population distribution of fluid particles in the $\mathbf{e}_i$ direction on the lattice node $\mathbf{x}$. For $D3Q19$: $i=0,1,...,18$, $\omega_i = 1/3$ $(i=0)$, $1/18$ $(i=1,2,...6)$, $1/36$ $(i=7,8,...18)$.

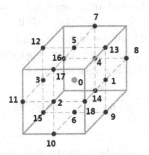

Fig. 1. Lattice mesh and nodes relations in the classical 3D LBM lattice structure ($D3Q19$): one resting particle (grey point in the cubic centre) and its 18 neighbours (black points)

Due to its particulate nature and local dynamics, LBM possesses advantages in simulating complex pore scale fluid flow and boundary conditions in porous media with detailed microstructures. However in the conventional LBM, there only can be either 0 or 1 particle on a lattice node moving in a lattice direction according to the microstructure features of porous media (i.e., only pores or impermeable solids). On the other hand, based on the scanned CT images, porous media structures and geometry with different kinds of materials can be observed and reconstructed, together with its physical characters (i.e., porosity). To extend applications of LBM in simulating porous media with diverse permeable material matrix across different scales, the following algorithm with force term is applied.

2.2   Force Model in Fractured Porous Media with Permeable Material Matrix

In porous media, fluid flow is always confronted with different kinds of forces: external forces such as gravity and buoyancy, interactions between the fluid particles, and the force derived from the porous matrix. Especially while the fluid is blocked by impermeable solids or flows through some permeable material matrix, the porosity and permeability will cause such forces to behave in a complicated way. Considering such force terms, Eq.(1) is further revised as below [12]:

$$f_i(\mathbf{x}+\mathbf{e}_i\Delta t,t+\Delta t)-f_i(\mathbf{x},t)=-\frac{1}{\tau}\left[f_i(\mathbf{x},t)-f_i^{eq}(\mathbf{x},t)\right]+\Delta t A_i \qquad (3)$$

A_i is the body force distribution of fluid flow in the $\mathbf{e}_i$ direction on the lattice node $\mathbf{x}$. As for seepage behaviour in porous media, both the linear and the non-linear influence of the porous matrix on fluid flow should be considered. The composition force of fluid flow in the porous media can be written as [9]:

$$\mathbf{F}=-\frac{\phi v_s}{k}\mathbf{u}-\frac{\phi F_\varepsilon}{\sqrt{k}}|\mathbf{u}|\mathbf{u}+\phi\mathbf{G} \qquad (4)$$

where v_s is fluid viscosity ($v_s=c_s^2(\tau-1/2)\Delta t$). As for one certain porous medium, F_ε is the average characteristic property of the porous medium related to the average porosity, $\mathbf{G}$ includes forces derived from particles and other external forces (i.e., gravity), and k is the average permeability of the porous medium. We take $F_\varepsilon^2=1.75^2/150\phi^3$, $k=\phi^3 d^2/150(1-\phi)^2$ in this paper, and d is the effective diameter of a solid particle of the porous medium. Especially considering the porous medium with more than one kind of permeable materials, the Eq.(4) can be modified as:

$$\mathbf{F}=-\frac{\phi_j v_s}{k_j}\mathbf{u}-\frac{\phi_j F_{\varepsilon j}}{\sqrt{k_j}}|\mathbf{u}|\mathbf{u}+\phi_j\mathbf{G} \qquad (5)$$

$F_{\varepsilon j}$ is the characteristic property of the jth permeable material matrix on each lattice node, ϕ_j and k_j represent the porosity and permeability of the jth permeable material matrix, respectively.

The macroscopic force $\mathbf{F}$ and velocity $\mathbf{u}$ in Eqs.(4) and (5) are continuous, which is different from the situation with other discrete variables in the Lattice Boltzmann equation. To build correlations between macroscopic continuum qualities and discrete pore scale values, the force model is embedded [9]. The discrete body force A_i is defined as:

$$A_i = \omega_i \rho \left[\frac{\mathbf{B} \cdot \mathbf{e}_i}{c_s^2} + \frac{\mathbf{C} : \left(\mathbf{e}_i \mathbf{e}_i - c_s^2 \mathbf{I} \right)}{2\phi_j c_s^4} \right], \tag{6}$$

where

$$\mathbf{B} = \left(1 - \frac{1}{2\tau} \right) \mathbf{F},$$

$$\mathbf{C} = \left(1 - \frac{1}{2\tau} \right) (\mathbf{uF} + \mathbf{Fu}),$$

$$\rho = \sum_i f_i, \qquad \rho \mathbf{u} = \sum_i \mathbf{e}_i f_i + \frac{\Delta t}{2} \rho \mathbf{F}.$$

Velocity $\mathbf{u}$ is obtained by combining Eqs.(5) and (6):

$$\mathbf{u} = \frac{\dfrac{\sum_i \mathbf{e}_i f_i}{\rho} + \dfrac{\phi_j \Delta t \mathbf{G}}{2}}{\dfrac{1}{2}\left(1 + \dfrac{\phi_j \Delta t \nu_s}{2k_j}\right) + \sqrt{\dfrac{1}{4}\left(1 + \phi_j \dfrac{\Delta t \nu_s}{2k_j}\right)^2 + \phi_j \dfrac{\Delta t F_{\varepsilon j}}{2\sqrt{k_j}}\left(\dfrac{\sum_i \mathbf{e}_i f_i}{\rho} + \dfrac{\phi_j \Delta t \mathbf{G}}{2}\right)}}. \tag{7}$$

According to the Champon-Enskog expansion, Eqs.(3), (5) and (6) are equivalently formulated with the following Navier-Stokes equations with force term $\mathbf{F}$:

$$\nabla \cdot \mathbf{u} = 0.$$

$$\frac{\partial \mathbf{u}}{\partial t} + (\mathbf{u} \cdot \nabla)\left(\frac{\mathbf{u}}{\phi} \right) = -\frac{1}{\rho} \nabla(\phi p) + \nu_e \nabla^2 \mathbf{u} + \mathbf{F}. \tag{8}$$

Based on this Enhanced Heterogeneous Porous Media Computational Model, there is no need to define boundary types for the border nodes that separate pores and the matrix, and thus no conventional bounce-back treatment required in our model.

2.3   Statistical Material Physics

In microscale material structures, material physics behaves a slight fluctuation. To describe these noise features, statistical material physics based on the Weibull analysis is employed and utilized here on different lattice nodes to slightly discriminate porosity distribution ϕ_{jp} instead of setting up a constant material porosity ϕ_j (0 for conventional LBM) at all the matrix nodes [12]. The porosity probability distribution meets the equation:

$$f(x) = \frac{\alpha}{\beta}\left(\frac{x}{\beta}\right)^{\alpha-1} \exp\left[-\left(\frac{x}{\beta}\right)^{\alpha}\right]. \tag{9}$$

According to the Weibull distribution, random number x for the matrix's physical property distribution on each LBM lattice node can be described as:

$$x = \beta\left(\ln\frac{1}{1-U}\right)^{\frac{1}{\alpha}}. \tag{10}$$

In Eqs.(9) and (10), $\alpha(>0)$ controls the shape of the distribution, while β stands for the scale (or the position, i.e., average porosity ϕ_j) of the distribution, and U is the uniform random number ranging from 0 to 1. Normally, when α is larger than 2, the distribution mimics Gaussian distribution, which is reasonable to be used for random distribution generation of certain matrix porosity. According to the embedded statistical material physics method mentioned above, Eqs.(3), (5) and (6) can be modified as:

$$f_i^{eq}(\mathbf{x},t) = \omega_i\rho\left[1 + \frac{\mathbf{e}_i\cdot\mathbf{u}}{c_s^2} + \frac{(\mathbf{e}_i\cdot\mathbf{u})}{2\phi_{jp}c_s^4} - \frac{\mathbf{u}^2}{2\phi_{jp}c_s^2}\right]. \tag{11}$$

$$\mathbf{F} = -\frac{\phi_{jp}v_s}{k_{jp}}\mathbf{u} - \frac{\phi_{jp}F_{\varepsilon j}}{\sqrt{k_{jp}}}|\mathbf{u}|\mathbf{u} + \phi_{jp}\mathbf{G} \tag{12}$$

$$A_i = \omega_i\rho\left[\frac{\mathbf{B}\cdot\mathbf{e}_i}{c_s^2} + \frac{\mathbf{C}:(\mathbf{e}_i\mathbf{e}_i - c_s^2\mathbf{I})}{2\phi_{jp}c_s^4}\right]. \tag{13}$$

2.4   Parallelism Implementation

To meet the computational resource needs of the complex 3D simulations, parallelism technology has been implemented into the code (written in C++) to improve computing performance of high resolution or large scale cases on supercomputers. Fig.2 demonstrates the code structure, including input preparation, parallel computing and data management, and post-processing such as data output and its visualization.

In the input preparation step, the control processor (processor 0) conducts I/O instructions, and both control parameters and geometry files are read for the program initialization. In the main computing step, message passing is conducted in this code with the Message Passing Interface (MPI) using domain decomposition, where Lattice data has been split as sub-blocks and broadcasted onto each processor. Each processor is responsible for performing density and velocity calculations by executing collision, propagation, and relaxation for its block portion of the simulation grids, as well as dealing with boundary conditions at the external edges of each block. A two-voxel-thick padding layer is employed to manage the most recently updated parameters exchanged from the edge of the neighbouring sub-blocks. In the output and visualization step, each processor outputs density and velocity data, the format (i.e., vtk) of which can be visualized by Paraview.

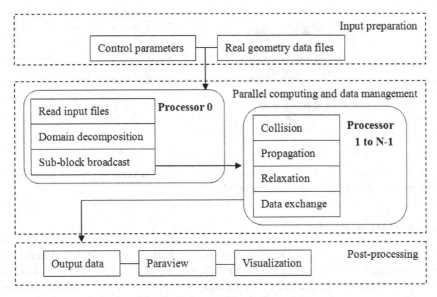

Fig. 2. Code structure of the model

The scalable code has been designed to maintain portability on different computer architectures, compilers, and MPI libraries, and a portable Makefile is programmed to enable easy compilation across different platforms. For this paper, the earth systems simulator, Savanna, at our centre is carried out for the calculation. Savanna is a 64bit SGI ICE 8200 EX Parallel Computer, with 128 Intel Xeon quad-core processors (512 cores at 2.8 GHz each), 2 TB memories, 14.4 TB disk on a Nexis 9000 NAS, and with peak speed reaching 5.7 TFlops. Based on Savanna, parallelization for both cases as above (Eqs.(1) and (3)) has been implemented with the message-passing paradigm of MPI library (v4.0.0.028) for running programs on the distributed-memory platforms, with the Intel C++ compiler employed. Up to 512 processors have been used, and up to 1GB time-varying volumetric output data can be produced in a single run.

3　Application Example

3.1　Porous Media Reconstruction of Sandstone Sample

Based on digital images, a pre-processing module is developed to convert such images into the lattice dataset, and then a 3D porous medium is reconstructed for computing over the discrete LBM lattice. In this paper, the scanned CT images of sandstone are taken as a 3D application example [7]. The sandstone core is a cube of 2.5mm in length, and a porous medium structure (331×331×331 voxels, Fig.3) has been generated according to 600 CT images.

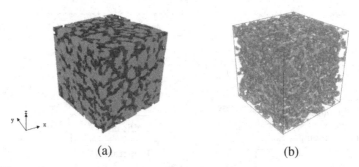

(a)   (b)

Fig. 3. The sandstone sample (331×331×331 voxels) to be analyzed. (a) matrix structure in grey, pores in black, (b) pore structure displayed in orange.

3.2   Simulation Results

The above sandstone (331×331×331 voxels) is used here to compare the influence of porous medium matrix characters on fluid flow behaviours. Two kinds of porous matrix (impermeable solids and permeable material matrix under Weibull distribution with an average porosity of 0.23) are calculated. The flow is driven from bottom to top under a constant dimensionless pressure (0.1/lattice grid) with all the other sides closed.

Detailed simulation results of flux features inside the sandstone are compared in Fig.4 and Fig.5. As it is shown, a transect is conducted right in the middle of y-direction to demonstrate the inside structure of sandstone (Fig.4(a) and Fig.5(a)). Although under the same boundary conditions, as well as same geometry and distribution of porous media, the velocity distributions are clearly different due to dissimilar properties of microscale matrix permeability (Figs.4(b)~(d), Figs.5(b)~(d)).

Figs.4(b)~(d) and Figs.5(b)~(d) have demonstrated the snapshots of fluid flow process to the steady situation under two different porous matrix conditions. Within the impermeable solids case, velocity distribution differs largely at pores and matrix (Figs.4(b)~(d)); while within the permeable material matrix case, the fluid flows through the matrix with a lower speed than through the fractures (Figs.5(b)~(d)). It is clearly shown in Figs.4(b)~(d) that the fluid flow is blocked by the impermeable solids (districts in black color), while the seepage phenomenon emerges clearly in the permeable material matrix (Figs.5(b)~(d)). In Fig.4, the conventional bounce-back boundary is arbitrarily utilized to simulate the pore/solid surfaces, while in Fig.5, our model is able to distinguish matrix and pores automatically according the different porosity description on each lattice node.

From the above comparison, our model can easily simulate the permeable behaviour of fluid passing through permeable material matrix, and distinguish complex solid/pore interfacial dynamics without arbitrarily defining solid/pore surfaces (e.g. bounce-back or slip walls). Additionally, it indicates that both porous matrix geometry and permeability of the matrix have significant effects on detailed fluid flow behaviours.

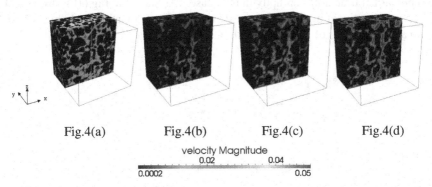

Fig.4(a)   Fig.4(b)   Fig.4(c)   Fig.4(d)

velocity Magnitude
0.02   0.04
0.0002   0.05

Fig. 4. (a) Transect of sandstone (solids in black, pores in grey), (b)~(d) Snapshots of velocity distribution within porous medium with impermeable solids (zero velocity in black)

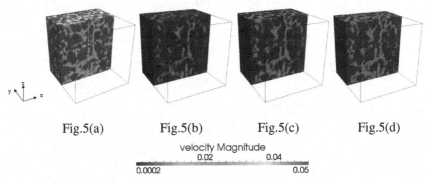

Fig.5(a)   Fig.5(b)   Fig.5(c)   Fig.5(d)

velocity Magnitude
0.02   0.04
0.0002   0.05

Fig. 5. (a) Transect of sandstone (permeable material matrix in brown, pores in grey), (b)~(d) Snapshots of velocity distribution within porous medium with permeable material matrix

It is highly applicable to combine the conventional LBM and Enhanced Heterogeneous Porous Media Computational Model to analyze fluid behaviours in fractured heterogeneous porous media across different scales. Upon completion of a solution step, the micro models based on the conventional LBM return with the upscale material properties as a local matrix character evaluation (i.e., average permeability) for macroscale analysis. The macroscale model distinguishing fractures, solids and other matrix with certain permeability by stating its physical property from the microscale simulation will be simulated to consider the multiscale heterogeneity.

3.3   Parallelization and Computation Performance

The sandstone sample (Fig.3) is discretized into 36 million cubes with a spatial resolution of $7\mu m$ and calculated using Savanna. The computing time and speedup are compared and shown in Fig.6 for both permeable material matrix and impermeable solids cases. Fig.6(a) shows curves of the logarithm of computational time vs. the number of processors. With the increasing CPU numbers up to 512 processors, computational time drops abruptly in both cases. As shown in Fig.6(b), strong scaling of the computational models has been demonstrated on Savanna, and the code scales up to a large number of processors with linear speedup.

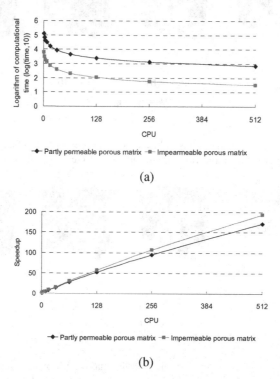

(a)

(b)

Fig. 6. Comparison of both cases on (a) computational time and (b) speedup

Additionally in Fig.6(a), it is demonstrated that extra about 40% computing time is consumed in the permeable material matrix case than the impermeable solids case while under the same resolution. The increase is mainly due to the additional calculation of complex forces term (i.e., Eqs.(3), (5) and (6)) for describing the permeable material matrix feature on each lattice node at each time step. However, if the conventional LBM model is applied to simulating such permeable mineral matrix conditions, it requires much higher resolution to distinguish detailed pore/solid structures, which will cost more computation time even using double resolution of that in the above permeable model (Fig.6(a)). Therefore, the proposed model is better for modeling complicated fluid flow in 3D fractured porous media across different scales.

4   Conclusions

With the development of advanced imaging technology, digital images are nowadays widely applied to describing the complicated structures of porous media. A LBM based Enhanced Heterogeneous Porous Media Computational Model is proposed to model the complicated fluid dynamics in the fractured porous media across different scales. The force model is extended and combined together with the statistical material physics and the parallel algorithm to effectively describe the feature changes while the fluid passes through the fractured porous media with diverse permeable material matrix on supercomputers. A 3D sandstone sample is successfully reconstructed with 36 million grids using the scanned CT images and characterized with different feature values at each grid to distinguish pores, impermeable solids and permeable material matrix by stating its local physical property. The calculation and comparison results with the conventional LBM demonstrate the advantages of our method in modeling complicated flow phenomena in fractured porous media with diverse permeable material matrix, and its sound computational performance which keeps the parallel speedup linearly with the number of processors. The proposed model will be used to explore complex multiscale fluid flow phenomena across different scales, which are widely observed in geo-science and geo-engineering. The model is also useful to deal with complex interfacial dynamics between solid/pore boundaries in porous media.

We gratefully acknowledge financial support from the Australian Research Council (Grant No. ARC DP066620, LP0560932, LX0989423 and DP110103024).

References

1. Sahimi, M.: Flow and Transport in Porous Media and Fractured Rock: From Classical Method to Modern Approaches. Wiley-VCH Verlag GmbH, New York (1995)
2. Nield, D.A., Bejan, A.: Convection in porous media. Springer (2006)
3. Nithiarasu, P., Seetharamu, K.N., Sundararajan, T.: Natural Convective Heat Transfer in a Fluid Saturated Variable Porosity Medium. Int. J. Heat Mass Transfer 40(16), 3955–3967 (1997)
4. Joseph, D.D., Nield, D.A., Papanicolaou, G.: Nonlinear Equation Governing Flow in a Saturated Porous Medium. Water Resources Research 18(4), 1049–1052 (1982)
5. Sukop, M.C., Thorne, D.T.: Lattice Boltzmann Modeling: An Introduction for Geoscientists and Engineers. Springer, Heidelberg (2006)
6. Qian, Y.H., D'Humières, D., Lallemand, P.: Lattice BGK Models for Navier-Stokes Equation. Europhys. Lett. 17(6), 479–484 (1992)
7. Hou, J., Li, Z., Zhang, S., Cao, X., Du, Q., Song, X.: Computerized Tomography Study of the Microscopic Flow Mechanism of Polymer Flooding. Transp. Porous Med. 79(3), 407–418 (2009)
8. Xing, H.L., Gao, J., Zhang, J., Liu, Y.: Towards An Integrated Simulator For Enhanced Geothermal Reservoirs. In: Proceedings World Geothermal Congress 2010, paper 3224, Bali, Indonesia (2010)

9. Guo, Z., Zhao, T.S.: Lattice Boltzmann Model for Incompressible Flows through Porous Media. Physical Review E 66, 036304 (2002)
10. Chen, H., Shen, S., Matthaeus, W.H.: Recovery of the Navier-Stokes equations using a lattice-gas Boltzmann method. Physical Review A 45, 5339–5342 (1992)
11. Swift, M.R., Orlandini, E., Osborn, W.R., Yeomans, J.M.: Lattice Boltzmann simulations of liquid-gas and binary fluid systems. Physical Review E 54, 5041 (1996)
12. Gao, J., Xing, H.: LBM simulation of fluid flow in fractured porous media with permeable matrix. Theor. Appl. Mech. Lett. 2, 032001 (2012), doi:10.1063/2.1203201
13. Xing, H., Liu, Y.: Automated quadrilateral mesh generation for digital image structures. Theor. Appl. Mech. Lett. 1, 061001 (2011), doi:10.1063/2.1106101
14. Liu, Y., Xing, H.L.: A boundary focused quadrilateral mesh generation algorithm for multi-material structures. Journal of Computational Physics 232, 516–528 (2012), doi:10.1016/j.jcp.2012.08.042

Parallel Scalability Enhancements
of Seismic Response and Evacuation Simulations
# of Integrated Earthquake Simulator

M.L.L. Wijerathne*, Muneo Hori, Tsuyoshi Ichimura, and Seizo Tanaka

Earthquake Research Institute, The University of Tokyo, Tokyo, Japan
{lalith,ichimura,hori,stanaka}@eri.u-tokyo.ac.jp

Abstract. We developed scalable parallel computing extensions for
Seismic Response Analysis (SRA) and evacuation simulation modules
of an Integrated Earthquake Simulator (IES), with the aim of simulating earthquake disaster in large urban areas. For the SRA module, near
ideal scalability is attained by introducing a static load balancer which
is based on the previous run time data. The use of SystemV IPC as
a means of reusing legacy seismic response analysis codes and its im-
pacts on the parallel scalability are investigated. For parallelizing the
multi agent based evacuation module, a number of strategies like communication hiding, minimizing the amount of data exchanged, virtual
CPU topologies, repartitioning, etc. are used. Preliminary tests on the
K computer produced above 94% strong scalability, with several million
agents and several thousand CPU cores. Details of the parallel computing strategies used in these two modules and their effectiveness are
presented.

Keywords: multi agent simulations, seismic response analysis, large urban area, HPC, scalability.

1   Introduction

Petascale super computers have opened new avenues for more reliable earthquake disaster predictions compared to the currently used simplified methods.
The current earthquake disaster predictions, which are based on the statistical
analysis of past events, are less reliable since the built environment has significantly changed since those decades old past events. It is possible to make more
reliable predictions by simulating large area earthquake disasters using cutting
edge numerical tools from many disciplines like seismology, earthquake engineering, civil engineering, social science, etc. High performance computing is vital to
meet the computational demand of simulating high fidelity models of large urban
areas, with high spatial and temporal resolutions. Further, the need of stochastic
modelling increases this high computational demand by severalfold. Stochastic

---

* Earthquake Research Institute, the University of Tokyo, 1-1-1, Yayoi, Bunkyo-ku,
  Tokyo 113-0032, Japan.

M. Daydé, O. Marques, and K. Nakajima (Eds.): VECPAR 2012, LNCS 7851, pp. 105–117, 2013.

modeling is required to improve reliability of the numerical predictions; what decision makers require is the high level of confidence in the predictions.

With the aim of developing a such a system for making for more reliable earthquake disaster predictions, a system called Integrated Earthquake Simulator (IES) is being developed[1]. The objective of IES is to seamlessly simulate earthquake hazards, disasters and aftermaths. Modules for simulating source to site seismic wave propagation, seismic response analysis (SRA) of buildings and underground structures, evacuation and recovery are being developed. Highly scalable parallel extensions are essential for IES, in order to realize more reliable disaster predictions using high fidelity models. While petascale machines provide the necessary hardware resources, it is a challenging task to develop scalable codes for simulating a large urban area with fine detailed models.

Prior to this work, a parallel computation extension for the IES's SRA module has been developed[2]. However, several bottlenecks seriously hinder its scalability to mere 32 CPU cores. The SRA module of IES consists of several simple to moderately advanced seismic response analysis methods. All these seismic response analysis models are implemented as serial programs. With task level parallelism and simple static load balancer, near ideal scalability is attained up to a limited number of CPUs. Runtime of some tasks are excessively large compared to the majority of tasks; large buildings involve much longer run time. These tasks with large runtime limit the scalability, and have to be parallelized to scale over a large number of CPUs.

The emergency evacuation module of IES is based on Multi-Agent Simulation (MAS), in which people are represented by agents that autonomously navigate, and interact with the neighbors and the environment. Even though MAS often advocates the use of simple agents, sophisticated and smart agents are necessary to model complex human behaviors. High performance computing enhancements are necessary to simulate the evacuation of large urban area with millions of smart agents. Most literature focuses on simulation of several ten thousands of simple agents on less than 50 CPU cores, and real time rendering; modeling virtual worlds for games and entertainment is the main objective of the existing studies. Consenza et al.[3] have demonstrated simulation of 100,000 agents on 64 CPUs, but it has limited scalability. With several strategies to hide communications and minimize the volume of data exchanged, our HPC enhanced MAS code attained 94% strong scalability up to 2048 cores on the K-computer, with two million agents.

Details of the parallel extension of the SRA module and scalability are presented in the section 2. Section 3 summarizes difficulties and the strategies used in parallelizing the multi-agent based evacuation module. Some concluding remarks are given in the last section.

2   Parallel Performance Enhancement of SRA Module

SRA module of IES has a series of simple to advance nonlinear SRA methods like discrete element method (DEM), one component model (OCM), fiber element model, etc. All these SRA methods are implemented as serial codes with FORTRAN 77. On the other hand, IES is developed with C++; extensive object oriented features make C++ a popular choice for projects developed by large groups. Though FORTRAN 77 is an outdated standard, the reuse of these SRA codes is unavoidable since rewriting in a latest standard, verification and validation require a significant effort and time.

A previous effort to parallelize SRA module [2] based on a master slave model produced much lower scalability than the required; some bottlenecks that hinder the scalability are extensive use of temporary files for inter-process communication and output data saving, uneven workloads assign to CPUs, and the large number of message passing. We eliminated these bottlenecks and achieved high scalability with task parallelism and a static load balancer. The rest of the section provides the details of these improvements.

2.1 Calling Legacy FORTRAN Codes

The existing FORTRAN 77 based SRA codes have to be converted to libraries, so that those can be reused in IES. This conversion is a complex and error prone process since the original codes have been developed under non-standard default compiler settings. The SRA codes have been developed with COMPAQ FORTRAN which used *save* semantics, which makes compiler to allocate all local variables in static storage and initialized to zero. The use of this special compiler setting is not recommended by most of the present day compilers including Intel FORTRAN, which is the successor of COMPAQ FORTRAN. To circumvent this problem, independent SRA executables can be called from IES using *system()* command while using temporary files for data exchnage. Though this is an acceptable solution in serial applications, using temporary files is a serious performance bottleneck in parallel applications.

It is necessary to find a simple and less error prone means of reusing old FORTRAN codes for the future needs. To this end, we explored the applicability of SystemV IPC[4] (commonly abbreviated SysV IPC) in parallel computing environment. In this setting, IES invokes the legacy FORTRAN codes as independent executable and use shared memory segments and semaphores of SysV-IPC to exchange data with IES. Compared with conversion to a library, this process involves fewer modifications to FORTRAN codes. As shown at the end of this section, the use of SysV-IPC is a good alternative to call legacy FORTRAN 77 codes in parallel environments.

One disadvantage of using SysV IPC shared memory segments is that those persist even after termination of the owner processes, unless explicitly cleaned. Under normal operations, the IPC classes introduced to IES take care of the cleaning of the used SysV IPC resources. However, if the main program dies prematurely, the IPC resources used must be manually cleaned, in each node; SysV IPC resources can be released with the command *ipcrm <-s/-m> <shmid/ semid>*. Unless the shared memory segments are manually cleaned, it becomes a system wide persistent memory leak. SysV IPC is a good solution for a cluster dedicated for IES. However, premature exits of IES may cause problems in clusters with multiple users and automated resource management. Therefore, this SysV IPC approach cannot be used in professional super computers; spawning new process is usually prohibited.

2.2 Saving Large Volume of Data with MPI-IO

When simulating a large urban area, SRA module produces a large volume of output data; around 10GB of binary data per 10,000 structures. IES should organize and save the output data in a ready-to-visualize format. We utilized the MPI-IO functionalities of MPI-2 standard to write the SRA output in ready to visualize format, thereby achieving higher parallel performance and reducing the visualization time. Compared to POSIX I/O, MPI-IO can deliver much higher I/O performance in parallel environment[5]. The level 3 access function *MPI_File_write_all()*, which allows non-contiguous collective access to a file, is used to write SRA output data in a ready to visualize format.

2.3   Load Balancing

The currently implemented task parallelism makes the SRA module an embarrassingly parallel problem. However, the difficulty of predicting the runtime of each task makes it difficult to attain a good load balance; run time for each structure depends on the location and magnitude of an earthquake, etc. Therefore, some form of dynamic load balancing is necessary. A hybrid solution with static load balancing for the majority of the buildings and switching to dynamic load balancing for the remaining is the best solution. However, only static load balancing is implemented in the modified version, since dynamic load balancing requires significant modifications to the present code of IES.

The static load balancer utilizes the runtime information recorded at previous executions to assign nearly equal workloads to each CPU. The simple static load balancer first distributes all the building information and previous run time data to all the CPUs. These data are grouped such that data of one or several GIS (Geographic Information System) tiles are in one set, and each data set is compressed to reduce message size. CPUs independently pick a subset of data so that each has nearly equal work load, estimated based on the previous runtime data. The use of static load balancer is an acceptable solution provided the runtime difference of a given building due to different input Strong Ground Motions (SGM's) of similar magnitudes would not be large. Irrespective of the number of input data files of building shapes and run times, this phase involves only 2 message broadcasts, provided the volume of building shape data and run time data is less than 4GB. Unlike a master slave load balancing, where master CPU decides and distributes the work, this communication independent load balancing does not degenerate the scalability.

2.4   Scalability of the New Parallel Extension of SRA Module

The modified parallel extension of SRA module involves message passing only at the very beginning and the end of a simulation: at the beginning to share some configuration files and building shapes and run time data of each GIS tile; at the end to save data with collective MPI-IO. In this setting, the modified parallel module should well scale with the number of CPUs, as long as previous run time based load estimation assigns equal workloads to CPUs.

To test the scalability of the new parallel extension, we simulated 125,500 buildings in the Kochi city of Japan, with the OCM model. The buildings are excited with the strong ground motion data observed during the 1995 Kobe earthquake. A DELL cluster with QLogic 12200 InfiniBand switch and 16 computation nodes, each with two hexa-core Intel Xeon X5680 CPUs and 47GB DDR3 memory, was used for these simulation. This cluster has hardware support for MPI-IO. However, it does not have a parallel file system supporting MPI-IO.

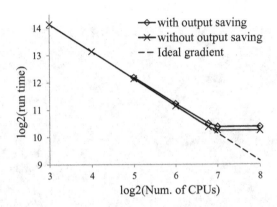

Fig. 1. Scalability of the SRA module

As shown in Fig. 1, the modified parallel extension of SRA module exhibits near ideal scalability, up to 110 CPUs. The same graph shows that saving 84 GB of output data with MPI-IO has little impact on the scalability, even though the cluster does not have supporting file system for MPI-IO. On the same hardware, another simulation is conducted to find the scalability issues due to the use of SysV IPC for inter-process communication; OCM model is called as an independent executable and SysV IPC resources are used for data exchange. It is found that the use of SysV IPC has no noticeable effect on the scalability. This confirms that SysV IPC resources can be used for calling legacy FORTRAN codes in parallel environments, without any impact on scalability. Surely, some work is necessary to introduce SysV IPS resources to FORTRAN codes. However, compared to conversion to a library, it is less error prone and requires much less time.

The sudden changes of the graphs in Fig. 1 to a constant, after 110 CPUs, indicates the inadequacy of task level parallelism. For complex SRA methods like OCM and fiber element method, run time for large buildings is 2-4 orders of magnitude larger compared to that for small buildings. Further, these complex

SRA methods involve hours of runtime for large buildings. One of the buildings in this demonstration involves 1211 seconds run time, which is almost equal to the run time with 110 CPUs. Hence the graphs in Fig. 1 flatten.

Parallelizing the SRA codes is necessary to break the above mentioned performance barrier. With parallelized SRA codes, CPU subgroups of different sizes have to be formed such that resources are optimally used. All the small buildings are executed as serial codes in one large CPU group while the other groups execute bigger buildings in parallel. Such strategy makes it possible to simulate a large urban area in a single simulation.

Figure 2 shows a snapshot of seismic response of 145,000 buildings in central Tokyo; assumed structural skeletons and material properties are used. Colors represent the magnitude of displacement vector; red color indicates displacements exceeding $0.5m$.

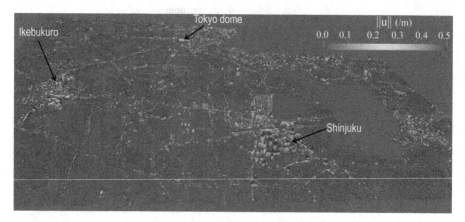

Fig. 2. Snapshot from a seismic response analysis demonstartion of central Tokyo

3   Parallel Performance Enhancements of Evacuation Module

To simulate emergency evacuation of millions of people in a large urban area like Tokyo, it is necessary to develop a scalable parallel extension of the evacuation module. The existing MAS based pedestrian simulation studies have reported low scalability which is limited to a few tens of CPU cores; the objective of majority of the existing studies is simulating virtual worlds, for which a few tens of CPU cores may be sufficient. With several strategies for hiding and minimizing communications, we attained near liner scalability at least up to 2048 CPU cores, with a few millions of agents. The rest of this section provides a short description of the abilities of current agents, difficulties in parallelization and main strategies used to attain high parallel scalability.

3.1   Complexity of Agents

Since the evacuation simulations are concerned with human lives, it is necessary to use smart agents which can reproduce the observed behaviors of real crowds. Modeling smart agents involves complex data structures consisting of a large number of variables. While the current agents are simple, we are implementing the required abilities that are important for tsunami triggered emergency evacuations. To model the heterogeneous real human crowds, a wide range of agents are implemented: agents with different physical abilities like speed, maximum sight distance, etc.; agents with different amount of information related to the environment and the emergency scenario; agents with different levels of responsibilities like police officers, fire fighters, volunteers and common residents. Figure 3 shows a part of the UML diagram for a resident agent. Another two types of agents, officials and non-residents, are implemented similarly by specializing the same _Agent base class.

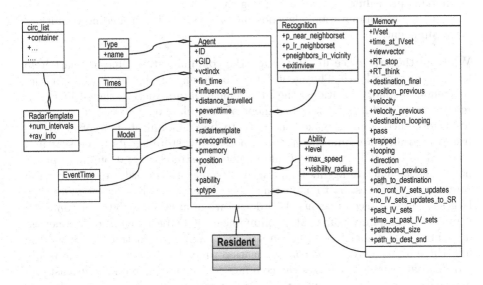

Fig. 3. Class hierachy of a resident agent

All the types of agents have common navigation behavior, though the different type of agents may take different actions depending on the role they play. As an example, official agents seek for the other agents requiring support while resident agents move to nearest evacuation center. With *See*() functionality, an agent makes a high resolution scan of his visible environment like a radar, and identify the boundary of visibility and visible neighboring agents. Next, with *Think*() functionality, he analyzes the area and the boundary of his visibility and identify the available paths and choose the closest path to his destination direction. Finally, with *Move*() functionality, he navigates avoiding collisions with neighboring agents and other obstacles.

3.2   Difficulties in Parallelization

The major steps in parallelizing the multi agent code are the same as that of other particle type simulations like SPH. However, compared to other particle type methods, parallelization of the multi agent code involves several additional difficulties, some of which are listed below.

1. Complexity of data structures and the amount of data involved
   (a) The volume of agent's data is 50 times or more compared to SPH.
   (b) Agents have dynamically growing data like memory of their experiences, etc.
   (c) Implementing smart agents requires the use of complex data structures like graphs, maps and trees, which grow in size.
2. Objects of different types of agents, like officials, residents, etc., have to be stored in non-contiguous locations in different vectors.
3. Require maintaining a wide ghost layer of thickness at least equal to the maximum visibility distance of an agent.
4. Amount of computations depends on the type and surrounding conditions of an agent.

While all the first three items increase the communication time, the last item leads to load imbalance. The agent data to be communicated are located in non-contiguous memory locations; the hierarchical data structure shown in Fig. 3 requires byte padding for the alignment of base class objects and for the sake of performance. Item 2 makes the data to be communicated become further fragmented and non-uniform; it is not possible to store agents to be sent and received to each CPU in a continuous memory stretches, and one may prefer not to delete inactive agents for performance reasons. In order to preserve the continuity, it is necessary to maintain a ghost layer of width at least equal to the largest sight distance of agents. In a dense urban area, ghost layer of $50m$ may contain a large number of people. Communicating a larger volume of fragmented data always is associated with increase in communication time[6]. Hence the first three items increase the communication time. The presence of dynamically growing data further increases the communication time; it requires at least two messages, memory allocation at the receiving end and packing and unpacking of data.

3.3   Strategies for Enhancing Scalability

The basic strategies used in parallelization of the multi agent code are more or less the same as that of other particle type simulations: the domain is decomposed such that each has equal work loads; ghost or overlapping layer is maintained and updated with the neighboring CPUs to preserve the continuity; agents are moved to neighbor CPUs when they enter the domain of a neighbor CPU; domain is repartitioned when the agent movements bring significant load imbalance. For domain decomposition, kd-tree is used. Although kd-tree does not minimize the volume of data being communicated, its simple geometry makes it possible to easily detect the movements of agents between different domains. In addition to these common strategies, the following strategies are used to deal with the above mentioned additional difficulties.

Virtual CPU Topologies. With 2D-tree based partitioning we cannot take the advantage of communication topologies of underlying hardware like hypercube, torus, etc.; the resulting communication patterns of kd-tree is too irregular to be mapped to these structured hardware topologies. Distributed graph topology interfaces of MPI-2.2 address this problem[7]. We used *MPI_Dist_graph_create()* to map MPI process ranks to processing elements, to better match the communication pattern of the partitions to the topology of the underlying hardware. However, we could not test the effectiveness of virtual topology due to the unavailability of a cluster supporting this feature.

Algorithm 1.

$comm\_freq = ghost\_update\_interval;$

for $k = 1$ **to** n **do**
  Execute send agents;
  if $(! (k\%comm\_freq))$ **then**
    Initialize ghost update;
  end
  Execute inner agents;
  if $(! (k\%comm\_freq))$ **then**
    Finalize ghost update;
  end
end

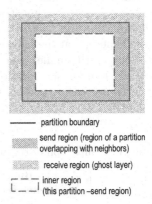

——— partition boundary

send region (region of a partition overlapping with neighbors)

receive region (ghost layer)

inner region (this partition –send region)

Fig. 4. Subdivisions of a partition

Hiding Communications and Minimizing Volume of Data Exchanged.

As mentioned in the Section 3.2, communication of large volume of fragmented data is time consuming. However, most of the communication time can be hidden behind calculations by processing the agents in a certain order (see Algorithm 1). To this end, agents in each CPU are divided into three sub groups (see Fig 4); agents to be received from other CPUs, agents to be sent to other CPUs and the rest of the active agents which are named inner most agents hereafter. To deal with the large amount of data to be communicated, only the necessary agent data is exchanged. The three agent sub groups are stored in *std::map<>* C++ containers instantiated with *std::pair< global ID, _Agent * >*. The map data structure makes it efficient to manage agent movements among CPUs and repartitioning.

The presence of dynamically growing data makes it difficult to eliminate the communication overhead completely. Exchanging all the dynamic data requires at least two messages and packing and unpacking of large amount of data. Instead, in a single message, only the newly added contents are exchanged when updating ghost boundaries; new updates are packed to a small temporary buffer in agent objects, sent with the static data members in one message, and unpacked at the receiving end. The explicit packing and unpacking makes it impossible to hide the communication overhead completely. However, a significant portion of communication overhead is hidden, making the code to attain high parallel scalability.

Reduction of the Frequency of Ghost Layer Updates. Even if the ghost update communications are hidden, packing and unpacking dynamically growing data, etc. incur some time. Therefore, further gain in scalability is possible by reducing the frequency of ghost layer updates. This introduces small error to the simulation. However, the error is negligibly small; the time increment used is $0.2\,s$. Table 1 shows that the reduction of ghost update frequency has a significant advantage only with smaller number of CPUs. With the increasing number of CPUs, its effect diminishes; the overhead of packing and unpacking data goes down with the decreasing number of agents. Therefore, this approach is effective only when CPUs have a large number of agents in the ghost regions.

Table 1. Comparison of runtimes with *ghost_update_interval*'s of 1 and 2

CPUs	Runtimes with /(s)		Difference
	comm_freq=1	*comm_freq=2*	
4	16701.7	15103.2	1598.5
8	7765.6	7025.6	740.0
16	3243.5	3170.5	73.0
32	1701.3	1694.4	6.9

Minimizing Data Exchange in Repartitioning. Migration of agents from a partition to another brings load imbalance. When a significant load imbalance occurs, repartition is necessary to maintain equal workloads. Repartitioning is an expensive step since sophisticated agents have large amount of data. With 2D-tree, it is observed that most of the agents remain in the same CPU even after repartitioning, unless *MPI_Dist_graph_create*() maps a partition to a different CPU. The repartitioning algorithm detects whether a partition is assigned to the same CPU and exchanges only the newly assign agents. This drastically reduces the communication overhead involved with repartitioning, effectively lowering any performance degeneration due to repartitioning. Table 2 compares run times without any repartitioning and with 4 repartitioning (once in 80 steps). As is seen, even with the current serial 2d-tree algorithm, the gain due to repartitioning is significantly increasing with the number of CPUs. Figure 6, about which a detailed explanation is given later, demonstrate the gain due to repartitioning. Instead of calling repartitioning at fixed intervals, load in each CPU has to be monitored and repartitioning has to be called based on the level of load imbalance.

Table 2. Effectiveness of repartitioning

CPUs	Runtime /(s)		Gain /(%)
	no repartition	4 repartitions	
32	1733.9	1656.7	4.6
64	939.1	866.5	8.4
128	459.8	434.1	5.9
256	250.9	220.6	13.7

3.4 Scalability

In order to test the effectiveness of the above major strategies, we conducted a series of simulations with 500,000 agents in a part of Kochi city environment. The same cluster used for the scalability test of the SRA module is used for these simulations. Ghost layer updating at each time step and repartitioning at an interval of 80 steps are considered for these simulations. As Fig. 5 indicates, the above strategies have produced near linear scalability; super linear behavior is due to the nonlinear time reduction of the neighbor search algorithm with the decreasing number of agents.

Further, preliminary tests in the K-computer produced 94% strong scalability with 2048 CPU cores; strong scalability is defined as $\frac{(\frac{T_m}{T_n})}{(\frac{n}{m})}$, where T_k is the time with k number of CPU cores and $n \geq 2m$. For the tests on the K-computer, 400 time steps with 2 million of agents are considered. Further the ghost boundaries are updated at each iteration, movements of agents between CPU cores are checked at each 10 iterations and domain is repartitioned at each 100 iterations for load balancing. Figures 6a and 6b show the runtime for 400 iterations, except the repartitioning time. These figures clearly show the significant performance gain due to repartitioning. As is seen, at each iteration, the run time with 2048 cores is nearly the half of that of 1024 cores.

Figure 7a shows the history of total run time with 2048 CPU cores. As is seen the major bottleneck in the current code is repartitioning, which is still a serial code. Further, as shown in Fig. 7b, migration of agents (detecting agent movements from the domain of one CPU to another and dispatch those agents to the new CPU) is relatively time consuming. In future developments, these two bottlenecks are to be addressed to further increase the scalability.

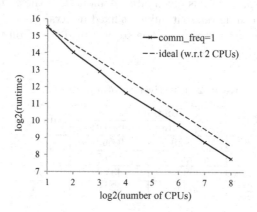

Fig. 5. Scalability of the multi agent code

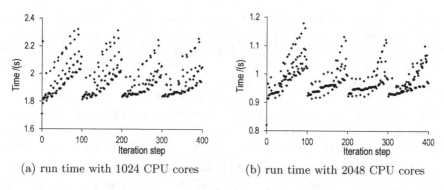

(a) run time with 1024 CPU cores

(b) run time with 2048 CPU cores

Fig. 6. History of run time with 1024 and 2048 cores. Repartitioning time is exculded.

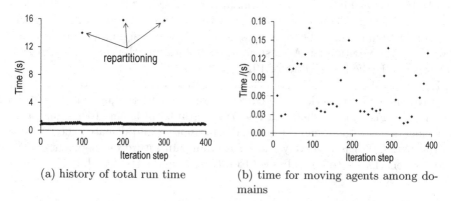

(a) history of total run time

(b) time for moving agents among domains

Fig. 7. History of total run time and agent migration time with 2048 CPU cores

4   Summary

Scalable parallel extensions for the seismic response analysis and multi-agent based evacuation modules of an Integrated Earthquake Simulator (IES) are being developed. Task parallelism is considered for the SRA module and near ideal scalability is attained with a static load balancer. In future, parallelization of the SRA codes is necessary to address the scalability limitations due to the presence of long time consuming tasks (i.e. large buildings). Further, out of a given number of CPUs, forming CPU subgroups of different sizes to execute the parallel SRA codes with minimum waste of CPU time and task scheduling have to be considered.

The multi agent based evacuation simulation module produced high strong scalability with a few thousands of CPU cores in the K-computer. In future, further scalability improvements of evacuation module are planned with direct remote memory access features of MPI. Such enhancements are necessary to cope with complexities arising from planned sophisticated agent features and introduction of different types of agents.

Acknowledgements. This work was supported by JSPS KAKENHI Grant Number 24760359. Part of the results is obtained by using the K computer at the RIKEN Advanced Institute for Computational Science.

References

1. Hori, M., Ichimura, T.: Current state of integrated earthquake simulation for earthquake hazard and disaster. J. of Seismology 12(2), 307–321 (2008)
2. Sobhaninejad, G., Hori, M., Kabeyasawa, T.: Enhancing integrated earthquake simulation with high performance computing. Advances in Engineering Software 42(5), 286–292 (2011)
3. Cosenza, B., Cordasco, G., De Chiara, R., Scarano, V.: Distributed Load Balancing for Parallel Agent-Based Simulations. In: 19th International Euromicro Conference on Parallel, Distributed and Network-Based Processing, pp. 62–69 (2011)
4. Richard Stevens, W.: UNIX Network Programming. Interprocess Communications 2 (1999) ISBN 0-13-081081-9
5. Latham, R., Ross, R., Thakur, R.: The impact of file systems on MPI-IO scalability. In: Kranzlmüller, D., Kacsuk, P., Dongarra, J. (eds.) EuroPVM/MPI 2004. LNCS, vol. 3241, pp. 87–96. Springer, Heidelberg (2004)
6. Balaji, P., Buntinas, D., Balay, S., Smith, B., Thakur, R., Gropp, W.: Nonuniformly Communicating Noncontiguous Data: A Case Study with PETSc and MPI. In: Proceedings of the 21th International Parallel and Distributed Processing Symposium (IPDPS 2007), Long Beach, March 26-30 (2007)
7. Hoefler, T., Rabenseifner, R., Ritzdorf, H., de Supinski, B.R., Thakur, R., Traff, J.L.: The Scalable Process Topology Interface of MPI 2.2. Concurrency and Computation: Practice and Experience 23(4), 293–310 (2010)

QMC=Chem: A Quantum Monte Carlo Program for Large-Scale Simulations in Chemistry at the Petascale Level and beyond

Anthony Scemama[1], Michel Caffarel[1], Emmanuel Oseret[2], and William Jalby[2]

[1] Lab. Chimie et Physique Quantiques, CNRS-Université de Toulouse, France
[2] Exascale Computing Research Laboratory, GENCI-CEA-INTEL-UVSQ
Université de Versailles Saint-Quentin, France

Abstract. In this work we discuss several key aspects for an efficient implementation and deployment of large-scale quantum Monte Carlo (QMC) simulations for chemical applications on petaflops infrastructures. Such aspects have been implemented in the QMC=Chem code developed at Toulouse (France). First, a simple, general, and fault-tolerant simulation environment adapted to QMC algorithms is presented. Second, we present a study of the parallel efficiency of the QMC=Chem code on the Curie machine (TGCC-GENCI, CEA France) showing that a very good scalability can be maintained up to 80 000 cores. Third, it is shown that a great enhancement in performance with the single-core optimization tools developed at Versailles (France) can be obtained.

1 Introduction

Quantum Monte Carlo (QMC) is a generic name for a large class of stochastic approaches solving the Schrödinger equation by using random walks. In the last forty years they have been extensively used in several fields of computational physics and are in most cases considered as state-of-the-art approaches. However, this is not yet the case in the important field of computational chemistry where the two "classical" computational methods are the Density Functional Theory (DFT) and post-Hartree-Fock methods. For a review of QMC and its status with respect to the standard approaches, see *e.g.* [1]. In the recent years several applications for realistic chemical problems have clearly demonstrated that QMC has a high potential in terms of accuracy and in ability of dealing with (very) large systems. However, and this is probably the major present bottleneck of QMC, simulations turn out to very CPU-expensive. The basic reason is that chemical applications are particularly demanding in terms of precision: the energy variations involved in a chemical process are typically several orders of magnitude smaller than the total energy of the system which is the quantity computed with QMC. Accordingly, the target relative errors are typically 10^{-7} or less and the Monte Carlo statistics needed to reach such a chemical accuracy can be tremendously large.

Now, the key point for the future is that this difficulty is expected to be largely overcome by taking advantage of the remarkable property of QMC methods (not valid for standard methods of chemistry) of being ideally suited to HPC and, particularly, to massive parallel computations. In view of the formidable development of computational

M. Daydé, O. Marques, and K. Nakajima (Eds.): VECPAR 2012, LNCS 7851, pp. 118–127, 2013.

platforms this unique property could become in the near future a definite advantage for QMC over standard approaches.

The stochastic nature of the algorithms involved in QMC enables to take advantage of today and tomorrow's computer architectures through the following aspects: i) Data structures are small inducing a fairly small memory footprint (less than 300 MiB per core for very large systems) and data accesses are organized to maximize cache usage: spatial locality (stride-one access) and temporal locality (data reuse), ii) Most of the computation can be efficiently vectorized making full use of the vector capabilities of recent processors, iii) Network communications can be made non-blocking, iv) Access to persistent storage is negligible and can be made non-blocking, v) Different parallel tasks can be made independent of each other so as to run asynchronously, vi) Fault-tolerance can be naturally implemented. All these features which have been implemented in the QMC=Chem code developed at Toulouse[2] are required to take advantage of large-scale computing grids[3] and to achieve a very good parallel efficiency on petascale machines.

In section 2 a short overview of the mathematical foundations of the QMC method employed here is presented. For a more detailed presentation the reader is referred to [1] and references therein. Section 3 is devoted to the presentation of the general structure of the simulation environment employed for running QMC=Chem on an arbitrary computational platform. A preliminary version of such an environment has been presented in [3]. Here, we present an improved implementation where the network communications are now fully handled by a client-server implementation and the computational part is isolated in multiple instances of a single-core Fortran program. These modifications allow the program to survive failures of some compute nodes. In section 4 the results of our study of the parallel efficiency of the QMC=Chem code performed on the Curie machine (TGCC-GENCI, France) thanks to a PRACE preparatory access[4] are presented. In section 5 the results of the optimization of the single-core performance are discussed. The optimization was performed after a static assembly analysis of the program and a decremental analysis.

2   Overview of a QMC Simulation

In the simulations discussed here, the basic idea is to define in the $3N$-dimensional electronic configuration space a suitable Monte Carlo Markov chain combined with a birth-death (branching) process to sample a target probability density from which exact (or high-quality) quantum averages can be evaluated. In our simulations we employ a variation of the Fixed-Node Diffusion Monte Carlo (FN-DMC) method, one of the most popular versions of QMC. In short, we aim at solving the electronic Schrödinger equation written as

$$\mathscr{H}\Psi_0(\mathbf{r}_1,\ldots,\mathbf{r}_N) = E_0\Psi_0(\mathbf{r}_1,\ldots,\mathbf{r}_N) \tag{1}$$

where $\mathscr{H}$ is the molecular Hamiltonian operator, $\Psi_0(\mathbf{r}_1,\ldots,\mathbf{r}_N)$ the N-electron wave function, and E_0 the total electronic energy.

To do that, the basic idea is to construct a stochastic process having the density

$$\pi(\mathbf{r}_1,\ldots,\mathbf{r}_N) = \frac{\Psi_0(\mathbf{r}_1,\ldots,\mathbf{r}_N)\Psi_T(\mathbf{r}_1,\ldots,\mathbf{r}_N)}{\int\ldots\int d\mathbf{r}_1\ldots d\mathbf{r}_N\Psi_0(\mathbf{r}_1,\ldots,\mathbf{r}_N)\Psi_T(\mathbf{r}_1,\ldots,\mathbf{r}_N)} \tag{2}$$

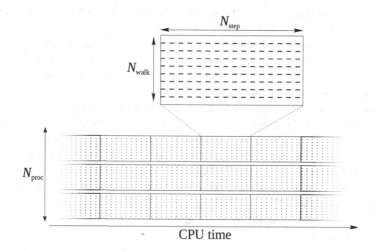

Fig. 1. Graphical representation of a QMC simulation. Each process generates blocks, each block being composed of N_{walk} walkers realizing N_{step} Monte Carlo steps.

as stationary density. Here, Ψ_T — called the trial wavefunction — is some good known (computable) approximation of Ψ_0. The role played by the trial wavefunction is central since it is used to implement the "importance sampling" idea, a fundamental point at the heart of any *efficient* Monte Carlo sampling of a high-dimensional space. In the important case of the total energy, it can be shown that the exact ground-state energy E_0 may be expressed as the average of the so-called local energy defined as $E_L(\mathbf{r}_1,\ldots,\mathbf{r}_N) \equiv \frac{\mathcal{H}\Psi_T}{\Psi_T}$ over the density π.

In brief, the stochastic rules employed are as follows:

1. Use of a standard Markov chain Monte carlo chain based on a drifted Brownian motion. The role of the drift term is to push the configurations (or "walkers" in the QMC terminology) towards the regions where the trial wavefunction takes its largest values (importance sampling technique).
2. Use of a birth-death (branching) process: the walkers are killed or duplicated a certain number of times according to the magnitude of the local energy (low values of the local energy are privileged).

It can be shown that by iterating these two rules for a population of walkers the stationary density π (Eq. 2) is obtained. Note that in the actual implementation in QMC=Chem, a variation of the FN-DMC method working with a *fixed* number of walkers is used[5].

Denoting as $\mathbf{X} = (\mathbf{r}_1,\ldots,\mathbf{r}_N)$ a walker in the $3N$-dimensional space, the random trajectories of the walkers differ from each other only in the initial electron positions $\mathbf{X}_0$, and in the initial random seed S_0 determining the entire series of pseudo-random numbers.

The main computational object is a *block*. In a block, N_{walk} independent walkers realize random walks of length N_{step}, and the energy is averaged over all the steps of each random walk. N_{step} is set by the user, but has to be taken large enough such that the positions of the walkers at the end of the block can be considered independent from their initial positions. A new block can be sampled using the final walker positions as X_0 and using the current random seed as S_0. The block-averaged energies are Gaussian distributed and the statistical properties can be easily computed. The final Monte Carlo result is obtained by super-averaging all the block-averages. If the block-averages are saved to disk, the final average can be calculated by post-processing the data and the calculation can be easily restarted at any time. As all blocks are completely independent, different blocks can be computed asynchronously on different CPU cores, different compute nodes and even in different data centers. Figure 1 shows a pictorial representation of three independent CPU cores computing blocks sequentially, each block having different initial conditions.

The core of QMC=Chem is a single-core Fortran executable that computes blocks as long as a termination event has not been received. At the end of each block the results are sent in a non-blocking way to a central server, as described in the next section.

3   The Client/Server Layer

In the usual MPI implementations, the whole run is killed when one parallel task will not be able to reach the *MPI_Finalize* statement. This situation occurs when a parallel task is killed, often due to a system failure. For deterministic calculations where the result of every parallel task is required, this mechanism is convenient since it immediately stops a calculation that will not give the correct result. In our case, as the result of the calculation of a block is a Gaussian random variable, removing the result of a block from the simulation is not a problem since doing that does not introduce any bias in the final result. Therefore, if one compute node fails, the rest of the simulation should survive.

A second disadvantage of using MPI libraries is that all resources need to be available for a simulation to start. In our implementation, as the blocks can be computed asynchronously we prefer to be able to use a variable number of cores during the simulation in order to reduce the waiting time in the batch queue.

These two main drawbacks lead us to write a lightweight TCP client/server layer in the Python language to handle all the network communications of the program and the storage of the results in a database. The Python program is divided into three distinct tasks shown in figure 2, the first and second tasks running only on the master compute node.

The first task is a *manager* that watches periodically the database associated with the running simulation. The manager computes the running averages and error bars, and checks if the stopping condition of the calculation is reached. The stopping condition can be for instance a threshold on the error bar of the energy, a condition on the total execution time, *etc.* If the stopping condition is reached, a stopping flag is set in the database.

The second task is a *data server*. This task contains two threads: one network thread and one I/O thread. The network thread periodically sends a UDP packet to the connected clients to update their stopping condition and to check that they are still running. The network thread also receives the block averages and puts them in a queue. Simultaneously, the I/O thread empties this queue by storing in the database the block averages.

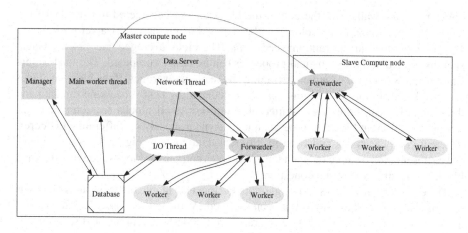

Fig. 2. The communication architecture of QMC=Chem

At any time, a new client can connect to the data server to request the input files and participate to a running calculation. Another way to increase the number of running cores is to submit another run reading and writing to the same database. As the managers read periodically the content of the database, each simulation is aware of the results obtained by the other simulations. This allows the use of multiple managers and data servers that can be submitted to the batch scheduler as independent jobs in order to gather more and more computing resources as they become available.

The third task is a *forwarder*. Each compute node (as well as the master node) has only one instance of the forwarder. The forwarder has different goals. The first goal is to spawn the computing processes (single-core Fortran executables) and to collect the results via Unix pipes after a block has been computed. Then, it sends the results to the data server while the computing processes are already computing the next block. If every compute node sends directly the results to the data server, the master node is flooded by small packets coming from numerous sources. Instead the forwarders are organized in a binary tree structure, and the second goal of a forwarder is to collect results from other forwarders and transfer them in a larger packet to its parent in the binary tree. Using this structure, the data server has much fewer connected clients, and receives much larger packets of data. All the nodes of the forwarder tree can possibly connect to all their ancestors if the parent forwarder does not respond.

On massively parallel machines an MPI launcher is used to facilitate the initialization step. The launcher sends, via the MPI library, the input files and Python scripts to all the slave nodes allocated by the batch scheduler. The files are written in a RAM disk on every node (the */dev/shm* location). The reason for this copy is to avoid too many simultaneous I/O on the shared file system, and also to avoid I/O errors if the shared file system fails, if a local disk fails or is full. The MPI launcher then forks to an instance of a forwarder that connects to the data server.

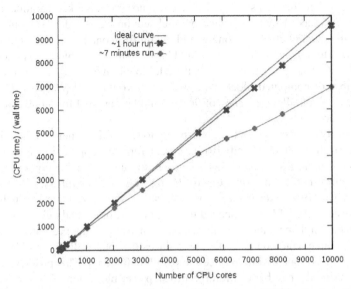

Fig. 3. Number of computed blocks as a function of the number of CPU cores for a fixed computation time

4 Parallel Efficiency

Using the design described in the previous section, the parallel section of the program is expected to display a parallel efficiency of about 100% since there is no blocking statement. In this section, we investigate in some detail how the initialization and finalization steps impact the parallel efficiency of QMC=Chem.

For that, a small benchmark was set up during a PRACE preparatory access on the Curie machine (TGCC-GENCI, France).[4] On this machine, all the compute nodes were equipped with the same processors, namely four Intel Nehalem 8-core sockets. Two important parameters are to be kept in mind. First, we chose to run a short simulation for which the average CPU time required to compute one block is 82 seconds.[1] Second, we have chosen to send the stopping signal after 300 seconds (during the fourth block). When the forwarders receive the stopping signal, they wait until all the working CPU cores finish their current block. Hence, in this study the ideal wall time for perfect scalability should be 328 seconds.

The additional time $T(N_{core})$ with respect to the ideal time can be expressed as

$$T(N_{core}) = W(N_{core}) - \frac{C(N_{core})}{N_{core}} \tag{3}$$

where N_{core} is the number of cores, $W(N_{core})$ is the wall time and $C(N_{core})$ is the CPU time, both measured for N_{core} cores. With 10 000 cores, 149 seconds are needed for the

---

[1] This is not representative of a real simulation since it is far too small: the time spent in network communications will be over-estimated compared to real simulations where the time to compute one block is typically much greater (10 minutes and more).

initialization and finalization steps. For this 7 minutes benchmark, a parallel efficiency of 69% was obtained. However, as the parallel section has an ideal scaling, one can extrapolate the parallel efficiency one would obtain for a one hour run. If the stopping signal occurs after one hour, each core would have computed 44 blocks. The total CPU time would be $\tilde{C}(N_{core}) = 11C(N_{core})$. As the additional time $T(N_{core})$ does not depend on the number of computed blocks, the wall time would be $\tilde{W}(N_{core}) = T(N_{core}) + \tilde{C}(N_{core})/N_{core}$. A parallel efficiency of 96% would be obtained for a one-hour run on 10 000 CPU cores (figure 3).

More recently, we were given the opportunity to test QMC=Chem on the thin nodes of the Curie machine (80 000 Sandy Bridge cores), for a real application on a biological molecule made of 122 atoms and 434 electrons (the largest application ever realized using all-electron Diffusion Monte Carlo). Using 51 200 cores for 3 hours, the parallel efficiency was 79%. After the runs were finished, we realized that for such a large molecular system, the CPU time needed to compute one block had quite large fluctuations due to the implementation the dense-sparse matrix product presented in the next section. This implementation considerably reduces the total wall time (which is what the end user wants), but slightly reduce the parallel efficiency. This problem has been solved by making the number of Monte Carlo steps per block non-constant. Nevertheless, these runs confirm that a good scaling can still be obtained for a real simulation.

5   Single-Core Efficiency

Our choice in the implementation of the QMC algorithms was to minimize the memory footprint. This choice is justified first by the fact that today the amount of memory per CPU core tends to decrease and second by the fact that small memory footprints allows in general a more efficient usage of caches. Today, the standard size of the molecular systems studied by QMC methods and published in the literature usually comprise less than 150 electrons. For a 158 electron simulation, the binary memory footprint (including code and data) per core is only 9.8 MiB. To check the memory footprint of much larger systems, a few Monte Carlo steps were performed successfully on a molecular system containing 1731 electrons; such a large system only required 313 MiB of memory per core. For a system beyond the largest systems ever computed with all-electron QMC methods, the key limiting factor is only the available CPU time and neither the memory nor disk space requirements. This feature is well aligned with the current trends in computer architecture for large HPC systems.

As the parallel scaling is very good, single-core optimization is of primary importance: the gain in execution time obtained on the single-core executable will also be effective in the total parallel simulation. The Fortran binary was profiled using standard profiling tools (namely gprof[6] and Scalasca[7]). Both tools exhibit two major hot spots in the calculation of a Monte Carlo step. The first hot spot is a matrix inversion, and the second hot spot is the product of a constant matrix A with five different matrices $B_1 \ldots B_5$. These two bottlenecks have been carefully optimized for the x86 micro-architectures, especially the for the AVX instruction set of the Sandy Bridge processors.

To measure the performance of the matrix inversion and the matrix products, small codelets were written. The final results are given in table 1, compared to the performance of the single core executable, with different molecular system sizes. Computational complexity (with respect to FP operations) of the matrix inversion is $\mathcal{O}(N_e^3)$ where N_e is the number of electrons. Exploiting the sparse structure of the right matrices in the matrix matrix products, computational complexity of such products is reduced to $\mathcal{O}(N_e^2)$ with a prefactor depending on N_{basis}, the size of the basis set used to describe the wave function.

The matrix inversion is performed in double precision (DP) using the Intel MKL library, an implementation of the LAPACK[9] and BLAS[10] APIs. To maximize MKL efficiency, arrays were padded to optimize array alignment (lined up on 256 bit boundaries) and the leading dimension of the array is chosen to be a multiple of 256 bits to ensure that all the column accesses are in turn properly aligned.

For the matrix matrix products, similar alignment/padding techniques were used. Loops were rearranged in order to use full vector length stride-one access on the left dense matrix and then blocked to optimize temporal locality (i.e. cache usage).

An x86_64 version of the MAQAO framework[11] was used to analyze the binary code and to generate best possible static performance estimates. This technique was used not only on the matrix matrix products but also on all of the hottest loops (i.e. accounting at least for more than 1% of the total execution time). This allowed us to detect a few compiler inefficiences and to fix them by hard coding loop bounds and adding up pragmas (essentially for allowing use of vector aligned instructions).

Then, the DECAN tool[12] was used to analyze performance impact of data access. For that purpose, for each loop, two modified binaries were automatically generated: i) FPISTREAM: all of AVX load instructions are replaced by PXOR instructions (to avoid introduction of extra dependencies) and all of the AVX store instructions were replaced by NOP instructions (issuing no operation but preserving the binary size). FPISTREAM corresponds to the ideal case where all of the data access are suppressed. ii) MISTREAM: all of the AVX arithmetic instructions were replaced by NOP instructions. By comparing cycle counts of FPISTREAM, MISTREAM binaries with the original binary, potential performance problems due to data access (essentially cache misses)

Table 1. Single core performance (GFlops/s) of the two hot routines: inversion (DP), matrix products (SP), and of the entire single-core executable. Measurements were performed on an Intel Xeon E31240, 3.30GHz, with a theoretical peak performance 52.8 GFlops/s (SP) and 26.4 GFlops/s (DP). The values in parenthesis are the percentages with respect to the peak. The turbo feature was turned off, and the results were obtained using Likwid performance tools.[8]. [1] As the matrix to invert is block-diagonal with two $N_e/2 \times N_e/2$ blocks, the inversion runs on the two sub-matrices.

System sizes	Matrix inversion[1]	Matrix products	Overall performance
$N_e = 158$, $N_{basis} = 404$	6.3 (24%)	26.6 (50%)	8.8 (23%)
$N_e = 434$, $N_{basis} = 963$	14.0 (53%)	33.1 (63%)	11.8 (33%)
$N_e = 434$, $N_{basis} = 2934$	14.0 (53%)	33.6 (64%)	13.7 (38%)
$N_e = 1056$, $N_{basis} = 2370$	17.9 (67%)	30.6 (58%)	15.2 (49%)
$N_e = 1731$, $N_{basis} = 3892$	17.8 (67%)	28.2 (53%)	16.2 (55%)

could be easily detected. Such an analysis revealed that for most of the hot loops, data access was accounting for less than 30% of the original time, indicating an excellent usage of the caches. Measurement of the binaries were performed directly with the whole application running allowing to take into account runtime context for the loops measured.

6 General Conclusion

In December 2011, GENCI gave us the opportunity to test our program on Curie (TGCC-GENCI, France) while the engineers were still installing the machine. At that time, up to 4 800 nodes (76 800 cores) were available to us for two sessions of 12 hours. As the engineers were still running a few benchmarks, our runs were divided into 3-hour jobs using 400 nodes (6 400 cores). In this way the engineers were able to acquire resources while our job was running. This aspect points out the importance of our flexible parallel model, but makes it impossible to evaluate rigorously the parallel efficiency. At some point, all the available nodes were running for our calculation during several hours. As we had previously measured a sustained performance of 200 GFlops/s per node for this run, we can safely extrapolate to a sustained value of $\sim$960 TFlops/s (mixed single/double-precision) corresponding to about 38 % of the peak performance of the whole machine for a few hours. As the machine was still in the test phase, we experienced a few hardware problems and maintenance shutdowns of some nodes during the runs. Quite interestingly, it turns out to be an opportunity for us to test the robustness of our program: it gave us the confirmation that our fault-tolerant scheme is indeed fully functional.

In this work we have presented a number of important improvements implemented in the QMC=Chem program that beautifully illustrate the extremely favorable computational aspects of the QMC algorithms. In view of the rapid evolution of computational infrastructures towards more and more numerous and efficient processors it is clear that such aspects could be essential in giving a definite advantage to QMC with respect to other approaches based on deterministic linear algebra-type algorithms.

Acknowledgments. AS and MC would like to thank ANR for support through Grant No ANR 2011 BS08 004 01. This work was possible thanks to the generous computational support from CALMIP (Toulouse) under the allocation 2011-0510, GENCI, CCRT (CEA), and PRACE through a preparatory access (No PA0356).

References

1. Caffarel, M.: Quantum monte carlo methods in chemistry. In: Encyclopedia of Applied and Computational Mathematics. Springer (2011),
   http://qmcchem.ups-tlse.fr/files/caffarel/qmc.pdf
2. See web site: Quantum Monte Carlo for Chemistry@Toulouse,
   http://qmcchem.ups-tlse.fr

3. Monari, A., Scemama, A., Caffarel, M.: Large-scale quantum monte carlo electronic structure calculations on the egee grid. In: Davoli, F., Lawenda, M., Meyer, N., Pugliese, R.,
   Wäglarz, J., Zappatore, S. (eds.) Remote Instrumentation for eScience and Related Aspects,
   pp. 195–207. Springer, New York (2012),
   http://dx.doi.org/10.1007/978-1-4614-0508-5_13
4. Caffarel, M., Scemama, A.: Large-scale quantum monte carlo simulations for chemistry.
   Technical Report PA0356, PRACE Preparatory Access Call (April 2011),
   http://qmcchem.ups-tlse.fr/files/scemama/Curie.pdf
5. Assaraf, R., Caffarel, M., Khelif, A.: Diffusion monte carlo methods with a fixed number of
   walkers. Phys. Rev. E 61 (2000)
6. The GNU profiler,
   http://sourceware.org/binutils/docs-2.18/gprof/index.html
7. Wolf, F.: Scalasca. In: Encyclopedia of Parallel Computing, pp. 1775–1785. Springer (October 2011)
8. Treibig, J., Hager, G., Wellein, G.: Likwid: A lightweight performance-oriented tool suite
   for x86 multicore environments. In: 39th International Conference on Parallel Processing
   Workshops (ICPPW), pp. 207–216 (September 2010)
9. Anderson, E., Bai, Z., Dongarra, J., Greenbaum, A., McKenney, A., Du Croz, J., Hammer-
   ling, S., Demmel, J., Bischof, C., Sorensen, D.: Lapack: a portable linear algebra library for
   high-performance computers. In: Proceedings of the 1990 ACM/IEEE Conference on Supercomputing, Supercomputing 1990, pp. 2–11. IEEE Computer Society Press, Los Alamitos
   (1990)
10. Dongarra, J.J., Du Croz, J., Hammarling, S., Hanson, R.J.: An extended set of fortran basic
    linear algebra subprograms. ACM Trans. Math. Softw. 14, 1–17 (1988)
11. Djoudi, L., Barthou, D., Carribault, P., Lemuet, C., Acquaviva, J.-T., Jalby, W.: MAQAO:
    Modular assembler quality Analyzer and Optimizer for Itanium 2. In: Workshop on EPIC
    Architectures and Compiler Technology, San Jose, California, United-States (March 2005)
12. Koliai, S., Zuckerman, S., Oseret, E., Ivascot, M., Moseley, T., Quang, D., Jalby, W.: A
    balanced approach to application performance tuning. In: Gao, G.R., Pollock, L.L., Cavazos,
    J., Li, X. (eds.) LCPC 2009. LNCS, vol. 5898, pp. 111–125. Springer, Heidelberg (2010)

Optimizing Sparse Matrix Assembly in Finite Element Solvers with One-Sided Communication

Niclas Jansson

School of Computer Science and Communication
KTH Royal Institute of Technology
SE-100 44 Stockholm, Sweden
njansson@csc.kth.se

Abstract. In parallel finite element solvers, sparse matrix assembly is often a bottleneck. Implemented using message passing, latency from message matching starts to limit performance as the number of cores increases. We here address this issue by using our own stack based representation of the sparse matrix, and a hybrid parallel programming model combining traditional message passing with one-sided communication. This gives an significantly faster insertion rate compared to state of the art implementations on a Cray XE6.

Keywords: UPC, PGAS, Hybrid Parallel Programming.

1   Introduction

In large scale finite element simulations a considerable amount of time is spent in sparse tensor assembly. These tensors are often represented as sparse matrices which can be frequently reassembled for time dependent problems. Efficient tensor assembly is therefore a key to obtain good performance. Sparse matrix formats are designed to have a low memory footprint and for good performance when rows are accessed consecutively, for example in sparse matrix vector multiplication. Finite element assembly on the other hand often inserts or adds data at random locations in the tensor, resulting in a poor access pattern which can have a tremendous impact on the assembly performance.

Unstructured meshes are excellent for accurate approximation of complex geometries. However, the lack of underlying structure implies an unstructured communication pattern, which can have a negative effect on the overall performance, in particular for the assembly stage on a large number of cores, as we have observed in our previous work [9].

The reason for this behaviour is partly due to the programming model used. Today, most scientific libraries and applications are parallelized using the *Message Passing Interface* (MPI) [13] or some hybrid incarnation combining MPI with OpenMP [14]. MPI is what is called two-sided, which means that each process can only communicate with send and receive operations, and since these have to be matched to each other it will unavoidably increase latency and synchronization costs in non-structured communication, as in sparse matrix assembly on unstructured meshes.

M. Daydé, O. Marques, and K. Nakajima (Eds.): VECPAR 2012, LNCS 7851, pp. 128–139, 2013.

In this paper we address this issue by replacing the linear algebra parts of the finite element library DOLFIN [11] with a one-sided communication linear algebra backend, based on the *Partitioned Global Address Space* (PGAS) programming model, resulting in a hybrid MPI/PGAS application. Using this hybrid model we observe a significant reduction in assembly time, especially for large core counts.

The outline of the paper is the following; in §2 we motivate our work and give a short background on related work. In §3 we present our sparse matrix format and in §4 we present our parallelization strategy and briefly describe DOLFIN's implementation. Performance results are discussed in §5, and we give conclusions and outline future work in §6.

2   Background

Most state of the art linear algebra packages, as for example PETSc [1], optimize the sparse matrix assembly by using non-blocking communication, overlapping computation and communication. It can be seen as a simulation of one-sided communication, since send and receive operations do not have to be matched directly. However, it requires the receiver to occasionally check for messages which introduces latency etc. Non-blocking communication also involves software buffering, either on the sending or receiving side which adds to the latency penalty. True one-sided communication is realized in MPI 2.0 with *Remote Memory Access* (RMA) operations, but unfortunately the API imposes a number of restrictions that limit the usability of these extensions. For a discussion of these constraints, see for example [3].

Today, at the dawn of exascale computing some concerns have been raised whether MPI is capable of delivering the needed performance. Since a tremendous amount of high quality scientific software has been written in MPI over the past decades, it is unreasonable to think that these are going to be rewritten or replaced with something written in for example PGAS. Therefore we argue that a reasonable way to prepare old legacy codes for exascale is to replace bits and pieces with more scalable one-sided communication, thus creating hybrid MPI/PGAS applications which to our knowledge is a quite unusual combination. Related work in this area is sparse, most of the work is focused on support in the runtime system [2][10] or the applicability of hybrid methods [5], with the exception of [16] which presents an entire large scale application rewritten using a hybrid PGAS/OpenMP approach.

In our previous work on sparse matrix formats [6], we have found that the most common format *Compressed Row Storage* (CRS) [17] is sub-optimal when it comes to sparse matrix assembly. CRS has an efficient access pattern for *Sparse Matrix Vector Multiplication* (SPMV) since data is stored in consecutive places in memory. On the fly insertion of elements can however be costly due to reallocation and data movement. Instead we propose a new way of using a stack-based representation. It is similar to the linked-list data structure where each row is represented by a linked list.

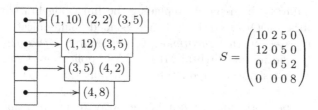

Fig. 1. An illustration of the stack-based representation of the matrix S

3   Stack-Based Representation

A stack-based representation of a sparse matrix is based around a long array A, with the same length as the number of rows in the matrix. For each entry in the array $A(i)$, we have a stack $A(i).rs$ holding tuples (c, v) representing the column index c and element value v, hereby referred to as a row-stack, illustrated in Fig. 1. Inserting an element into the matrix is now straightforward. Namely, find the corresponding row-stack and push the new (c, v) tuple on the top. Adding a value to an already inserted element is also straightforward. Find the corresponding row-stack, perform a linear search until the correct (c, v) tuple is found and add the value to v, as illustrated in Algorithm 1 below:

Algorithm 1. *Matrix update $(A_{i,c} + = v)$:*
   for $j = 1 : length(A(i).rs)$ **do**
     if $A(i).rs(j).c == c$ **then**
       $A(i).rs(j).v+ = v$
      return
     end if
    end for
    push (c, v) onto the row-stack $A(i).rs$

We argue that this representation is more efficient for matrix assembly than the CRS, in particular for random insertion such as finite element assembly, foremost since the indirect addressing to find the corresponding start of a row is removed. Secondly we do not require these stacks to be ordered. Thus we could push new elements regardless of the column index. In the case of adding a value to an already inserted element, the linear search will still be efficient since each row-stack has a short length equal to the number of non-zeros in the corresponding row.

4   Parallelization Strategy

The finite element framework DOLFIN is written in C++ and parallelized using MPI. Our HPC branch is based on a fully distributed mesh approach, where everything from preprocessing, assembly of linear systems and postprocessing is performed in parallel, without representing the entire problem or any pre-/postprocessing step on one single core. Each core is assigned a whole set of

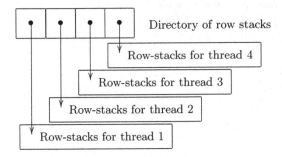

Fig. 2. An illustration of the representation of row stacks as a directory of objects

elements, defined initially by the graph partitioning of the corresponding dual graph of the mesh. Since whole elements are assigned to each core, assembly of linear systems based on evaluation of element integrals can be performed in a straightforward way with low data dependency. Furthermore, we renumber all degrees of freedom such that a minimal amount of communication is required when modifying entries in the sparse matrix. A more detailed description of DOLFIN's parallelization can be found in [7].

The parallelization of the stack-based linear algebra backend is based on a row wise data distribution. Each core is assigned a continuous set of rows, such that the first core is assigned rows $0 \ldots n$, the second $n+1 \ldots m$, and so forth. For the implementation we used *Unified Parallel C* (UPC) [18], a C like language that extends ISO C99 with PGAS constructs. In UPC the memory is partitioned into a private and a global space. Memory can then either be allocated in the private space as usual or in the global space using UPC provided functionality. Once memory is allocated in the global space it can be accessed in the same manner on all threads.

4.1 Directory of Objects Representation

In theory the stack-based (or CRS based) representation can easily be implemented in UPC by allocating the entire list of row-stacks in global space. However, this is not possible. For performance reasons memory in UPC is allocated in blocks with affinity to a certain thread. Hence we would either waste memory or force the matrix dimension to be even divisible by the number of threads.

The solution to this problem is to use a technique called directory of objects, where a list of pointers is allocated in the global space such that each thread has affinity to one pointer. Each pointer then points to a row-stack, allocated in the global space with affinity to the same thread as the pointer. This technique enables us to have unevenly distributed global memory, and each piece can grow or shrink independently of each other, as illustrated in Fig. 2.

Algorithm 2. *Lock free matrix update:*
>**if** $owner(A(i).rs)! = threadid$ **then**
>>$stage(owner(A(i).rs), c, v)$
>
>**else**
>>**for** $j = 1 : length(A(i).rs)$ **do**
>>>**if** $A(i).rs(j).c == c$ **then**
>>>>$A(i).rs(j).v+ = v$
>>>>*return*
>>>
>>>**end if**
>>
>>**end for**
>>$push\ (c, v)$ *onto the row-stack* $A(i).rs$
>
>**end if**

4.2   Parallel Matrix Operations

Operating on the matrix in parallel is almost the same as for the serial implementation. The big difference is how matrix add/insert operations handle non thread-local elements. One solution is to allow the thread which updates an entry to update it in the row-stack on the thread to which the entry has affinity to. The possibility of data races makes this approach error prone, which of course can be solved by adding locks around certain regions, and pay a certain latency fee for acquiring the locks.

Instead, we use a lock-free approach where Algorithm 1 is modified such that if an entry does not belong to a thread it is placed in a staging area (offthread row-stack), one for each thread, as illustrated in Algorithm 2. After all entries have been added or inserted into the matrix, each thread copies the relevant staging areas from each other using the remote memory copy functionality in UPC, and add/insert them into the local row-stack. Possible communication contention is reduced by pairing UPC threads together, such that each thread copies data from the thread with number mod (threadid + i, nthreads), where threadid is the thread's own id and nthreads is the total number of UPC threads. For static sparsity patterns, we use the initial assembly to gather dependency information such that consecutive assemblies can be optimized, only fetching data from threads in the list of dependencies, as illustrated in Algorithm 3.

4.3   Hybrid Interface

Mixing different programming languages in scientific code has always been a cause for headache and portability issues. Since there is no C++ version of UPC we had to use an interface that did not expose the UPC specific data structures to DOLFIN's C++ code. To overcome this problem we access the UPC data types from DOLFIN as opaque objects [15]. On the C++ side we allocate memory for an object of the same size as the UPC data type and the object is never accessed by the C++ code. All modifications are done through the interface to

Algorithm 3. *Finalization of matrix assembly:*

> *if* Initial assembly *then*
>> *for* $i = 1 : nthreads$ *do*
>>> $src = mod(threadid + i, nthreads))$
>>> *if* $length(staging\_area(src, threadid)) > 0$ *then*
>>>> $data = memget(staging\_area(src, threadid))$
>>>> *for* $j = 1 : length(data)$ *do*
>>>>> *add* $data(j)$ *to matrix*
>>>> *end for*
>>>> *add* src *to list of dependencies (dep)*
>>> *end if*
>> *end for*
> *else*
>> *for* $i = 1 : length(dep)$ *do*
>>> $src = dep(i)$
>>> $data = memget(staging\_area(src, threadid))$
>>> *for* $j = 1 : length(data)$ *do*
>>>> *add* $data(j)$ *to matrix*
>>> *end for*
>> *end for*
> *end if*

```
typedef struct {
    shared [] row_stack *shared *a_dir;
    row_stack *rs;
    ...

} jp_mat_t;

int jp_mat_init(jp_mat_t *restrict A,
                uint32_t m,
                uint32_t n);
```
(a) UPC side

```
extern "C" {
    int jp_mat_init(char *restrict A,
                    uint32_t m,
                    uint32_t n);

}

char A[144]; /* sizeof(jp_mat_t) */

jp_mat_init(A, M, N);
```
(b) C++ side

Fig. 3. An illustration of the hybrid interface

the UPC library. For example, the code given in Fig. 3(a) illustrates how an UPC matrix (`jp_mat_t`) and its initialization function (`jp_mat_init`) are defined. As mentioned above, since shared pointers are illegal in C++, we have to redefine the function on this side as illustrated in Fig. 3(b). This technique enables us to access exotic UPC types from C++ with minor portability issues, except for determining the size of the data type for each new platform the library is compiled on. During runtime we use a flat model and map MPI ranks to UPC threads one-to-one. How this mapping is created and managed is left to the runtime system. Also, in order to minimize possible runtime problems we ensured that no UPC and MPI communication overlaps.

5   Performance Analysis

We conducted a performance study in order to determine if sparse matrix assembly, in particular for finite element solvers, can gain anything from one-sided communication, and also to determine if the hybrid MPI/PGAS model is feasible for optimizing legacy MPI codes. For our experiments we chose four partial differential equations (PDE), all with different kinds of communication and computational costs. Two of them are more synthetic with large mesh sizes and the other two are more application oriented with realistic mesh sizes. We measured the time to recompute the stiffness matrix, thus assuming that the sparsity pattern is known a priori. Insertion rate r (per core) was also measured, computed as

$$r = \frac{N}{t \cdot c} \qquad (1)$$

where N is the number of measured matrix updates, t the time it took to assemble the matrix and c the number of cores.

Benchmark A: Laplace Equation in 3D. In the first benchmark we compute the stiffness matrix corresponding to the continuous linear Lagrange FEM discretization of Laplace equation:

$$-\Delta u = 0 \qquad (2)$$

This benchmark corresponds to a worst case scenario, since the stiffness matrix can be computed with minimal work. Hence, this benchmark tests the communication cost more than the insertion rate. For this experiment we use an unstructured tetrahedral mesh of the unit cube consisting of 317M elements.

Benchmark B: Convection-diffusion in 2D. The second benchmark computes the matrix for the convection-diffusion equation:

$$\dot{u} + \nabla \cdot (\beta u) + \alpha u - \nabla \cdot (\epsilon \nabla u) = f \qquad (3)$$

on a 214M element continuous linear Lagrange FEM discretization of the unit square. In this benchmark, the assembly cost for each element is significantly higher. Since we are using vector elements more data is also inserted per element. Hence, this benchmark tests both the communication and the insertion rate for a more balanced problem.

Benchmark C: Linear Elasticity in 3D. The third benchmark assembles the matrix for the linear elasticity equation:

$$\nabla \cdot \sigma = f \qquad (4)$$

on a 14M element continuous linear Lagrange FEM discretization of a three dimensional gear. This problem has an order of magnitude smaller mesh than the previous problems, and is more within the range of a typical application than the extreme mesh sizes of benchmark A and B. The problem has a balanced communication and computational cost (similar to benchmark B) if the number of cores are kept small. If the latter is increased the communication cost will start to dominate and the problem will be more similar to benchmark A.

Benchmark D: Navier-Stokes Equations in 3D (Momentum part). Our last benchmark computes the momentum matrix for the three dimensional Navier-Stokes equation:

$$\dot{u} + (u \cdot \nabla)u + \nabla p - \nu \Delta u = f \tag{5}$$

on a 80M element unstructured discretization of a wind tunnel using continuous linear Lagrange elements. As in the previous benchmark, the mesh size reflects the size of a real application. The momentum part of the Navier-Stokes equations (5), is the most computational expensive problem of all benchmarks, and with its moderately sized mesh, it should also have a more balanced communication cost than benchmark C, behaving more like a combination of benchmarks A and B.

5.1 Experimental Platform

This work was performed on a 1516 node Cray XE6 called Lindgren, located at PDC/KTH. Each node consists of two 12-core AMD "Magny-Cours" running at 2.1 GHz, equipped with 32GB of RAM. The Cray XE6 is especially well suited for our work since its Gemini interconnect provides hardware accelerated PGAS support. We used the Cray Compiler Environment (CCE) version 8.0.6 to compile everything. For our UPC based sparse matrix library JANPACK [8], we used the C compiler with UPC enabled (-hupc) and compiled a library. For DOLFIN we used the C++ compiler to compile another library for the finite element framework, but without any PGAS flag. Finally, we linked them together (now with the -hupc flag) forming a true hybrid MPI/PGAS application. DOLFIN supports several different linear algebra backends. For this work we chose the Cray provided PETSc 3.2.0 as the base line for our experiments.

5.2 Results

The benchmark problems were run on 384 up to 12288 cores, reassembling the matrix several times and measuring the mean. The experiments were also repeated several times in order to remove possible effects of poor node placement within the machine (due to fragmentation of available nodes). In Fig. 4 and Table 1 we present the reassembly time. In Table 2 insertion rates (normalized to one core) are presented.

In Fig. 4 we can see that for all benchmarks, JANPACK inserts entries faster than PETSc. Part of this is due to the less costly communication in UPC, but also due to our own sparsity format which has proven to outperform PETSc's in

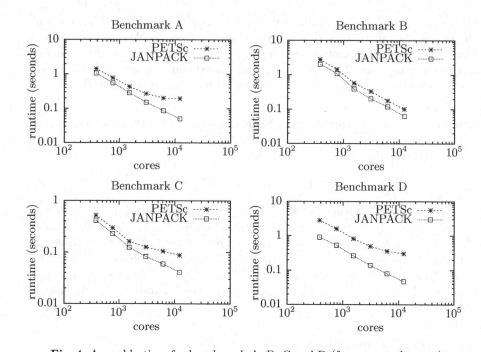

Fig. 4. Assembly time for benchmark A, B, C and D (from top to bottom)

previous single core studies [6]. Furthermore, the result also confirms our claim that one-sided communication has a lower latency than traditional message passing. As it can be seen for benchmark A, C and D, PETSc's performance flattens out when the number of elements per core becomes small, while the less latency affected UPC code continues to perform well. For a problem with a more balanced computation/communication ratio, benchmark B, the result is different. JANPACK still performs better than PETSc, but the performance degradation due to higher latency cost on large core counts is not visible, since the latency cost is completely hidden behind the computational cost. Furthermore, benchmarks C and D also demonstrate the ability of PGAS to run small problems efficiently on a large amount of cores, despite the small amount of elements assigned per core.

To summarize we see that PGAS, and in particular UPC, can be faster than MPI, in contrary to the conclusions in [4][12]. However, it might be fair to mention that the evaluations in these references were performed on platforms without hardware accelerated one-sided communication. Besides the performance improvements, the hybrid approach also has some drawbacks when it comes to productivity. We have to acknowledge that PGAS offers great potential when it comes to ease of programming and expressiveness. However, this has not yet influenced currently available development tools, which could have a direct impact on both performance and productivity. Furthermore, the lack of support for celestial linking of MPI and UPC libraries in most application development

Table 1. Assembly time (in seconds) for all benchmarks

cores	Benchmark A PETSc	Benchmark A JANPACK	Benchmark B PETSc	Benchmark B JANPACK	Benchmark C PETSc	Benchmark C JANPACK	Benchmark D PETSc	Benchmark D JANPACK
384	1.420	1.050	2.800	2.040	0.515	0.411	2.800	0.908
768	0.772	0.547	1.430	1.080	0.294	0.228	1.590	0.531
1536	0.429	0.285	0.571	0.390	0.161	0.124	0.822	0.263
3072	0.268	0.150	0.330	0.202	0.132	0.082	0.495	0.139
6144	0.200	0.086	0.177	0.119	0.104	0.059	0.357	0.079
12288	0.194	0.050	0.101	0.063	0.087	0.041	0.303	0.047

Table 2. Insertion rate (Mega entries/second) for all benchmarks

cores	Benchmark A PETSc	Benchmark A JANPACK	Benchmark B PETSc	Benchmark B JANPACK	Benchmark C PETSc	Benchmark C JANPACK	Benchmark D PETSc	Benchmark D JANPACK
384	9.31	12.6	7.16	9.83	10.9	13.6	10.8	33.2
768	8.56	12.1	7.01	9.28	9.51	12.3	9.49	28.4
1536	7.70	11.6	8.78	12.8	8.68	11.3	9.18	28.7
3072	6.16	11.0	7.60	12.4	5.30	8.52	7.62	27.1
6144	4.13	9.60	7.08	10.5	3.36	5.92	5.28	23.9
12288	2.13	8.26	6.20	9.96	2.01	4.26	3.11	20.1

environments also limits the portability of our approach. Hopefully, these are things that will improve in the future.

6 Summary and Future Work

In this work we have investigated the feasibility of optimizing finite element solvers with one-sided communication. Our results show that one can gain a significant speedup by switching from traditional message passing sparse matrix assembly to a one-sided communication using the PGAS programming model. The approach of developing a hybrid MPI/PGAS application also demonstrates that old legacy code can be optimized further with PGAS without rewriting the entire application.

For many applications good SPMV performance is equally or even more important than matrix assembly. Since this is the first step towards solving non-linear transient problems, where assembly performance matters since matrices are reassembled for each time step, tuning the stack based representation for good SPMV performance is left as future work. Also, we acknowledge that replacing PETSc in our solvers is a serious undertaking. Future work also includes the implementation of Krylov solvers and efficient preconditioners in UPC such that we can use our sparse matrix library for solving large industrial flow problems.

Acknowledgments. The author would like to acknowledge the financial support from the Swedish Foundation for Strategic Research and is also grateful for the large amount of computer time given on Lindgren during its test phase.

The research was performed on resources provided by the Swedish National Infrastructure for Computing (SNIC) at PDC - Center for High-Performance Computing.

References

1. Balay, S., Buschelman, K., Gropp, W.D., Kaushik, D., Knepley, M.G., McInnes, L.C., Smith, B.F., Zhang, H.: PETSc Web page (2009),
   http://www.mcs.anl.gov/petsc
2. Blagojević, F., Hargrove, P., Iancu, C., Yelick, K.: Hybrid PGAS runtime support for multicore nodes. In: Proceedings of the Fourth Conference on Partitioned Global Address Space Programming Model, PGAS 2010, pp. 3:1–3:10. ACM, New York (2010)
3. Bonachea, D., Duell, J.: Problems with using MPI 1.1 and 2.0 as compilation targets for parallel language implementations. Int. J. High Perform. Comput. Networking 1, 91–99 (2004)
4. Coarfă, C., Dotsenko, Y., Mellor-Crummey, J., Cantonnet, F., El-Ghazawi, T., Mohanti, A., Yao, Y., Chavarría-Miranda, D.: An Evaluation of Global Address Space Languages: Co-Array Fortran and Unified Parallel C. In: Proceedings of the Tenth ACM SIGPLAN Symposium on Principles and Practice of Parallel Programming, PPoPP 2005, pp. 36–47. ACM, New York (2005)
5. Dinan, J., Balaji, P., Lusk, E., Sadayappan, P., Thakur, R.: Hybrid parallel programming with MPI and unified parallel C. In: Proceedings of the 7th ACM International Conference on Computing Frontiers, CF 2010, pp. 177–186. ACM, New York (2010)
6. Jansson, N.: Data Structures for Efficient Sparse Matrix Assembly. Technical Report KTH-CTL-4013, Computational Technology Laboratory (2011),
   http://www.publ.kth.se/trita/ctl-4/013/
7. Jansson, N.: High performance adaptive finite element methods for turbulent fluid flow. Licentiate thesis, Royal Institute of Technology, School of Computer Science and Engineering, TRITA-CSC-A 2011, 02 (2011)
8. Jansson, N.: JANPACK (2012), http://www.csc.kth.se/~njansson/janpack
9. Jansson, N., Hoffman, J., Nazarov, M.: Adaptive Simulation of Turbulent Flow Past a Full Car Model. In: State of the Practice Reports, SC 2011, pp. 20:1–20:8. ACM, New York (2011)
10. Jose, J., Luo, M., Sur, S., Panda, D.K.: Unifying UPC and MPI runtimes: experience with MVAPICH. In: Proceedings of the Fourth Conference on Partitioned Global Address Space Programming Model, PGAS 2010, pp. 5:1–5:10. ACM, New York (2010)
11. Logg, A., Wells, G.N.: DOLFIN: Automated finite element computing. ACM Trans. Math. Softw. 37(2), 20:1–20:28 (2010)
12. Mallón, D.A., Taboada, G.L., Teijeiro, C., Touriño, J., Fraguela, B.B., Gómez, A., Doallo, R., Mouriño, J.C.: Performance Evaluation of MPI, UPC and OpenMP on Multicore Architectures. In: Ropo, M., Westerholm, J., Dongarra, J. (eds.) EuroPVM/MPI 2009. LNCS, vol. 5759, pp. 174–184. Springer, Heidelberg (2009)
13. MPI Forum. Message Passing Interface (MPI) Forum Home Page,
   http://www.mpi-forum.org/
14. OpenMP Architecture Review Board. Openmp application program interface (2008), http://www.openmp.org/mp-documents/spec30.pdf

15. Pletzer, A., McCune, D., Muszala, S., Vadlamani, S., Kruger, S.: Exposing Fortran Derived Types to C and Other Languages. Comput. Sci. Eng. 10(4), 86–92 (2008)
16. Preissl, R., Wichmann, N., Long, B., Shalf, J., Ethier, S., Koniges, A.: Multi-threaded Global Address Space Communication Techniques for Gyrokinetic Fusion Applications on Ultra-Scale Platforms. In: Proceedings of 2011 International Conference for High Performance Computing, Networking, Storage and Analysis, SC 2011, pp. 1–5. ACM, New York (2011)
17. Saad, Y.: Iterative Methods for Sparse Linear Systems, 2nd edn. Society for Industrial and Applied Mathematics, Philadelphia (2003)
18. UPC Consortium. UPC Language Specifications, v1.2. Technical Report LBNL-59208, Lawrence Berkeley National Lab (2005)

Implementation and Evaluation of 3D Finite Element Method Application for CUDA

Satoshi Ohshima, Masae Hayashi, Takahiro Katagiri, and Kengo Nakajima

The University of Tokyo, 2-11-16 Yayoi, Bunkyo-ku, Tokyo, Japan
{ohshima,masae,katagiri,nakajima}@cc.u-tokyo.ac.jp

Abstract. This paper describes a fast implementation of a FEM application on a GPU. We implemented our own FEM application and succeeded in obtaining a performance improvement in two of our application components: Matrix Assembly and Sparse Matrix Solver. Moreover, we found that accelerating our Boundary Condition Setting component on the GPU and omitting CPU–GPU data transfer between Matrix Assembly and Sparse Matrix Solver slightly further reduces execution time. As a result, the execution time of the entire FEM application was shortened from 44.65 sec on only a CPU (Nehalem architecture, 4 cores, OpenMP) to 17.52 sec on a CPU with a GPU (TeslaC2050).

1 Introduction

The performance of GPUs is rapidly improving, attracting greater and greater attention. Through various projects, numerous numerical applications and libraries have been optimized for GPUs. On the other hand, there are many applications which are not suitable for GPUs, and some kinds of applications are suitable for both CPUs and GPUs. Therefore, it is necessary to implement and evaluate various real applications on GPUs.

Our specific interest is in investigating the acceleration of numerical applications though the use of GPUs and creating numerical libraries based on our results. Our current aim to accelerate our 3D finite element method (FEM) application on a NVIDIA GPU using CUDA[1]. We proposed and implemented some optimization techniques specific to our FEM application for a GPU and demonstrated that better performance can be attained than for a multi-core CPU.

The remainder of this paper is as follows. Section 2 describes the abstractions of CUDA and our target FEM application. Section 3 describes special characteristics of our FEM application and proposes three types of optimization techniques, which we implement and evaluate the effectiveness of. Section 4 is the Conclusion section.

2 CUDA and FEM Application

2.1 CUDA and NVIDIA GPU

CUDA is an architecture and application development framework of NVIDIA GPUs. It provides a GPGPU program development environment using an

M. Daydé, O. Marques, and K. Nakajima (Eds.): VECPAR 2012, LNCS 7851, pp. 140–148, 2013.
© Springer-Verlag Berlin Heidelberg 2013

extension of the C/C++ programming language and a corresponding compiler. There is only a slight difference between C/C++ and CUDA in terms of language specification, but CUDA has unique specifications in its hardware model, memory model, and execution model. Therefore, the program optimization techniques for CPUs and GPUs (CUDA) are greatly different.

Some optimization techniques and strategies for GPU programming are already well known[2]. For example, GPU programs must have high parallelism in order to hide latencies of memory access. Also, reducing the number of branch instructions and random memory accesses is important because the associated performance penalties are larger for GPUs than for CPUs.

2.2 FEM Application

FEM is widely used in various scientific applications, and the acceleration of FEM is important and in high demand. There are many FEM applications and various libraries are used to accelerate these. Generally, Sparse Matrix Solver and Matrix Assembly take up a significant portion of the execution time of FEM applications.

Our target FEM application is based on the GeoFEM program[3], which is an existing FEM program for CPUs. The target problem is a 3D solid mechanics problem. This application has been parallelized for a multi-core CPU and PC cluster by using OpenMP and MPI. We use the modified OpenMP version [4] as a target of GPU acceleration.

Figure 1 shows the structure of our target FEM application, which has five parts. The execution time breakdown is shown in Figure 2. The execution environment is described in Table 1. This environment has a Xeon W3520 CPU (Nehalem architecture, 4 cores, 2.67 GHz) and a Tesla C2050 GPU (Fermi architecture, 448 SPs, 1.15 GHz). The number of elements is 512,000 ($=80 \times 80 \times 80$). The most time-consuming part is Sparse Matrix Solver, for both sequential execution (90.36%) and parallel execution (80.15%). In this part, this program uses a Conjugate Gradient Solver (CG solver) with a simple block diagonalization preconditioner as a sparse matrix solver. The second most time-consuming part is Matrix Assembly. Therefore, we focused mainly on trying to accelerate Sparse Matrix Solver and Matrix Assembly, which are described in Section 3.

Our FEM program has a special memory structure. Our matrix format is similar to the Compressed Row Storage (CRS) and blocked CRS formats. Matrices are partitioned into 3×3 blocks, and also divided into diagonal matrix, upper matrix, and lower matrix parts (Figure 3). This is based on physical problem setting that is stress strain. This 3×3 blocks structure is effective in terms of the memory requirements. Moreover, because 3×3 blocks structure improves cache hit rate, execution time is shorter than non-3×3 blocks structure.

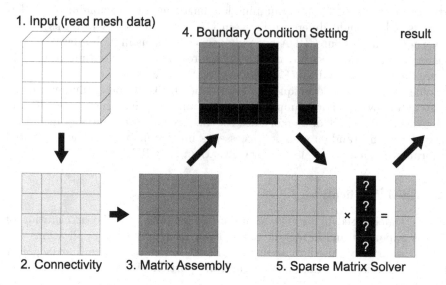

Fig. 1. Structure of our target FEM program

2.3   Related Works

Accelerating a CG solver requires speeding up matrix and vector calculations —
for example, summation and multiplication of a vector and scalar, and multiplication of a matrix and vector — and these calculations are easy to parallelize.
Also, these calculations are suitable for GPUs. Thus, there have been many studies of sparse matrix solvers (CG solvers) for GPUs over the years [5] [6] [7]. Also,
some libraries for executing a CG solver on a GPU have been published. For
example, the CUSP library[8] provides a fast CG solver and some useful data
structures and calculation methods as a C++ class library.

One example of matrix assembly on a GPU is Cris[9].

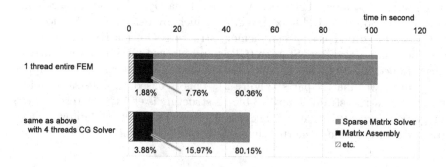

Fig. 2. Breakdown of execution time on a CPU

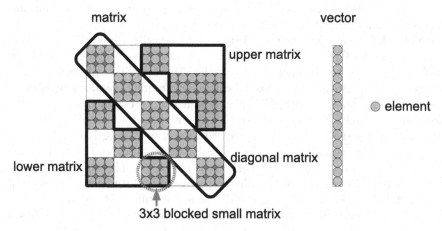

Fig. 3. Matrix structure

Table 1. Evaluation environment

CPU	Intel Xeon W3530 (Nehalem, 4cores, 2.67GHz)
Main memory	PC3-10600(DDR3-1333) 12GB
GPU	NVIDIA Tesla C2050 (Fermi, 448SPs, 1.15GHz)
Video memory	DDR5 3GB
connection	PCI-Express x16 (Gen 2)
OS	CentOS5.7 (kernel 2.6.18)
Compiler	gcc version 4.4.4 20100726 (Red Hat 4.4.4-13) (GCC)
	Cuda compilation tools, release 4.0, V0.2.1221

Our work has a specific target application and focuses on optimization techniques for that application. There are no existing reports of accelerating GeoFEM-based FEM applications on a GPU. We only are trying to accelerate a 3×3-block CG solver. Few studies have considered the implementation of boundary condition setting on a GPU or evaluating the CPU–GPU data transfer time between matrix assembly and sparse matrix solver components.

3 Implementation

3.1 Optimization of Sparse Matrix Solver

In this section, we describe three kinds of optimization strategies and implementations. The first is the optimization of Sparse Matrix Solver.

As described in Section 2, for our FEM application, Sparse Matrix Solver, which uses a CG solver with a matrix partitioned into 3×3 blocks, is the most time-consuming part. The CG solver involves only a few types of calculation, the most time-consuming of which is sparse matrix–vector multiplication (SpMV).

Therefore, in this section, we mainly focusing on describing acceleration on a GPU for a matrix partitioned into 3×3 blocks.

SpMV calculation with the CRS format is very easy to parallelize and accelerate, whether for CPU or GPU calculation, because the calculation of each row is independent and can be executed in parallel. However, it is difficult to obtain very good performance because SpMV calculation requires random memory accesses. A GPU can execute SpMV quickly and simply by dividing the matrix based on rows and assigning them to CUDA Threads. Our problem is how to divide and assign the 3×3 block partitioned matrix to CUDA Threads.

A simple strategy is dividing the block based on rows and assigning each block to CUDA ThreadBlock and each row to a CUDA Thread (Figure 4(a)). Although this strategy can obtain almost the same performance as SpMV calculation without 3×3 blocking on a GPU, it is not sufficiently optimized. This strategy cannot perform coalesced memory access and does not create sufficient parallelism.

Our optimization strategy is to assign each 3×3 block to three CUDA Threads. Each CUDA Thread reads matrix data in coalesced rule (Figure 4(b)) and has data in the shared memory, and these CUDA Threads calculate multiplication and addition using SharedMemory. Moreover, parallelism is increased by assigning multiple 3×3 blocks to each CUDA TreadBlock. These strategies are simple and effective.

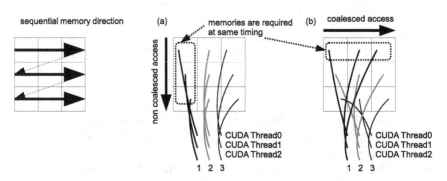

Fig. 4. Optimization strategy for SpMV (coalesced memory access)

Figure 5(a) shows the SpMV performance. The evaluation environment and problem setting are same as in Section 2. As a result, 3×3 blocked SpMV obtained 3.20 times better performance on a GPU than on a CPU. The performance ratio is much smaller than that for FLOPS (515 GFLOPS/42.7 GFLOPS = 12.06). Rather, it is closer to the memory bandwidth ratio (144 GB/s / 32 GB/s = 4.50). Moreover, by applying the same strategies to vector summation, multiplication, and dot product calculation, the execution time of the entire CG solver was shortened from 39.90 sec to 14.15 sec (Figure 5(b)).

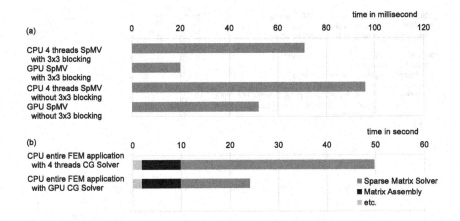

Fig. 5. Performance evaluation 1 (SpMV calculation and entire FEM application)

3.2 Optimization of Matrix Assembly

As shown in Section 2, Matrix Assembly was the second most time-consuming part of the FEM application. As the result of accelerating the CG solver on a GPU, the relative amount of time used by Matrix Assembly increased. In this subsection, we describe the acceleration of Matrix Assembly.

Figure 6 shows the structure (flow of source code) of Matrix Assembly. The flow of Matrix Assembly is more complicated than SpMV, and a hierarchical loop structure (loopA, loopB, and loopC in Figure 6) is characteristic. The problem of how to assign calculations to the GPU is more important than the problem of Sparse Matrix Solver. Because Matrix Assembly has dependencies between each matrix element, coloring computation is used to obtain parallelism. In this study, we make the CPU execute coloring computation (multicolor method) and make the GPU execute parallel calculation after coloring.

We tried to implement the calculation in two styles. The first style (data strategy) performs the entire calculation in one GPU kernel, and the second style (data+task strategy) divides the calculation into three parts. While the data+task strategy can make each GPU kernel simple and small, the CPU and GPU have to synchronize their kernels, which may degrade performance. According to the results of our implementation and evaluation, the data+task strategy obtained better performance than the data strategy (Figure 7).

3.3 Optimization of Entire FEM Application

The execution time of the FEM application is shortened by accelerating Matrix Assembly and Sparse Matrix Solver on a GPU. However, some parts of the FEM

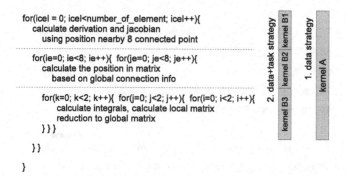

```
for(icel = 0; icel<number_of_element; icel++){
    calculate derivation and jacobian
        using position nearby 8 connected point

    for(ie=0; ie<8; ie++){ for(je=0; je<8; je++){
        calculate the position in matrix
            based on global connection info

        for(k=0; k<2; k++){ for(j=0; j<2; j++){ for(i=0; i<2; i++){
            calculate integrals, calculate local matrix
            reduction to global matrix
        }}}

    }}

}
```

Fig. 6. Assignment of Matrix Assembly to the GPU

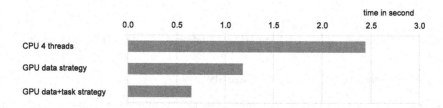

Fig. 7. Performance evaluation 2 (Matrix Assembly)

application are still executed on the CPU, but the execution times of these parts are not so time-consuming. However, we think that accelerating more parts on the GPU is important in order to obtain the best performance in the CPU with a GPU environment.

Of the five parts of the FEM application, Sparse Matrix Solver and the latter half of Matrix Assembly are already implemented to execute on the GPU. Here, we implement Boundary Condition Setting on the GPU. Because our FEM application has a simple boundary condition, the computation time for the boundary condition is small and we can implement it easily on the GPU. However, if we implement Boundary Condition Setting on the GPU, the data transfer between CPU and GPU after Matrix Assembly and before Sparse Matrix Solver can be omitted and the performance may improve. In order to obtain correct result, we modify Matrix Assembly and Sparse Matrix Solver to omit the data management computation. Therefore, the CPU only performs control computations, such as kernel launching of the GPU and loop control in the CG solver from after the coloring procedure of Matrix Assembly to the end of Sparse Matrix Solver (Figure 8).

Figure 9 shows the resulting execution times. The middle bar shows the results of the above-described optimization. It is true that the effect of this optimization is not large, but it is significant: a 5.14% performance improvement.

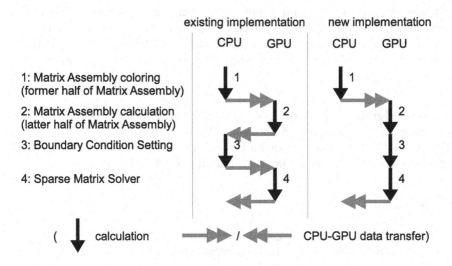

Fig. 8. Assignment more parts to the GPU

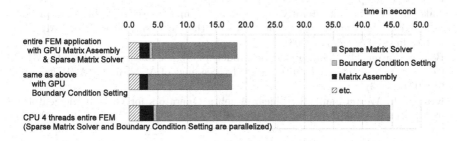

Fig. 9. Performance evaluation 3 (entire FEM application)

4   Conclusion

In this paper, we described the acceleration of a FEM application on a NVIDIA GPU with CUDA. We implemented three components of the application, Sparse Matrix Solver, Matrix Assembly, and Boundary Condition Setting, on a GPU. The execution time of Sparse Matrix Solver was shortened from 39.30 sec to 14.15 sec, and the execution time of Matrix Assembly was shortened from 2.44 sec to 0.65 sec. By implementing the Boundary Condition Setting on the GPU and omitting the CPU–GPU data transfer, the execution time of the entire FEM application was reduced. The most important technique for accelerating execution was memory assignment. Exact assignment obtained good performance. As a result, the execution time of entire FEM application was shortened from 44.65 sec on only a CPU (Nehalem architecture, 4 cores, OpenMP) to 17.52 sec on a CPU with a GPU (TeslaC2050).

There remains room for improvement and some challenges for FEM applications on a GPU. For example, coloring computation, complex preconditioners, and complex boundary conditions are difficult for a GPU to accelerate. Utilizing multiple GPUs is also an advanced topic. These remain as future work of our project.

Acknowledgment. This work is partially supported by "Framework and Programming for Post Petascale Computing (FP3C)" (JST, ANR) and "ppOpen-HPC: Open Source Infrastructure for Development and Execution of Large-Scale Scientific Applications with Automatic Tuning" (JST CREST).

References

1. NVIDIA: NVIDIA Developer Zone (CUDA ZONE),
   http://developer.nvidia.com/category/zone/cuda-zone
2. NVIDIA: NVIDIA CUDA C Programming Guide
3. Research Organization for Information Science & Technology (RIST): GeoFEM Homepage, http://geofem.tokyo.rist.or.jp/
4. Nakajima, K.: Parallel iterative solvers of geofem with selective blocking preconditioning for nonlinear contact problems on the earth simulator. In: ACM/IEEE Proceedings of SC 2003 (2003)
5. Bolz, J., Farmer, I., Grinspun, E., Scheróder, P.: Sparse Matrix Solvers on the GPU: Conjugate Gradients and Multigrid. In: Proceedings of ACM SIGGRAPH 2003, pp. 917–924 (2003)
6. Krüger, J., Westermann, R.: Linear Algebra Operators for GPU Implementation of Numerical Algorithms. In: Proceedings of ACM SIGGRAPH 2003, pp. 908–916 (2003)
7. Cevahir, A., Nukada, A., Matsuoka, S.: High performance conjugate gradient solver on multi-gpu clusters using hypergraph partitioning. Computer Science - Research and Development 25, 83–91 (2010)
8. cusp-library: Generic Parallel Algorithms for Sparse Matrix and Graph Computations, http://code.google.com/p/cusp-library/
9. Cecka, C., Lew, A.J., Darve, E.: Assembly of finite element methods on graphics processors. International Journal for Numerical Methods in Engineering 85(5) (2011)

Evaluation of Two Parallel Finite Element Implementations of the Time-Dependent Advection Diffusion Problem: GPU versus Cluster Considering Time and Energy Consumption

Alberto F. De Souza[1], Lucas Veronese[1], Leonardo M. Lima[2], Claudine Badue[1], and Lucia Catabriga[1]

[1] Departamento de Informática, Universidade Federal do Espírito Santo, Vitória, Brazil
[2] Instituto Federal de Educação, Ciência e Tecnologia do Espírito Santo, Vitória, Brazil

Abstract. We analyze two parallel finite element implementations of the 2D time-dependent advection diffusion problem, one for multi-core clusters and one for CUDA-enabled GPUs, and compare their performances in terms of time and energy consumption. The parallel CUDA-enabled GPU implementation was derived from the multi-core cluster version. Our experimental results show that a desktop machine with a single CUDA-enabled GPU can achieve performance higher than a 24-machine (96 cores) cluster in this class of finite element problems. Also, the CUDA-enabled GPU implementation consumes less than one twentieth of the energy (Joules) consumed by the multi-core cluster implementation while solving a whole instance of the finite element problem.

1 Introduction

The advances of numerical modeling in the past decades have allowed scientists to solve problems of increasing complexity. Frequently, these problems require the solution of very large systems of equations at each time step and/or iteration. Because of that, a great effort has been made on the development of more efficient and optimized solution algorithms. But, along the past few decades, the underlying hardware for running these algorithms has changed significantly. A recent important development was the advent of the Compute Unified Device Architecture (CUDA) [12].

CUDA is a new Graphics Processing Unit (GPU) architecture that allows general purpose parallel programming through a small extension of the C programming language. The Single Instruction Multiple Thread (SIMT [12]—it is similar to SIMD, but more flexible on the use of resources) architecture of CUDA-enabled GPUs allows the implementation of scalable massively multithreaded general purpose C+CUDA code. Currently, CUDA-enabled GPUs possess arrays of hundreds of cores (called stream processors) and peak performance surpassing 1 Tflop/s. More than 200 million CUDA-enabled GPUs have been sold [10], which makes it the most successful high performance parallel computing platform in computing history and, perhaps, up to this point in time, one of the most disruptive computing technologies of this century—many relevant programs have been ported to C+CUDA and run orders of magnitude faster in CUDA-enabled GPUs than in multi-core CPUs.

M. Daydé, O. Marques, and K. Nakajima (Eds.): VECPAR 2012, LNCS 7851, pp. 149–162, 2013.

In this paper, we analyze two parallel finite element implementations of the 2D time-dependent advection diffusion problem: one for multi-core clusters and one for CUDA-enabled GPUs [12]. We also compare their performances in terms of time and energy consumption.

The finite element method is one of the most used numerical techniques for finding approximated solutions of partial differential equations (PDE). In this method, the solution approach is based either on rendering the PDE into an approximating system of ordinary differential equations, which are then numerically integrated using standard techniques, such as the Euler's method [6].

The finite element formulation requires the solution of linear systems of equations involving millions of unknowns that are usually solved by Krylov space iterative update techniques [13], from which the most used is the Generalized Minimum Residual method (GMRES). One of the most time consuming operations of this solution strategy is the matrix-vector product, which can be computed on data stored according to global and local schemes. The most well known global scheme is the compressed storage row (CSR) [13], while the most well known local schemes are the element-by-element (EBE) and edge-based data structure (EDS) [2, 4]. The code for CSR is easily parallelized in different computer architectures. This type of implementation is often preferred to local schemes—matrix-vector products computed on EBE or EDS can be memory intensive, needing more operations than on CSR. However, particularly for large-scale nonlinear problems, EBE and EDS schemes have been very successful because they handle large sparse matrices in a simple and straightforward manner.

In this work, we consider the parallel finite element formulation of the 2D time-dependent advection diffusion equation. To solve the system of ordinary differential equations that results from the finite element formulation, we employ the well known implicit predictor/multicorrector scheme [15]. The sparse linear system of each time-step (stored in a Compressed Storage Row (CSR) scheme in both implementations) is solved by the GMRES method.

We implemented one code for multi-core clusters and, from that, one code for CUDA-enabled GPUs, and run them in a 24-machine (96 cores) cluster and in a 4-GPU desktop machine. Both implementations were written in C and use the MPI library for inter-core communication. Our simulations show that a desktop computer with a single GPU can outperform a 24-machine (96 cores) cluster of the same generation and that a 4-GPU desktop can offer more than twice the cluster performance. Also, with four GPUs, the CUDA-enabled implementation consumes less than one twentieth of the energy (Joules) consumed by the multi-core cluster implementation while solving a whole instance of the finite element problem. These results show that, currently, considering the benefits of shorter executing times, smaller energy consumption, smaller dimensions and maintenance costs, Multi-GPU desktop machines are better high performance computing platforms than small clusters without GPUs, even though they are somewhat harder to program.

2   Governing Equations and Finite Element Formulation

Let us consider the following time-dependent boundary-value problem defined in a domain $\Omega \in \Re^2$ with boundary Γ:

$$\frac{\partial u}{\partial t} + \boldsymbol{\beta}.\nabla u - \nabla.(\boldsymbol{\kappa}\nabla u) = f \quad \text{(time-dependent advection-diffusion equation)} \quad (1)$$

$$u = g \quad \text{on } \Gamma_g \quad \text{(essential boundary condition)} \quad (2)$$

$$\boldsymbol{n}.\boldsymbol{\kappa}\nabla u = h \quad \text{on } \Gamma_h \quad \text{(natural boundary condition)} \quad (3)$$

$$u(\boldsymbol{x},0) = u_o(\boldsymbol{x}) \quad \text{on } \Omega \quad \text{(initial condition)} \quad (4)$$

where u represents the quantity being transported (e.g. concentration), $\boldsymbol{\beta}$ is the velocity field, and $\boldsymbol{\kappa}$ is the volumetric diffusivity. g and h are known functions of $\boldsymbol{x} = (x,y)$ and t, $\boldsymbol{n}$ is the unit outward normal vector at the boundary, and Γ_g and Γ_h are the complementary subsets of Γ where boundary conditions are prescribed.

Consider a finite element discretization of Ω into elements $\Omega_e, e = 1, \ldots, n_{el}$, where n_{el} is the number of elements. Let the standard finite element approximation be given as

$$u^h(\boldsymbol{x}) \cong \sum_{i=1}^{nnodes} N_i(\boldsymbol{x})u_i, \quad (5)$$

where $nnodes$ is the number of nodes, N_i is a shape function corresponding to node i, and u_i are the nodal values of u. Then, applying this approximation on the variational form of Equation (1), we arrive at a system of ordinary differential equations:

$$\boldsymbol{Ma} + \boldsymbol{Kv} = \boldsymbol{F}, \quad (6)$$

where $\boldsymbol{v} = \{u_1, u_2, \ldots, u_{nnodes}\}^t$ is the vector of nodal values of u, $\boldsymbol{a}$ is its time derivative, $\boldsymbol{M}$ is the "mass" matrix, $\boldsymbol{K}$ is the "stiffness" matrix, and $\boldsymbol{F}$ is the "load" vector [6]. In this work, we approximate the domain Ω using linear triangular elements. Thus, the global interpolation of Equation (5) is restricted to an element by

$$u^e(\boldsymbol{x}) \cong \sum_{i=1}^{3} N_i(\boldsymbol{x})u_i, \quad (7)$$

where the superscript e means that u is restricted to an element, and N_1, N_2 and N_3 are the conventional shape functions [6]. Proceeding in the standard manner, matrices $\boldsymbol{M}$ and $\boldsymbol{K}$ and vector $\boldsymbol{F}$ are built from element contributions and it is convenient to identify their terms as:

$$\boldsymbol{M} = \underset{e=1}{\overset{nel}{\boldsymbol{A}}}(\boldsymbol{m}^e), \quad \boldsymbol{K} = \underset{e=1}{\overset{nel}{\boldsymbol{A}}}(\boldsymbol{k}^e) \quad \text{and} \quad \boldsymbol{F} = \underset{e=1}{\overset{nel}{\boldsymbol{A}}}(\boldsymbol{f}^e) \quad (8)$$

where $\boldsymbol{A}$ is the assembling operator and $\boldsymbol{m}^e$, $\boldsymbol{k}^e$ and $\boldsymbol{f}^e$ are the local contributions.

3   Solution Algorithm

To solve the time-dependent advection diffusion problem numerically employing the approach described in the previous section, we just have to solve the system of ordinary differential equations stated in Equation (6) towards a final time t_{final}. To do that, we use the Algorithm 1, which implements the well known implicit predictor/multicorrector solution scheme [15]. The algorithm: receives as input the initial values of v and a (see Equation (6)), t_{final}, Δt, the maximum number of multicorrection attempts, n_{max}, and the tolerance of the multicorrection phase, ϵ; and returns the values of v and a at t_{final}.

Algorithm 1. Predictor/multicorrector

1: Data: v_0 and a_0, t_{final}, Δt, n_{max}, ϵ
2: $t = 0$, $n = 0$
3: $M^* = M + \alpha \Delta t K$
4: **while** $t \leq t_{final}$ **do**
5:    $i = 0$
6:    $v_{n+1}^{(i)} = v_n + (1 - \alpha)\Delta t a_n$
7:    $a_{n+1}^{(i)} = 0$
8:    $norm_d = 0$
9:    **while** $i \leq n_{max}$ and $\|a_{n+1}^{(i)}\| \geq \epsilon \times norm_d$ **do**
10:      $b = F - M a_{n+1}^{(i)} - K v_{n+1}^{(i)}$
11:      Solve $M^* d = b$
12:      $a_{n+1}^{(i+1)} = a_{n+1}^{(i)} + d$
13:      $v_{n+1}^{(i+1)} = v_{n+1}^{(i)} + \alpha \Delta t d$
14:      $i = i + 1, norm_d = \|d\|$
15:    **end while**
16:    $a_{n+1} = a_{n+1}^{(i)}$
17:    $v_{n+1} = v_{n+1}^{(i)}$
18:    $t = t + \Delta t, n = n + 1$
19: **end while**

In Algorithm 1, the prediction phase (lines 5 to 8) calculates an initial guess of the nodal values v and a at iteration $n + 1$, where n denotes a time step, and the multicorrection phase (lines 9 to 15) iteratively calculates new nodal approximations until a convergence criteria (line 9) is reached. The most time consuming step of the algorithm is solving the linear system derived from Equation (6), lines 10 and 11. In this linear system, M^* is denoted the effective matrix, b is the residual vector, and d is the correction of the nodal values of a from one multicorrection iteration to the next. Matrix M^* is constant in time and is computed in line 3. The residual vector b, however, must be computed in every multicorrection step (line 10).

Apart from the solution of the linear system in line 11, the other time consuming operations of the algorithm are the matrix vector product in line 10, and the saxpy vector update operations of lines 6 (the number of multicorrection iterations is small, but one always have to remember the Amdhal's Law), 12 and 13. We solve the linear system of line 11 using GMRES [13]. The most time consuming operations of GMRES are a matrix vector product per iteration, and several saxpy and vector inner products. Therefore, the most time consuming operations of the whole predictor/multicorrector algorithm are matrix vector products, saxpy and vector inner products. For more on the predictor/multicorrector algorithm see [15].

4 Parallel Implementations

To solve our problem in parallel, it is necessary to code all matrices and vectors in Algorithm 1 in a way that allows parallel access, and to calculate their most time consuming operations in parallel. In order to achieved this, we create a partition of non-overlapping sets of elements, Ω_e. For that, we discretized the domain into a mesh composed of linear triangular elements, $\mathsf{T} = \Omega_e$, where $\{\mathsf{T}_1, \mathsf{T}_2, \cdots, \mathsf{T}_p\}$ represents a partition of the triangulation in subdomains, p is the number of subdomains, and $\bigcup_{i=1}^{p} \mathsf{T}_i = \mathsf{T}$ and $\mathsf{T}_i \cap \mathsf{T}_j = \emptyset$ when $i \neq j$. By dividing the computation domain into p subdomains, it is possible to spread the workload between p different cores. That is, by partitioning the matrices M, K and M^*, and the vectors v, a and d (see Equation (6) and Algorithm 1) independently over p cores (with core i working only on subdomain T_i), one can spread the workload among the p different cores.

We rewrite all matrices and vectors presented in Algorithm 1 into block matrix and block vector forms employing the well known Schur complement decomposition [13], as suggested by Jimack and Touheed [8]. By doing that, a generic vector u (representing v, a or d) can be ordered in the following way:

$$u = (\underline{u}_1, \underline{u}_2, \cdots, \underline{u}_i, \cdots, \underline{u}_p, \underline{u}_S)^T. \tag{9}$$

The nodes of the linear triangular elements of the mesh T can be classified into interior nodes, interface nodes and boundary nodes of the domain.

Figure 1 illustrates a mesh with 50 nodes and 74 triangular elements, where the domain was partitioned into 4 subdomains to be assigned to four cores. In this mesh, nodes I and J are interior nodes of cores 3 and 4, respectively, while node K is an interface node of cores 1, 3 and 4.

In Equation (9), the sub-vector $\underline{u}_i$ is associated with the interior nodes in T_i, $i = 1, 2, \cdots, p$; while u_S, in turn, is defined as $\underline{u}_S = \bigcup_{i=1}^{p} \underline{u}_{s(i)}$, an assembly of others sub-vectors that are associated with the interface nodes of each subdomain T_i, $i = 1, 2, \cdots, p$. That is, each sub-vector $u_{s(i)}$ holds the interface nodes of T_i. Boundary nodes are not unknowns and need not be represented in u. Also following the approach suggested by Jimack and Touheed [8], a generic matrix A (M, K and M^*) can be written in a block matrix form as:

$$A = \begin{bmatrix} A_1 & & & & B_1 \\ & A_2 & & & B_2 \\ & & \ddots & & \vdots \\ & & & A_p & B_p \\ C_1 & C_2 & \cdots & C_p & A_S \end{bmatrix} \tag{10}$$

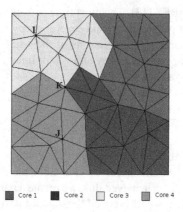

Core 1   Core 2   Core 3   Core 4

Fig. 1. Example of the partitioned mesh with 4 subdomains

where the block-arrowhead structure of the new matrix comes from the local support of the finite element basis functions. In Equation (10), the sub-matrices A_i, B_i, C_i and A_S are sparse matrices that are stored using a CSR data structure.

The sub-matrix A_i stores the contribution of the interior nodes of core i on the interior nodes of core i. The matrix B_i stores the contribution of the interior nodes of core i on the interface nodes of core i. The sub-matrix C_i stores the contribution of the interface nodes of core i on the interior nodes of core i. Finally, the sub-matrix A_S, an assembly of a set of blocks distributed over the p cores, is defined as $A_S = \bigcup_{i=1}^{p} A_{s(i)}$, where the sub-matrix $A_{s(i)}$ stores the contribution of the interface nodes of core i on the interface nodes of core i.

With this approach, each of the sub-vectors $\underline{u}_i$ and $\underline{u}_{s(i)}$, and each of the sub-matrix A_i, B_i, C_i, $A_{s(i)}$ may be computed entirely by core i, for $i = 1, 2, \cdots, p$. One can also observe that core i will work only on the elements of its own subdomain T_i. Assuming that the partition T is built in such way that each core deals with approximately the same number of elements and the number of vertices lying on the partition boundary is as small as possible, the amount of calculations performed by each core i will be balanced and the amount of communication will be minimized.

Following the same procedure explained above for a generic vector u and a generic matrix A, we rewrite all the matrices (M, K and M^*) and vectors (v, a or d) of Algorithm 1 in a block matrix form and execute the most time consuming operations of the whole predictor/multicorrector algorithm—matrix vector product, saxpy vector update and vector inner product—in parallel.

Using the domain partitioning presented above, a matrix-vector product, $v = Au$, can be computed in parallel by computing both expressions on Equation (11) below (see also Equations (9) and (10))

$$\underline{v}_i = A_i \underline{u}_i + B_i \underline{u}_{s(i)} \quad \text{and} \quad \underline{v}_{s(i)} = A_{s(i)} \underline{u}_{s(i)} + C_i \underline{u}_i \qquad (11)$$

on each core $i = 1, 2, \cdots, p$. Also, using the domain partitioning presented, a saxpy vector update, $v = v + \lambda u$, can be formulated as

$$\underline{v}_i = \underline{v}_i + \lambda \underline{u}_i \quad \text{and} \quad \underline{v}_{s(i)} = \underline{v}_{s(i)} + \lambda \underline{u}_{s(i)} \tag{12}$$

for $i = 1, 2, \cdots, p$, where λ is a real number. Finally, using the domain partitioning presented, a vector inner product, $scalar = \boldsymbol{u} \cdot \boldsymbol{v}$, can be computed on each core as

$$scalar = \sum_{i=1}^{p} (\underline{u}_i \cdot \underline{v}_i + \underline{u}_{s(i)} \cdot \underline{v}_{s(i)}) \tag{13}$$

for $i = 1, 2, \cdots, p$. It is important to note that this last operation requires a global communication because its result is a scalar that always must be known by all cores. This communication is a global reduction, which computes the sum of the contributions to the inner product coming from each core, and then provides each core with a copy of this sum.

In addition to global reductions required by inner products, our Multi-Core Cluster implementation performs core-to-core communication before every matrix vector product (lines 10 and 11 of Algorithm 1) in order to communicate the value of the interface nodes—we use MPI_send and MPI_Recv for that. Thanks to the assembly presented in Equation 4, the data that needs to be communicated is clearly specified (interface nodes). The partitioning of the work between the cores is made before the whole computation using METIS [9]. Please refer to our internal technical report for more details about our multi-core cluster implementation (http://www.lcad.inf.ufes.br/~alberto/techrep01-11.pdf).

The CUDA-enabled GPU parallel version was derived from the Multi-Core Cluster parallel version and, therefore, follows the same principles described above. It was implemented in C+CUDA and, as we wanted to run it in multi-core desktop computers with multiple GPUs (or clusters of multi-core machines each of which with one or more GPUs), it takes advantage of the multiple cores for distributing the domain (or subdomains in the case of a cluster) between multiple GPUs (one subdomain per GPU) and employs MPI for inter-core communication. We choose to do this way (i) to avoid large modifications in the Multi-Core Cluster version in the process of morphing it into the C+CUDA version, and (ii) to transform our multi-core cluster code into a code that runs in clusters of multi-core machines each of which with multiple GPUs. For this process, we basically moved the main functions of the Multi-Core Cluster version into CUDA kernels and optimized the use of the GPU memory hierarchy.

The main strategy adopted in the design of the C+CUDA version was (i) to identify the most time consuming operations of the predictor/multicorrector (Figure 1) and GMRES algorithms, (ii) to parallelize and optimize these operations, and (iii) to try and avoid data transfer between the CPU and GPU memories as much as possible.

We identified the most time consuming operations of the predictor/multicorrector and GMRES algorithms—the matrix-vector product, $\boldsymbol{v} = \boldsymbol{A}\boldsymbol{u}$, and the vector inner product, $scalar = \boldsymbol{u} \cdot \boldsymbol{v}$—using gprof. Please refer to our internal technical report for details about the C+CUDA implementation (http://www.lcad.inf.ufes.br/~alberto/techrep01-11.pdf).

5 Experimental Evaluation

5.1 Hardware

The Multi-Core Cluster implementation was run on the Enterprise 3 cluster of the *Laboratório de Computação de Alto Desempenho* (LCAD) at UFES. Enterprise 3 is a 24-node cluster of 24 quad-core Intel 2 Q6600 machines (96 cores), with 2.4GHz clock frequency, 4MB L2 and 4GB of DRAM, interconnected with a 48-Port 4200G 3COM Gigabit Ethernet switch. The C+CUDA implementation was run on LCAD's BOXX Personal Supercomputer, which is a quad-core AMD Phenon X4 9950 of 2.6GHz, with 2MB L2, 8GB of DRAM, and four GPU NVIDIA Tesla C1060 PCIE boards, with 240 1.3GHz CUDA cores and 4GB DRAM each.

5.2 Rotating Cone Problem

In our experimental evaluation we solved a standard test problem for transient dominated advection flow, named rotating cone problem. The problem (Figure 2(a)) considers a cosine hill profile advected in a two-dimensional rotating flow field (see [1] for details). The homogeneous Dirichlet boundary conditions is imposed zero everywhere on the external boundaries and the initial condition is a hill profile. In our experiments, the velocity field is $\beta = (-y, x)^T$ and the diffusivity is $\kappa = kI$, where $k = 10^{-6}$. The exact solution consists of a rigid rotation of a cone about the center of the square domain $[-5, 5] \times [-5, 5]$. Figure 2(b) shows the solution obtained after 7 seconds of simulation.

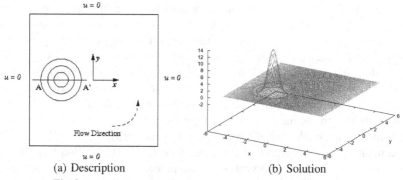

(a) Description   (b) Solution

Fig. 2. Description and solution of the rotating cone problem

To evaluate the performances in terms of time of the machines on the solution of a large size problem, we consider the rotating cone problem in a regular mesh of 1024×1024 cells, totalizing $2,097,152$ elements, $1,050,625$ nodes and $1,046,529$ unknowns with $\Delta t = 10^{-2}$, the $t_{final} = 7$, GMRES and predictor-multicorrector tolerances equal to 10^{-3}; and number of restart vectors for GMRES equal to 10. The observed number of GMRES iterations for each correction was around 15.

5.3 Performance in Terms of Time

Figure 3(a) shows the time it takes to solve this problem with the Multi-Core Cluster implementation running on the Enterprise 3 configured with 1, 4, 8, 12, 16, 24, 32, 48, 64 and 96 cores, while Figure 3(b) shows the speedups obtained with 1, 4, 8, 12, 16, 24, 32, 48, 64 and 96 cores.

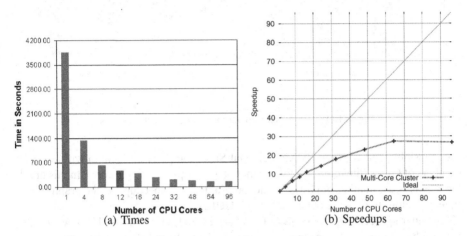

(a) Times   (b) Speedups

Fig. 3. Enterprise 3 times and speedups

In the graph of Figure 3(a), the x-axis is the number of cores, while the y-axis is the time it takes to solve the problem in seconds. As the graph shows, there is an almost linear reduction of the time it takes to solve the problem as the number of processors increases from 1 to 8. However, the performance gains obtained increasing the number of cores from 8 onwards decreases as the number of cores increases. This can be more easily appreciated by examining the graph of Figure 3(b). In this graph, the x-axis is the number of cores, while the y-axis is the speedup. As the graph of Figure 3(b) shows, although the speedup starts augmenting linearly, as the number of cores increases, the speedup levels of—there is no gain as one goes from 64 to 96 cores. This is to be expected because, as the number of cores increases, the amount of inter-machine communication increases, while the amount of compute work per core decreases. So, the time spent waiting for data transfer (communication) ends up surpassing the time doing computation.

Figure 4(a) shows the time it takes to solve this problem with the C+CUDA implementation running on the BOXX Personal Supercomputer configured with 1, 2 and 4 GPUs, while Figure 4(b) shows the speedups obtained with 1, 2 and 4 GPUs—these speedups were computed against a single Enterprise 3 core.

In the graph of Figure 4(a), the x-axis is the number of GPUs, while the y-axis is the time it takes to solve the problem in seconds. As the graph shows, the time it takes to solve the problem decreases as the number of GPUs increases, but not linearly. This is to be expected since the multi GPU C+CUDA implementation uses the PCI Express bus to transfer interface nodes data between the multi-core CPU and the GPUs and, as the number of GPUs increases, this bus becomes a bottleneck. Figure 4(b) presents the

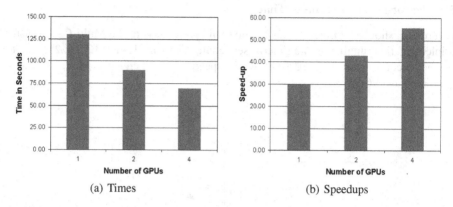

(a) Times   (b) Speedups

Fig. 4. BOXX Personal Supercomputer times and speedups

speedups obtained with the BOXX Personal Supercomputer configured with different numbers of GPUs (the reference time is that of a Enterprise 3 single core). In this graph, the x-axis is the number of GPUs, while the y-axis is the speedup. As the graph shows, speedups close to 60 were obtained with C+CUDA.

To better appreciate the benefits of CUDA-enabled GPUs and C+CUDA, we plot on the graph of Figure 5 the speedups obtained with the BOXX Personal Supercomputer against the best performing Enterprise 3 cluster (96 cores). In the graph of Figure 5, the x-axis is the number of GPUs, while the y-axis is the time it takes to solve the problem with Enterprise 3 divided by the time it takes to solve the problem with the BOXX Personal Supercomputer configured with different numbers of GPUs. As the graph of Figure 5 shows, a desktop machine with a single GPU can outperform a 24-machine cluster (96 cores). Also, a desktop machine with four GPUs can deliver more the twice the performance of a 24-machine cluster (96 cores).

5.4 Performance in Terms of Energy Consumption

To compare the performance of our Multi-Core Cluster implementation with that of our C+CUDA implementation in terms of energy consumption, we run the rotating cone problem in a regular mesh of 2048×2048 cells in both machines and measured the total current drained by each at 10-second intervals using a Digital Clamp Meter Minipa, Model ET-3880, while measuring the voltage. Figure 6 shows the measurement setup employed with each machine (voltage measurement not shown).

By numerically integrating the current $\times$ voltage (power in Watts) required by the machines in the period of time they took to solve the rotating cone problem, we were able to estimate the total energy (in Joules) consumed by each machine. The amount of Joules consumed by Enterprise 3 (all 96 cores) was equal to approximately 5,545,530 Joules (45 Amperes $\times$ 114 Volts $\times$ 1,081 Seconds). The amount of Joules consumed by the BOXX Personal Supercomputer on equivalent circumstances (127 Volts, but different currents and times for each number of GPUs) was measured for 1, 2 and 4 GPUs. Figure 7 shows the energy consumed by each machine configuration.

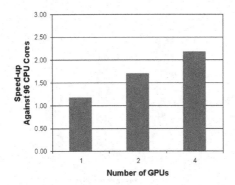

Fig. 5. BOXX Personal Supercomputer speedups: C+CUDA x Multi-Core Cluster

As Figure 7 shows, the amount of Joules decreases as the number of GPUs increases. This is to be expected, since the time to solve the problem diminishes. Note that we did not remove the unused GPU boards during these experiments and, even when not doing useful computation, the GPUs consume a significant amount of energy. Note also that the energy consumed by the whole machine was measured in all cases, and the ratio computation/energy consumption becomes worth with fewer GPUs doing useful work.

Finally, Figure 8 presents a comparison between the amount of energy consumed by Enterprise 3 versus (divided by) the amount of energy consumed by the BOXX Personal Supercomputer while solving the rotating cone problem with 1, 2 and 4 GPUs. As the graph of Figure 8 shows, the BOXX Personal Supercomputer consumes more than 20 times less energy than the Enterprise 3 cluster while solving the same problem. This result shows that, currently, considering the benefits of shorter executing times, smaller energy consumption, and smaller size and maintenance costs, Multi-GPU desktop machines are better high performance computing platforms than small clusters without GPUs such as Enterprise 3, even though they are somewhat harder to program. It is important to note that our C+CUDA code runs unmodified in clusters of multi-core machines each of which with multiple GPUs (it is, in fact, a C+CUDA+MPI code).

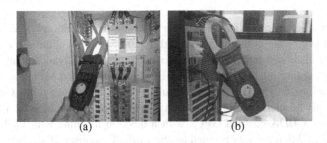

Fig. 6. Power (current) measurement setup. (6(a)) Cluster setup: the total current consumed by Enterprise 3 was measured on the neutral wire of its power distribution panel. (6(b)) BOXX Personal Supercomputer setup: the total current consumed by it was measured on the neutral wire of its power cord.

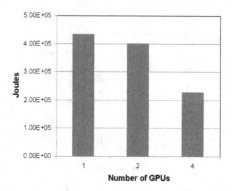

Fig. 7. Joules consumed while running the rotating cone problem with the BOXX Personal Supercomputer with 1, 2 and 4 GPUs. The unused GPU boards were not removed during the experiments.

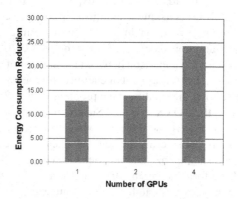

Fig. 8. Energy reduction observed while solving the rotating cone problem in the BOXX Personal Supercomputer for 1, 2 and 4 GPUs when compared with the 96-core Enterprise 3 Cluster

6   Related Work

Since the introduction of CUDA, a number of works have demonstrated that the use of GPUs can accelerate computational fluid dynamics (CFD) simulations ([3, 11, 14, 16]). Recently, Jacobsen et al. [7] have exploited some of the advanced features of MPI and CUDA programming to overlap both GPU data transfer and MPI communications with computations on the GPU. Their results demonstrated that multi-GPU clusters can substantially accelerate CFD simulations. In this work, we compared a Multi-Core Cluster without CUDA-enabled GPUs with a desktop machine with CUDA-enabled GPUs and showed that the way pointed by the work of Jacobsen et al. and others [7] is perhaps the current only way forward in the high performance CFD simulation field.

Little research has been conducted on the evaluation of energy consumption of GPUs against that of clusters. Huang et al. [5] analyzed two parallel implementations of a biological code that calculates the electrostatic properties of molecules—a multithreaded

CPU version (for a single multi-core machine) and a GPU version—and compared their performance in terms of execution time, energy consumption, and energy efficiency. Their results showed that the GPU version performs the best in all three aspects. In this work, we showed that a parallel CUDA-enabled GPU implementation consumes considerably less energy (Joules) than a parallel multi-core cluster implementation while solving a whole instance of the finite element problem.

7 Conclusions

We used a finite element formulation to solve the 2D time-dependent advection diffusion equation in Multi-Core Clusters and CUDA-enabled GPUs. Our experimental results have shown that a desktop computer with a single GPU can outperform a 24-machine cluster of the same generation and that a 4-GPU desktop can offer more than twice the cluster performance (performance in terms of time to compute a solution). Our experimental results have also shown that a 4-GPU desktop can consume less than one twentieth of the energy (Joules) consumed by a 24-machine cluster while solving a whole instance of this relevant finite element problem. The techniques we employed for the problem tackled in this paper can be employed in much harder problems. In future works, we will examine multidimensional compressible problems governed by the Navier-Stokes equations.

Acknowledgments. We thank CNPq-Brazil (grants 552630/2011-0, 309831/2007-5, 314485/2009-0, 309172/ 2009-8) and FAPES-Brazil (grant 48511579/2009) for their support to this work.

References

1. Brooks, A.N., Hughes, T.J.R.: Streamline upwind/Petrov-Galerkin formulations for convection dominated flows with particular emphasis on the incompressible Navier-Stokes equations. Computer Methods in Applied Mechanics and Engineering 32, 199–259 (1982)
2. Catabriga, L., Coutinho, A.L.G.A.: Implicit SUPG solution of Euler equations using edge-based data structures. Computer Methods in Applied Mechanics and Engineering 191, 3477–3490 (2002)
3. Cohen, J.M., Molemaker, M.J.: Cohen and M. Jeroen Molemaker. A fast double precision CFD code using CUDA. In: Proceedings of the 21st Parallel Computational Fluid Dynamics, Monffett Fiel, California (2010)
4. Coutinho, A.L.G.A., Martins, M.A.D., Alves, J.L.D., Landau, L., Moraes, A.: Edge-based finite element techniques for nonlinear solid mechanics problems. International Journal for Numerical Methods in Engineering 50, 2053–2068 (2001)
5. Huang, S., Xiao, S., Feng, W.: On the energy efficiency of graphics processing units for scientific computing. In: Proceedings of the IEEE International Symposium on Parallel & Distributed Processing, pp. 1–8 (2009)
6. Hughes, T.J.R.: The Finite Element Method. Linear Static and Dynamic Finite Element Analysis. Prentice-Hall, Englewood Cliffs (1987)
7. Jacobsen, D.A., Thibault, J.C., Senocak, I.: An MPI-CUDA implementation for massively parallel incompressible flow computations on multi-GPU clusters. In: Proceedings of the 48th AIAA Aerospace Sciences Meeting, Orlando, Florida (2010)

8. Jimack, P.K., Touheed, N.: Developing parallel finite element software using mpi. In: Topping, B.H.V., Lammer, L. (eds.) High Performance Computing for Computational Mechanics, pp. 15–38. Saxe-Coburg Publications (2000)

9. Karypis, G., Kumar, V.: Multilevel k-way partioning scheme for irregular graphs. Technical Report 95-064, Department of Computer Science, University of Minnesota (1995)

10. Kirk, D.B., Hwu, W.W.: Programming massively parallel processors: a hands-on approach. Elsevier (2010)

11. Klockner, A., Warburton, T., Bridge, J., Hesthaven, J.S.: Nodal discontinuous Galerkin methods on graphics processors. J. Comput. Phys. 228, 7863–7882 (2009)

12. NVIDIA. NVIDIA CUDA 3.0 - Programming Guide. NVIDIA Corporation (2010)

13. Saad, Y.: Iterative Methods for Sparse Linear Systems. PWS Publishing, Boston (1996)

14. Senocak, I., Thibault, J., Caylor, M.: Rapid-response urban CFD simulations using a GPU computing paradigm on desktop supercomputer. In: Proceedings of the Eighth Symposium on the Urban Environment. Phoenix, Arizona (2009)

15. Tezduyar, T.E., Hughes, T.J.R.: Finite element formulations for convection dominated flows with particular emphasis on the compressible Euler equations. In: Proceedings of AIAA 21st Aerospace Sciences Meeting, AIAA Paper 83-0125, Reno, Nevada (1983)

16. Thibault, J.C., Senocak, I.: CUDA implementation of a Navier-Stokes solver on multi-GPU desktop platforms for incompressible flows. In: Proceedings of the 7th AIAA Aerospace Sciences Meeting Including The New Horizons Forum and Aerospace Exposition, Orlando, Florida (2009)

A Service-Oriented Architecture for Scientific Computing on Cloud Infrastructures*

Germán Moltó, Amanda Calatrava, and Vicente Hernández

Instituto de Instrumentación para Imagen Molecular (I3M). Centro mixto CSIC
Universitat Politècnica de València CIEMAT, camino de Vera s/n,
46022 Valencia, España
{gmolto,vhernand}@dsic.upv.es, amcaar@ei.upv.es

Abstract. This paper describes a service-oriented architecture that eases
the process of scientific application deployment and execution in IaaS
Clouds, with a focus on High Throughput Computing applications. The
system integrates i) a catalogue and repository of Virtual Machine Images, ii) an application deployment and configuration tool, iii) a meta-
scheduler for job execution management and monitoring. The developed
system significantly reduces the time required to port a scientific application to these computational environments. This is exemplified by a
case study with a computationally intensive protein design application
on both a private Cloud and a hybrid three-level infrastructure (Grid,
private and public Cloud).

Topics. Parallel and Distributed Computing.

1 Introduction

With the advent of virtualization techniques, Virtual Machines (VM) represent
a key technology to provide the appropriate execution environment for scientific applications. They are able to integrate the precise hardware configuration,
operating system version, libraries, runtime environments, databases and the application itself in a Virtual Machine Image (VMI) which can be instantiated into
one or several runnable entities commonly known as Virtual Appliances. With
this approach, the hardware infrastructure is decoupled from the applications,
which are completely encapsulated and self-contained. This has paved the way
for Cloud computing [1,2], which enables to dynamically provision and release
computational resources on demand.

The efficient and coordinated execution of scientific applications on Cloud infrastructures requires, at least: (i) the dynamic provision and release of computational resources (ii) the configuration of VMs to offer the appropriate execution
environment required by the applications and (iii) the allocation and execution of

---

* The authors wish to thank the financial support received from the Generalitat Va-
lenciana for the project GV/2012/076 and to the Ministerio de Economía y Com-
petitividad for the project CodeCloud (TIN2010-17804).

M. Daydé, O. Marques, and K. Nakajima (Eds.): VECPAR 2012, LNCS 7851, pp. 163–176, 2013.

the jobs in the virtualised computational resources. This requires the coordination of different Cloud-enabling technologies in order to automate the workflow required to execute scientific application jobs on the Cloud. To that end, we envision a system where users express their application requirements via declarative procedures and the burden of its deployment, execution and monitoring on an IaaS (Infrastructure as a Service) Cloud is automated. There are previous studies that aim at using Cloud computing for scientific computing [3,4]. However, as far as the authors are aware, there is currently no generic platform that provides automated deployment of scientific applications on IaaS Clouds which deals with VMI management, configuration of VMs and the meta-scheduling of jobs to the virtual computing resources. This represents the whole life cycle of scientific application execution on the Cloud.

For that, the main contribution of this paper is to present a service-oriented architecture integrated by the following developed components: i) a generic catalogue and repository system that indexes VMIs together with the appropriate metadata describing its contents (operating system, capabilities and applications), ii) a contextualization system that allows to deploy scientific applications together with its dependences, iii) a meta-scheduler to manage and monitor the execution of jobs inside VMs and to access the generated output data of the jobs with support for computational steering. The usage of such a system would significantly reduce the time required to migrate an application to be executed on the Cloud. The integration of the different components of the architecture enables to abstract many of the details that arise when interacting with Cloud platforms. This would reduce the entry barrier to incorporate the Cloud as a new source of computational power for scientific applications. This way, scientists would focus on the definition of the jobs and delegate on the proposed platform the orchestration of the components to execute the jobs on the provisioned virtualised infrastructure on the Cloud.

The remainder of the paper is structured as follows. First, section 2 introduces the architecture and details the features of the principal components. Later, section 3 addresses a case study for the execution of a protein design scientific application using the aforementioned system. Finally, section 5 summarises the paper and points to future work.

2   Architecture for Scientific Application Execution on the Cloud

Many scientific applications require the execution of batch jobs, where each job basically consists of an executable file that processes some input files (or command line arguments) and produces a set of files (or data to the standard output) without the user intervention. This is the case of many parameter sweep studies and Bag of Tasks (BoT) applications commonly found in High Throughput Computing (HTC) approaches, where the jobs share common requirements. For these applications, the benefits of the Cloud are two-fold. Firstly, computational resources can be provisioned on demand according to the number of jobs to be

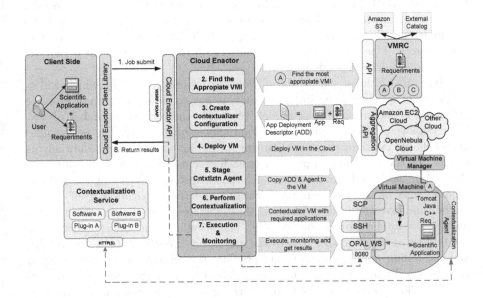

Fig. 1. Scientific applications execution on IaaS Cloud via the Cloud Enactor

executed (and the budget of the user in the case of a public Cloud). Secondly, the provisioned VMs can be configured for the precise hardware and software configuration required by the jobs. This means that VMs can be reused to perform the execution of multiple jobs.

Figure 1 summarises the main interactions between a user and the proposed architecture. The user employs the client-side API to describe each task to be executed (executable file or source code, and required input files) together with the hardware (i.e. CPU architecture, RAM, etc.) and software requirements (OS, applications, system packages, etc.). The jobs might optionally include budget information, since the underlying Cloud infrastructure could require a pay-per-use access to resources. These jobs are submitted (step 1) to the Cloud Enactor (CE) which is the central manager that orchestrates all the components.

The CE checks whether the job could be executed on one of the already deployed (if any) VMs. For the jobs that cannot be executed on the currently deployed VMs, the CE queries the Virtual Machine image & Repository Catalogue (VMRC) [5] with the job's requirements to find the most appropriate VMI to execute the application (step 2). The VMRC, a software that we previously developed, implements matchmaking capabilities to offer a ranked list of suitable VMIs to the Cloud Enactor. The VMRC discards the VMIs that do not satisfy the mandatory requirements (i.e., different OS or CPU architecture) and it ranks the resulting VMIs according to the degree of satisfaction with respect to the optional requirements (mainly, software applications). The CE computes the deviation from the current state of the most appropriate VMI found and the desired state for the job execution in order to create the Application Deployment

Descriptor (ADD) for the contextualization software (step 3). The ADD specifies the deployment process of the application so that the contextualization software can unattendedly perform the installation of the application and its software dependences. This will be executed inside the VM at boot time to deploy the application and its dependences.

Next, the CE must decide the deployment strategy of VMs, which will be in charge of executing the jobs. For that, it has to consider a mixture of performance, economic and trust models to decide the optimum number of VMs to be deployed, together with their Cloud allocation strategy. The performance model should consider the execution time of the jobs (which can be initially estimated by the user but computed after each execution), the deployment time of the VM in the Cloud infrastructure, the time invested in deploying the software requirements of the job (contextualization) and the application itself, as well as the time invested in data transfer, that is, staging out the generated output data of the application inside the VM. The economical model should consider the budget of the user allocated to the execution of each job (or a set of jobs), and the billing policies of the Cloud provider (i.e. hourly rates, economic time zones, etc.). Finally, the trust model plays an important role on scenarios with multiple Cloud providers (Sky Computing), where reputation and the ability of a provider to systematically fulfill the Service Level Agreement (SLA) must be considered. The trust model would be employed to rank a Cloud provider according to its adherence to SLA and the Quality of the Service it offered along the time, among other possible characteristics. For example, a Cloud provider that systematically violates its own SLA should be ranked lower than a provider that has always fulfilled the terms of conditions. The user would express the precise rank function according to the aforementioned categories, as performed in other meta-scheduling softwares such as GridWay.

Therefore, the CE decides to fire up a new VM (or a group of them). This is achieved by delegating on a Virtual Infrastructure Manager (VIM), which deploys the VM on top of a physical infrastructure (step 4). Notice that the CE could use elasticity rules in order to enlarge or shorten the number of VMs dynamically assigned for the allocation of jobs, depending on the budget and the deadline constraints imposed by the user.

When the VM has booted, the CE stages the contextualization agent and the ADD into the VM using SSH (step 5). The VMRC service stores the login name and the private key (or the password) of an account in the VM as part of the metadata stored for a VMI. Then, the contextualization process is started, where software dependences are retrieved from the Contextualization Service and then installed. Next, the scientific application is deployed and a Web services (WS) wrapper is automatically created and deployed into an application server, which is finally started (step 6). This WS wrapper enables to remotely start and monitor the application running inside the VM. All this automated process results in a VA fully configured for the execution of the scientific application.

Once the VA is up and running, the meta-scheduler can perform the execution of the jobs inside the VAs (step 7). This involves managing and monitoring the execution of the jobs inside the VM during their lifetime. For efficiency purposes each VM would be in charge of the execution of several jobs. In the case of parameter sweep studies and BoT applications commonly found in HTC approaches, the jobs share common requirements and, therefore, they can be executed in the same contextualized VM. In addition, scientific applications might require a periodical access of the generated output data during their executions, mainly for computational steering purposes. Once the application inside the VM has finished executing, then its output data must be retrieved so that another job (with the same requirements) can execute inside the VM.

After all the executions have been carried out, the VAs can be gracefully shutdown which is achieved by the VIM. Notice that it is possible to catalogue the resulting VMI (after the contextualization process) together with the metadata information concerning the new applications installed. Therefore, this would minimize the contextualization time for subsequent executions of that scientific application, since no additional software should have to be installed. This streamlined orchestration of components enables the user to simply focus on the definition of the jobs and thus delegate to the central manager the underlying details of interacting with the Cloud technologies for computational resource provisioning and scientific application execution.

This Service-Oriented Architecture relies on several interoperable services that can be orchestrated by the Cloud Enactor due to the usage of standard protocols and interfaces (WS, WSRF, XML). Concerning the software employed, we have relied on the GMarte meta-scheduler [6], which provides execution management capabilities of scientific tasks on computational Grid infrastructures. By incorporating the functionality to access Cloud infrastructures in this software we can simultaneously schedule jobs on both Grid and Cloud infrastructures. In fact, once the virtual infrastructure of computational resources has been provisioned, other job dispatchers could be fit within the proposed architecture, such as Condor or GridWay. The WS Wrapper for the application is created by the Opal 2 Toolkit [7], which has been integrated in the lightweight contextualization software that we previously developed. Other tools for software configuration, such as Puppet or Chef could also be employed within this architecture.

2.1   The Virtual Machine Catalogue and Repository

In a previous work we introduced an early version of the Virtual Machine Catalogue and Repository system (VMRC) [5], whose main capabilities are explained in this section for the sake of completeness. This paper also describes novel features recently included in the system, such as multi-user support by means of Access Control Lists (ACLs) in order to introduce certain levels of security and prevent malware distribution in the VMs and the development of a web-based GUI. In addition, the integration with the Cloud Enactor module is unique in this paper, which can be seen as a practical usage of its functionality.

The main goal of VMRC is to enable users to upload, store and catalogue their VMIs so that they can be indexed. This way, others can search and retrieve them, thus leveraging sharing and collaboration. For that, we have used industry standards such as the Open Virtualization Format (OVF) [8] to describe the VMIs in an hypervisor-agnostic manner, and Web Services to develop the core of the VMRC service.

Each VMI can be catalogued together with appropriate metadata including information about the hardware configuration (memory, architecture, disk size, etc.), the operating system (type of OS, version and release) and the applications currently installed (application name and version). Linking metadata to the VMI enables the development of matchmaking algorithms to retrieve the most appropriate VMI to execute a job considering its requirements.

The following snippet of code summarises the declarative language employed to query the catalogue considering the job's requirements. This example queries the catalogue for a Linux-based VMI, preferably an Ubuntu 11.10 or greater, created for the KVM hypervisor which must have MySQL 5.0 and Tomcat 7.0.22 and it would be desirable to have also the Java Development Kit version 1.5 or greater.

```
vm.type="kvm"
os.name="linux"
os.name="linux" && os.flavour="ubuntu" &&
        os.version>="11.10", soft, 20
app.name="org.mysql" && app.version="5.0"
app.name="org.apache.tomcat" && app.version="7.0.22"
app.name="org.oracle.java-jdk" &&
        app.version>="1.5", soft, 40
```

This language, inspired by the Condor classads language [9], differentiates between the hard requirements, which should be met by a VMI to be considered a potential candidate, and the soft ones, which can be ranked by the client. Certain applications might be considered soft requirements since the client might rely on proper deployment software to delegate the installation of these software on the VM. The inclusion of matchmaking capabilities in the catalogue is a key differential aspect with other catalogues of VMIs.

It is important to point out that the usage of preconfigured VMIs as base images for other VMIs involves security concerns that should be addressed, such as the distribution of malware among images. This can be alleviated by enforcing access control to images [10]. Therefore, we have included multi-user support in the VMRC. The VMRC has an administrator account that has privileges to create new users. The user that registers a VMI can optionally specify the list of users (or give public access) that can perform a given operation on its VMI (search, download, modify). This allows having public images in the catalogue, downloadable by everyone, and private images which might be shared by a collection of users. This is of importance for a research collaboration that might require the usage of a set of VMI, with their specific requirements for their scientific applications.

The VMRC features a web-based GUI which enables authentication via user and password in order to list and download the VMIs together with its metadata that the user can access. Therefore, the catalogue can currently be used via its Web Service API, through the Java bindings for programmatic access, and also using the web based GUI. This allows seamless access to the VMIs. Notice that since the VMRC is a generic component, and it has no specific bindings with a particular Cloud infrastructure, it can be deployed as a central VMI sharing module in a Cloud deployment in order to foster sharing and collaboration. This software is open source and it is available online[1].

2.2 Contextualization of Scientific Virtual Appliances

As stated earlier, the process of configuring a VM to obtain a VA can be referred to as contextualization, a term initially employed in [11] for the configuration of virtual machines to create virtual clusters. This term is employed in this paper for application contextualization, i.e., providing the application with the appropriate execution environment (mainly software dependences) to guarantee its execution. An application with a reduced number of external dependencies can be perfectly contextualized at the time the VM is deployed by the VIM. This way, it is possible to start from a base VM, that only includes the operating system and common use libraries, and to perform the application deployment and contextualization when the VM boots, before executing the application.

However, applications with a large number of dependencies on third-party software are not candidate to perform the contextualization at the time of deployment. In some cases, the time required for contextualization might represent an important overhead, depending on the total execution time of the applications running on the VA. As an example, the compilation and installation of the Globus Toolkit 4 [12], a toolkit for deploying Grid services can take several hours. Additionally, in most cases, performing automatic contextualization requires a considerable complexity from a technical point of view. For these cases, a practical approach consists in performing the installation of the most complex software components by the user, in order to produce a pool of partially contextualized VMIs which are stored on the VMRC. These VMs would then be completely contextualized at boot time in order to create the appropriate environment required for the execution of the scientific application.

Automatic Deployment of Scientific Applications. In order to avoid manual installation procedures when the VMs are allocated by the VIM, we developed a tool (called *cntxtlzr*) that enables to automate the flow of deploying scientific applications. The main goal is to perform the main steps required when deploying a scientific application (packaging, configuration, compilation, execution) without the user intervention. This way, instead of manually configuring the VM via SSH, application inoculation into the VM with minimal user intervention is achieved.

---

[1] http://www.grycap.upv.es/vmrc

This tool supports a small declarative language based on XML employed to create an Application Deployment Descriptor (ADD) which specifies the common actions employed to deploy a scientific application, together with its software requirements.

These are the typical steps involved in the deployment of a scientific application which are addressed by the developed tool:

1. Package Installation. Installs the software packages that the application depends on. It resolves dependencies with other software components and installs those dependencies first. The software packages can be made accessible to the contextualization software via an URL, an installable system package via *yum* or *apt-get* or simply staged into the VM together with the contextualization tool.
2. Configuration. Enables the user to detail the configuration process of the software package. This is achieved by specifying common actions such as copying files, changing properties in configuration files, declaring environment variables, etc.
3. Build. Compiles the software package using the appropriate build system (Configure + Make, Apache Ant, SCons, etc.)
4. Opal-ize. Creates the configuration required by the Opal toolkit, the Web services wrapper for the application. It then installs Opal and its requirements (Tomcat + Java), deploys the scientific application and, finally, starts Tomcat. This causes the scientific application to be deployed and the jobs ready to be started by the Cloud Enactor.

The following snippet of code shows a simplified version of an ADD. It describes an application called *gBiObj* that requires the MPICH Message-Passing Interface (MPI) library, the GNU C compiler and the *make* utility. Its source code is available in a compressed TAR file called *gBiObj.tgz*. We want the application to be accessible via the Opal WS Wrapper so we specify the Opalize XML element. In addition, we want to modify the Makefile of the scientific application to point to where the MPICH library has been installed. Notice that dependencies are installed before the application.

```
<DeployableApp name="gBiObj" requires="mpich gcc make">
  <Package name="gBiObj" file="gBiObj.tgz"/>
  <Opalize exec_file="gBiObj"
           default_args="--gra1 @gBiObj#INSTALL_PATH@/energy.gra"
<Configuration>
 <ReplaceInFile file="@gBiObj#INSTALL_PATH@/Makefile"
                from="mpicc" to="@mpich#INSTALL_PATH@/bin/mpicc"/>
</Configuration>
  <Build type="make"/>
</DeployableApp>
```

The contextualization tool relies on plugins, in the shape of other ADDs, to deploy specific software. This way application developers can specify the installation procedure required by their applications. Thus, it is possible to integrate different software installation descriptions in order to perform complex installations. There currently exists plugins for commonly used software such as Java, Globus Toolkit 4 WS-Core, etc.

The *cntxtlzr* tool currently consists of a highly portable Python script that processes the XML ADD and performs the required actions. This script is staged into the VM and started so that the contextualization process starts. The plugins (or ADDs) and the packages for the software dependencies can be stored in a separate web server. Therefore, the tool can download all the required information at runtime inside the VM in order to perform the contextualization. This lightweight approach to application contextualization only requires Python support in the VM, which is commonly found in the pristine installations of many GNU/Linux distribution.

We plan to combine our tool with other software configuration tools such as Puppet or Chef in order to take advantage of their software deployment approaches. Our approach would complement these software since we use a high level XML-based declarative description of the deployment process which targets at the specific workflow required for the deployment of scientific applications.

2.3   Application Management and Monitoring Inside the VM

Starting and monitoring the execution of the jobs inside the VMs is far from being a trivial task because it requires the deployment of a special agent inside the VM in charge of starting and cancelling the application, and which provides information about the appropriate states of the job (running, finished, etc.). For that, we have relied on the Opal 2 Toolkit [7], which is a tool that wraps scientific applications as Web services so that they can be managed via remote invocations.

Opal requires the user to write an Application Configuration File (ACF) which provides metadata information about the application, such as the location of the executable file and the command-line arguments together with its description. It also accepts advanced features such as the execution method (either locally, inside the VM or delegating the execution to another component such as Globus or Condor).

Then, Opal generates a Web service wrapper and deploys the application into an application server such as Apache Tomcat. The WS front-end to the application allows starting, stopping and monitoring the application that runs in the VM. Different executions of the application can be concurrently carried out within the same VM, since separate folders are employed to generate output data files. It also allows to obtain a list of generated output files. An interesting point is that the output files can easily be accessed from outside the VM via the HTTP protocol, since they are generated inside the Tomcat deployment folder. This allows for computational steering capabilities, where scientific applications performing long simulations periodically generate output data. These data can be retrieved while the computation takes place, thus being able to steer the execution depending on the intermediate results. As an example, if a certain job in a Bag of Tasks submission takes longer than the expected time, it can be cancelled and resubmitted by the Cloud Enactor.

3   Case Study

In order to test the suitability of the Cloud infrastructure as a computational source for scientific applications, two case studies were performed. They involve a scientific application that designs proteins with targeted properties via a computationally intensive process based on Monte Carlo Simulated Annealing (MCSA) [13]. The application is developed in the C programming language and it depends on common build tools available in Linux (configure, make and a C compiler). It also requires the MPICH 2 library.

For the first case study, we used a fixed number of 8 jobs (an appropriate number for our test infrastructure) and we analysed the total execution time. This time includes from the beginning of the task allocation process until the last job has been executed and its output results have been retrieved. Each job requires the initial configuration of the protein and the matrix that indicates the energetic interactions among the different rotamers of the protein. This amounts to a total of 172 MBytes per job. The job outputs the results of the optimization process to the standard output. This computationally intensive application is typically CPU-bound, but we configured the executions to periodically read the energy matrix from the disk (as part of the optimization process) so that I/O would also be significant in the total runtime.

The test infrastructure is based on four dual-processor Intel Xeon QuadCore with 16 GBytes of RAM Blade servers, with a total of 32 cores, managed by OpenNebula 2.2 and the KVM hypervisor. Two nodes were exclusively used for this particular case study. In order to focus on the execution time, the case study was carried out on pre-started VMs where all the contextualization process had finished and the VMs were ready to receive the execution of the jobs. The allocation of tasks to VMs is achieved by the GMarte meta-scheduler. The current configuration controls that only one job is executed inside a single VM. Therefore, using N VMs allows the concurrent execution of up to N tasks. Other jobs are executed as soon as free VMs are available.

In addition, since the architecture can simultaneously schedule jobs to Grid and both private and public Cloud infrastructures, the second case study executes 30 protein design jobs on a hybrid infrastructure composed by resources from a Grid, the aforementioned private Cloud and the Amazon EC2 public Cloud. This demonstrates its ability to scale out computations on demand as long as resources from different infrastructures become exhausted.

3.1   Results

The solid line in Figure 2.a depicts the global execution time of the first case study. As expected, the global execution time decreases when the number of VMs increases, since more computational resources are available to carry out jobs. The plateau in the execution time seen between 4 and 7 VMs is explained by the fact that only one job is executed in each VM and the execution time of each job is expected to be quite similar. Therefore, the executions are actually carried out in groups. As an example, with 5 VMs there is a first group of 5 jobs that are concurrently executed. When they finish, the meta-scheduler allocates the remaining 3 jobs to the free VMs. This would take a similar time as the allocation of the 8 jobs into 7 VMs, which carries out 7 concurrent jobs and a final single job when spare computational resources are available.

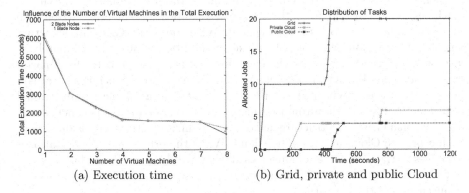

(a) Execution time   (b) Grid, private and public Cloud

Fig. 2. Global execution time (a) of the case study, considering two different distributions of VMs. Allocated jobs (b) on an infrastructure composed of Grid, private Cloud and public Cloud resources.

The dotted line in Figure 2.a compares the degree of scalability of the Blade servers since it shows the global execution time of the case study when all the VMs are running inside a single node. It can be seen that a similar execution time is achieved except for the case of using 8 VMs, where a minor difference is noticed. Since each node features a dual quad-core processor, it appears that scalability issues are only noticeable starting from the 8-th VM in a single node, where the usage of shared resources such as memory and disk start affecting the execution of the applications. These results suggest that VM consolidation in few physical nodes might still deliver good performances for computationally intensive applications, depending on resource consumption.

Concerning the performance improvement gained using the Cloud infrastructure, the results show that up to an speed up of 7.13 is achieved with 8 VMs evenly distributed among the two physical nodes. The global execution time of the case study reduces from a total 6041 seconds in a single VM to just 847 seconds using the aforementioned 8 VMs. Therefore, the usage of virtualised resources from a Cloud as a provider of computational power can deliver a significant improvement for resource-starved scientific applications.

For the second case study we used 10 Grid nodes (from a local resource inte-grated in the Spanish National Grid Initiative), 4 provisioned VMs for the private Cloud and 4 for the public Cloud. The provisioned VMs where contextualized at boot time in order to deploy the application. We used the Free Usage Tier provided by Amazon EC2 to provision low-performance VMs, thus requiring a noticeably larger time to execute the jobs

The task allocation of the 30 jobs is shown on Figure 2.b, further detailed in [14]. The system starts submitting jobs to the Grid infrastructure until all the execution slots are used (approximately at instant 31 in the figure). Since there are pending jobs to be executed, a virtual infrastructure composed of 4 virtual machines is provisioned from the private Cloud provider in order to be able to submit additional jobs to be executed. When both the Grid infrastructure and the private Cloud are not able to execute additional jobs (approximately at instant 258 in the figure) then the computations are scaled out to the Amazon EC2 public Cloud provider. Therefore, 4 additional virtual machines on a pay-per-use basis are provisioned in order to enlarge the available computational infrastructure. Notice that from that moment on, the jobs are being concurrently executed on a Grid infrastructure and on virtual infrastructures provisioned from both a private Cloud and a public Cloud. When the provisioned computational resources of the Cloud are no longer used, they will be shut down. This enables to dynamically adjust the size of the virtual infrastructure to the computational requirements of the case study.

Therefore, the developed system allows to simultaneously harvest computa-tional power from three different infrastructures, in order to reduce the execution time of HTC-based applications.

4   Related Work

This paper aims at abstracting the details of scientific applications execution on Cloud platforms. The literature reveals research efforts into this area.

In a work related to the Nimbus project [15], the authors offer the Workspace Service, which enables to publish different VMIs ready to be used for the execu-tion of certain applications. Therefore, each VMI must be properly configured in advance with the hardware parameters and software dependencies required for the execution of the application. A different approach is offered by the Swarm project [16] which is a task scheduler that acts over three kind of infrastructures (Grid, Windows Server Cluster and Cloud). However, the task execution on the Cloud requires the VMs deployed in the Cloud configured by means of a Hadoop cluster. It uses the MapReduce execution model for the execution of tasks.

SAGA [17] allows to remotely execute applications on top of Grid and Cloud infrastructures. The SAGA libraries and its dependences need to be deployed in advance into the VM, but the main advantage over the previous approaches is that it allows basic VM contextualization once it has been deployed in the Cloud. This includes package installation and minor application configuration during VM startup. There also exists the Cloud Scheduler[2], which is a cloud-enabled distributed resource manager. It provides part of the functionality of a VIM but uses the Condor scheduler [18] to delegate the scheduling decisions for jobs. The user can reference VMIs stored either in Nimbus (via its URL) or Amazon EC2 (via the name of the Amazon Machine Image (AMI)), the same IaaS providers currently available with this tool.

In [19] the authors propose a system to deploy and invoke science applications in the Cloud with minimal user effort. They address the principal challenges when porting an application to the Cloud: application deployment, application execution and data transfer from and into the Cloud. They propose several pre-defined application runtime environments which can be staged into the VM, and an execution framework to start the application. However, being implemented in Windows Azure [20], their approach only targets Windows platforms. In addition, their approach focuses on self-contained applications (binaries and libraries), which are assumed to seamlessly run on the target VM. Therefore, they do not consider the intricacies of deploying complex scientific applications.

5 Conclusion

This paper has introduced a software architecture that abstracts the details of application deployment and execution on IaaS Clouds. The system features the provision of computational virtualised resources, the configuration of these resources to support the execution of the applications, the cataloguing of virtual machine images and, finally, the job execution management on the virtual infrastructure. The benefits of the proposed architecture have been exemplified by the execution of a protein design case study on both a private Cloud infrastructure and a hybrid infrastructure (Grid, private and public Cloud). The automated deployment and execution of scientific applications fosters the widespread adoption of Cloud technologies by the scientific community. This way, Clouds deliver important benefits for scientific computing in terms of the ability to provision computational resources and the customizability of the execution environments.

Therefore, the main contribution of this work to the state-of-the-art is the development of generic components and an architecture to integrate them all in order to ease the process of executing scientific applications on the Cloud. In addition, some of the components of the architecture, such as the VMRC system, have been released to the community as open source.

References

1. Vaquero, L.M., Rodero-Merino, L., Caceres, J., Lindner, M.: A break in the clouds. ACM SIGCOMM Computer Communication Review 39(1), 50 (2008)
2. Armbrust, M., Fox, A., Griffith, R., Joseph, A.: Above the clouds: A berkeley view of cloud computing. Technical report, UC Berkeley Reliable Adaptive Distributed Systems Laboratory (2009)

---

[2] http://www.cloudscheduler.org

3. Rehr, J., Vila, F., Gardner, J., Svec, L., Prange, M.: Scientific computing in the cloud. Computing in Science 99 (2010)
4. Keahey, K., Figueiredo, R., Fortes, J., Freeman, T., Tsugawa, M.: Science Clouds: Early Experiences in Cloud Computing for Scientific Applications. In: Cloud Computing and its Applications (2008)
5. Carrión, J.V., Moltó, G., De Alfonso, C., Caballer, M., Hernández, V.: A Generic Catalog and Repository Service for Virtual Machine Images. In: 2nd International ICST Conference on Cloud Computing (CloudComp 2010) (2010)
6. Moltó, G., Hernández, V., Alonso, J.: A service-oriented WSRF-based architecture for metascheduling on computational Grids. Future Generation Computer Systems 24(4), 317–328 (2008)
7. Krishnan, S., Clementi, L., Ren, J., Papadopoulos, P., Li, W.: Design and Evaluation of Opal2: A Toolkit for Scientific Software as a Service. In: 2009 IEEE Congress on Services (2009)
8. Distributed Management Task Force (DMTF): The Open Virtualization Format Specification (Technical report)
9. Raman, R., Livny, M., Solomon, M.: Matchmaking: Distributed Resource Management for High Throughput Computing. In: Proceedings of the Seventh IEEE International Symposium on High Performance Distributed Computing, pp. 28–31 (1998)
10. Wei, J., Zhang, X., Ammons, G., Bala, V., Ning, P.: Managing security of virtual machine images in a cloud environment. ACM Press, New York (2009)
11. Keahey, K., Freeman, T.: Contextualization: Providing One-Click Virtual Clusters. In: Fourth IEEE International Conference on eScience, pp. 301–308 (2008)
12. Foster, I.: Globus toolkit version 4: Software for service-oriented systems. Journal of Computer Science and Technology 21(4), 513–520 (2006)
13. Moltó, G., Suárez, M., Tortosa, P., Alonso, J.M., Hernández, V., Jaramillo, A.: Protein design based on parallel dimensional reduction. Journal of Chemical Information and Modeling 49(5), 1261–1271 (2009)
14. Calatrava, A.: In: Use of Grid and Cloud Hybrid Infrastructures for Scientific Computing (M.Sc. Thesis in Spanish), Universitat Politècnica de València (2012)
15. Keahey, K., Freeman, T., Lauret, J., Olson, D.: Virtual workspaces for scientific applications. Journal of Physics: Conference Series 78(1), 012038 (2007)
16. Pallickara, S., Pierce, M., Dong, Q., Kong, C.: Enabling Large Scale Scientific Computations for Expressed Sequence Tag Sequencing over Grid and Cloud Computing Clusters. In: Eigth International Conference on Parallel Processing and Applied Mathematics (PPAM 2009), Citeseer (2009)
17. Merzky, A., Stamou, K., Jha, S.: Application Level Interoperability between Clouds and Grids. In: 2009 Workshops at the Grid and Pervasive Computing Conference, pp. 143–150 (2009)
18. Thain, D., Tannenbaum, T., Livny, M.: Distributed computing in practice: the Condor experience. Concurrency and Computation: Practice and Experience 17(2-4), 323–356 (2005)
19. Simmhan, Y., van Ingen, C., Subramanian, G., Li, J.: Bridging the Gap between Desktop and the Cloud for eScience Applications. In: 2010 IEEE 3rd International Conference on Cloud Computing, pp. 474–481. IEEE (2010)
20. Chappell, D.: Introducing windows azure. Technical report (2009)

Interactive Volume Rendering Based on Ray-Casting for Multi-core Architectures

Alexandre S. Nery[1], Nadia Nedjah[2], Felipe M.G. França[1], and Lech Jozwiak[3]

[1] LAM – Computer Architecture and Microelectronics Laboratory
Systems Engineering and Computer Science Program, COPPE
Universidade Federal do Rio de Janeiro, Brazil
[2] Department of Electronics Engineering and Telecommunications
Faculty of Engineering – Universidade do Estado do Rio de Janeiro, Brazil
[3] Department of Electrical Engineering – Electronic Systems
Eindhoven University of Technology, The Netherlands

Abstract. The Volume Ray-Casting rendering algorithm, often used to produce medical imaging, is a well-known algorithm and the underlying computation can be easily executed in parallel. This is due to the fact that the huge number of rays, used to sample the volumetric data, can be processed independently. However, the algorithm's performance may drop substantially when the complexity/size of the volumetric dataset increases. In this paper, we present three implementations of our parallel volume ray-casting algorithm in different multi-core architectures, such as CMPs, GPUs and MPSoCs. Furthermore, we show that using multi-GPUs, that perform in parallel, we can almost halve the rendering time. The performance and aspects of the three implementations are discussed.

1 Introduction

High performance visualization of 3-D datasets has always been one of the main goals in Computer Graphics. For 3-D volumetric datasets, such as those acquired by *Computer Tomography* (CT), the rendering process is generally known as Volume rendering. The volumetric dataset is usually composed of several stacked parallel slices (images) that form a 3-D volumetric dataset. There are different techniques to render 3-D volumetric datasets [9,2]. For instance, the *Marching Cubes* algorithm [11] is one approach to turn voxels samples into polygonal data, in order to create an actual set of 3-D primitives that can be rendered by regular GPUs pipeline. On the other hand, such technique may lead to a poor quality polygonal representation of the volume, because of the approximations that are performed to create the polygonal data. Thus, the Volume Ray-Casting algorithm is a better candidate for producing more accurate results [10,4]. Essentially, this algorithm samples equidistant points along the ray, inside the volumetric 3-D dataset. Each sample, i.e. Voxel (*volumetric pixel*), corresponds to a given color and opacity in one of the parallel slices of the dataset.

M. Daydé, O. Marques, and K. Nakajima (Eds.): VECPAR 2012, LNCS 7851, pp. 177–186, 2013.

The interpolated colors and opacities are merged through *compositing* to yield the color of the view-plane pixel through which a primary ray has been traversed. For instance, the algorithm can show specific parts of a human body volumetric dataset, such as bones or internal organs.

Interactive visualization of volumetric datasets is often difficult. The volume ray-casting performance can drop significantly as more complex datasets are used. On the other hand, volume ray-casting has a very high parallelization potential, as each ray can be processed independently, producing one corresponding pixel information. Therefore, there are consistent approaches to accelerate volume ray-casting with custom parallel architectures in hardware. In [6], a pipelined application-specific integrated circuit (ASIC) was created, fabricated in 0.35 μ technology and running at 125MHz. Such ASIC is capable of producing interactive frame-rates at some degree, since there are limitations regarding the size of the dataset (256^3 voxels). GPUs have recently become a good option for massively parallel processing of floating point data [8]. Thus, there are also approaches to accelerate volume ray casting using GPUs [5,3]. However, most of the volume ray-casting algorithms on GPU strongly depend on optimizations to achieve real-time rendering performance. For example, using *texture* or *constant* memories of the GPU to store frequently-used data can substantially improve the given algorithm performance, because of their much lower latency [8].

In this paper, we propose and discuss the implementations of our interactive, un-optimized and flexible parallel volume ray-casting algorithm with *supersampling* on three different multi-core architectures: Chip Multiprocessor (CMP), Graphics Processing Unit (GPU) and Multiprocessor System on Chip (MPSoC). The CMP implementation of the algorithm uses OpenMP, while the GPU implementation is CUDA-based. The MPSoC-based implementation on FPGA uses the shared DDR memory for synchronization. We extensively compare performance results of the GPU and OpenMP implementations, showing that the GPU implementation can reach interactive visualization, especially when a multi-GPU configuration is used. We also compared and analyzed the advantages of using multi-GPU configuration over a single-GPU configuration for varying workloads (number of primary rays). Finally, the MPSoC-based implementation on FPGA (Xilinx Virtex-5) shows the portability and scalability of the volume ray-casting algorithm, as several microprocessors (MicroBlaze [13] cores) can be mapped on the FPGA and run the algorithm in parallel. All the implementations have not been optimized to use any special features of the corresponding architectures.

The rest of this paper is organized as follows: Sections 2, 3 and 4 briefly explains the parallel Volume Ray-Casting algorithm in CMP, GPU and MPSoC. Then, Section 5 presents extensive performance results for the three implementations and compare them, while Section 6 draws the conclusion of this work.

2   Parallel Volume Ray-Casting in CMP

The OpenMP-based parallel volume ray-casting technique is presented in Algorithm 1, where we use a *for work-sharing construct*, that splits the execution of the parallel section among the group of threads. Thus, iterations of the *for loop* are split across the group of threads. Therefore, in Algorithm 1, groups of rays are assigned to groups of threads for execution, leading to parallelization of rays. Each thread has its own private variables (i, j and s) that are used to control the loop iterations assigned to each thread in the beginning of the parallel section. Also, if *supersampling* is enabled, then each ray spawns a given number of neighbor sampling rays (i.e. in the vicinity of the primary ray), that are executed by the same thread. Thus, the color information of each pixel is measured from all the sampling rays, improving the overall quality of the resulting image.

3 Parallel Volume Ray-Casting in GPU

The CUDA-based parallel volume ray-casting is presented in Algorithm 2. In the CUDA programming model, a thread is actually a *lightweight thread*, because of their simplicity and faster context switching mechanism when compared to regular threads. Throughout this section, we refer to threads in CUDA as *lightweight threads*. In addition, the CUDA-based implementation in Algorithm 2 has not been optimized for GPU execution. For example, the kernel do not make use of *shared memory* or *texture memory*, that are usually employed to avoid global memory long latency penalties.

Algorithm 1. Volume Ray-Casting with OpenMP

Require: rays, uniform grid structure, 3-D dataset
Ensure: image
1: # pragma omp parallel for private(i,j,s)
2: **for** $i = 0$ to WIDTH **do**
3:      **for** $j = 0$ to HEIGHT **do**
4:          color pixel;
5:          **for** $s = 0$ to N_SAMPLES **do**
6:              ray ry $\Leftarrow$ get_ray(i,j,s);
7:              color aux $\Leftarrow$ intersectGrid(grid, ry, dataset);
8:              pixel $\Leftarrow$ pixel + aux;
9:          pixel $\Leftarrow$ pixel / N_SAMPLES;
10:         image[i][j] $\Leftarrow$ pixel ;

Modern general purpose GPUs are capable of executing many thousands of threads in parallel [8]. Thus, each thread can be assigned to a primary ray that crosses a pixel of the view-plane. The result is that a portion of the final image is going to be produced by a *block of threads* (one pixel per thread). The corresponding *CUDA Kernel* is presented in Algorithm 2, considering that all data transfers between the host and the GPU have been already performed. If *supersampling* is enabled, the thread will execute as many sampling rays as required, as shown in line 5 of Algorithm 2. The sampling rays are addressed in column chunks, as shown in line 6.

Algorithm 2. Volume Ray-Casting CUDA–kernel

Require: rays, uniform grid structure, 3-D dataset
Ensure: image
 1: ray ry;
 2: i ⇐ blockDim.x * blockIdx.x + threadIdx.x;
 3: j ⇐ blockDim.y * blockIdx.y + threadIdx.y;
 4: color pixel;
 5: **for** *samples* = 0 to N_SAMPLES **do**
 6:    ry ⇐ rays[i][j+samples];
 7:    color aux ⇐ intersectGrid(uniform grid, ry, dataset);
 8:    pixel ⇐ pixel + aux;
 9: pixel ⇐ pixel / N_SAMPLES;
10: image[i][j] ⇐ c; {corresponding pixel color}

3.1 CUDA-Based Implementation Using Multi-GPUs

In this implementation, the same kernel shown in Algorithm 2 is executed by each GPU. However, the input rays are split among the GPUs, increasing even more the parallel processing of rays. In order to use two GPUs, a separate thread must be created to access each GPU, because one thread cannot control both GPUs at the same time. For that reason, we use OpenMP to create two threads, each one controlling one GPU. The same idea can be extended for more than two GPUs, if available. In the end, the results from both GPUs are merged by the host process into one single image.

4   Parallel Volume Ray-Casting in MPSoC

The MPSoC architecture consists of up to four Xilinx MicroBlaze [13] microprocessors running in parallel at 125MHz. They are connected to a shared DDR memory via a Xilinx Multi-Port Memory Controller (MPMC) [12]. One of the microprocessors is connected to a few communication peripherals, to enable input/output data transmission between the MPSoC and a host machine, as well as to enable access to the FPGA's flash memory. Thus, all the microprocessors must wait until the whole 3-D volume data is available for computation.

The parallel volume ray-casting implementation is presented in Algorithm 3, where iterations of the *for loop* are split across the microprocessors, as shown in line 2. Therefore, in Algorithm 3, groups of rays are assigned to different microprocessors, since rays can be processed independently from the others. Each microprocessor knows which data to read and to write, according to its own identification number (CPU_ID= 0, 1, 2 or 3) and also according to the total number of enabled microprocessors (N_CPU= 1, 2, 3 or 4), as shown in line 2 of Algorithm 3. Finally, at each loop iteration, an image pixel is produced, as shown in line 9 of Algorithm 3.

Algorithm 3. Volume Ray-Casting with MicroBlaze

Require: rays, uniform grid structure, 3-D dataset
Ensure: image
1: **for** $(i = 0;\ i < \text{IMG\_WIDTH};\ \text{i++})$ **do**
2:　**for** $j = \text{CPU\_ID};\ j < \text{IMG\_HEIGHT};\ \text{j} \Leftarrow \text{j} + \text{N\_CPU})$ **do**
3:　　color pixel;
4:　　**for** $s = 0$ to N_SAMPLES **do**
5:　　　ray ry $\Leftarrow$ get_ray(i,j,s);
6:　　　color aux $\Leftarrow$ intersectGrid(grid, ry, dataset);
7:　　　pixel $\Leftarrow$ pixel + aux;
8:　　pixel $\Leftarrow$ pixel / N_SAMPLES;
9:　　image[i][j] $\Leftarrow$ pixel ;

5　Experimental Results

In this section we present the experimental results on different datasets for each multi-core architecture implementation. The CUDA-based implementation was compiled using the CUDA Toolkit 4.0, while the OpenMP-based implementation was compiled in GCC 4.4.4. Up to two NVIDIA GTX 470 GPU were used for execution of the algorithm in CUDA, while a Core i7 960 Intel Multiprocessor (at 3.2 GHz) was used for the algorithm execution in OpenMP. The MPSoC-based architecture was synthesized in Xilinx EDK 13.1 for a Virtex-5 XC5VLX50T FPGA and the parallel algorithm implementation was compiled using MicroBlaze gcc compiler 4.1.2. All the execution time results are measured in seconds and the volumetric dataset (Fig. 1) used in this work is available in [1].

Fig. 1. Images produced by the proposed parallel volume ray-casting algorithm

Table 1. High-resolution execution times for eight different datasets

Data	Sampling rays, OpenMP Core i7						Sampling rays, single-GPU						Sampling rays, dual-GPU					
	1	2	4	8	16	32	1	2	4	8	16	32	1	2	4	8	16	32
foot	2.54	4.68	9.02	18.30	34.58	69.75	0.03	0.08	0.17	0.34	0.67	1.35	0.03	0.06	0.12	0.24	0.48	0.96
skull	2.07	3.94	7.22	15.08	30.36	55.45	0.04	0.09	0.19	0.38	0.77	1.54	0.02	0.05	0.09	0.19	0.38	0.76
engine	2.14	4.12	8.63	16.81	33.42	66.45	0.02	0.04	0.09	0.18	0.37	0.74	0.02	0.03	0.07	0.14	0.28	0.56
aneurism	2.69	5.43	11.20	21.41	41.19	81.48	0.03	0.07	0.15	0.29	0.59	1.18	0.03	0.06	0.12	0.24	0.49	0.97
bonsai	2.23	4.19	7.41	14.65	30.16	55.78	0.03	0.06	0.14	0.27	0.53	1.11	0.03	0.05	0.11	0.22	0.44	0.87
teapot	2.58	4.57	8.72	16.84	38.36	73.00	0.03	0.06	0.13	0.26	0.53	1.06	0.02	0.05	0.09	0.19	0.38	0.77
aorta	6.19	12.22	24.08	47.94	95.76	192.46	0.08	0.19	0.39	0.80	1.62	3.26	0.06	0.12	0.23	0.45	0.88	1.76
backpack	6.36	12.51	24.97	49.45	99.22	198.32	0.12	0.29	0.58	1.20	2.43	4.91	0.08	0.17	0.33	0.66	1.32	2.62

5.1   High-Resolution Performance Results

For each volumetric dataset, the volume ray-casting algorithm was executed for 1280×800 primary rays, producing high-resolution images. In addition, the algorithm was executed with *supersampling* enabled, varying from 1 to 32 sampling rays per pixel. Therefore, up to 32 sampling rays were cast around the region of the primary ray pixel, producing smoother edges in the resulting image. The performance results are summarized in Table 1, for the OpenMP and the CUDA based implementations, using 1 and 2 GPUs, respectively. The MPSoC does not support high-resolution volume ray-casting processing because of memory limitations. Thus, its results are not included in Table 1.

The OpenMP-based implementation uses 8 parallel threads, since the Core i7 microprocessor can execute up to eight parallel processes. The results in Table 1 show that even for one sampling ray, the performance is still not enough to ensure interactive visualization of the datasets. However, good results, i.e. image quality and interactive visualization, can still be obtained at lower resolutions, as fewer primary rays are processed. This will be shown in Section 5.2.

On the other hand, the GPU-based implementation results show that interactive visualization of volumetric datasets is possible even for high-resolution volume ray-casting. As depicted in Fig. 2a and 2b, the volume ray-casting execution time for every dataset is still below one second if up to 4 sampling rays are used, which corresponds to processing $1280 \times 800 \times 4$ rays, in total. Thus, more than one image, a.k.a frame, can be produced in one second, especially if less than four sampling rays are used. The dual-GPU implementation is around 90 times faster than the OpenMP implementation. Comparing the algorithm execution results using one and two GPUs, the performance is almost two times faster when two GPUs are employed instead of one. Also, observe that as more sampling rays are used, the performance gap increases, making the dual-GPU configuration a better candidate for high-quality interactive volume ray-casting, especially for complex datasets such as *aorta* and *backpack*, as shown in Fig. 3.

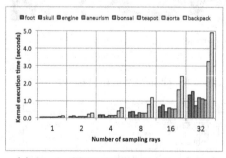

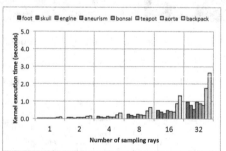

(a) Single-GPU, CUDA-based results.   (b) Dual-GPU, CUDA-based results.

Fig. 2. GPU performance results in CUDA

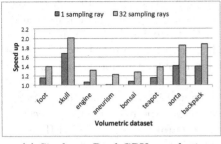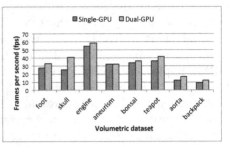

(a) Single vs. Dual GPU speed up.   (b) GPU frame rate (1 sampling ray).

Fig. 3. Acceleration rate using two GPUs and frames per second rate

5.2   Lower-Resolution Performance Results

Lower resolution volume ray-casting can still provide a good trade-off between image quality and performance. In this section, we present some experimental results for the *foot* and *backpack* datasets, rendered in lower resolutions. The performance results are presented in Fig. 4, for one sampling ray.

It is clear that the GPU-based implementation can easily achieve real-time visualization (30 fps) of volumetric datasets when the resulting image resolution is decreased, which means that fewer primary rays are used to sample the volume data. For a simple dataset (*foot*), interactive visualization (around 60 fps) can be achieved even for higher resolutions, as in Fig.4a. On the other hand, the *backpack* dataset can achieve interactive visualization performance for very-low resolutions only, as depicted in Fig.4b. Moreover, the OpenMP-based implementation cannot provide real-time or interactive rendering yet. Thus, optimizations are necessary in order to improve the algorithm performance in OpenMP, as shown in [7].

5.3   MPSoC Synthesis and Performance Results

The MPSoC-based implementation results are shown in Fig. 5. Because of memory limitations of the FPGA we could render images of 640 × 480 pixels. Also,

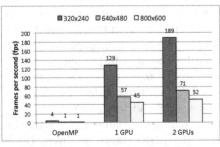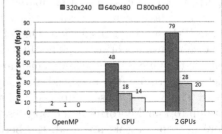

(a) *Foot* dataset low-resolution *fps*.   (b) *Backpack* dataset low-resolution *fps*.

Fig. 4. Frames per second rendering rate for lower resolutions

we could not fit the *aorta* and *backpack* datasets in memory. In Fig. 5a, one can observe that almost all the FPGA slices are being used (82%), as well as the available BlockRAMs (95%). Because of that, we could only fit up to 4 microprocessors running in parallel. The high usage of BlockRAMs is due to the MPMC implementation of FIFOs for each input/output memory port, in order to improve timing and performance [12].

Performance and scalability results are shown in Fig. 5b. For most datasets the parallel algorithm execution time improves as more MicroBlaze microprocessors (Processing Elements - PEs) are being used in parallel. The MPMC FIFOs for the fourth microprocessor are implemented using shift register lookup tables instead of BlockRAMs, which can contribute to create stalls in the datapath and, hence, worsen the overall performance of the microprocessor.

Finally, it is clear that interactive performance is not yet achieved. However, an Application-Specific Integrated Circuit (ASIC) implementation of such application-specific MPSoC design, instead of FPGA, could most probably run faster, with lower area and power consumption, as in [6].

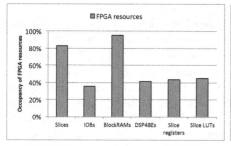

(a) FPGA area occupancy (4 PEs).

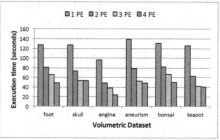

(b) Execution times for 640 × 480 res.

Fig. 5. MPSoC synthesis and scalability, for up to 4 parallel microprocessors

6   Conclusions

In this paper, three un-optimized implementations of our volume ray-casting algorithm are discussed and compared. The GPU-based implementation is up to 90 times faster when a dual-GPU configuration is used, in comparison to the OpenMP-based implementation. One of the reasons for such speed up gain is because thousands of lightweight threads can be executed in parallel on GPU, while in the OpenMP-based implementation only 8 threads are executing in parallel. Furthermore, the overhead of changing between threads in GPU is much lower. The MPSoC-based implementation on a single Virtex-5 FPGA can execute up to four MicroBlaze microprocessors in parallel, running at 125MHz. As more microprocessors are used, the better is the performance achieved. However, interactive performance is not yet achieved, although in ASIC technology it could most probably run at higher frequencies, with more dedicated hardware.

Summing up, we demonstrated that the un-optimized GPU implementation of our volume ray-casting algorithm is able to deliver a high performance, between 60 and 90 times higher than that of our OpenMP-based implementation. For most datasets, high-resolution interactive visualization is achievable. Also, if we would make use of the texture and constant memories of the GPU, we would very likely achieve much higher frame rates, since the latency of these memories is much lower than the global memory latency. On the other hand, interactive performance may only be achieved in OpenMP unless several optimizations are applied to the algorithm. Furthermore, since the GPU implementation introduces more hardware overhead comparing to an MPSoC-based ASIC implementation, a MPSoC-based ASIC is expected to have lower area/power consumption.

References

1. Bartz: Volvis – volume library (2005), http://www.volvis.org/ (last access May 2012)
2. Bhaniramka, P., Demange, Y.: Opengl volumizer: a toolkit for high quality volume rendering of large data sets. In: Proceedings of the 2002 IEEE Symposium on Volume Visualization and Graphics, VVS 2002, pp. 45–54. IEEE Press, Piscataway (2002)
3. Cox, G., et al.: Exploring parallelism in volume ray casting: understanding the programming issues of multithreaded accelerators. In: Proceedings of the 2012 International Workshop on Programming Models and Applications for Multicores and Manycores, PMAM 2012, pp. 64–73. ACM, New York (2012)
4. Wald, I., et al.: Faster isosurface ray tracing using implicit kd-trees. IEEE Transactions on Visualization and Computer Graphics 11(5), 562–572 (2005)
5. Mensmann, J., et al.: An advanced volume raycasting technique using gpu stream processing. In: GRAPP: International Conference on Computer Graphics Theory and Applications, pp. 190–198. INSTICC Press, Angers (2010)
6. Hanspeter, P., et al.: The volumepro real-time ray-casting system. In: Proceedings of the 26th Annual Conference on Computer Graphics and Interactive Techniques, SIGGRAPH 1999, pp. 251–260. ACM Press/Addison-Wesley Publishing Co., New York (1999)
7. Lee, V., et al.: Debunking the 100x gpu vs. cpu myth: an evaluation of throughput computing on cpu and gpu. In: Proceedings of the 37th Annual International Symposium on Computer Architecture, ISCA 2010, pp. 451–460. ACM, New York (2010)
8. Kirk, D., Hwu, W.M.: Programming Massively Parallel Processors: A Hands-on Approach. Morgan Kaufmann Publishers Inc., San Francisco (2010)
9. Lacroute, P., Levoy, M.: Fast volume rendering using a shear-warp factorization of the viewing transformation. In: Proceedings of the 21st Annual Conference on Computer Graphics and Interactive Techniques, SIGGRAPH 1994, pp. 451–458. ACM, New York (1994)
10. Levoy, M.: Efficient ray tracing of volume data. ACM Trans. Graph. 9, 245–261 (1990)

11. Lorensen, W., Cline, H.: Marching cubes: A high resolution 3d surface construction algorithm. SIGGRAPH Comput. Graph. 21, 163–169 (1987)
12. Xilinx. Logicore ip multi-port memory controller (mpmc) v6.03.a, http://www.xilinx.com/support/documentation/ip_documentation/mpmc.pdf (last access May 2012)
13. Xilinx. Microblaze reference, http://www.xilinx.com/support/documentation/ sw_manuals/xilinx13_1/mb_ref_guide.pdf (last access May 2012)

Automatic Generation of the HPC Challenge's Global FFT Benchmark for BlueGene/P

Franz Franchetti[1], Yevgen Voronenko[2], and Gheorghe Almasi[3]

[1] Carnegie Mellon University, ECE Department, Pittsburgh, PA 15213, USA
franzf@ece.cmu.edu
[2] Accuray, Inc., Sunnyvale, CA 94089, USA
yvoronen@gmail.com
[3] IBM Research, T.J. Watson Research Center, Yorktown Heights, NY 10598, USA
gheorghe@us.ibm.com

Abstract. We present the automatic synthesis of the HPC Challenge's Global FFT, a large 1D FFT across a whole supercomputer system. We extend the Spiral system to synthesize specialized single-node FFT libraries that combine a data layout transformation with the actual on-node FFT computation to improve the network performance through enabling all-to-all collectives. We run our optimized Global FFT benchmark on up to 128k cores (32 racks) of ANL's Blue-Gene/P "Intrepid" and achieved 6.4 Tflop/s, outperforming ANL's 2008 HPC Challenge Class I Global FFT run (5 Tflop/s). Our code was part of IBM's winning 2010 HPC Challenge Class II submission. Further, we show first single-thread results on BlueGene/Q.

1 Introduction

The HPC Challenge (HPCC) [1] has been developed to provide more in-depth benchmarking of supercomputers beyond the HPL benchmark used for the TOP500 ranking [2]. HPCC contains seven benchmarks: HPL, STREAM, RandomAccess, PTRANS, FFT, DGEMM, and b_eff. In this paper we focus on the Global FFT benchmark, which computes a large 1D FFT across a large distributed-memory machine, using FFTE [3]. Given network bandwidth and node (CPU) performance developments, Global FFT is on large machines dominated by the machine's cross-sectional bandwidth and thus the three global transposes required to compute a 1D FFT where input and output are in natural order.

An optimized Global FFT implementation must combine an optimized communication library (e.g., the vendor MPI library) with an optimized FFT library (e.g., the vendor FFT library). However, performance does not compose. To obtain the best performance within each of these two libraries, incompatible data formats are required: (1) The MPI library typically is best optimized to send large messages through collective communication functions (e.g., MPI all-to-all), which requires all data for the same destination processor to be packed into one contiguous memory area. (2) Conversely, FFT libraries usually require the data for FFTs to be contiguous in the node memory to obtain best performance. However, the Global FFT algorithm requires sending neighboring data elements to different target processors. Thus, one either needs to convert

M. Daydé, O. Marques, and K. Nakajima (Eds.): VECPAR 2012, LNCS 7851, pp. 187–200, 2013.
© Springer-Verlag Berlin Heidelberg 2013

the data between the storage formats in between library calls to the two libraries or one needs to resort to less efficient library functions (e.g., MPI all-to-allv instead of all-to-all). In either case the overhead can be significant.

Contribution. In this paper we present a novel distributed memory 1D FFT algorithm for large processor counts. Our algorithm blocks the FFT's three global transposes so that for each transpose every processor sends only one large message (that is contiguous in memory) to each other processor, using the most optimized collective communication call. The necessary data scrambling before sending and after receiving and the twiddle factor scaling becomes part of modified node FFT libraries. We extend the program generation and autotuning framework Spiral to automatically generate and optimize these modified FFT libraries, and use UPC as communication layer. Our optimized Global FFT reaches 6.4 Tflop/s on 128k cores (32 racks) of ANL's BlueGene/P while FFTE (the original HPCC Global FFT) reached 5 Tflop/s on the same machine in the 2008 Class I HPC Challenge award. Finally, we make a first step towards targeting BlueGene/P's successor—the BlueGene/Q system—and demonstrate Spiral's ability to automatically generate highly optimized single threaded code taking full advantage of the new QPX SIMD vector instruction set.

Related Work. One and multi-dimensional FFT algorithms for distributed memory are an extensively studied topic. Most 1D FFT algorithms are building on the *Six Step FFT Algorithm* [4], which breaks a large 1D FFT into two stages of local FFTs on contiguous data plus a twiddle stage and three global transposes. FFTW [5] provides a well-optimized open source single node and MPI FFT library. FFTE is specifically designed for large 1D distributed memory FFTs [3], and the reference implementation of the HPC Challenge's FFT benchmark (Global FFT) [1]. In this work we extend the program generation system Spiral [6, 7], and builds on Spiral's code generation for BlueGene/L's *Double FPU* [8], and multicore CPUs [9]. Our system uses IBM's UPC runtime [10] as communication layer. We are building on Spiral's experimental MPI FFT code generation for fixed problem-size and small and fixed processor count (up to 16) [11]. We are extending these concepts to automatically generate the whole FFT computation required for the Global FFT HPCC benchmark, which needs to be a single library working for any problem size and processor count without recompilation.

2   Background

In this section we discuss the necessary background for this paper. We review the Kronecker product formalism to describe fast Fourier transform (FFT) algorithms, the autotuning and program generation system Spiral, and the BlueGene/P supercomputer.

Fast Fourier Transform. Given n real or complex inputs $x_0, \ldots, x_{n-1}$, the discrete Fourier transform (DFT) is defined as

$$y_k = \sum_{0 \le \ell < n} \omega_n^{k\ell} x_\ell, \quad 0 \le k < n, \tag{1}$$

with $\omega_n = \exp(-2\pi i/n)$, $i = \sqrt{-1}$. Stacking the x_ℓ and y_k into vectors $x = (x_0, \ldots, x_{n-1})^T$ and $y = (y_0, \ldots, y_{n-1})^T$ yields the equivalent form of a matrix-vector product:

$$y = \text{DFT}_n\, x, \quad \text{DFT}_n = [\omega_n^{k\ell}]_{0 \le k, \ell < n}. \tag{2}$$

Computing the DFT by its definition (2) requires $\Theta(n^2)$ many operations. FFT algorithms reduce the runtime to $O(n \log(n))$ and can be written in the form of structured sparse matrix factorization using the Kronecker product formalism [6, 12]. The two-point FFT is given by the butterfly matrix,

$$\text{DFT}_2 = \begin{bmatrix} 1 & 1 \\ 1 & -1 \end{bmatrix}. \tag{3}$$

In the following, we use I_n to denote an $n \times n$ identity matrix, and

$$A \otimes B = [a_{k\ell} B], \quad A = [a_{k\ell}]$$

for the tensor (Kronecker) product of matrices. It replaces every entry $a_{k,\ell}$ of A by the matrix $a_{k,\ell} B$. Most important for FFTs are the cases where A or B is the identity matrix. As examples consider

$$I_4 \otimes \text{DFT}_2 = \begin{bmatrix} 1 & 1 & & & & & & \\ 1 & -1 & & & & & & \\ & & 1 & 1 & & & & \\ & & 1 & -1 & & & & \\ & & & & 1 & 1 & & \\ & & & & 1 & -1 & & \\ & & & & & & 1 & 1 \\ & & & & & & 1 & -1 \end{bmatrix} \quad \text{and} \quad \text{DFT}_2 \otimes I_4 = \begin{bmatrix} 1 & & & & 1 & & & \\ & 1 & & & & 1 & & \\ & & 1 & & & & 1 & \\ & & & 1 & & & & 1 \\ 1 & & & & -1 & & & \\ & 1 & & & & -1 & & \\ & & 1 & & & & -1 & \\ & & & 1 & & & & -1 \end{bmatrix}.$$

Further we introduce the stride permutation matrix defined by

$$L_m^{mn} : jn + i \mapsto im + j, \quad 0 \le i < n, \; 0 \le j < m.$$

L_m^{mn} can be seen as transposing a $n \times m$ matrix which is stored in row-major order and is derived from reshaping a mn-dimensional vector into a $n \times m$ matrix. As example consider

$$L_2^8 = \begin{bmatrix} 1 & & & & & & & \\ & & 1 & & & & & \\ & & & & 1 & & & \\ & & & & & & 1 & \\ & 1 & & & & & & \\ & & & 1 & & & & \\ & & & & & 1 & & \\ & & & & & & & 1 \end{bmatrix}.$$

Equation (4) shows the general mixed-radix Cooley-Tukey FFT algorithm:

$$\text{DFT}_{mn} = \left(\text{DFT}_m \otimes I_n \right) T_n^{mn} \left(I_m \otimes \text{DFT}_n \right) L_m^{mn}. \tag{4}$$

In (4), T_n^{mn} is a complex diagonal matrix [12]. Using (3) and (4), an 8-point FFT can be derived by two recursive applications:

$$\text{DFT}_8 = (\text{DFT}_2 \otimes I_4) T_4^8 \left((\text{DFT}_2 \otimes I_2) T_2^4 (I_2 \otimes \text{DFT}_2) L_2^4 \right) L_2^8.$$

Formulas can be manipulated using formula identities like

$$I_n \otimes (BC) = (I_n \otimes B)(I_n \otimes C) \tag{5}$$
$$(BC) \otimes I_n = (B \otimes I_n)(C \otimes I_n) \tag{6}$$
$$\left(L_m^{mn} \right)^\top = L_n^{mn} \tag{7}$$
$$(BC)^\top = C^\top B^\top \tag{8}$$
$$A \otimes B = L_m^{mn}(B \otimes A) L_n^{mn} \tag{9}$$
$$(A \otimes B)^\top = A^\top \otimes B^\top \tag{10}$$
$$L_n^{kmn} = (L_n^{kn} \otimes I_m)(I_k \otimes L_n^{mn}) \tag{11}$$
$$L_{km}^{kmn} = (I_k \otimes L_m^{mn})(L_k^{kn} \otimes I_m) \tag{12}$$
$$L_k^{kmn} = L_{km}^{kmn} L_{kn}^{kmn} . \tag{13}$$

In (5)–(10), A is a $m \times m$ matrix and B and C are $n \times n$ matrices. Further we denote the conjugation of a matrix A by a permutation P by $A^P = P^\top A P$ and note that for permutation matrices $P^{-1} = P^\top$.

```
void FFT8(_Complex double *Y, _Complex double *X) {
    __alignx(16,Y);
    __alignx(16,X);
    _Complex double s34, s35, s36, s37, s38, t100, t101, t102
        , t103, t104, t94, t95, t96, t97, t98, t99;
    t94 = (*(X) + *((X + 4)));
    t95 = (*(X) - *((X + 4)));
    t96 = (*((X + 2)) + *((X + 6)));
    s34 = (__I*(*((X + 2)) - *((X + 6))));
    t97 = (t94 + t96);
    t98 = (t94 - t96);
    t99 = (t95 + s34);
    t100 = (t95 - s34);
    t101 = (*((X + 1)) + *((X + 5)));
    t102 = (*((X + 1)) - *((X + 5)));
    t103 = (*((X + 3)) + *((X + 7)));
    s35 = (__I*(*((X + 3)) - *((X + 7))));
    t104 = (t101 + t103);
    s36 = (__I*(t101 - t103));
    s37 = ((0.70710678118654757 + __I * 0.70710678118654757)*(t102 + s35));
    s38 = ((-0.70710678118654757 + __I * 0.70710678118654757)*(t102 - s35));
    *(Y) = (t97 + t104);
    *((Y + 4)) = (t97 - t104);
    *((Y + 1)) = (t99 + s37);
    *((Y + 5)) = (t99 - s37);
    *((Y + 2)) = (t98 + s36);
    *((Y + 6)) = (t98 - s36);
    *((Y + 3)) = (t100 + s38);
    *((Y + 7)) = (t100 - s38);
}
```

Fig. 1. 8-point FFT, using complex C99 data types and the IBM XL C dialect

Spiral. Recursive application of rules like (3) and (4) yields many different algorithms for a FFT size. Spiral [6] uses this fact to search for the fastest on a given platform. A user-specified transform (like DFT_{256}) is expanded by Spiral using rules into a formula,

Table 1. Compiling SPL into code is done by recursively using the above correspondences. x denotes the input and y the output vector. We use Matlab-like notation: $x[b:s:e]$ denotes the subvector of x starting at b, ending at e, and extracted at stride s. T_km_k is a array of pre-computed constants.

SPL construct	code
$y = (A_n B_n)x$	```t[0:1:n-1] = B(x[0:1:n-1]);``` ```y[0:1:n-1] = A(t[0:1:n-1]);```
$y = (I_m \otimes A_n)x$	```for (i=0;i<m;i++)``` ``` y[i*n:1:i*n+n-1] = A(x[i*n:1:i*n+n-1]);```
$y = (A_m \otimes I_n)x$	```for (i=0;i<m;i++)``` ``` y[i:n:i+m-1] = A(x[i:n:i+m-1]);```

$y = L_k^{km} x$	```for (i=0;i<k;i++)```
	``` for (j=0;j<m;j++)```
	``` y[i+k*j]=x[m*i+j];```

$y = T_k^{km} x$	```for (i=0;i<k*m;i++)``` ``` y[i]=T_km_k[i]*x[i];```

which is then translated into a C program by a special formula compiler. The formula compiler is based on a translation table similar to Table 1 and uses traditional compiler techniques like unrolling, array scalarization, constant folding, and strength reduction to produce high quality fixed-size FFT functions from a given formula. The runtime of the program is measured and fed into a search module, which triggers, in a feedback loop, the generation of a modified formula based on a search strategy. Upon termination, Spiral out the fastest program found. Figure 1 shows an 8-point FFT generated by Spiral for BlueGene/P, using the complex data type extension of C99.

For sizes too large to be implemented as a single basic block, Spiral is automatically generating a recursive mixed-radix FFT library [7] similar to FFTW [5]. Spiral employs a rewriting system to symbolically expand breakdown rules like (4) to find a closure of recursive functions that is needed to implement the recursive FFT library. It then automatically implements these recursive functions as well as recursion leafs (codelets) for a sufficiently large set of sizes. At runtime, a planner autotunes the recursive decomposition of the FFT in an one-time setup effort. After tuning, a fast FFT library call for the respective problem size is available.

The key insight is that a straightforward implementation of (4) suggests four steps corresponding to the four factors, where two steps call smaller DFTs. However, to improve locality, the initial permutation L_m^{mn} is usually not performed but interpreted as data access for the subsequent computation, and the twiddle diagonal T_n^{mn} is fused with the subsequent DFTs. This strategy is chosen, for example, in the library FFTW 2.x and the code can be sketched as shown in Figure 2. A simplified description of performing this process by hand can be found in [13].

BlueGene/P. BlueGene/P is the second generation BlueGene architecture from IBM, succeeding BlueGene/L [14]. In its compute nodes BlueGene/P uses four PowerPC 450 cores operating at 850 MHz with a double precision, dual pipe floating point unit per

```
void dft(int n, complex *y, complex *x) {
   int k = choose_factor(n);
   // t1 = (I_k tensor DFT_m)L(n,k)*x
   for(int i=0; i < k; ++i)
      dft_iostride(m, k, 1, t1 + m*i, x + m*i);
   // y = (DFT_k tensor I_m) diag(d(j))
   for(int i=0; i < m; ++i)
      dft_scaled(k, m, precomp_d[i], y + i, t1 + i);
}

// DFT variants needed
void dft_iostride(int n, int istride, int ostride, complex *y, complex *x);
void dft_scaled(int n, int stride, complex *d, complex *y, complex *x);
```

Fig. 2. Recursive FFT implementation in the style of FFTW 2.X.

core. Each node has 13.6 Gflop/s peak performance (3.4Gflop/s per core) and 2 GB
RAM with 13.6 GB/s memory bandwidth. Each core has a private 32 kB L1 cache and
the four cores of a node share an 8 MB L3 cache. The compute nodes are connected
with multiple interconnection networks including a 3-D torus (used for standard messaging), a global collective network (used for reductions), and a global barrier network.
Each node has six bi-directional network links supporting 425 MB/s in each direction
into the torus network leading to 5.1 GB/s bidirectional bandwidth per node. The Blue-
Gene/P system "Intrepid" installed at Argonne National Laboratory (ANL) consists of
40 BlueGene/P racks. Each rack contains 1,024 compute nodes (32 node cards, each
holding 32 compute nodes), and each compute node four cores (one quad-core CPU).

BlueGene/P Messaging Layer. We use the IBM UPC runtime system as messaging
layer. It provides an equivalent to the MPI all-to-all collective operation that fully utilizes BlueGene/P's 3D torus interconnection network. To achieve best performance,
exactly one large message of equal size that is contiguous in the node memory should
be sent from every processor to every other processor.

BlueGene/Q. BlueGene/Q is the third generation BlueGene architecture from IBM,
succeeding BlueGene/P [15]. The Blue Gene/Q Compute chip [16] is a system-on-a-
Chip (SOC) ASIC with 16 user-accessible 4-way SMT (Symmetric Multi Threading)
A2 cores clocked at 1.6 GHz. A quad floating unit implementing the QPX instruction
set is associated with each core. The BlueGene/Q A2 chip achieves 204.8 Gflop/s peak
performance. At 1024 chips per rack, the 48 rack ANL "Mira" systems achieves a
peak performance of 10 Pflop/s and the 96 rack LLNL "Sequoia" system 20 Pflop/s,
respectively.

3   Global FFT Algorithm

We now derive our novel 1D Global FFT algorithm, which is a variant of the Six Step
FFT algorithm. Like the Six Step FFT algorithm, it has three global data exchanges.
However, we block the global transpositions so that exactly one pair of large messages
that are contiguous in memory are exchanged between every pair of processors in every
communication step. This can be mapped efficiently to collective communication functions (all-to-all). We formally merge the ensuing data scrambling necessary to produce

consume the contiguous messages with the on-node FFT computations and derive modified FFT libraries (working on custom scrambled data format) that perform the reordering at no extra cost compared to standard FFT libraries. We use the Kronecker product formalism to derive the algorithm and use Spiral to automatically build the modified node FFT libraries from the Kronecker product specification. Finally, we show pseudocode for the top-level parallel (single program multiple data, SPMD) function that calls the modified node FFT libraries.

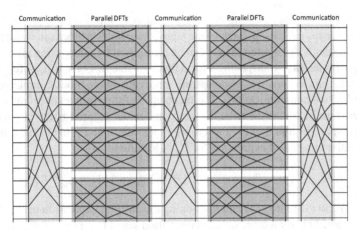

$$L_4^{16}\left(I_4 \otimes ((\mathrm{DFT}_2 \otimes I_2) T_2^4 (I_2 \otimes \mathrm{DFT}_2) L_2^4)\right) L_4^{16}\, T_4^{16}\left(I_4 \otimes ((\mathrm{DFT}_2 \otimes I_2) T_2^4 (I_2 \otimes \mathrm{DFT}_2) L_2^4)\right) L_4^{16}$$

Fig. 3. Six-step FFT for $n = 2^4$ and $k = m = \sqrt{n} = 4$

Algorithm Derivation. Using (5)–(13), the Six Step FFT algorithm can be derived from (4):

$$\mathrm{DFT}_{mn} = L_m^{mn}\left(I_n \otimes \mathrm{DFT}_m \right) L_n^{mn}\, T_n^{mn}\left(I_m \otimes \mathrm{DFT}_n \right) L_m^{mn}. \tag{14}$$

By flipping both tensor products into their parallel form ($I_n \otimes \mathrm{DFT}_m$ and $I_m \otimes \mathrm{DFT}_n$), the algorithm is guaranteed to perform all DFT computations within the local memory of each node. This is achieved by reshaping the data vector of length mn into a $n \times m$ matrix and explicitly transposing it back and forth (a total of three transpositions is required). Typically, choosing $m \approx \sqrt{mn}$ and $n = mn/m$ gives the so-called "square-root decomposition" which maximizes the number of processors that the DFT can be run on in parallel and provides good load balancing. A visual (data flow) representation of the 16-point six-step FFT is shown in Fig. 3. Details can be found in [17].

Below assume p processors and $p \mid m, n$. Using (5)–(13) and associativity and distributivity the stride permutation L_m^{mn} can be expressed as three permutation stages. First we use (12) to obtain

$$L_m^{mn} = \left(I_p \otimes L_{m/p}^{mn/p} \right)\left(L_p^{np} \otimes I_{m/p} \right). \tag{15}$$

Next we use (5) to obtain

$$L_p^{np} = \left(L_p^{p^2} \otimes I_{n/p} \right) \left(I_p \otimes L_p^n \right). \tag{16}$$

Inserting (16) into (15) yields

$$L_m^{mn} = \left(I_p \otimes L_{m/p}^{mn/p} \right) \left(L_p^{p^2} \otimes I_{mn/p^2} \right) \left(I_p \otimes L_p^n \otimes I_{m/p} \right) \tag{17}$$

after further simplification.

Equation (17) describes treating the $n \times m$ matrix as block matrix of $p \times p$ blocks with block size $nm/p \times nm/p$. Each of the p processor holds p blocks in its local memory. (17) states that a distributed matrix can be transposed by transposing all blocks (p^2 local transpositions, each transposing a local block of size $nm/p \times nm/p$) followed by transposing the blocks (one $p \times p$ transposition moving whole blocks, implemented as all-to-all collective communication). The mechanics of the Kronecker product formalism requires three factors to describe the two steps. In our algorithm derivation we also require a transposed version of (17) where we first swap m and n in (17) and the apply (7) to obtain the transposed expression for L_m^{mn}, leading to

$$L_m^{mn} = \left(I_p \otimes L_{m/p}^m \otimes I_{n/p} \right) \left(L_p^{p^2} \otimes I_{mn/p^2} \right) \left(I_p \otimes L_m^{mn/p} \right). \tag{18}$$

Inserting (17) and (18) into (14) and regrouping the ensuing expression using (5) leads to the final algorithm (for a more detailed derivation see [11, 18]),

$$\mathrm{DFT}_{mn} = \underbrace{\left(I_p \otimes (L_{m/p}^m \otimes I_{n/p}) \right)}_{\text{local transpose}} \underbrace{\left(L_p^{p^2} \otimes I_{mn/p^2} \right)}_{\text{all-to-all}} \underbrace{\left(I_p \otimes (\mathrm{DFT}_m \otimes I_{n/p}) \right)}_{\text{inplace FFT library call}}$$

$$\underbrace{\left(L_p^{p^2} \otimes I_{mn/p^2} \right)}_{\text{all-to-all}} \underbrace{(T_n^{mn})^{(I_p \otimes L_{m/p}^m \otimes I_{n/p})} \left(I_p \otimes (L_p^m \otimes I_{n/p})(I_{m/p} \otimes \mathrm{DFT}_n) L_{m/p}^{mn/p} \right)}_{\text{out-of-place scaled FFT library call}}$$

$$\underbrace{\left(L_p^{p^2} \otimes I_{mn/p^2} \right)}_{\text{all-to-all}} \underbrace{\left(I_p \otimes (L_p^n \otimes I_{m/p}) \right)}_{\text{local transpose}}. \tag{19}$$

Eq. (19) makes the minimal necessary changes to the Six Step algorithm to make it compatible to highly optimized all-to-all communication calls, and to allow for specialized high-performance local recursive FFT libraries. Reading (19) from right to left, first each processor performs local data scrambling (*local transpose*) in their own memory space to produce the first set of contiguous messages. This cannot be folded into any FFT library call but could be merged with computation that produces the input data. Next all processors invoke *all-to-all* collective communication; all p processors send one message of size mn/p^2 to every of the p processors (including themselves). Then a modified node FFT—the *out-of-place scaled FFT library*—is called to perform the local FFT computation on scrambled data and performs twiddle scaling. Next the same *all-to-all* call is invoked a second time, followed by the second modified node FFT library, an *inplace FFT library* operating on scrambled data. Note that too make this stage inplace, one needs to chose (17) and (18) carefully. Lastly, the same *all-to-all* collective communication is called a third time to redistribute the data to the target processor and a final *local transpose* unscrambling phase puts the data back into natural order. This final scrambling cannot be merged with any of the modified libraries but could be merged with the code consuming the transformed data.

Specialized FFT Node Libraries. Our derivation extracted the formal definition of two modified node FFT libraries that are invoked independently but in parallel on all p processors. The first node library is specified as

$$(\mathrm{T}_{n,i}^{mn})^{(\mathrm{L}_{m/p}^m \otimes \mathrm{I}_{n/p})} \left((\mathrm{L}_p^m \otimes \mathrm{I}_{n/p})(\mathrm{I}_{m/p} \otimes \mathrm{DFT}_n) \; \mathrm{L}_{m/p}^{mn/p} \right) \tag{20}$$

with $\mathrm{T}_{n,i}^{mn}$ being the global FFT twiddle factors for processor i. The library specified by (20) performs an out-of-place batch FFT (m/p FFTs of size n) plus twiddle scaling on a block-matrix data format. The second library is specified as

$$\mathrm{DFT}_m \otimes \mathrm{I}_{n/p} \tag{21}$$

and performs an inplace strided batch FFT (n/p FFTs of size m) that can be viewed as column FFT. The modified node FFT libraries are automatically generated from the specification using Spiral's general size library generation framework [7].

To turn the algorithmic advantage into a performance advantage, the automatically generated libraries need to be of equivalent performance as FFTW or the vendor library ESSL. Since we are targeting Global FFT for 128k processors, the largest FFT sizes are up to $mn = 2^{38}$, and thus m and n can be up to 2^{19}. Thus, the node FFT libraries built from the specifications (20) and (21) need to provide good performance for batches of large FFTs. The generated libraries must perform all state-of-the-art optimizations including SIMD vectorization for the Double FPU [8,9] and must be parallelized across the four cores of a BlueGene/P node [9] when running in SMP mode. Further, aggressive memory hierarchy optimizations like buffering and vector recursion need to be applied [5,9]. All these optimizations need to be performed fully automatically [7].

Full Global FFT Code. In Figure 4 we show the full HPCC Global FFT algorithm using a partitioned global address space (PGAS) abstraction similar to Unified Parallel C (UPC). The data vectors x and y are block distributed (mn/p elements reside in the local memory of each of the p nodes) and all parallel for loops are run across p nodes of the parallel machine. For simplicity, on-node threading and SIMD vectorization is omitted.

4   Experimental Results

We experimentally evaluated our optimized Global FFT benchmark on BlueGene/P configurations from one node card (32 quadcore nodes or 128 cores) up to 32 racks (32k quadcore nodes or 128k cores), with one process per node. We used the IBM UPC runtime for process and thread management and as messaging layer. The benchmark is executed as UPC program that calls external (C/C++) libraries for the on-node FFT computation. UPC uses IBM's XL C compiler as backend, and our generated synthesized on-node libraries were compiled with IBM's XL C compiler and options "-O3 -qarch=440d".

```
// HPCC Gloabal FFT
// data is block distributed on p processors
// the p iterations of parallel for loops are executed across p nodes

// all to all data exchange
//
// implements a block transpose of a p x p matrix on vectors of n elements
// exchange between all pairs of p processors packets of size n
//
// input/output: x[p*p*n], block distributed across p processors
// x := L^{p^2}_p (x) I_n * x
//
void all_to_all(int p, int n, _Complex double *x) {
    int i, j;

    par_forall (i=0; i<p; i++)
        for (j=i+1; j<p; j++)
            SENDRECV(i, j, x+n*(i*p+j), x+n*(i+j*p), n);
}


// local transpose
//
// transposes a n x m matrix of vectors of v complex elements,
// stored in row major order in local node memory
// x := L^{mn}_m (x) I_v * x
//
void transpose(int mn, int m, int v, _Complex double x) {
    int i, j, k, n = mn/m;

    for (i=0; i<n; i++)
        for (j=i; j<m; j++)
            for (k=0; k<v; k++)
                SWAP(x[v*(i*m+j)+k], x[v*(i+j*n)+k]);
}


// global FFT of size m*n on p processors
// y = DFT_mn * x
//
// input: x[m*n], block distributed across p processors
// output: y[m*n], block distributed across p processors
//
void global_fft(int m, int n, int p, _Complex double y, _Complex double x) {
    int i, j, k;

    par_forall (i=0; i<p; i++)
        transpose(n, p, m/p, x+i*m*n/p);

    all_to_all(p, m*n/(p*p), x);

    par_forall (i=0; i<p; i++)
        fft_scaled(n, m, p, n/p, m*n, m/p, m*n/p, x+i*m*n/p, y+i*m*n/p);

    all_to_all(p, m*n/(p*p), x);

    par_forall (i=0; i<p; i++)
        fft_inplace(m, m*n/p, m, n/p, y+i*m*n/p);

    all_to_all(p, m*n/(p*p), x);

    par_forall (i=0; i<p; i++)
        transpose(m, m/p, n/p, y+i*m*n/p);
}
```

Fig. 4. HPCC Global FFT implementation using a UPC-like PGAS syntax, implementing (19)

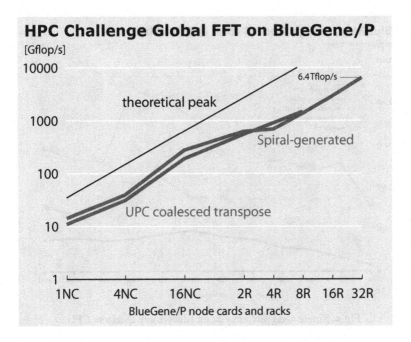

Fig. 5. Performance of the HPC Challenge Global FFT Benchmark on BlueGene/P from 128 cores (1 node card) up to 128k cores (32 racks)

We implemented a baseline Global FFT version that uses IBM's BlueGene/P ESSL for local FFTs and UPC coalesced transpose (the equivalent of MPI all-to-all) for messaging. This implementation requires explicit data reordering between the UPC messaging and the invocation of ESSL but provides best performance for the FFT computation and the messaging in separation. This implementation is part of IBM's winning 2010 HPC Challenge Class II UPC submission.

Figure 5 summarizes the performance results. We run the UPC+ESSL baseline benchmark on the IBM T.J. Watson BlueGene/P system for up to eight racks. We run our Spiral-generated library from one node card to 2 racks on the T.J. Watson machine and on ANL's "Intrepid" from 4 racks to 32 racks. The Spiral-generated Global FFT generally outperforms the UPC+ESSL baseline which shows that (a) Spiral's automatically generated node libraries offer performance competitive with ESSL, and (b) the memory traffic savings obtained by merging data scrambling with the node-libraries improves performance. Finally, the Spiral-generated Global FFT reaches 6.4 Tflop/s on 32 racks of "Intrepid". The winning 2008 ANL HPC Challenge Class I submission reported 5 Tflop/s Global FFT performance on the same machine. Thus, the combination of algorithmic optimization and library generation improved the Global FFT on "Intrepid" by 1.4 Tflop/s or 28%.

Single Node Performance. Figure 6 shows single node performance on the Blue-Gene/P quadcore PowerPC 450D. We compare the GNU Scientific Library (GSL) [19] to Spiral-generated sequential and multi-threaded scalar and Double FPU-vectorized code. Spiral-generated scalar single-core code significantly outperforms the GSL for

BlueGene/P Single Node (4 cores @ 850 MHz)

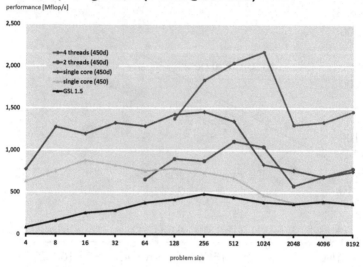

Fig. 6. Single node performance on BlueGene/P quadcore CPU

Spiral FFT: BlueGene/Q Single Thread @ 1.6 GHz

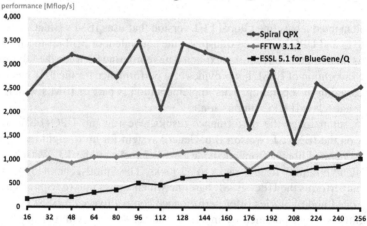

Fig. 7. BlueGene/Q QPX single thread performance

in-cache sizes and performs equally to the GSL for memory-bound sizes, demonstrating the quality of Spiral's base line code generation on BlueGene/P. Spiral's Double FPU two-way SIMD vector code provides between 50% and 2x speed-up on top of the scalar base-line. Using all four cores of the BlueGene/P multicore CPU yields speed-up of 2x–2.5x except for the smallest sizes where parallelization overhead makes sequential code the fastest choice. Using only two threads is never a winning strategy.

Towards Global FFT on BlueGene/Q. We are in the process of porting the Spiral Global FFT code generation to the next generation BlueGene machines, BlueGene/Q. One major difference is that BlueGene/Q features a new 4-way SIMD vector unit called QPX that is twice as wide as the Double FPU of BlueGene/P. In Figure 7 we show first performance results of Spiral-generated QPX code run on a single thread of a Blue-Gene/Q node. We observe that for small FFT sizes Spiral-generated code substantially outperforms both FFTW and ESSL. We are currently porting and adapting the remaining two levels of parallelism of the Global FFT (intra-node threading and inter-node message passing) to BlueGene/Q.

5 Conclusion

The increased complexity and performance levels of high performance and supercomputing systems makes the automatic generation and tuning of performance libraries for the petascale and beyond a promising alternative to hand-tuning. In this paper we present an novel 1D Global FFT algorithm for the HPC Challenge. Extending the Spiral system, we automatically generate specialized node FFT libraries that support the data layout required by the messaging layer while providing FFT performance of the native FFT data layout. The resulting reduction in memory traffic and high node performance enabled us to reach 6.4 Tflop/s on 128k cores of ANL's BlueGene/P system, improving performance by 28% over the previously reported Global FFT Class I benchmark. Finally, we show first single-node results on BlueGene/Q in which we significantly outperform FFTW and ESSL.

Acknowledgement. The authors acknowledge support by NSF through awards 0702386 and 1116802, and by the U.S. Army through contract W911NF-10-1-0004. Yevgen Voronenko was partially supported by a Kauffman Entrepreneur Postdoctoral Fellowship. The authors wish to thank Kalyan Kumaran, Scott Parker and Vitali Morozov of Argonne National Laboratory for access and help with ANL's BlueGene/P "Intrepid" and BlueGene/Q "VEAS"/"VESTA".

References

1. Luszczek, P., Bailey, D., Dongarra, J., Kepner, J., Lucas, R., Rabenseifner, R., Takahashi, D.: The HPC Challenge (HPCC) benchmark suite. In: SC 2006 Conference Tutorial (2006)
2. Meuer, H.W.: The top500 project: Looking back over 15 years of supercomputing experience (2008)
3. Takahashi, D.: An implementation of parallel 1-D FFT using SSE3 instructions on dual-core processors. In: Kågström, B., Elmroth, E., Dongarra, J., Waśniewski, J. (eds.) PARA 2006. LNCS, vol. 4699, pp. 1178–1187. Springer, Heidelberg (2007)
4. Bailey, D.H.: FFTs in external or hierarchical memory. J. Supercomputing 4, 23–35 (1990)
5. Frigo, M., Johnson, S.G.: The design and implementation of FFTW3. Proceedings of the IEEE 93(2), 216–231 (2005); special issue on Program Generation, Optimization, and Adaptation

6. Püschel, M., Moura, J.M.F., Johnson, J., Padua, D., Veloso, M., Singer, B.W., Xiong, J., Franchetti, F., Gačić, A., Voronenko, Y., Chen, K., Johnson, R.W., Rizzolo, N.: SPIRAL: Code generation for DSP transforms. Proceedings of the IEEE 93(2), 232–275 (2005); special issue on Program Generation, Optimization, and Adaptation

7. Voronenko, Y., de Mesmay, F., Püschel, M.: Computer generation of general size linear transform libraries. In: Proc. Code Generation and Optimization (CGO), pp. 102–113 (2009)

8. Franchetti, F., Kral, S., Lorenz, J., Püschel, M., Überhuber, C.W.: Automatically tuned fFTs for blueGene/L's double FPU. In: Daydé, M., Dongarra, J., Hernández, V., Palma, J.M.L.M. (eds.) VECPAR 2004. LNCS, vol. 3402, pp. 23–36. Springer, Heidelberg (2005)

9. Franchetti, F., Püschel, M., Voronenko, Y., Chellappa, S., Moura, J.M.F.: Discrete Fourier transform on multicore. IEEE Signal Processing Magazine, special issue on "Signal Processing on Platforms with Multiple Cores" 26(6), 90–102 (2009)

10. UPC Consortium: UPC language specifications, v1.2, Lawrence Berkeley National Lab. Tech Report LBNL-59208 (2005)

11. Bonelli, A., Franchetti, F., Lorenz, J., Püschel, M., Uberhuber, C.W.: Automatic performance optimization of the discrete fourier transform on distributed memory computers. In: Guo, M., Yang, L.T., Di Martino, B., Zima, H.P., Dongarra, J., Tang, F. (eds.) ISPA 2006. LNCS, vol. 4330, pp. 818–832. Springer, Heidelberg (2006)

12. Van Loan, C.: Computational Framework of the Fast Fourier Transform. SIAM (1992)

13. Chellappa, S., Franchetti, F., Püschel, M.: How to write fast numerical code: A small introduction. In: Lämmel, R., Visser, J., Saraiva, J. (eds.) Generative and Transformational Techniques in Software Engineering II. LNCS, vol. 5235, pp. 196–259. Springer, Heidelberg (2008)

14. Alam, S., Barrett, R., Bast, M., Fahey, M.R., Kuehn, J., McCurdy, C., Rogers, J., Roth, P., Sankaran, R., Vetter, J.S., Worley, P., Yu, W.: Early evaluation of IBM BlueGene/P. In: Proceedings of the 2008 ACM/IEEE Conference on Supercomputing, SC 2008, pp. 23:1–23:12. IEEE Press, Piscataway (2008)

15. Team, T.B.G.: Blue Gene/Q: by co-design. Computer Science - Research and Development, 1–9 (2012)

16. Haring, R., Ohmacht, M., Fox, T., Gschwind, M., Satterfield, D., Sugavanam, K., Coteus, P., Heidelberger, P., Blumrich, M., Wisniewski, R., Gara, A., Chiu, G., Boyle, P., Chist, N., Kim, C.: The ibm blue gene/q compute chip. IEEE Micro 32(2), 48–60 (2012)

17. Franchetti, F., Püschel, M.: Fast Fourier Transform. In: Encyclopedia of Parallel Computing. Springer (2011)

18. Chellappa, S.: Computer Generation of Fourier Transform Libraries for Distributed Memory Architectures. PhD thesis, Electrical and Computer Engineering, Carnegie Mellon University (2010)

19. Galassi, M., Davies, J., Theiler, J., Gough, B., Jungman, G., Alken, P., Booth, M., Rossi, F.: GNU Scientific Library Reference Manual (v1.12), 3rd edn. Network Theory Ltd. (2009)

Matrix Multiplication on Multidimensional Torus Networks

Edgar Solomonik and James Demmel

Division of Computer Science
University of California at Berkeley, CA, USA
{solomon,demmel}@cs.berkeley.edu

Abstract. Blocked matrix multiplication algorithms such as Cannon's algorithm and SUMMA have a 2-dimensional communication structure. We introduce a generalized 'Split-Dimensional' version of Cannon's algorithm (SD-Cannon) with higher-dimensional and bidirectional communication structure. This algorithm is useful for torus interconnects that can achieve more injection bandwidth than single-link bandwidth. On a bidirectional torus network of dimension d, SD-Cannon can lower the algorithmic bandwidth cost by a factor of up to d. With rectangular collectives, SUMMA also achieves the lower bandwidth cost but has a higher latency cost. We use Charm++ virtualization to efficiently map SD-Cannon on unbalanced and odd-dimensional torus network partitions. Our performance study on Blue Gene/P demonstrates that a MPI version of SD-Cannon can exploit multiple communication links and improve performance.

1   Introduction

Torus interconnects can scale to hundreds of thousands of nodes because they achieve good bisection bandwidth while maintaining bounded router degree on each node. Additionally, many scientific simulation and physically structured codes can be mapped to exploit locality on torus networks. In particular, 3-dimensional (3D) tori have been widely deployed in networks (e.g. IBM Blue Gene/L, Blue Gene/P, and the Cray XT series). The newest generation of high-end supercomputer networks is beginning to move to higher dimensionality (e.g. IBM Blue Gene/Q is 5D [6], K computer is 6D [19]). This transition is natural since the minimal-cost (bisection bandwidth with respect to number of pins) topology for a network of 100,000 nodes is 3D, while for 1,000,000 nodes it is 5D or 6D [7]. Higher-dimensional interconnects motivate the design of algorithms that can use such networks efficiently. In this paper, we adapt a classical matrix multiplication algorithm to exploit full injection bandwidth on a torus network of any dimension.

Cannon's algorithm [5] is a parallel algorithm for matrix multiplication ($C = A \cdot B$) on a square ($\sqrt{p}$-by-$\sqrt{p}$) processor grid. After staggering the initial matrix layout, Cannon's algorithm performs $\sqrt{p}$ shifts of A and B along the two dimensions of the processor grid. The algorithm can be done in-place and all

M. Daydé, O. Marques, and K. Nakajima (Eds.): VECPAR 2012, LNCS 7851, pp. 201–215, 2013.

communication is efficiently expressed in the form of near-neighbor data passes. Given n-by-n matrices, each processor must send $O(n^2/\sqrt{p})$ words of data in $O(\sqrt{p})$ messages along each dimension. The number of words and messages sent by each node in Cannon's algorithm is asymptotically optimal [12,3] assuming minimal memory usage. However, since each node sends messages to nearest neighbors in 2 dimensions, at most 2 network links can be saturated per node. However, a d-dimensional bidirectional torus network has $2d$ outgoing links per node that can be utilized.

It is known that a different algorithm, SUMMA [1,17], can utilize all $2d$ links and send a minimal number of words. For matrix multiplication of n-by-n matrices, SUMMA sends $O(n^2/\sqrt{p})$ data in the form of n outer-products, which can be pipelined or blocked. Each update requires a broadcast along a row or column of processors. If a higher-dimensional torus is flattened into each row and column of the mapping, rectangular collective algorithms [18,9,15] can utilize all dimensions of the network. Rectangular algorithms subdivide and pipeline the messages into edge-disjoint spanning trees formed by traversing the network in different dimensional orders. However, SUMMA typically sends more messages since it does $O(\sqrt{p})$ broadcasts, rather than the $O(\sqrt{p})$ near-neighbor sends in Cannon's algorithm.

Cannon's algorithm does not employ communication collectives, so it cannot utilize rectangular collectives. We design a generalization of Cannon's algorithm, Split-Dimensional Cannon's algorithm (SD-Cannon), that explicitly sends data in all dimensions of the network at once. This algorithm does not need topology-aware collectives and retains all the positive features of the classical Cannon's algorithm. However, like Cannon's algorithm, SD-Cannon is difficult to generalize to non-square processor grids. We get around this challenge by using a virtualization framework, Charm++ [14]. Our performance results on Blue Gene/P (Intrepid, located at Argonne National Lab) demonstrate that SD-Cannon outperforms Cannon's algorithm and can match the performance of SUMMA with rectangular collectives. The virtualized version of Cannon's algorithm does not incur a high overhead but our Charm++ implementation is unable to saturate all networks links at once.

The rest of the paper is structured as follows, Section 2 introduces the SD-Cannon algorithm, Section 3 gives a cost analysis of the SD-Cannon algorithm, Section 4 studies the performance of MPI and Charm++ implementations of SD-Cannon, and Section 5 concludes.

2   Previous Work

Matrix multiplication computes $C = A \cdot B$ where A is m-by-k, B is k-by-n, and C is m-by-n. Our algorithms target the case where $m \approx n \approx k$. Other optimizations or different algorithms (e.g. 1D blocking) may be worth considering when the matrices are rectangular but are out of the scope of this paper. The algorithms are designed for a l-ary d-cube bidirectional torus network (a d dimensional network of $p = l^d$ processors). The algorithms require that the torus network is of even dimension ($d = 2i$ for $i \in \{1, 2, ...\}$). Virtualization will be used to extend our approach to odd-dimensional networks and rectangular processor grids.

2.1 Matrix Layout

A matrix is a 2D array of data. To spread this data in a load balanced fashion, we must embed the 2D array in the higher-dimensional torus network. We do not consider algorithms that replicate data (e.g. 3D matrix multiplication [8,1,2,4,13]. However, the extension is natural, given the replication approach presented in [16].

Any l-ary d-cube Π^{dD}, where d is a multiple of 2, can be embedded onto a square 2D grid. Each of two dimensions in this square grid is of length $l^{d/2}$ and is formed by folding a different $d/2$ of the d dimensions. For simplicity we fold the odd $d/2$ dimensions into one of the square grid dimensions and the $d/2$ even dimensions into the other square grid dimension. The algorithms below will assume the initial matrix layout follows this ordering. In actuality, the ordering is irrelevant since a l-ary d-cube network is invariant to dimensional permutations. We define a dimensional embedding for a processor with a d-dimensional index $I^d \in \{0, 1, \ldots, l-1\}^d$, to a two dimensional index (i, j) as

$$G^{dD \to 2D}[I^d] = \left(\sum_{i=0}^{d/2-1} l^i I^d[2i], \sum_{i=0}^{d/2-1} l^i I^d[2i+1] \right).$$

We denote the processor with grid index I^d as $\Pi^{dD}[I^d]$.

2.2 SUMMA Algorithm

Algorithm 1. $[C] = \text{SUMMA}(A, B, C, n, m, k, \Pi^{2D})$

Input: $m \times k$ matrix A, $k \times n$ matrix B distributed so that $\Pi^{2D}[i,j]$ owns $\frac{m}{\sqrt{p}} \times \frac{k}{\sqrt{p}}$ sub-matrix $A[i,j]$ and $\frac{k}{\sqrt{p}} \times \frac{n}{\sqrt{p}}$ sub-matrix $B[i,j]$, for each $i, j \in [0, \sqrt{p} - 1]$

Output: square $m \times n$ matrix $C = A \cdot B$ distributed so that $\Pi^{2D}[i,j]$ owns $\frac{m}{\sqrt{p}} \times \frac{n}{\sqrt{p}}$ block sub-matrix $C[i,j]$, for each $i, j \in [0, \sqrt{p} - 1]$

//In parallel with all processors
for all $i, j \in [0, \sqrt{p} - 1]$ **do**
    for $t = 1$ *to* $t = \sqrt{p}$ **do**
        Multicast $A[i,t]$ along rows of Π^{2D}
        Multicast $B[t,j]$ along columns of Π^{2D}
        $C[i,j] := C[i,j] + A[i,t] \cdot B[t,j]$
    end
end

The SUMMA algorithm [1,17] (Algorithm 1), utilizes row and column multicasts to performs parallel matrix multiplication. The algorithm is formulated on

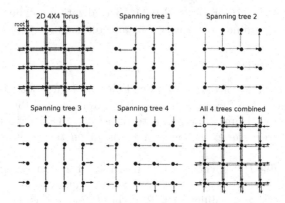

(a) 2D rectangular multicast spanning trees

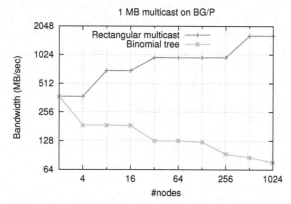

(b) Performance of rectangular multicasts on BG/P

Fig. 1. Rectangular collectives

a 2D grid, with each process owning a block of the matrices A, B, and C. At each step, the algorithm performs an outer product of parts of A and B. While SUMMA can be done with k rank-one outer products, latency can be reduced by performing $\sqrt{p}$ rank-$(k/\sqrt{p})$ outer-products. The latter case yields an algorithm where at each step every process in a given column of the processor grid multicasts its block of A to all processors in its row. Similarly, a row of processors multicasts B along columns.

The SUMMA algorithm performs all communication via multicasts [1]. Therefore, the communication-performance of the algorithm is dictated by the performance of a multicast on the architecture. The most efficient way to perform multicasts on a torus network architecture is to employ rectangular collectives [18,9,15]. Rectangular collectives use edge-disjoint spanning trees of

---

[1] SUMMA can be modified to communicate C rather than A or B via reductions, which is useful if C is the smaller than A or B. A and B can also be communicated via all-gathers among rows and columns rather than multicasts.

the entire processor grid. On a bidirectional torus network of dimension l, $2l$ such trees can be constructed (see Figure 1(a) for the $l = 2$ case). The message is then subdivided and pipelined in chunks to each of the spanning trees, which allows the saturation of all torus links simultaneously.

The BG/P architecture has a low level implementation of rectangular collectives. The performance of these optimized topology-aware collectives far exceeds that of regular collectives, e.g. binomial multicast tree. Further, Figure 1(b) demonstrates that the rectangular collectives actually become faster on more processors, the opposite of the behavior of most communication primitives. This feature is due to the fact that the BG/P partitions become higher-dimensional as they grow, so there are more links to exploit, which allow more rectangular trees. Thus, by employing these collectives the SUMMA algorithm can exploit all links of a torus network.

2.3   Cannon's Algorithm

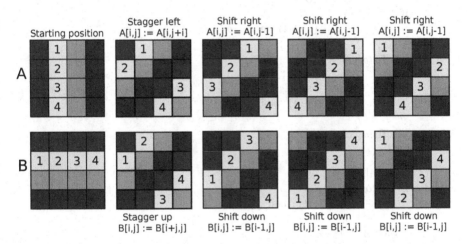

Fig. 2. Cannon's algorithm, stagger and shift. A and B blocks of the same color must be multiplied together. Notice that the colors (blocks that need to be multiplied) align after each shift.

Cannon's algorithm is a parallel matrix multiplication algorithm that uses shifts blocks among columns and rows of a processor grid. The algorithm starts by staggering the blocks of A and B leftwards and upwards, respectively. Then the A and B blocks are shifted rightwards and downwards, respectively. We describe Cannon's algorithm on a $\sqrt{p}$-by-$\sqrt{p}$ grid (Π^{2D}) (Algorithm 2). The procedure is demonstrated in Figure 2, where each color corresponds to an outer product. One of the outer products (the yellow blocks) is numbered, and we see that after each shift, different blocks are multiplied, and overall all sixteen distinct block multiplies are performed for that outer product (this also holds for the other 3 outer products).

Once we embed the dD grid onto a 2D grid, we can run Cannon's algorithm with the matrix distribution according to the ordered 2D processor grid. However, in this embedded network, Cannon's algorithm will only utilize $1/d$ of the links, since two messages are sent at a time by each processor and there are $2d$ links per node.

Algorithm 2. $[C] = \textbf{Cannon}(A,\ B,\ C,\ n,\ m,\ k,\ p,\ \Pi^{2D})$

Input: $m \times k$ matrix A, $k \times n$ matrix B distributed so that $\Pi^{2D}[i,j]$ owns $\frac{m}{\sqrt{p}} \times \frac{k}{\sqrt{p}}$ sub-matrix $A[i,j]$ and $\frac{k}{\sqrt{p}} \times \frac{n}{\sqrt{p}}$ sub-matrix $B[i,j]$, for each $i,j \in [0, \sqrt{p} - 1]$

Output: square $m \times n$ matrix $C = A \cdot B$ distributed so that $\Pi^{2D}[i,j]$ owns $\frac{m}{\sqrt{p}} \times \frac{n}{\sqrt{p}}$ block sub-matrix $C[i,j]$, for each $i,j \in [0, \sqrt{p} - 1]$

//In parallel with all processors
for all $i,j \in [0, \sqrt{p} - 1]$ **do**
   for $t = 1$ *to* $\sqrt{p} - 1$ **do**
      if $t \le i$ **then**
         $A[i,j] \leftarrow A[i,((j+1) \mod \sqrt{p})]$
         /** */
         [f]*stagger A*
      end
      if $t \le j$ **then**
         $B[i,j] \leftarrow B[((i+1) \mod \sqrt{p}), j]$
         /** */
         [f]*stagger B*
      end
   end
   $C[i,j] := A[i,j] \cdot B[i,j]$
   for $t = 1$ *to* $\sqrt{p} - 1$ **do**
      $A[i,j] \leftarrow A[i,((j-1) \mod \sqrt{p})]$
      /** */
      [f]*shift A rightwards*
      $B[i,j] \leftarrow B[((i-1) \mod \sqrt{p}), j]$
      /** */
      [f]*shift B downwards*
      $C[i,j] := C[i,j] + A[i,j] \cdot B[i,j]$
   end
end

---

3 Split-Dimensional Cannon's Algorithm

We can formulate another version of Cannon's algorithm by using more dimensional shifts. A shift can be performed with a single message sent over a single link from each processor to the next. Since the shifts will be done along dimensions of the l-ary d-cube network, $2d$ links will be available. Algorithm 3 performs this dimensional shift. Split-dimensional (SD) Cannon's algorithm will use exclusively this shift for communication. In fact, all shifts operate on a static message size. Therefore, communication cost can be calculated by counting shifts. The algorithm achieves complete utilization on any l-ary d-cube network during the shift stage. We specify the algorithm for a bidirectional network as those are much more common. However, the algorithms can be trivially simplified to the unidirectional case.

Algorithm 3. Shift$<$ dim, dir $>$(l, M, p, Π^{dD}, I^d)

Input: $\Pi^{dD}[I^d]$ owns sub-matrix M.

$S^d \leftarrow I^d$

if $dir = +1$ then

$\quad S^d[\text{dim}] = (S^d[\text{dim}] + 1) \mod l$

end

if $dir = -1$ then

$\quad S^d[\text{dim}] = (S^d[\text{dim}] - 1) \mod l$

end

Send M to $\Pi^{dD}[S^d]$.

/** */

$[f]\Pi^{dD}[I^d]$ sends to $\Pi^{dD}[S^d]$

SD-Cannon on a 3-ary 6-cube

Each color corresponds to an outer product

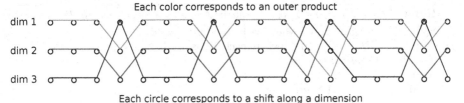

Each circle corresponds to a shift along a dimension

Fig. 3. Snapshot of network usage in SD-Cannon. A and B use a disjoint set of 3 network dimensions in the same fashion, so it suffices to pay attention to 3.

Algorithm 4 describes how the stagger step is done inside the SD-Cannon algorithm. A different shift is done on each sub-panel of A and B concurrently. These calls should be done asynchronously and ideally can fully overlap. One interpretation of this stagger algorithm is that sub-panels of each matrix are being staggered recursively along $d/2$ disjoint dimensional orders.

Algorithm 5 is a recursive routine that loops over each dimension performing shifts on sub-panels of A and B. The order of the shifts is permuted for each sub-panel. Each sub-panel is multiplied via a recursive application of Cannon's algorithm over a given dimensional ordering.

Algorithm 4. SD-Stagger(A, B, n_b, m_b, k_b, l, p, Π^{dD}, I^d)

Input: $\Pi^{dD}[I^d]$ owns $m_b \times k_b$ sub-matrix A and $k_b \times n_b$ sub-matrix B.
Split $A = [A_0, A_1, \ldots, A_d]$ where each A_h is $m_b \times k_b/d$.
Split $B^T = [B_0^T, B_1^T, \ldots, B_d^T]$ where each B_h is $k_b/d \times n_b$.
//At each level, apply index shift
for $level \in [0, d/2 - 1]$ **do**
    //To stagger must shift up to $l - 1$ times
    for $t = 1$ **to** $l - 1$ **do**
        for all $dh \in [0, d/2 - 1]$ **do**
            $h \leftarrow (dh + \text{level}) \mod (d/2)$ //Shift the ordering
            if $t \leq I^d[2h + 1]$ **then**
                Shift$< (2 * h), +1 >(l, A_h, p, \Pi^{dD}, I^d)$
            end
            if $t \leq l - I^d[2h + 1]$ **then**
                Shift$< (2 * h), -1 >(l, A_{h+d/2}, p, \Pi^{dD}, I^d)$
            end
            if $t \leq I^d[2h]$ **then**
                Shift$< (2 * h + 1), +1 >(l, B_h, p, \Pi^{dD}, I^d)$
            end
            if $t \leq l - I^d[2h]$ **then**
                Shift$< (2 * h + 1), -1 >(l, B_{h+d/2}, p, \Pi^{dD}, I^d)$
            end
        end
    end
end

Algorithm 5. SD-Contract< level >(A, B, C, n_b, m_b, k_b, l, p, Π^{dD}, I^d)

Input: $\Pi^{dD}[I^d]$ owns $m_b \times k_b$ sub-matrix A and $k_b \times n_b$ sub-matrix B.
Split $A = [A_0, A_1, \ldots, A_d]$ where each A_h is $m_b \times k_b/d$.
Split $B^T = [B_0^T, B_1^T, \ldots, B_d^T]$ where each B_h is $k_b/d \times n_b$.
//Shift and contract l times
for $t = 0$ **to** $l - 1$ **do**
    if $level = d/2 - 1$ **then**
        $C \leftarrow C + A \cdot B$
    else
        SD-Contract< level $+ 1$ >(A, B, C, $\frac{n}{\sqrt{p}}$, $\frac{m}{\sqrt{p}}$, $\frac{k}{\sqrt{p}}$, l, p, Π^{dD},
        I^d)
    end
    for all $dh \in [0, d/2 - 1]$ **do**
        $h \leftarrow (dh + \text{level}) \mod (d/2)$
        /** */
        [f]*Shift the ordering*
        Shift$< (2 * h), +1 >(l, A_h, p, \Pi^{dD}, I^d)$
        Shift$< (2 * h + 1), +1 >(l, B_h, p, \Pi^{dD}, I^d)$
        Shift$< (2 * h), -1 >(l, A_{h+d/2}, p, \Pi^{dD}, I^d)$
        Shift$< (2 * h + 1), -1 >(l, B_{h+d/2}, p, \Pi^{dD}, I^d)$
    end
end

Algorithm 6. SD-Cannon(A, B, C, n, m, k, l, p, Π^{dD}, $G^{dD \to 2D}$)

Input: $m \times k$ matrix A, $k \times n$ matrix B distributed so that $\Pi^{dD}[I]$ owns $\frac{m}{\sqrt{p}} \times \frac{k}{\sqrt{p}}$

sub-matrix $A[G^{dD \to 2D}[I]]$ and $\frac{k}{\sqrt{p}} \times \frac{n}{\sqrt{p}}$ sub-matrix $B[G^{dD \to 2D}[I]]$

Output: square $m \times n$ matrix $C = A \cdot B$ distributed so that $\Pi^{dD}[I]$ owns
$\frac{m}{\sqrt{p}} \times \frac{n}{\sqrt{p}}$ block sub-matrix $C[G^{dD \to 2D}[I]]$

//In parallel with all processors
for all $I^d \in \{0, 1, \ldots, l-1\}^d$ **do**
  $(i, j) \leftarrow G^{dD \to 2D}[I^d]$
  SD-Stagger($A[i,j]$, $B[i,j]$, $\frac{n}{\sqrt{p}}$, $\frac{m}{\sqrt{p}}$, $\frac{k}{\sqrt{p}}$, l, p, Π^{dD}, I^d)
  SD-Contract$< 0 >$($A[i,j]$, $B[i,j]$, $C[i,j]$, $\frac{n}{\sqrt{p}}$, $\frac{m}{\sqrt{p}}$, $\frac{k}{\sqrt{p}}$, l, p, Π^{dD},
  I^d)
end

Figure 3 demonstrates how different network dimensions of a 3-ary 6-cube are used by SD-Cannon. A and B get shifted along 3 of 6 dimensions, so Figure 3 records usage along 3 dimensions (corresponding to one of A or B). Each outer product (corresponding to a color) is shifted (each shift corresponds to a circle) along a different dimension at any given step.

Note that the local multiplication call is the same as in Cannon's algorithm. The granularity of the sequential work does not decrease in the SD-Cannon algorithm but only changes its ordering. This is a virtue of splitting into outer-products that accumulate to the same buffer.

The control flow of SD-Cannon is described in Algorithm 6. The algorithm can be elegantly expressed with one-sided communication since the sends should be asynchronous (puts). Our MPI SD-Cannon code uses one-sided put operations and is compact (a few hundred lines of C).

4 Analysis

We analyze the communication costs of Cannon's algorithm, SUMMA, and SD-Cannon. We consider bandwidth cost, as the total volume of data sent by each process, and latency cost, as the number of messages sent by each process. As before, we assume a l-ary d-cube bidirectional torus network.

4.1 Bandwidth Cost

We can analyze the bandwidth cost of these algorithms by the embedding of the algorithm onto the physical l-ary d-cube network. The bandwidth cost of the algorithm is proportional to the number of shifts along the critical path.

In traditional Cannon's algorithm we shift $2\sqrt{p}$ blocks along the critical path ($\sqrt{p}$ times for stagger and $\sqrt{p}$ times for contraction) of size mk/p and nk/p. Given the ordering of the embedding, we can always find a link which has to communicate $mk/\sqrt{p}$ values and a link that has to communicate $nk/\sqrt{p}$ values. Therefore the bandwidth cost is

$$W_{\text{Cannon}} = O\left(\frac{\max(m,n)\cdot k}{\sqrt{p}}\right).$$

In the bidirectional, split-dimensional algorithm, all shifts in the communication inner loops (in Algorithm 4 and Algorithm 5) can be done simultaneously (so long as the network router can achieve full injection bandwidth). So the communication cost is simply proportional to the number of inner loops, which, for staggering (Algorithm 4) is $N_T = l \cdot d/2$. For the recursive contraction step (Algorithm 5), the number of these shift stages is

$$N_S = \sum_{i=1}^{d/2} l^i \le 2\sqrt{p}.$$

If the network is bidirectional, at each shift stage we send asynchronous messages of sizes $mk/(d \cdot p)$ and $kn/(d \cdot p)$ values. Ignoring the lower-order stagger term in SD-Cannon we have a cost of

$$W_{\text{SD-Cannon}} = O\left(\frac{\max(m,n)\cdot k}{d \cdot \sqrt{p}}\right).$$

So the bandwidth cost of SD-Cannon, $W_{\text{SD-Cannon}}$, is d times lower than that of Cannon's algorithm, W_{Cannon}. In SUMMA, throughout the algorithm A of size mk and B of size kn are multicast along two different directions. An optimal multicast algorithm would utilize d links for the multicasts of A and B respectively. So, the bandwidth cost of SUMMA is

$$W_{\text{SUMMA}} = O\left(\frac{\max(m,n)\cdot k}{d \cdot \sqrt{p}}\right),$$

which is asymptotically the same as the bandwidth cost of SD-Cannon.

4.2   Latency Cost

The latency overhead incurred by these algorithms will differ depending on the topology and collective algorithms for SUMMA. The SD-Cannon algorithm sends more messages than Cannon's algorithm, but into different links, so it incurs more sequential and DMA overhead, but no extra network latency overhead. However, both Cannon's algorithm and SD-Cannon will have a lower latency cost than SUMMA on a typical network. In each step of SUMMA, multicasts are done along each dimension of the processor grid. So, on a torus network, a message must travel $l \cdot d/2$ hops at each step, rather than 1 hop as in SD-Cannon. The Blue Gene/P machine provides efficient multicast collectives that work at a fine granularity and incur little latency overhead [9]. However, on a machine without this type of topology-aware collectives, SD-Cannon would have a strong advantage, as messages would need to travel fewer hops.

If we count latency as the number of hops a message must travel on the network and assume processes can send multiple messages at once, we can derive definite latency costs. For Cannon's algorithm, which does $O(\sqrt{p})$ near neighbor sends, the latency cost is unambiguously

$$S_{\text{Cannon}} = O\left(\sqrt{p}\right).$$

For SD-Cannon, if we assume messages can be sent simultaneously into each dimension, the latency cost is

$$S_{\text{SD-Cannon}} = O\left(\sqrt{p}\right).$$

However, the on-node messaging overhead goes up by a factor of $O(d)$. For SUMMA, there are again $\sqrt{p}$ steps, but at each step a multicast happens among $\sqrt{p}$ processes. On a torus network, the most distant processor would be $l \cdot d/2$ hops away, giving a hop-messaging cost of

$$S_{\text{SUMMA}} = O\left(l \cdot d \cdot \sqrt{p}\right).$$

This latency overhead is higher than Cannon and SD-Cannon, though this cost reflects the number of hops travelled not the number of synchronizations. However, generally it is reasonable to state that a multicast incurs a larger latency overhead than near-neighbor sends, so our qualitative conclusion is valid.

5   Results

We implemented version of SD-Cannon in MPI [10] and Charm++ [14]. Both versions work on matrices of any shape and size, but we only benchmark square matrices. Both versions assume the virtual decomposition is a k-ary n-cube. In MPI, the process grid is a k-ary n-cube, while in Charm++ we get a k-ary n-cube of chare objects. We use Charm++ to explicitly map the objects onto any unbalanced process grid we define at run time. While we explicitly define the mapping function to fold the chare array onto a smaller processor grid, the Charm++ run-time system manages how the sequential work and messaging get scheduled.

The MPI version uses MPI put operations for communication and barriers for synchronization. The Charm++ version uses the underlying run-time system for messaging between chares, and is dynamically scheduled (no explicit synchronization).

We benchmarked our implementations on a Blue Gene/P (BG/P) [11] machine located at Argonne National Laboratory (Intrepid). We chose BG/P as our target platform because it uses few cores per node (four 850 MHz PowerPC processors) and relies heavily on its interconnect (a bidirectional 3D torus with 375 MB/sec of achievable bandwidth per link).

Since the BG/P network only has three dimensions, the benefit of SD-Cannon is limited to trying to exploit the backwards links and the third dimensional links.

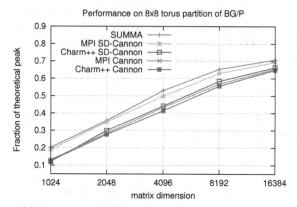

(a) Matrix multiplication on a 2D torus partition

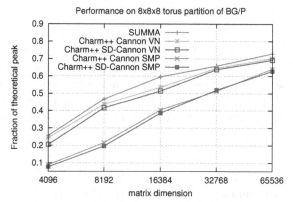

(b) Matrix multiplication on a 3D torus partition

Fig. 4. Performance of SD-Cannon on BG/P

The MPI version of our code is limited to even-dimensional tori, so it could only exploit 4 of 6 links. We study relative performance of the MPI version of SD-Cannon on an 8-by-8 torus partition of BG/P. [2]

Figure 4(a) details the performance of MPI versions of SUMMA, Cannon's algorithm, and SD-Cannon on an 8x8 node 2D torus partition. SD-Cannon improves upon Cannon's algorithm as it can utilize the backwards as well as the forwards links simultaneously. The one-dimensional rectangular multicasts used by SUMMA achieve the same effect. We see that the performance of SD-Cannon is higher than Cannon's algorithm (up to 1.5x) and slightly worse than SUMMA. The performance difference between SUMMA and SD-Cannon is due to the extra cost of the initial stagger, which SUMMA does not need. The Charm++ versions

---

[2] The partitions allocated by the BG/P scheduler are only toroidal if they have 512 or more nodes. So, we allocated a 512 node partition and worked on the bottom 64 node slice.

were executed with 1 chare per node. In this case, we see that Charm++ has a small overhead and we see a small benefit from bidirectionality.

Figure 4(b) shows the performance on a 3D 512 node partition. Since this partition is odd-dimensional, we cannot efficiently map the MPI version of Cannon or SD-Cannon. We execute the Charm++ codes in two modes, one with 1 process per node and 8 chares per process (SMP), and one with 4 processes per node and 2 chares per process (VN). Using multiple processes per node improves the performance of the Charm++ codes, because it is more efficient to perform a separate multiplication on each core, rather than execute each multiplication in sequence across all cores. While the VN-mode version performs almost as well as SUMMA, neither version benefits from multidimensional shifts. Our Charm++ implementations use two-sided communication, while the MPI version uses one-sided. It is likely that a Charm++ implementation with one-sided sends would successfully exploit all of the links.

6   Conclusion

SD-Cannon is an improvement on top of Cannon's algorithm. While Cannon's algorithm has some nice properties, SUMMA has seen more wide-spread adoption. In this paper, we demonstrate how SD-Cannon can get closer to the performance of the SUMMA algorithm, and how virtualization can be used to map SD-Cannon and Cannon's algorithm efficiently. On the Blue Gene hardware it still does not make sense to use SD-Cannon over SUMMA, but SD-Cannon has advantages that could prove to be faster than SUMMA on other hardware. In particular, on networks without optimized collectives or with higher latency cost, the near-neighbor sends performed by Cannon's algorithm and SD-Cannon are preferable to SUMMA's multicasts.

More generally, our work demonstrates how algorithmic design can couple with topology-aware mapping and virtualization. These techniques are already important on modern supercomputers with 3D interconnects as demonstrated by our performance results. As the scale and dimensionality of high performance networks grow, topology-aware mapping and communication-avoidance are becoming pivotal to scalable algorithm design.

Acknowledgements. The first author was supported by a Krell Department of Energy Computational Science Graduate Fellowship, grant number DE-FG02-97ER25308. Research was also supported by Microsoft (Award #024263) and Intel (Award #024894) funding and by matching funding by U.C. Discovery (Award #DIG07-10227). This material is supported by U.S. Department of Energy grants numbered DE-SC0003959, DE-SC0004938, and DE-FC02-06-ER25786. This research used resources of the Argonne Leadership Computing Facility at Argonne National Laboratory, which is supported by the Office of Science of the U.S. Department of Energy under contract DE-AC02-06CH11357.

References

1. Agarwal, R.C., Balle, S.M., Gustavson, F.G., Joshi, M., Palkar, P.: A three-dimensional approach to parallel matrix multiplication. IBM J. Res. Dev. 39, 575–582 (1995)
2. Aggarwal, A., Chandra, A.K., Snir, M.: Communication complexity of PRAMs. Theoretical Computer Science 71(1), 3–28 (1990)
3. Ballard, G., Demmel, J., Holtz, O., Schwartz, O.: Minimizing communication in linear algebra. SIAM J. Mat. Anal. Appl. 32(3) (2011)
4. Berntsen, J.: Communication efficient matrix multiplication on hypercubes. Parallel Computing 12(3), 335–342 (1989)
5. Cannon, L.E.: A cellular computer to implement the Kalman filter algorithm. Ph.D. thesis, Bozeman, MT, USA (1969)
6. Chen, D., Eisley, N.A., Heidelberger, P., Senger, R.M., Sugawara, Y., Kumar, S., Salapura, V., Satterfield, D.L., Steinmacher-Burow, B., Parker, J.J.: The IBM Blue Gene/Q interconnection network and message unit. In: Proceedings of 2011 International Conference for High Performance Computing, Networking, Storage and Analysis, SC 2011, pp. 1–2. ACM, New York (2011)
7. Dally, W.: Performance analysis of k-ary n-cube interconnection networks. IEEE Transactions on Computers 39(6), 775–785 (1990)
8. Dekel, E., Nassimi, D., Sahni, S.: Parallel matrix and graph algorithms. SIAM Journal on Computing 10(4), 657–675 (1981)
9. Faraj, A., Kumar, S., Smith, B., Mamidala, A., Gunnels, J.: MPI collective communications on the Blue Gene/P supercomputer: Algorithms and optimizations. In: 17th IEEE Symposium on High Performance Interconnects, HOTI 2009 (2009)
10. Gropp, W., Lusk, E., Skjellum, A.: Using MPI: portable parallel programming with the message-passing interface. MIT Press, Cambridge (1994)
11. IBM Journal of Research and Development staff: Overview of the IBM Blue Gene/P project. IBM J. Res. Dev. 52, 199–220 (2008)
12. Irony, D., Toledo, S., Tiskin, A.: Communication lower bounds for distributed-memory matrix multiplication. Journal of Parallel and Distributed Computing 64(9), 1017–1026 (2004)
13. Johnsson, S.L.: Minimizing the communication time for matrix multiplication on multiprocessors. Parallel Comput. 19, 1235–1257 (1993)
14. Kale, L.V., Krishnan, S.: CHARM++: a portable concurrent object oriented system based on C++. In: Proceedings of the Eighth Annual Conference on Object-Oriented Programming Systems, Languages, and Applications, OOPSLA 1993, pp. 91–108. ACM, New York (1993)
15. Solomonik, E., Bhatele, A., Demmel, J.: Improving communication performance in dense linear algebra via topology aware collectives. In: Supercomputing, Seattle, WA, USA (November 2011)
16. Solomonik, E., Demmel, J.: Communication-optimal parallel 2.5D matrix multiplication and LU factorization algorithms. In: Jeannot, E., Namyst, R., Roman, J. (eds.) Euro-Par 2011, Part II. LNCS, vol. 6853, pp. 90–109. Springer, Heidelberg (2011)

17. Van De Geijn, R.A., Watts, J.: SUMMA: scalable universal matrix multiplication algorithm. Concurrency: Practice and Experience 9(4), 255–274 (1997)
18. Watts, J., Van De Geijn, R.A.: A pipelined broadcast for multidimensional meshes. Parallel Processing Letters 5, 281–292 (1995)
19. Yokokawa, M., Shoji, F., Uno, A., Kurokawa, M., Watanabe, T.: The k computer: Japanese next-generation supercomputer development project. In: International Symposium on Low Power Electronics and Design, ISLPED 2011, pp. 371–372 (August 2011)

High Performance CPU Kernels
for Multiphase Compressible Flows

Babak Hejazialhosseini, Christian Conti, Diego Rossinelli,
and Petros Koumoutsakos

Computational Science and Engineering Lab, ETH Zurich, Zurich, Switzerland
{hbabak,cconti,diegor,petros}@mavt.ethz.ch

Abstract. We develop efficient CPU kernels for multiphase compressible flows and evaluate different optimization strategies. The presented software achieves up to 48% of the peak performance on shared memory architectures, outperforming by 9-14X what is considered to be state-of-the-art. On 48-core CPUs we observe speedups of 40-45X and measure up to 360 GFLOP/s over 840 GFLOP/s of the peak.

1 Introduction

Simulations of multiphase compressible flows are a critical testbed for high performance computing as they face performance issues that are often encountered in other branches of computational science. Such simulations are essential for studies of shock wave lithotripsy and combustion, problems that do not easily render themselves to experimental studies. Numerical investigations have therefore become an established approach in studying such complex flows [1]. Among the most modern techniques, we find adaptive mesh refinement (AMR) [2] and wavelet-based adaptive grids [3], which concentrate the computation on regions of interest. Although these techniques provide substantial algorithmic improvements, their performance impact in terms of GFLOP/s and hardware utilization (i.e. fraction of the peak performance) has not been extensively reported except for a few cases limited to heterogenous multicore/GPUs platforms [4].

In contrast to synthetic benchmarks, for real world applications harnessing the full potential of current state-of-the-art multicores has been shown to be hardly possible as they reach only about 1% of the peak [5]. A primary cause of this issue is the imbalance between the peak performance and the system memory bandwidth (1-10 TFLOP/s versus 0.1-1 TB/s) meaning that applications should employ kernels exhibiting ratios of 10 FLOP/B (bytes of off-chip memory traffic) or more to achieve peak performance. As discussed in this work, such high ratios in the context of multiphase compressible flows are not always realistic and, in the few cases they are, reaching them requires revisiting both algorithms and memory layouts.

Optimization techniques have been proposed to address this widespread challenge ranging from software autotuning tools [6] to automated code generation [7] and hardware/software co-design for domain specific problems [8]. Other successful optimization techniques [9] are based on the roofline model [10].

M. Daydé, O. Marques, and K. Nakajima (Eds.): VECPAR 2012, LNCS 7851, pp. 216–225, 2013.
© Springer-Verlag Berlin Heidelberg 2013

A number of open issues can be identified in the context of fast simulations for compressible flows. Firstly, there is poor analysis on the performance expectation, which helps contextualizing the measurements. Secondly, it is not clear how the performance of AMR solvers would compare to a highly optimized uniform resolution solver. Thirdly, to the best of our knowledge, for simulations of compressible flows there are no standard packages to assist the software development (such as "High-Performance Linpack" (HPL) [11]) whose performance can reach a significant fraction of the peak: reported per-core performance of the most compute intensive kernel (6000 FLOP/grid point) of the fastest existing software for compressible flows indicate results in the range of 2-3% of the peak per-core performance [12,13].

In this work, we evaluate a set of optimization techniques for the simulation of multiphase compressible flows that is tailored to state-of-the-art multicore platforms and discuss the measured performance. Furthermore, we provide an a-priori analysis on the expected performance. The paper is organized as follows. In Section 2 we describe the governing equations and the considered numerical schemes. In Section 3 we identify the performance bottlenecks in the solver and we discuss the data structures and techniques adopted in the software used to mitigate these barriers. In Section 4 we analyze and discuss the resulting performance for the simulations of the shock-bubble interaction.

2   Governing Equations and Numerical Methods

We model an inviscid multiphase compressible flow described by the Euler equations using the one-fluid formulation [14]:

$$\frac{\partial \rho}{\partial t} = -\nabla \cdot (\rho \mathbf{u}), \tag{1}$$

$$\frac{\partial \rho \mathbf{u}}{\partial t} = -\nabla \cdot (\rho \mathbf{u}\mathbf{u}^T - p\mathbb{I}), \tag{2}$$

$$\frac{\partial E}{\partial t} = -\nabla \cdot ((E + p)\mathbf{u}), \tag{3}$$

$$\frac{\partial \phi}{\partial t} = -\mathbf{u} \cdot \nabla \phi, \tag{4}$$

with ρ being the density, $\mathbf{u}$ the velocity vector, p the pressure, E the total energy of the fluid and ϕ the interface marker function. To close the system of equations, we assume that the fluid follows the ideal gas equation of state,

$$p = (\gamma - 1)(E - \frac{1}{2}\rho|\mathbf{u}|^2), \tag{5}$$

with γ being the ratio of specific heats of each phase.

We use the finite volume discretization and solve the integral form of Equations (1-4) by reconstructing flow quantities and computing the numerical flux on the finite volume cell interfaces. The computation and summation of fluxes are referred to as the computation of the right hand side (RHS) and are coupled to the low-storage third-order TVD Runge-Kutta time stepping scheme.

One simulation step consists of the following stages: estimation of the time step, conversion of conserved to primitive quantities, computation of the RHS and update of the conserved quantities. The RHS computation is itself composed of a series of substeps consisting in 5th order WENO reconstructions [15], evaluations of the HLLE numerical fluxes [16] and summation of the fluxes.

3   Design and Techniques

Previous works indicate that the computation of the RHS is the most expensive part of a simulation step [4]. Furthermore, its *Operational Intensity* (OI), i.e. the ratio of FLOP to bytes of off-chip memory transfers, is low and therefore it implies that in order to observe decent performance, we need to increase data locality. A low OI means that the OI is below the ratio of peak performance to peak memory bandwidth and indicates that the kernel performance is bound by the memory bandwidth of the system that runs it.

Our compressible flow simulations rely on a "software stack" with two software layers, namely core and node layers. This modular structure separates the optimization "domains" as it is directly related to the underlying hardware and increase flexibility and code reusability. These separations in turn facilitate the software development for new applications, techniques and hardware. Optimizations that affect the operational intensity are in general applied to the core layer whereas optimizations related to thread level parallelism and communication are covered in the node layer.

Data Structures. The governing equations considered in this work involve 6 unknowns, ρ, u, v, w, E and ϕ, which are organized into an Array of Structures (AoS) format referred to as *grid point*, to maintain maximum software flexibility. To increase the data locality we introduce an intermediate data structure, called hereafter *block*, which contains 16-32 grid points per dimension. In order to further enforce data locality, Morton space-filling is used to index the blocks. A secondary advantage of introducing blocks is that any technique developed here can be employed in block-based AMR solvers as well. The main downside comes from the extra overhead of replicating some grid points (ghosts) required to process the blocks.

Core Layer. The core layer maps to the processor cores and is thus responsible for Data Level Parallelism (DLP) and Instruction Level Parallelism (ILP).

As we expect the performance of the kernels to be memory-bound, all data structures are properly 16- or 32-byte aligned. Our design is also oriented towards data and computation reordering: temporal locality is improved by processing blocks slice-by-slice and the memory footprint is reduced by computing on ring-like buffers composed of a few slices, analogously to [17].

Table 1. Arithmetic and operational intensities

Kernel	AI [FLOP/B]	OI [FLOP/B]
dt	1.7	5.1
RHS	1.5	45.4
Update	0.2	0.2

Intermediate data structures for RHS computation are represented in Structures of Arrays (SoA) format since it is required by some kernels to benefit from vectorization, which in our solver is implemented with SSE and AVX instructions. The performance is further increased by the use of division and square root operations which are less accurate (1.5/2.5 ulps) [18] for the computation of the WENO values.

In order to increase the *Arithmetic Intensities* (AI), i.e. ratios of FLOP over total amount of bytes transferred, the AoS/SoA conversion is "fused" with the conversion of flow quantities from conserved to primitive (Figure 1). Similarly, copying back the RHS from SoA to AoS is fused with the first stage of the low-storage Runge-Kutta time stepper. We further replace the conditional branches with conditional moves in the HLLE fluxes.

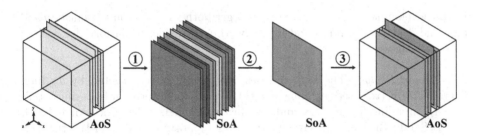

Fig. 1. Slice-wise processing of one block with ring buffers made of two slices (left to right). The two slices undergo a conversion stage (denoted as stage 1) from AoS to SoA format (colors differentiate the 6 flow quantities). This data is then processed by the computing kernels (denoted as stage 2) and the computed RHS (blue) is written back (stage 3) to the associated slice (gray) in the block.

The core layer is composed of three kernels: the RHS, the dt (which computes the time step based on the maximum characteristic velocity) and the update (which updates the conserved quantities) kernels.

The respectives AIs and OIs for the three kernels considered in this work are shown in Table 1. The RHS kernel has an OI that is dependent on the block size and is of 45.4 FLOP/B for a block size of 32, meaning that potentially it could reach peak performance. The update and dt kernel, on the other hand, have fixed OIs of 0.17 and 5 FLOP/B respectively. Without the use of blocks, the OI of the RHS kernel decreases 3.6 FLOP/B, whereas the other two kernels are not affected. The AIs for RHS, dt and update are 1.5, 1.7 and 0.17 FLOP/B respectively, which means that they cannot reach peak performance without exploiting caches.

Node Layer. This layer relies on the core layer and runs simulations on a single computing node, i.e. a full multiprocessor node by using the block size as the parallel grain size. The choice of the concurrency level, i.e. the number of logical threads, is not obvious since one of the considered platforms features multiple hardware threads per core. Since computing the RHS in our solver is limited by the memory bandwidth, we choose to have a fully-subscribed node as we expect that the resource contention is hidden by the memory access costs.

Load imbalance in our solver is not as dramatic as that encountered by spatially adaptive solvers, as the number of blocks contained in our grid remains constant in time. It is however an existing concern due to the presence of ghosts in conjunction with the use of NUMA architectures which make the computation asymmetrical with respect to the cores. To mitigate this issue, we employ OpenMP to maximize the thread-level parallelism (TLP), and by way of static scheduling to initialize the data on ccNUMA platforms, which are subject to the "first-touch" policy.

4   Results and Analysis

In the development of our software we target both supercomputing clusters and high-end desktop machines and in this work we consider a platform for each of these two classes.

Intel Sandy Bridge. The high-end desktop system is represented by an Intel Core i7-2600K (released in January 2011), a quad-core processor featuring Intel's Hyper-Threading Technology and running at 3.4 GHz. The processor is based on the Sandy Bridge micro-architecture and supports the AVX instruction set. The machine has 16 GB of DRAM memory, its measured peak performance is 213 GFLOP/s in single precision and its measured peak DRAM bandwidth is 20.5 GB/s.

AMD Magny-Cours. The supercomputing cluster node is an AMD Magny-Cours (released in March 2010), a 4P 12-cores AMD Opteron 6174 based on the Barcelona micro-architecture with 48 cores at 2.2 GHz and support for the SSE instruction set. One socket contains two ccNUMA nodes (exa-cores), each with one dedicated dual-channel memory controller. The node features 16 GB of DRAM memory per socket for a total of 64 GB. The measured peak performance is 842 GFLOP/s in single precision and the measured peak DRAM bandwidth is 96 GB/s.

Computational Settings. We consider a 3D computational domain size of (1,0.5,0.5) for the simulation of a shock-bubble interaction. A helium bubble with an initial radius of 0.05 is centered at (0.15,0.25,0.25) and a shock wave of $M = 3$ is placed at $x = 0.075$ in air. The specific heat ratios of air and helium are $\gamma = 1.4$ and $\gamma = 1.667$ respectively, whereas the helium to air density ratio is set to 0.138.

Figure 2 shows the isosurface of the helium/air interface location at $\tilde{t} = 2$ (left) and the volume rendering of the density field for this simulation at the same time (right). The shock passage compresses the bubble and deposits vorticity on the helium/air interface. The counter-rotating vorticity pair advects the bubble downstream, turns it inwards and forms a primary vortex ring (PVR) and a smaller secondary vortex ring (SVR).

 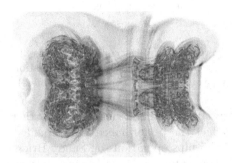

Fig. 2. Shock-Bubble interactions at Mach 3: Isosurface rendering of the interface location (left). Volume rendered image of the vorticity magnitude field for a simulation with 2 billion grid cells (right).

Core Layer. The core layer results presented in the following are obtained by running single precision computations of the same kernels on all cores, in order to capture possible resource contentions. On Magny-Cours, each kernel is explicitly vectorized with SSE whereas on Sandy Bridge AVX is used.

Since the RHS kernel is the most expensive, the studies on the core layer reported are focused on this kernel. We investigate the performance impact of the block size, which represents a tradeoff between spatial locality and memory footprint of the working data set. The left picture of Figure 3 shows the per-core RHS performance versus the block size. We observe that this kernel reaches 35-45% of the peak. Secondly, we note that the performance on Sandy Bridge system is 3-4X than on Magny-Cours, which delineates an important difference between the two platforms: the former's computational power is given by a few cores at high frequency and vector width while the latter enjoys the performance from many cores running at lower frequency. Using a block size of 32 with respect to a block size of 16, we observe an improvement of 4-12% in performance.

With the best block size found in the first performance benchmark, we examine the performance gains provided by the individual optimization techniques applied, namely the use of ring buffers, vectorization and 1.5/2.5 ulps operations. As illustrated in Figure 3 (right), the overall performance gain over the baseline is 14X on Sandy Bridge and 9X on Magny-Cours. The vectorization yields the largest contribution: 9X with AVX and 5X with SSE. Ring buffers contribute another extra 1.2-3.4X. A comparison between the columns for the block-based

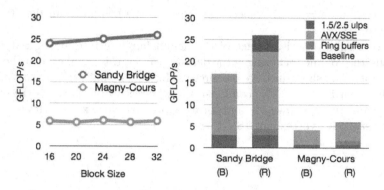

Fig. 3. Performance of the core layer versus block size for the RHS computation (left) and improvements brought by the optimization techniques ('B' indicates block-based memory layout and 'R' stands for ring buffers) (right).

memory layout and the ring buffers further shows that ring buffers also positively affect the performance gain provided by vectorization. We observe that the 1.5/2.5 ulps operations on Sandy Bridge provide a further improvement of 2.3X, whereas for the Magny-Cours platform the gain is only 30%.

We further assess the performance of the kernels by comparing it with the range of the achievable performance estimated with the roofline model. For each kernel, the maximum is computed with its OI (see Table 1), which in our case represents an optimal use of caches, whereas the lower bound is estimated by its AI (see Table 1), which represents a case where the memory hierarchy is not exploited.

As shown in Table 2, we clearly see that, for all kernels, the measured performance is within the estimated intervals. The column "efficiency" denotes the fraction of the peak achievable performance reached by the kernel given its OI, whereas "HU", which stands for hardware utilization, represents the fraction of absolute peak performance.

The RHS and dt kernels reach 48% and 42% of the peak on the Sandy Bridge, with the former kernel being compute bound and the latter being memory bound. Due to its low OI, the update kernels is strongly bound by memory bandwidth and only reaches 2% of the peak performance, although it shows a 100% efficiency. However, because of its simplicity and its total lack of temporal locality, the update kernel has no space for improvement.

Node Layer. We investigate the additional costs incurred by the necessary copying of ghost values between blocks, by the increased memory footprint required to run a simulation and by potential load imbalance issues.

We consider a weak scaling benchmark in terms of GFLOP/s for a system size of $512 \times 256 \times 256$, shown in Figure 4. On the Magny-Cours, the thread placement scheme for the benchmark was chosen so that NUMA nodes are filled first and thread-data affinity is maintained. This causes the piecewise behavior of the (OI-based) predicted performance shown by the solid line: each step represents the

Table 2. Measured performance of the core layer and predicted range, efficiency w.r.t. the maximum achievable performance, nominal peak and fraction of the peak (HU)

Kernel	Platform	Measured [GFLOP/s]	Range [GFLOP/s]	Efficiency [-]	Peak Perf. [GFLOP/s]	HU [-]
RHS	Sandy Bridge	26.0	7.8 - 54.4	48%	54.4	48%
	Magny-Cours	6	3 - 17.6	34%	17.6	34%
update	Sandy Bridge	0.9	0.9 - 0.9	100%	54.4	2%
	Magny-Cours	0.3	0.3 - 0.3	100%	17.6	2%
dt	Sandy Bridge	22.8	8.7 - 26	88%	54.4	42%
	Magny-Cours	4.6	3.4 - 10.2	45%	17.6	26%

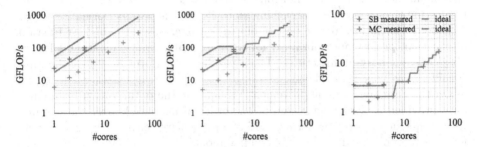

Fig. 4. Weak scaling plot for the RHS (left), dt (center) and update (right) kernels

addition of a NUMA node, and therefore an increase in bandwidth available, to the group of nodes on which the test runs.

On Sandy Bridge we observe speedups up to 3.7X (over 4 cores), whereas on Magny-Cours the speedup reaches 45X (over 48 cores). As illustrated in Figure 4 (left), on Magny-Cours the RHS performance scales almost linearly, attaining 246 GFLOP/s. On Sandy Bridge we observe similar results, with the addition of a point representing the performance obtained by running two hardware threads per core, which increases the performance from 80 GFLOP/s to 92 GFLOP/s, leading to 42% of the peak. A similar behavior is observed for the dt kernel (Figure 4, middle). On Sandy Bridge, the kernel reaches 81 GFLOP/s, corresponding to 37% of the peak, whereas for the same kernel Magny-Cours yields 362 GFLOP/s, corresponding to 43% of the peak performance. We note that the expected peak performance on the two platforms changes from compute bound to memory bound as the number of threads increases, due to contention of the memory bandwidth which decreases the resources available per core.

Figure 4 (right) shows the measured performance for the update kernel. On Sandy Bridge the bandwidth is already saturated with a single thread and the performance reached is around 3.5 GFLOP/s. With two threads the performance of the kernel is slightly above the prediction: we attribute this to caching effects in the LLC. On Magny-Cours the measured performance is 2.3 GFLOP/s with one NUMA node and 16 GFLOP/s with 8 nodes.

For the considered simulation of SBI, the time spent on Magny-Cours for a single time step is measured to be 2.9 seconds with 98% of the time spent in the RHS kernel and 1% in the update and the dt kernels, which further justifies the effort spent on the RHS kernel.

5   Conclusions and Outlook

We have developed efficient CPU kernels for compressible multiphase flows and performed an a-priori analysis to estimate the maximum achievable performance using the roofline model. We have presented a high performance uniform resolution simulation software for multiphase compressible flows tailored to state-of-the-art multicores, capable of reaching up to 48% of the peak performance on the considered platforms. To the best of our knowledge, this is the highest fraction of the peak performance obtained for a compressible flow simulation software on shared memory architectures. These results outperform by 9-14X the ones considered to be state-of-the-art.

On the core layer, the improvement brought by the optimization techniques have provided a performance gain of one order of magnitude over the baseline C++ implementation. Furthermore, all of our measurements are within the performance bounds predicted by the roofline model.

On Magny-Cours, we have reported a speedup of 40-45X over 48 cores, reaching 360 GFLOP/s which corresponds to 42% of the peak performance and of 3.7X over 4 cores on a single-socket Sandy Bridge platform.

Because of the block-based nature of the present solver, the optimizations considered here are also applicable to other frameworks such as AMR. These are currently being applied to wavelet-based adaptive solvers [4] for compressible flow simulations with and without solid boundaries.

Acknowledgments. The authors wish to thank Dr. Olivier Byrde, Teodoro Brasacchio and Adrian Ulrich of the Brutus cluster support team at ETH Zurich for their crucial assistance.

References

1. Quirk, J., Karni, S.: On the dynamics of a shock-bubble interaction. Journal of Fluid Mechanics 318, 129–163 (1996)
2. Colella, P., Graves, D.T., Ligocki, T.J., Martin, D.F., Mondiano, D., Serafini, D.B., Van Straalen, B.: Chombo software package for amr applications design document. Technical report, Lawrence Berkeley National Laboratory (2003)
3. Alam, J.M., Kevlahan, N.K.R., Vasilyev, O.V.: Simultaneous space-time adaptive wavelet solution of nonlinear parabolic differential equations. Journal of Computational Physics 214(2), 829–857 (2006)
4. Rossinelli, D., Hejazialhosseini, B., Spampinato, D., Koumoutsakos, P.: Multicore/Multi-GPU Accelerated Simulations of Multiphase Compressible Flows Using Wavelet Adapted Grids. SIAM J. Scientific Computing 33(2) (2011)

5. Cameron, K., Ge, R., Feng, X.: High-performance, power-aware distributed computing for scientific applications. Computer 38(11), 40–47 (2005)
6. Luk, C.K., Newton, R., Hasenplaugh, W., Hampton, M., Lowney, G.: A Synergetic Approach to Throughput Computing on x86-Based Multicore Desktops. IEEE Softw. 28, 39–50 (2011)
7. Puschel, M., Moura, J., Johnson, J., Padua, D., Veloso, M., Singer, B., Xiong, J., Franchetti, F., Gacic, A., Voronenko, Y., Chen, K., Johnson, R., Rizzolo, N.: SPIRAL: Code Generation for DSP Transforms. Proceedings of the IEEE 93(2), 232–275 (2005)
8. Shalf, J., Quinlan, D., Janssen, C.: Rethinking hardware-software codesign for exascale systems. IEEE Computer 44(11), 22–30 (2011)
9. Chen, G., Chacón, L., Barnes, D.C.: An efficient mixed-precision, hybrid CPU-GPU implementation of a fully implicit particle-in-cell algorithm. ArXiv (2011)
10. Williams, S., Waterman, A., Patterson, D.: Roofline: an insightful visual performance model for multicore architectures. Commun. ACM 52, 65–76 (2009)
11. Petitet, A., Whaley, R.C., Dongarra, J., Cleary, A.: HPL - A Portable Implementation of the High-Performance Linpack Benchmark for Distributed-Memory Computers
12. Yelick, K., Semenzato, L., Pike, G., Miyamoto, C., Liblit, B., Krishnamurthy, A., Hilfinger, P., Graham, S., Gay, D., Colella, P., Aiken, A.: Titanium: a high-performance Java dialect. CCPE 10(11-13), 825–836 (1998)
13. Van Straalen, B., Shalf, J., Ligocki, T., Keen, N., Yang, W.S.: Scalability challenges for massively parallel amr applications. In: IEEE International Symposium on Parallel Distributed Processing, pp. 1–12 (2009)
14. Prosperetti, A., Tryggvason, G. (eds.): Computational Methods for Multiphase Flow. Cambridge University Press, Cambridge (2007)
15. Jiang, G., Shu, C.: Efficient implementation of weighted ENO schemes. Journal of Computational Physics 126(1), 202–228 (1996)
16. Wendroff, B.: Approximate Riemann solvers, Godunov schemes and contact discontinuities. In: Toro, E.F. (ed.) Godunov Methods: Theory and Applications, pp. 1023–1056. Kluwer Academic/Plenum Publ. (2001)
17. Datta, K., Murphy, M., Volkov, V., Williams, S., Carter, J., Oliker, L., Patterson, D., Shalf, J., Yelick, K.: Stencil computation optimization and auto-tuning on state-of-the-art multicore architectures. In: SC 2008, pp. 4:1–4:12. IEEE Press (2008)
18. AMD Inc.: Software Optimization Guide for the AMD 15h Family (2011)

Efficient Algorithm for Linear Systems Arising in Solutions of Eigenproblems and Its Application to Electronic-Structure Calculations

Yasunori Futamura[1], Tetsuya Sakurai[1,2], Shinnosuke Furuya[3],
and Jun-Ichi Iwata[3]

[1] Department of Computer Science, University of Tsukuba, 1-1-1 Tennodai,
Tsukuba-shi, Ibaraki 305-8573, Japan
futamura@mma.cs.tsukuba.ac.jp, sakurai@cs.tsukuba.ac.jp
[2] JST CREST, 4-1-8 Hon-cho, Kawaguchi-shi, Saitama 332-0012, Japan
[3] Department of Applied Physics, The University of Tokyo, 7-3-1 Hongo,
Bunkyo-ku, Tokyo 113-8656, Japan
furuya@comas.t.u-tokyo.ac.jp, iwata@ap.t.u-tokyo.ac.jp

Abstract. We consider an eigenproblem derived from first-principles electronic-structure calculations. Eigensolvers based on a rational filter require solutions of linear systems with multiple shifts and multiple right hand sides for transforming the spectrum. The solutions of the linear systems are the dominant part of the eigensolvers. We derive an efficient algorithm for such linear systems, and develop implementation techniques to reduce time-consuming data copies in the algorithm. Several experiments are performed on the K computer to evaluate the performance of our algorithm.

The first-principles electronic-structure calculation based on the density functional theory (DFT) is currently one of the best choices for understanding and predicting phenomena in material sciences. In terms of parallelization on distributed parallel computers, the real-space method in which the basic equation of DFT, the Kohn-Sham equation, is solved as a finite-difference equation is a promising idea due to small communication costs comparing to the other method such as the reciprocal-space method with fast Fourier transformations [1, 2, 3]. A real-space first-principles calculation of a large system, which consists of about 100,000 Si atoms, was performed on the K computer [3]. In such calculations, we have to solve eigenproblems of large Hermitian matrices called the Hamiltonians self consistently, because the Hamiltonian depends on the charge density that is constructed from its eigenvectors correspond to a certain number of the smallest eigenvalues. Thus it is an exterior eigenproblem.

By using the self-consistent charge density, various physical quantities can be calculated. The electronic band structure around the Fermi energy is important information on the electric current flow of the system. In order to get the band structure, eigenproblems of the self-consistent Hamiltonian with several different parameters need to be solved. However, we need only the eigenvalues within a certain range in this case. Thus it is an interior eigenproblem. In

M. Daydé, O. Marques, and K. Nakajima (Eds.): VECPAR 2012, LNCS 7851, pp. 226–235, 2013.

the self-consistent calculation, the exterior eigenproblem with a large number of eigenpairs is inevitable. On the other hand, the band structure calculation is just the interior eigenproblem with relatively small number of eigenpairs, and we can reduce much computational time by utilizing the eigensolver suitable for interior eigenproblems.

In this study, we solve this interior eigenproblem in band structure calculations with the Sakurai-Sugiura (SS) method [4]. The SS method is proposed as a solver for interior eigenproblem. In the method, solutions of linear systems with multiple shifts and multiple right hand sides (RHSs) are required. Ohno et al. [5] and Mizusaki et al. [6] solve them by a conjugate gradient (CG) type methods for linear systems with multiple shifts. They consider a restricted case of single RHS. We extend these approaches to deal with both multiple shifts and multiple RHSs on an additional degree of freedom. In addition, we introduce implementation techniques to reduce time-consuming data copies of the dominant part in our approach.

The present paper is organized as follows. Section 1 describes a brief introduction of the Sakurai–Sugiura method and linear systems solved in the method. We propose an algorithm of a CG type method for multiple shifts and multiple RHSs in Section 2. We also describe implementation techniques to reduce time-consuming data copies for the algorithm. In Section 3, we show the performance evaluation of our algorithm on the K computer. Conclusions and future work are presented in Section 4.

1   Linear Systems in the Sakurai–Sugiura Method

The Sakurai–Sugiura method is an eigensolver which seeks eigenvalues in specified closed curve and their corresponding eigenvectors. Let $A = A^H \in \mathbb{C}^{n \times n}$. Let us describe the Rayleigh-Ritz projection type method for standard eigenproblem. In the SS method, we calculate matrices

$$S_k \equiv \frac{1}{2\pi i} \int_\Gamma z^k (zI - A)^{-1} V \mathrm{d}z$$

where, $V \in \mathbb{C}^{n \times L}$ is an arbitrary nonzero matrix, Γ is an Jordan curve, i is the imaginary unit, I is the n dimensional unit matrix, L is called block size usually $L << n$, $z \in \mathbb{C}$ and $k = 0, 1, \ldots, M-1$. Assume that L is greater than maximum multiplicity of eigenvalues in Γ, $L \times M$ is greater than number of eigenvalues in Γ and $M \leq N$. So as to calculate S_k numerically, the N-point trapezoidal rule is applied, and we approximate S_k by

$$\hat{S}_k \equiv \sum_{j=0}^{N-1} w_j \zeta_j^k (z_j I - A)^{-1} V, \tag{1}$$

where z_j and w_j are a quadrature point and a weight, respectively. The condition for ζ_j and w_j is given in [5]. Let $S \equiv [\hat{S}_0, \hat{S}_1, \ldots, \hat{S}_{m-1}] \in \mathbb{C}^{n \times (LM)}$ and

$U \in \mathbb{C}^{n \times (\ell m)}$ be an unitary matrix given by the QR decomposition of $\hat{S}$. Let $\tilde{\lambda}$ and $\tilde{x}$ be eigenvalues and corresponding eigenvectors that obtained by diagonalizing $U^H A U$. Eigenpairs of A are approximated by $\lambda \approx \tilde{\lambda} x \approx U\tilde{x}$. To calculate (1), we should solve linear systems with multiple shifts and multiple RHSs. This process often become the most time-consuming part of the SS method.

2   Solver for Linear Systems with Multiple Shifts and Multiple Right Hand Sides

In the SS method, the linear systems with multiple shifts and multiple RHSs should be solved. From this section, we refer to the target shifted linear systems as

$$(A + \sigma_j I)X_j = B, \quad j = 0, 1, \ldots, N - 1. \tag{2}$$

Here, $X_j, B \in \mathbb{C}^{n \times L}$. We refer $AX = B$ as the seed system. Ohno et al. [5] and Mizusaki et al. [6] solve them by conjugate gradient (CG) type methods in case of $L = 1$. They compare the SS method with a widely used method, the Lanczos method, and found that the methods are comparable. When seed system is Hermitian, the linear systems with multiple shifts can be solved by the shifted CG method [7] even if σ_j are complex numbers [5]. Using the shift invariance of the Krylov subspace, the update of solution vectors for shifted systems can be performed without time-consuming matrix-vector products, i.e. matrix-vector products are only required for the seed system. In this study, we deal with multiple RHSs in addition to multiple shifts to reduce the iteration count by exploiting this additional degree of freedom. A GMRES algorithm for both multiple shifts and multiple RHSs was proposed by Darnell et al. [8]. Since we consider the case that the seed system is Hermitian, we choose the CG method as the base method. Thus, we propose the CG method for multiple shifts and multiple RHSs. We refer to the approach shown in [5] as the conventional approach.

2.1   Shifted Block CG-rQ Method

We derive the CG method for multiple shifts and multiple RHSs by extending the block CG method [9] for shifted systems. The block CG method solves systems with multiple RHSs by using the block Krylov subspace [10]. In the block CG method, the search space is extended by L basis per iteration. The block CG method often requires fewer iteration count than the CG method. Several techniques and variants to stabilize the block CG method are presented in [9, 11, 12]. Dubrulle [12] showed that a variant BCGrQ (we refer this as the block CG-rQ method) is the best variant in terms of execution time by numerical experiments. Therefore we choose the block CG-rQ method as the base method of extension for shifted systems. In a similar way as the standard Krylov subspace, the block Krylov subspace also has the shift invariance. Thus there is a relation between the orthonormalized residual matrix Q_k in the block CG-rQ method for the seed system and the residual matrix of $R_k^{\sigma_j}$ in the block CG method for the shifted systems, that is $R_k^{\sigma_j} = \xi_k^{\sigma_j} Q_k$. Here, $\xi_k^{\sigma_j} \in \mathbb{C}^{L \times L}$. By using this relation, an algorithm of the block CG-rQ method for multiple shifts can be obtained. We refer to this algorithm as the shifted block CG-rQ (SBCGrQ) method . The pseudo code of the algorithm is shown in Fig.1. Note that the time-consuming matrix-vector products with $(A + \sigma_j I)$ do not appear in the algorithm of the SBCGrQ method. The computational cost for the SBCGrQ method is much smaller than that of the case that block CG-rQ method is applied for each shifted system.

If a preconditioner is applied, preconditioned coefficient matrices of shifted linear systems are no longer shifted matrices in general. Thus applicable preconditioners are limited (e.g. the incomplete LU preconditioner can not be applied) for block Krylov subspace methods that use the shift invariance. For this reason we omit considering preconditioners in this study.

1: $X_0^{\sigma_j} = O_{n \times L}, \xi_{-1}^{\sigma_j} = \alpha_{-1} = I_L,$
2: $Q_0 \rho_0 = \mathrm{qr}(B)$
3: $\xi_0^{\sigma_j} = \Delta_0 = \rho_0, P_0^{\sigma_j} = P_0 = Q_0$
4: **for** $k = 0, 1, \dots$ until solutions converge **do**
5:    $\alpha_k = (P_k{}^H A P_k)^{-1}$
6:    $X_{k+1} = X_k + P_k \alpha_k \Delta_k$
7:    $Q_{k+1} \rho_{k+1} = \mathrm{qr}(Q_k - A P_k \alpha_k)$
8:    $\Delta_{k+1} = \rho_{k+1} \Delta_k$
9:    $P_{k+1} = Q_{k+1} + P_k \rho_{k+1}^H$
10:   **for** $j = 0, 1, \dots, N - 1$ **do**
11:      $\xi_{k+1}^{\sigma_j} = \rho_{k+1} \left[I_L + \sigma_j \alpha_k + \left\{ \rho_k - \xi_k^{\sigma_j} (\xi_{k-1}^{\sigma_j})^{-1} \right\} (\alpha_{k-1})^{-1} \rho_k^H \alpha_k \right]^{-1} \xi_k^{\sigma_j}$
12:      $\alpha_k^{\sigma_j} = \alpha_k (\rho_{k+1})^{-1} \xi_{k+1}^{\sigma_j}$
13:      $\beta_k^{\sigma_j} = \alpha_k (\rho_{k+1})^{-1} \xi_{k+1}^{\sigma_j} (\xi_k^{\sigma_j})^{-1} (\alpha_k)^{-1} \rho_{k+1}^H$
14:      $X_{k+1}^{\sigma_j} = X_k^{\sigma_j} + P_k^{\sigma_j} \alpha_k^{\sigma_j}$
15:      $P_{k+1}^{\sigma_j} = Q_{k+1} + P_k^{\sigma_j} \beta_k^{\sigma_j}$
16:   **end for**
17: **end for**

Fig. 1. Pseudo code of the SBCGrQ method. $O_{n \times L}$ is the $n \times L$ dimensional zero matrix. I_L is the L dimensional unit matrix. $\mathrm{qr}(C)$ indicates the QR decomposition of matrix C.

To implement the SBCGrQ method for distributed parallel computers, we introduce the row-wise distribution. We implement our distributed parallel code with Message Passing Interface (MPI). In row-wise distribution, matrix-matrix product with a Hermitian transpose matrix in the third line and the QR decomposition in the 7th line are performed with MPI_Allreduce to sum local results. The parallel implementation for the matrix-vector products $A P_k$ depends on the application. The calculations in lines 8,11-13 are replicated. Other lines can be executed without MPI communications.

2.2 Efficient Implementation with Recurrence Unrolling

In the SS method, a number of shifted systems should be solved. In such a case, computational cost for lines 11-15 becomes dominant. Especially lines 14,15 are the most time-consuming part of the algorithm. In addition, the computational cost of lines 14,15 increases $O(L^2)$ with increasing L. We reduce execution time for this computation by following techniques. Fig.2 shows an naive implementation of the 9th line. Note that we reuse the memory area of the variables with subscript k for corresponding variables with subscript $k + 1$. We use simplified notations of the two BLAS subroutines ZGEMM and ZCOPY. Here, ZGEMM(A,B,C) operates $C \leftarrow AB + C$ and ZCOPY(A,B) operates $B \leftarrow A$. To

$$
\begin{array}{l}
\text{ZGEMM}(P_k^{\sigma_j}, \alpha_k^{\sigma_j}, X_{k+1}^{\sigma_j}) \\
\text{ZCOPY}(Q_{k+1}, T) \\
\text{ZGEMM}(P_k^{\sigma_j}, \beta_k^{\sigma_j}, T) \\
\text{ZCOPY}(T, P_{k+1}^{\sigma_j})
\end{array}
$$

Fig. 2. Naive implementation. $T \in \mathbb{C}^{n \times L}$ is a temporary variable.

exploit the efficiency of the cache blocking of ZGEMM, we operate the products $P_k^{\sigma_j}\alpha_k^{\sigma_j}$ and $P_k^{\sigma_j}\beta_k^{\sigma_j}$ in block as $P_k^{\sigma_j}[\alpha_k^{\sigma_j}, \beta_k^{\sigma_j}]$. The drawback of this approach is that additional 2 ZCOPY calls for X^{σ_j} are required. We reduce the total number of ZCOPY calls by unrolling the recurrences for X_{k+1} and P_{k+1}. The recurrences can be unrolled as

$$
X_{k+1}^{\sigma_j} = X_{k-u}^{\sigma_j} + \sum_{h=0}^{u-1} Q_{k-h}\gamma_h^{\sigma_j} + P_{k-u}\gamma_u^{\sigma_j}
$$

and

$$
P_{k+1}^{\sigma_j} = Q_{k+1} + \sum_{h=0}^{u-1} Q_{k-h}\delta_h^{\sigma_j} + P_{k-u}^{\sigma_j}\delta_u^{\sigma_j}.
$$

Here,

$$
\begin{cases}
\gamma_0^{\sigma_j} = \alpha_k^{\sigma_j} \\
\gamma_h^{\sigma_j} = \alpha_{k-h}^{\sigma_j} + \beta_{k-h}^{\sigma_j}\gamma_{h-1}^{\sigma_j}
\end{cases},
$$

$$
\begin{cases}
\delta_0^{\sigma_j} = \beta_k^{\sigma_j} \\
\delta_h^{\sigma_j} = \beta_{k-h}^{\sigma_j}\delta_{h-1}^{\sigma_j}
\end{cases}
$$

and

$$
\theta_h^{\sigma_j} = [\gamma_h^{\sigma_j}, \delta_h^{\sigma_j}].
$$

Fig.3 shows the implementation which uses these relations. By this implementation, the total number of ZCOPY calls is reduced from $2K$ to $4K/u$ when $u > 2$ since ZCOPY is only called every u iterations. Here K is the number of iterations which is required to satisfy the stopping criterion. Simular to the implementation in Fig.2, we reuse the memory area of the variables with subscript $k - u$ for corresponding variables with subscript $k + 1$. The problem is that the implementation shown in Fig.3 requires an additional memory requirement, mainly that of Q_{k-h} $(h = 0, 1, \ldots, u - 1)$. Note that this memory requirement is comparable with that of $X_k^{\sigma_j}$ and $P_k^{\sigma_j}$ $(j = 0, 1, \ldots, N - 1)$ when $u \approx N$.

---

if $\mathrm{mod}(k + 1, u + 1) = 0$ **then**
   $\mathrm{ZCOPY}(X_{k-u}^{\sigma_j}, T_2(:,1{:}L))$
   $\mathrm{ZCOPY}(Q_{k+1}, T_2(:,L + 1{:}2L))$
   for $h = 0, 1, \ldots, u - 1$ **do**
      $\mathrm{ZGEMM}(Q_{k-h}, \theta_h, T_2)$
   end for
   $\mathrm{ZGEMM}(P_{k-u}^{\sigma_j}, \theta_u, T_2)$
   $\mathrm{ZCOPY}(T_2(:,1{:}L), X_{k+1}^{\sigma_j})$
   $\mathrm{ZCOPY}(T_2(:,L + 1{:}2L), P_{k+1}^{\sigma_j})$
end if

Fig. 3. Implementation with recurrence unrolling. $T_2 \in \mathbb{C}^{n \times 2L}$ is a temporary variable.

3 Numerical Experiments

In this section, we perform numerical experiments to evaluate the efficiency of the SBCGrQ method and the recurrence unrolling technique described in the previous section. In the experiments, all examples are performed on the K computer. The K computer is a distributed memory supercomputer system which has more than 80,000 compute nodes. It is currently under development at the RIKEN Advanced Institute for Computational Science as a Japanese national project. A SPARC64TM VIIIfx CPU which has eight cores is equipped for a compute node. The clock frequency and the peak performance of the CPU are 2 GHz and 128 giga-flops, respectively. The target system is a silicon nanowire which consists of 9924 Si atoms [3]. The dimension of the Hamiltonian matrix A is $n = 8,719,488$. Our code is compiled with Fujitsu Fortran Compiler. We describe common parameter setting for all experiments as follows. The contour pass for the SS method is a circle with a center of 0.05 and a radius of 0.01. The number of quadrature points is $N = 32$. The RHS vectors are generated by random numbers. We executed the experiments with 768 MPI processes and each MPI process had 8 OpenMP threads. Note that the results of the numerical experiments are tentative since they are obtained by early access to the K computer.

First, we evaluate the execution time of the SS method, the number of eigenvalues that can be obtained by the SS method, and the iteration count and the execution time for the SBCGrQ method. The results of experiments are shown in Table 1. The parameter u is set to $u = 32$. Table 1 shows the elapsed time for the SS method is mostly occupied by the solutions of the linear systems with the SBCGrQ method in all cases. Large $\#eig$ is obtained by large L. This result is predictable since large subspace is given by large number of RHSs. The remarkable thing is that although the number of linear systems to be solved increase L-fold, $linsol\_time$ does not. This trend is mainly supported by the behavior that $\#iter$ decreases with increasing L as is the case in the block CG method [11]. We have succeeded to extend this feature for multiple shifts by developing the SBCGrQ method. Note that the case $L = 1$ and the conventional approach described in [5] are equivalent except that scaling of the vectors are different and the conventional approach was not implemented with recurrence unrolling. Thus, we can find in the column *Speed-up* for the case $L = 32$ that the SBCGrQ method is more than five times faster than the case that the shifted CG method is sequentially applied to each RHS if these is no significant difference in the iteration count for different RHSs.

Table 1. $\#iter$ and $linsol\_time$ are iteration count and elapsed time for SBCGrQ method, respectively. $\#eig$ is the number of eigenvalues derived in contour pass with relative residuals less than 1e−2. SS_time is elapsed time for the SS method. $Speed\_up$ is the speed-up ratio of average elapsed time for one RHS comparing to $L = 1$, i.e. $(128.2 \times L)$ / $linsol\_time$.

L	1	2	4	8	16	32	64
$\#iter$	10626	10560	9999	8382	6501	4455	4026
$\#eig$	10	21	43	82	159	212	271
SS_time [sec]	131.8	197.7	247.0	395.2	442.5	721.1	1714.6
$linsol\_time$ [sec]	128.2	195.3	246.3	349.3	432.8	698.1	1600.5
$Speed$-up	1	1.31	2.08	2.93	4.74	5.87	5.12

Next we see the detailed data that support the remarkable results described above. Fig.4 shows the results of experiments to see the behaviors of the dominant parts of the SBCGrQ method with increasing L. *Matvec* is the elapsed time of the matrix-vector products with A in the 5th line of Fig.1. *QR* is the elapsed time of the QR decomposition for the 7th line of Fig.1. *Shift* is the elapsed time of the calculations for lines 11-15 of Fig.1. Note that the time data are average data for one RHS of one iteration. *Matvec* slightly decreases with increasing L since latency for communication was reduced by sending or receiving L-fold data at once. *QR* increases with increasing L since the computational cost increases $O(L^2)$. The most time-consuming item *Shift* decreases until $L = 16$. This result indicates that the efficiency of cache blocking of ZGEMM hides the growth of the computational cost. However, *Shift* increases when $L = 32, 64$ due to the high

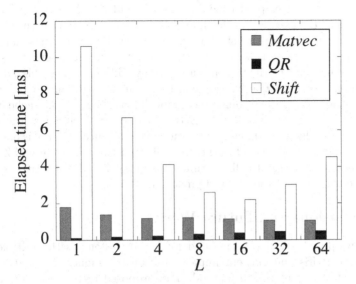

Fig. 4. Details of elapsed time for *linsol_time*

complexity. Fig.5 shows the results of experiments to see the behaviors of the dominant parts of *Shift* with increasing u of the recurrence unrolling technique. The number of RHSs is fixed to $L = 32$. *Square* is the elapsed time for calculations that involve L dimensional square matrices in lines 11-13 of Fig.1. *ZCOPY*

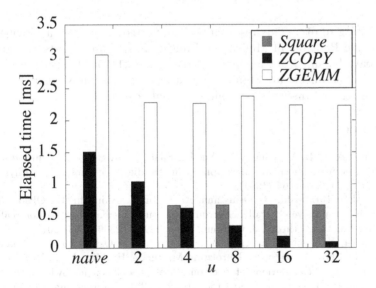

Fig. 5. Details of elapsed time for *Shift*

and *ZGEMM* are the elapsed time for ZCOPY and ZGEMM in Fig.2 or Fig.3 that implement the calculations for lines 14-15 of Fig.1. Note that the time data indicates average data for one shift of one iteration. The computational cost for *Square* other than *naive* is larger than that of *naive* due to calculations for θ_h. Practically, the elapsed time of all cases rarely different since this additional computational cost is negligible. We can find that elapsed time for ZGEMM is reduced by the recurrence unrolling technique. This is because the cache hit ratio is improved by merging two calls of ZGEMM into once. Moreover the elapsed time for ZCOPY decreases linearly with increasing u, since ZCOPY is only called every u iterations. We can find in these details that the efficient use of ZGEMM and the reduction of total call for time-consuming ZCOPY contribute to the remarkable efficiency of the SBCGrQ method.

4   Conclusions and Future Work

We have proposed a CG type method for linear systems with multiple shifts and multiple RHSs and efficient implementation techniques that reduce time-consuming data copies in the method. The proposed method can be used for linear systems that arise in solutions of eigenproblems by an interior eigensolver such as the SS method. We utilized the proposed method for the electronic-structure calculation of a large system which consists of about 10,000 Si atoms. We have found that the proposed method solves the linear systems more than five times faster than the conventional approach and have shown how much our implementation techniques contribute to efficiency of the proposed method. For future work, we will apply the proposed method in unprecedented simulations to clarify important physical properties.

Acknowledgments. This research was supported in part by a Grant-in-Aid for Scientific Research of Ministry of Education, Culture, Sports, Science and Technology, Japan, Grant number: 21246018 and 21105502. The results of the numerical experiments are obtained by early access to the K computer at the RIKEN Advanced Institute for Computational Science.

References

1. Chelikowsky, J.R., Troullier, N., Wu, K., Saad, Y.: Higher-order finite-difference pseudopotential method: An application to diatomic molecules. Phys. Rev. B 50(16), 11355–11364 (1994)
2. Iwata, J.I., Takahashi, D., Oshiyama, A., Boku, T., Shiraishi, K., Okada, S., Yabana, K.: A massively-parallel electronic-structure calculations based on real-space density functional theory. J. Comput. Phys. 229(6), 2339–2363 (2010)
3. Hasegawa, Y., Iwata, J.I., Tsuji, M., Takahashi, D., Oshiyama, A., Minami, K., Boku, T., Shoji, F., Uno, A., Kurokawa, M., Inoue, H., Miyoshi, I., Yokokawa, M.: First-principles calculations of electron states of a silicon nanowire with 100,000 atoms on the K computer. In: Proceedings of 2011 International Conference for High Performance Computing, Networking, Storage and Analysis, SC 2011, pp. 1:1–1:11. ACM, New York (2011)

4. Sakurai, T., Sugiura, H.: A projection method for generalized eigenvalue problems using numerical integration. J. Comput. Appl. Math. 159(1), 119–128 (2003)
5. Ohno, H., Kuramashi, Y., Sakurai, T., Tadano, H.: A quadrature-based eigensolver with a krylov subspace method for shifted linear systems for Hermitian eigenproblems in lattice QCD. JSIAM Letters 2, 115–118 (2010)
6. Mizusaki, T., Kaneko, K., Honma, M., Sakurai, T.: Filter diagonalization of shell-model calculations. Phys. Rev. C 82(2), 024310 (2010)
7. Jegerlehner, B.: Krylov space solvers for shifted linear systems. arXiv:hep-lat/9612014v1 (1996)
8. Darnell, D., Morgan, R.B., Wilcox, W.: Deflated GMRES for systems with multiple shifts and multiple right-hand sides. Linear Algebra Appl. 429(10), 2415–2434 (2007)
9. O'Leary, P.D.: The block conjugate gradient algorithm and related methods. Linear Algebra Appl. 29, 293–322 (1980); Special Volume Dedicated to Alson S. Householder
10. Gutknecht, M.H., Schmelzer, T.: The block grade of a block Krylov space. Linear Algebra Appl. 430(1), 174–185 (2009)
11. Nikishin, A.A., Yeremin, A.Y.: Variable block CG algorithms for solving large sparse symmetric positive definite linear systems on parallel computers, I: General iterative scheme. SIAM J. Matrix Anal. Appl. 16, 1135–1153 (1995)
12. Dubrulle, A.A.: Retooling the method of block conjugate gradients. Electron Trans. Numer. Anal. 12, 216–233 (2001)

Control Formats for Unsymmetric and Symmetric Sparse Matrix–Vector Multiplications on OpenMP Implementations

Takahiro Katagiri[1], Takao Sakurai[2], Mitsuyoshi Igai[3], Satoshi Ohshima[1],
Hisayasu Kuroda[4], Ken Naono[2], and Kengo Nakajima[1]

[1] Information Technology Center, The University of Tokyo,
2-11-16 Yayoi, Bunkyo-ku, Tokyo 113-8658, Japan
[2] Central Research Laboratory, Hitachi, Ltd.
292 Yoshida-cho, Totsuka-ku, Yokohama, Kanagawa 244-0817, Japan
[3] Hitachi ULSI Systems Co., Ltd.
292 Yoshida-cho, Totsuka-ku, Yokohama, Kanagawa 244-0817, Japan
[4] Graduate School of Science and Engineering, Ehime University
3 Bunkyo-cho, Matsuyama, Ehime 790-8577, Japan
katagiri@cc.u-tokyo.ac.jp

Abstract. In this paper, we propose "control formats" to obtain better thread performance of sparse matrix–vector multiplication (SpMV) for unsymmetric and symmetric matrices. By using the control formats, we established the following maximum speedups of SpMV in 16-thread execution on one node of the T2K Open Supercomputer: (1) 7.14× for an unsymmetric matrix by using the proposed Branchless Segmented Scan compared to the original Segmented Scan method; (2) 12.7× for a symmetric matrix by using the proposed Zero-element Computation-free method compared to a simple SpMV implementation.

Keywords: Sparse Matrix–Vector Multiplication (SpMV), Control Formats, Zero-element Computation-free, Branchless Segmented Scan.

1   Introduction

Current computer architectures are very complex due to their hierarchical caches and unsymmetric memory accesses. With the increasingly pervasive use of multicore CPUs, highly threaded parallelism is required. To solve this hardware complexity, many numerical libraries with an auto-tuning (AT) facility have been studied and developed [1][2][3][4].

In this paper, we focus on a sparse iterative solver for linear equations. The main part of the solver is sparse matrix–vector multiplication (SpMV). The performance of SpMV depends on the computer architecture, the number of parallel threads, and the locations of non-zero elements in the input sparse matrix. Hence, it is desired to adapt AT technologies.

M. Daydé, O. Marques, and K. Nakajima (Eds.): VECPAR 2012, LNCS 7851, pp. 236–248, 2013.
© Springer-Verlag Berlin Heidelberg 2013

Before actually adapting these AT technologies, it is important to know the variants of SpMV implementation, since the total performance of sparse iterative solvers affects the AT performance with respect to SpMV. In particular, controlling the thread parallelism for CPUs is crucial in current multicore architectures. This is because the number of parallel threads can reach more than 100.

The objective of this paper is to obtain better performance in thread execution. We accomplish this by introducing "control formats" for SpMV. The control format, which is a different concept from the sparse matrix format, controls sequential and parallel (thread) optimizations for symmetric and unsymmetric SpMVs.

The contributions of this paper are summarized as follows. First, we propose a new control format for a symmetric SpMV. Second, we evaluate three kinds of control formats for symmetric and unsymmetric cases on one node of a parallel machine.

This paper is organized as follows. Section 2 explains the control formats and their SpMV implementations. Section 3 is a performance evaluation of control formats on one node of the T2K supercomputer (U. Tokyo). Section 4 shows related work. Finally, we present our conclusions about this research.

2   Control Formats of SpMV

2.1   Definition of Computation

SpMV is defined as

$$y = A\,x, \tag{1}$$

where y and x are dense vectors in $\Re^n$, and A is a sparse matrix in $\Re^{n \times n}$. Since A is a sparse matrix, we need to reduce the amount of memory. We can use several formants to represent sparse matrices. For example, Compressed Row Storage (CRS), Coordinate (COO), Ellpack (ELL), and DIA (Diagonal) are widely used. We focus on the CRS format because it is an easy and widely used format in several numerical libraries.

The CRS format uses the following three arrays to represent a space matrix: IRP(1:N+1) for row index pointers of the matrix, ICOL(1:NNZ) for indexes of the columns of the matrix, and VAL(1:NNZ) for the values of non-zero elements, where N is the total number of rows, and NNZ is the total number of non-zero elements. If the matrix is symmetric, lower elements of A are *not* stored in order to reduce the amount of memory. This is a restriction of the design policy of Xabclib.

2.2   Control Formats for SpMV

First we define the control format for SpMV. To represent sparse matrices, we should select a sparse matrix format, such as CRS.

SpMV from the viewpoint of computational components is organized according to the following formula (2).

$$SpMV := \textit{Sparse Matrix Format} + \textit{Control Format.} \qquad (2)$$

As seen in formula (2), we need an additional format to control computations in SpMV to implement effective computation of SpMV with a sparse matrix format. We call the additional format for computations the control format.

For example, the control format keeps its own row indexes for each thread for the sparse matrix to increase thread parallelism. The control format to manage the computational complexity of SpMV has other potential abilities. Increasing sequential efficiency is also a crucial factor to be considered.

In the next paragraphs, we will show examples of control formats.

Control Format to Establish Load Balancing
With respect to thread parallelization, load imbalance occurs when the total number of rows is not equally divisible by the number of threads. Fig. 1 shows an example of this load imbalance.

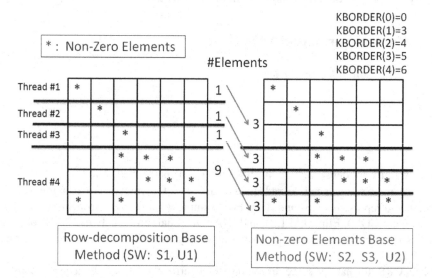

Fig. 1. An example of a load imbalance. The number of threads is 4, the number of sparse rows is 6, and the number of non-zero elements (NNZ) is 12. The U1 and U2 are SpMV implementations for non-symmetric matrices. The S1, S2, and S3 are SpMV implementations for symmetric matrices.

Fig. 1 shows that a heavy load imbalance is caused by the division (Number of rows)/(Number of threads) = floor(6/4) = 1. Thread #4 owns all of the remaining rows, which are three rows in this case. Consequently, a heavy load imbalance for execution time is also caused by OpenMP parallelization. We call this conventional parallelization method the *Row-decomposition Base* method.

To solve the load imbalance problem in the Row-decomposition Base method, we reallocate its load according to the number of non-zero elements per thread. In Fig. 1,

NNZ is 12; hence, perfect load balancing is established with NNZ/(Number of threads) = 12/4 = 3 elements per thread. That is, thread #1 owns three rows while the others own one row in this example. We call this the *Non-zero Element Base* method.

To establish load balancing, we introduce a new control format. Let KBORDER(0:4) be the control format that keeps its own rows for each thread. For example, the number of rows to be owned for thread #1 is represented by KBORDER(1). Determining the number of rows to maintain good load balancing is one of the crucial issues. We implemented a method to find better splitting points by utilizing the divide-and-conquer approach [5].

Branchless Control Format
The Segmented Scan (SS) method [6] is known as an efficient implementation for vector computers. In the control format of SS, a flag array, FLAG(1:NNZ), is used to know the row ends. However, one disadvantage of SS is inefficient execution of the innermost loop, which has an IF-line to prevent branch prediction. The IF-line prevents several optimizations of current cache machines (both compiler and hardware).

To solve this problem, we propose a pair of new control formats, which are JFSTART(1:JL) and MFLAG(1:*). We call computation of SpMV with these new formats the *Branchless Segmented Scan* (BSS) method.[5] Fig. 2 shows the control formats.

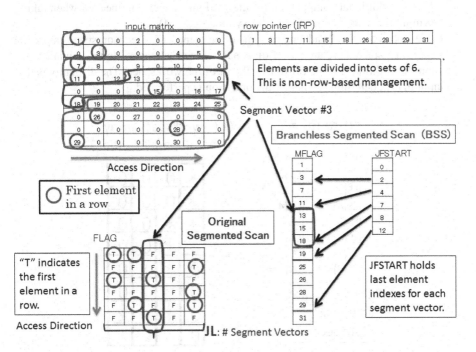

Fig. 2. Control formats of Segmented Scan (SS) and Branchless Segmented Scan (BSS). FLAG(1:NNZ) is the flag to know row ends. JFSTART(1:JL) and MFLAG(1:*) are stride information to know the loop start and loop end.

JFSTART holds the last element indexes for each row of MFLAG. MFLAG holds row-continuous information. This information shows continuous access in the innermost loop, which means the access information of the computation for each segment vector with respect to the CRS format. By using these two control formats, we can establish an "IF-line-free" version of SS. This is why we refer to this method, BSS, as Branchless SS.

In Fig. 2, segment vector #3 starts with the third element in the fourth row. To know the row end, the original SS holds information about the row end in the third row of the array of FLAG. The BSS holds this information in the array of MFLAG. To know the MFLAG information, we refer to the array of JFSTART.

Zero-Element Computation-Free Control Format

If the matrix is symmetric, a workspace is required to maintain parallelism on an OpenMP implementation. One easy way to remove the workspace is to copy elements from the upper part to the lower part. Then, an unsymmetric SpMV is performed. This is against our library design policy.

This procedure raises another issue. The memory space for the working space depends on the number of threads if we use symmetricity for the sparse matrix format. Using symmetricity makes additional computations, such as additions to all workspace areas. The computational costs increase according to the number of threads; hence, this can degrade the parallelization performance. Once again, please note that this additional computation for the working space is not needed when taking an unsymmetric sparse matrix format for symmetric SpMV.

If the matrix is a band matrix, most elements of the workspace must be zero-elements, since there are no off-diagonal non-zero elements. In this situation, we can omit the additions for the off-diagonal part of the workspace. The difficulty is that the input matrix determines whether the omission is performed.

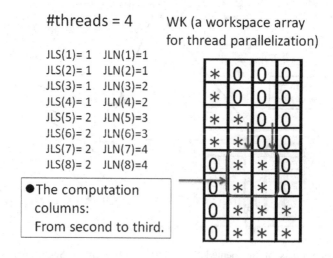

Fig. 3. An example of removing zero-element computations in symmetric SpMV

To address this issue, we introduce a new control format, the *Zero-element Computation-free* method. With respect to the locations of non-zero elements of *A*, we determine the locations of the additions before calling the symmetric SpMV. Fig. 3 gives an example.

As shown in Fig. 3, JLS(1:n) represents the start rows of the index for non-zero elements. JLN(1:n) represents the end rows of the index for non-zero elements.

By using these two control formats, we can compute non-zero elements only for the working space. This is very crucial if the input matrix is a stencil or a band matrix.

2.3   Implementation Details of SpMV on OpenMP

Fig. 4 shows the SpMV implementation of unsymmetric matrices using the control format KBORDER, explained in Section 2.2.

```
!$OMP PARALLEL DO PRIVATE(S, J_PTR, I)
<1>DO K=1,NUM_SMP
<2>   DO I=KBORDER(K-1)+1,KBORDER(K)
<3>     S=0.0D0
<4>     DO J_PTR=IRP(I),IRP(I+1)-1
<5>       S=S+VAL(J_PTR)*X(ICOL(J_PTR))
<6>     END DO
<7>     Y(I)=S
<8>   END DO
<9>END DO
!$OMP END DO PARALLEL
```

Fig. 4. SpMV for unsymmetric matrices with KBORDER

With KBORDER, the computation load from lines <4> to line <6> is almost balanced if it works well. Fig. 5 shows a BSS implementation with JFSTART and MFLAG.

In Fig. 5, by using MFLAG, no IF-line is needed in the innermost loop of the BSS. (See lines <4> to <6>.)

Fig. 6 also shows a simple SpMV implementation for symmetric matrices. Lines <9> to <14> in Fig. 6 cannot be parallelized since a data dependency exists.

Fig. 7 shows the implementation of the SpMV for symmetric matrices with the Non-zero Element Base and Zero-element Computation-free methods.

```
!$OMP PARALLEL DO
          PRIVATE(K,K1,S,I)
<1>DO J=1,JL
<2>  DO K=JFSTART(J),
          JFSTART(J-1)+1,-1
<3>    K1=K+1; S=0.0D0;
<4>    DO I=MFLAG(K1)-1,
          MFLAG(K),-1
<5>      S=VAL(I)*X(ICOL(I))+S
<6>    END DO
<7>    VALSS(K)=S
<8>  END DO
<9>END DO
!$OMP END DO PARALLEL
* Make spanning array
!$OMP PARALLEL DO
<10>DO J=2,JL
<11>  IF(JSFLAG(J).EQV.
         .FALSE.) THEN
<12>    SUM(J-1)=
         VALSS(JFSTART(J-1)+1)
<13>  END IF
<14>END DO
!$OMP END DO PARALLEL
```

```
* Sum up SUM array
<15>DO J=JL-1,1,-1
<16>  IF(PRESENT(J+1) .EQV.
          .FALSE.) THEN
<18>    SUM(J)=SUM(J)+SUM(J+1)
<19>  END IF
<20>END DO
*  Spanning sum
!$OMP PARALLEL DO
<21>DO J=1,JL
<22>  VALSS(JFSTART(J))=
         VALSS(JFSTART(J))+SUM(J)
<23>END DO
!$OMP END PARALLEL
*  Make output
!$OMP PARALLEL DO
          PRIVATE(IN,JN,I)
<24>DO J=1,JL
<25>  IN=JFSTART(J);
<26>  JN=JSY(J);
<27>  DO I=JYN(J),1,-1
<28>    Y(JN)=VALSS(IN)
<29>    IN=IN-1; JN=JN-1;
<30>  END DO; END DO;
!$OMP END DO PARALLEL
```

Fig. 5. SpMV for unsymmetric matrices of BSS with JFSTART and MFLAG

```
!$OMP PARALLEL DO PRIVATE
(S,JJ,JC,I)
<1>DO I=1,N
<2> S=0.0D0
<3> DO JC=IRP(I),IRP(I+1)-1
<4>   JJ=ICOL(JC)
<5>   S=S+VAL(JC)*X(JJ)
<6> ENDDO
<7> Y(I)=S
<8>ENDDO
!$OMP END DO PARALLEL
```

```
<9>DO I=1,N
<10> DO JC=IRP(I)+1,
          IRP(I+1)-1
<11>   JJ=ICOL(JC)
<12>   Y(JJ)=Y(JJ)
          +VAL(JC)*X(I)
<13> ENDDO
<14>ENDDO
```

Fig. 6. Simple SpMV for symmetric matrices

```
!$OMP PARALLEL DO PRIVATE          <14>      Y(I)=S
(S, XDIAG, AA, JJ, I, JC)          <15>   ENDDO
<1>DO K=1,NUM_SMP                  <16>ENDDO
<2>  DO I=KMBORDER(K-1)+1, N       !$OMP END DO PARALLEL
<3>     WK(I,K)=0.0D0
<4>  ENDDO                         !$OMP PARALLEL DO PRIVATE(S)
<5>  DO I=KMBORDER(K-1)+1,         <17>DO K=1, NUM_SMP
        KMBORDER(K)                <18>  DO I=KWBORDER(K-1)+1,
<6>     XDIAG=X(I)                         KWBORDER(K)
<7>     S=VAL(IRP(I))*XDIAG        <19>     S=0.0D0
<8>     DO JC=IRP(I)+1,            <20>     DO J=JLS(I), JLN(I)
           IRP(I+1)-1             <21>        S=S+WK(I,J)
<9>        JJ=ICOL(JC)            <22>     ENDDO
<10>       AA=VAL(JC)             <23>     Y(I)=Y(I)+S
<11>       S=S+AA*X(JJ)           <24>  ENDDO
<12>       WK(JJ,K)=WK(JJ,K)      <25>END DO
              +AA*XDIAG           !$OMP END DO PARALLEL
<13>    ENDDO
```

Fig. 7. SpMV for symmetric matrices with control formats KMBORDER, JLS, and JLN

In Fig. 7, lines <20> to <22> indicate the kernel of the Zero-element Computation-free method with JLS and JLN.

3   Performance Evaluation

We used the T2K Open Supercomputer (HITACHI HA8000) installed at the Information Technology Center at the University of Tokyo. Each node contains four sockets with AMD Opteron 8356 processors (Quad core, 2.3 GHz). The L1 cache is 64 KB/core, the L2 cache is 512 KB/core, and the L3 cache is 2 MB/4 cores. The memory on each node is 32 GB with 667 MHz DDR2. The theoretical peak is 147.2 GFLOPS/node.

We used Intel Fortran Compiler Professional Version 11.0 with the options "-O3 -m64 -openmp -mcmodel=medium." We used 20 types of symmetric matrices and 22 types of asymmetric matrices from the University of Florida sparse matrix collection (referred to hereinafter as UF collection) [7].

We implemented the SpMV with the proposed control formats in OpenATLib [8]. In OpenATLib, the AT of the SpMV is implemented as the first call of the SpMV routine to survey the performances of the implemented SpMVs. In the current version, three kinds of SpMVs were implemented with an AT switch for symmetric and unsymmetric matrices.

The switch of implementations of the SpMVs is summarized as follows: (1) simple SpMV (U1 for unsymmetric, S1 for symmetric); (2) Non-zero Element Base (U2 for unsymmetric, S2 for symmetric with a simple workspace for each thread); (3) BSS (U3, unsymmetric only); (4) original SS (U4, unsymmetric only); and (5) Zero-element Computation-free with Non-zero Element Base (S3, symmetric only). In this section, we evaluate these implementations of SpMV to know the AT effect of OpenATLib.

3.1   Performance of Unsymmetric SpMV

Fig. 8 shows the performance of the unsymmetric SpMV.

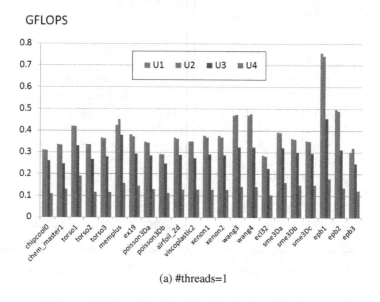

(a) #threads=1

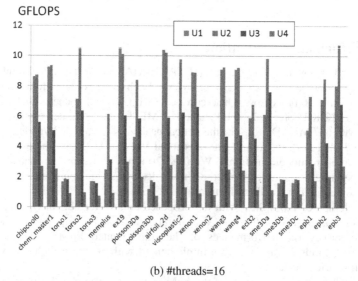

(b) #threads=16

Fig. 8. Performance of unsymmetric SpMV on one node of T2K

According to Fig. 8, we can establish the following speedups compared to U4 (SS): (1) #thread=1: from 1.72× (torso1) to 2.57× (epb1); (2) #threads=16: from 1.63× (epb1) to 7.14× (xenon1) with the BSS control format. The average number of non-zero elements per row (NZEPR) for the UF collection of xenon1 is 24.3. Its NNZ is 1,181,120. One of reasons why we can obtain such a high performance for xenon1 is the decreased cache miss–hit ratio by adapting the branchless control format, since xenon1 is almost a stencil matrix, which is suitable for access optimization for the right-hand-side vector b.

Moreover, we can establish the following speedups compared to U1 with the Non-zero Element Base control format: #threads=16: from 0.95× (ex19) to 2.81× (viscoplastic2). The average number of NZEPR for viscoplastic2 is 11.6. The derivation ratio of NZEPR is 13.9. Since the derivation ratio is very close to the average number of NZEPR for this matrix, a load imbalance easily occurs with highly threaded execution; hence, U2 is very crucial in this situation.

3.2   Performance of Symmetric SpMV

Fig. 9 shows the performance of the symmetric SpMV.

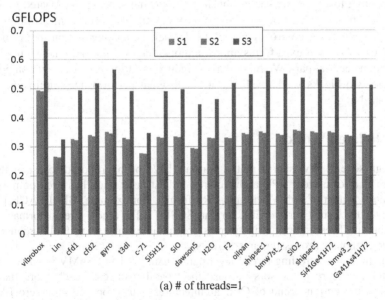

(a) # of threads=1

Fig. 9. Performance of symmetric SpMV on one node of T2K

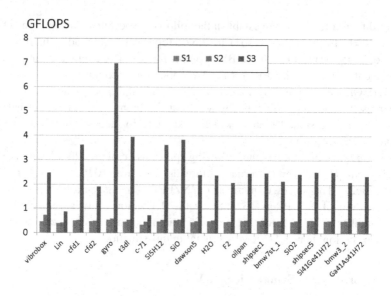

(b) # of threads=16

Fig. 9. (*continued*)

According to Fig. 9, we can establish the following speedups compared to S1: (1) #thread=1: from 1.22× (Lin) to 1.61× (gyro); (2) #threads=16: from 2.14× (c-71) to 12.7× (gyro) with the control format of the Zero-element Computation-free method.

The UF collection gyro forms almost a block diagonal matrix, but two big blocks appear in its central part. We think that the reduction of the zero-element computation for these two blocks works well. As a result, great performance improvement was established in gyro.

4   Related Work

Although several new sparse matrix formats have been proposed to establish high performance for current CPU architectures, almost no method has been proposed for the control format of SpMV with the CRS format. For example, BELLPACK [9] is a new blocking format for the ELL format. SCSR [10] is a new stream format for the CRS format. These new matrix formats are suitable for current CPU architectures, including graphics processing units (GPUs).

In particular, controlling thread level parallelism (TLP) of SpMV is one of crucial tasks for multicore processors because their parallelism can reach more than 100 threads in the current trend of CPU architectures. Our proposed Non-zero Element Base method to increase TLP is a new control format that supports TLP.

Moreover, the conventional Segmented Scan (SS) method has a drawback for optimization of instruction level parallelism (ILP) in CPUs. To increase ILP, our proposed new control format, BSS, enables us to remove IF-lines from the computational kernel of the original SS. We think that BSS is the first control format for SpMV with CRS that is aimed at the increase of ILP in CPUs.

Several approaches focusing primarily on TLP to increase parallelism in GPUs are also in demand for SpMVs.

5   Conclusion

In this paper, we proposed new "control formats" to obtain better thread performance in computations of SpMV for unsymmetric and symmetric matrices. Especially, the control format for symmetric matrices to reduce computations of zero-elements is a contribution of this paper.

By using these control formats, we established maximum speedups in 16-thread execution on one node of the T2K Open Supercomputer: (1) 7.14× for an unsymmetric matrix using the BSS method compared to the original SS method; (2) 12.7× for a symmetric matrix using the Zero-element Computation-free method compared to a simple symmetric SpMV implementation. These speedups are crucial for applications with SpMVs.

We implemented these control formats for a sparse iterative solver with an AT facility, named Xabclib [8]. The AT in Xabclib was implemented with a run-time selection function for all implementations of the SpMV in the first call of the library. The effects of AT, hence, depend on the performance of the SpMV.

Evaluating Xabclib with the SpMV by utilizing the proposed control formats in real applications is important future work.

References

1. Whaley, R.C., Petitet, A., Dongarra, J.J.: Automated Empirical Optimizations of Software and The ATLAS Project. Parallel Computing 27(1-2), 3–35 (2001)
2. Frigo, M., Johnson, S.G.: FFTW: An Adaptive Software Architecture for the FFT. In: Proceedings IEEE International Conference on Acoustics, Speech, and Signal Processing, vol. 3, pp. 1381–1384. IEEE Press, Los Alamitos (1998)
3. Katagiri, T., Kise, K., Honda, H., Yuba, T.: ABCLib_DRSSED: A Parallel Eigensolver with An Auto-tuning Facility. Parallel Computing 32(3), 231–250 (2006)
4. Vuduc, R., Demmel, J.W., Yelick, K.A.: OSKI: A Library of Automatically Tuned Sparse Matrix Kernels. In: Proceedings of SciDAC, Journal of Physics: Conference Series, vol. 16, pp. 521–530 (2005)
5. Sakurai, T., Naono, K., Katagiri, T., Nakajima, K., Kuroda, H., Igai, M.: Sparse Matrix-Vector Multiplication Algorithm for Auto-Tuning Interface "OpenATLib". IPSJ SIG Notes, vol. 2010-HPC-125(2), pp. 1–8 (2010) (in Japanese)
6. Blelloch, G.E., Heroux, M.A., Zagha, M.: Segmented Operations for Sparse Matrix Computation on Vector Multiprocessors. Carnegie Mellon University, Pittsburgh (1993)
7. The university of Florida sparse matrix collection,
   http://www.cise.ufl.edu/research/sparse/matrices/

8. Xabclib and OpenATLib, http://www.abc-lib.org/Xabclib/index.html
9. Chop, J.W., Singh, A., Vuduc, R.: Model-driven Autotuning of Sparse Matrix-vector Multiply on GPUs. In: Proc. ACM SIGPLAN Symp. Principles and Practice of Parallel Programming (PPoPP) (2010)
10. Guo, D., Gropp, W.: Optimizing Sparse Data Structures for Matrix-vector Multiply. International Journal of High Performance Computing Applications 25(1), 115–131 (2011)

Sparsification on Parallel Spectral Clustering*

Sandrine Mouysset[1] and Ronan Guivarch[2]

[1] University of Toulouse, IRIT-UPS, France
[2] University of Toulouse, INP(ENSEEIHT)-IRIT, France

Abstract. Spectral clustering is one of the most relevant unsupervised method able to gather data without a priori information on shapes or locality. A parallel strategy based on domain decomposition with overlapping interface is reminded. By investigating sparsification techniques and introducing sparse structures, this parallel method is adapted to treat very large data set in fields of Pattern Recognition and Image Segmentation.

1 Introduction

Spectral clustering selects dominant eigenvectors of a parametrized affinity matrix in order to build a low-dimensional data space wherein data points are grouped into clusters [1]. This method based on eigendecomposition of affinity matrix is used in Pattern Recognition or image segmentation to cluster non-convex domains without a priori on the shapes. The main difficulties of this method could be summarized by the two following questions: how to automatically separate clusters one from the other and how to perform clustering on large dataset, for example on image segmentation. This means that we look for some full-unsupervising process with parallelization. Several studies exist for defining a parallel implementation which exploits linear algebra [3], [4] for the affinity computation of the whole data set [2]. But the input parameters which are the affinity parameter and the number of clusters limit these methods. To address this limitation, a fully unsupervised parallel strategy based on domain decomposition was proposed in [6] which preserves the quality of global partition thanks to overlapping interface. From the first results, we have observed that the main part of the time is spent in the spectral clustering step and we encountered memory limitation with large problems.

In this paper, we study the robustness of the parallel spectral clustering with overlapping interface presented in [6] by investigating sparsification techniques and introducing sparse structures and adapted eigensolvers in order to treat larger problems. Then we test this improvements on geometrical examples and image segmentations.

2 Parallel Spectral Clustering

Let consider a data set $S = \{x_i\}_{i=1..n} \in \mathbb{R}^p$. Assume that the number of targeted clusters k is known. First, the spectral clustering consists in constructing the

---

* This work was performed using HPC resources from CALMIP (Grant 2012-p0989).

M. Daydé, O. Marques, and K. Nakajima (Eds.): VECPAR 2012, LNCS 7851, pp. 249–260, 2013.

affinity matrix based on the Gaussian affinity measure between points of the dataset S. After a normalization step, the k largest eigenvectors are extracted. So every data point x_i is plotted in a spectral embedding space of $\mathbb{R}^k$ and the clustering is made in this space by applying $K - means$ method. Finally, thanks to an equivalence relation, the final partition of data set is defined from the clustering in the embedded space. Algorithm 1 presents the different steps of spectral clustering.

Algorithm 1. Spectral Clustering Algorithm

Input: data set S, number of clusters k

1. Form the affinity matrix $A \in \mathbb{R}^{n \times n}$ defined by:

$$A_{ij} = \begin{cases} \exp\left(-\frac{\|x_i - x_j\|^2}{(\sigma/2)^2}\right) & \text{if } i \neq j, \\ 0 \text{ otherwise,} \end{cases} \tag{1}$$

2. Construct the normalized matrix: $L = D^{-1/2}AD^{-1/2}$ with $D_{i,i} = \sum_{j=1}^{n} A_{ij}$,
3. Assemble the matrix $X = [X_1 X_2 .. X_k] \in \mathbb{R}^{n \times k}$ by stacking the eigenvectors associated with the k largest eigenvalues of L,
4. Form the matrix Y by normalizing each row in the $n \times k$ matrix X,
5. Treat each row of Y as a point in $\mathbb{R}^k$, and group them in k clusters via the K-means method,
6. Assign the original point x_i to cluster j when row i of matrix Y belongs to cluster j.

This spectral clustering method could be adapted for parallel implementation [6] as a fully unsupervised method (see Figure 1). This avoid extracting the largest eigenvectors of a fully affinity matrix which complexity is of $O(n^3)$ [5].

The principle is based on domain decomposition with overlaps. By dividing the data set S in q sub-domains, each processor applies independently the spectral clustering algorithm on the subsets and provide a local partition. For each subdomain, a quality measure which exploits the block structure of indexed affinity matrix per cluster is used to determine the number of clusters. This heuristic avoids us to fix the targeted of clusters k. The final number of clusters k will be provided after the grouping step. The gathering step is dedicated to link the local partitions from the sub-domains thanks to the overlapping interface and the following transitive relation: $\forall x_{i_1}, x_{i_2}, x_{i_3} \in S$,

$$\text{if } x_{i_1}, x_{i_2} \in C^1 \text{ and } x_{i_2}, x_{i_3} \in C^2 \text{ then } C^1 \cup C^2 = P \text{ and } x_{i_1}, x_{i_2}, x_{i_3} \in P \tag{2}$$

where S is a data set, C^1 and C^2 two distinct clusters and P a larger cluster which includes both C^1 and C^2. By applying this transitive relation (2) on the overlapping interface, the connection between subsets of data is established and

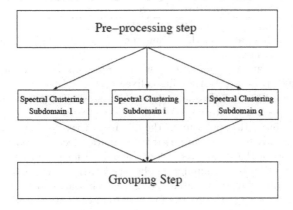

Fig. 1. Principle of the parallel spectral clustering

provides a global partition. We can summarize this Master-Slave implementation with Algorithm 2 and Algorithm 3.

We can notice that when we split the original data set into overlapping sub-pieces of data set, we gain on two aspects:

- *memory consumption:* the local spectral clustering analysis of each sub-piece involves the creation of a local affinity matrix. The size of the matrix is n^2, n being the cardinal of the data subset. The sum of the memory needs for all these local affinity matrix is much less than that needed for the affinity matrix covering the global data set. The consequence is that we can manage bigger data set, data set whose size cannot permit us to run with only one processor.
- *floating point operations:* the analysis of each subproblem is made from the extraction of eigenvectors in the scaled affinity sub-matrix: one extracted eigenvector for each identified cluster of the data subset. In that respect, the parallel approach enables us to decrease drastically the cost of this eigenvector computation: each subproblem will include a number of clusters much less than the total number of clusters in the whole data set.

Nevertheless, as we want to be able to consider larger and larger data sets, as, for instance, in image segmentation (see 4.2) or genomic applications, we still encounter memory limitation when the number of points in a local data subset is too much for the memory capacity of one processor.

3   Sparsification of Spectral Clustering

Despite the domain decomposition, the most time consuming is dedicated to the spectral clustering algorithm. To address this limitation and the memory consumption ones, we investigate a thresholding as sparsification technique.

Algorithm 2. Parallel Algorithm: Master

1: Pre-processing step
   1.1 Read the global data and the parameters
   1.2 Split the data into q subsets
   1.3 Compute the affinity parameter σ with the formula given in paper [6];
     the bandwidth of the overlapping is fixed to $3 \times \sigma$
2: Send the sigma value and the data subsets to the other processors (MPI_SEND)
3: Perform the Spectral Clustering Algorithm on its subset
   3.1 Computation of the spectrum of the affinity matrix (1): classical routines
     from LAPACK library [7] are used to compute selected eigenvalues,
     eigenvectors of the normalized affinity matrix A for its subset of data points
   3.2 Number of clusters: the number of clusters k with the heuristic [6]
   3.3 Spectral embedding: the centers for K-means initialization in the spectral
     embedding are chosen to be the furthest from each other along a direction
4: Receive the local partitions and the number of clusters from each processor
   (MPI_RECV)
5: Grouping Step
   5.1 Gather the local partitions in a global partition thanks to the transitive relation
     given in paper [6]
   5.2 Output a partition of the whole data set S and the final number of clusters k

---

Algorithm 3. Parallel Algorithm: Slave

1: Receive the sigma value and its data subset from the Master processor (MPI CALL)
2: Perform the Spectral Clustering Algorithm on its subset
3: Send the local partition and its number of clusters to the Master processor (MPI CALL)

3.1 Theoretical Interpretation

From the definitions of both the Gaussian affinity A_{ij} between two data points x_i and x_j and the Heat kernel $K_t(x) = (4\pi t)^{-\frac{p}{2}} \exp\left(-\|x\|^2/4t\right)$ in free space $\mathbb{R}_+^* \times \mathbb{R}^p$, we can interpret the gaussian affinity matrix as discretization of heat kernel by the following equation:

$$A_{ij} = \left(2\pi\sigma^2\right)^{\frac{p}{2}} K_t\left(\sigma^2/2, x_i - x_j\right). \tag{3}$$

So, we can prove that eigenfunctions for bounded and free space Heat equation are asymptotically close [8]. With Finite Elements theory, we can also prove that the difference between eigenvectors of A and discretized eigenfunctions of K_t is of an order of the distance between points include inside the same cluster. This means that applying spectral clustering into subdomains resumes in restricting the support of these L^2 eigenfunctions which have a geometrical property: their supports are included in only one connected component. In fact, the domain decomposition by overlapping interface does not alter the global partition because the eigenvectors carry the geometrical property and so, the clustering property.

Let now interpret a thresholding of the affinity matrix on the clustering result. This leads to restrict the approximation to the finite elements which satisfy homogeneity mesh condition in the interpretation. In other words, this means that it strengthens the piece-wise constancy of the dominant eigenvectors from the normalized Gaussian affinity matrix. But the threshold should be well-chosen and should be coherent according to the data distribution. So it should be defined function of both dimension of the data and number of data as defined in [8].

(a) Without thresholding   (b) With thresholding

Fig. 2. Thresholding of the weighted adjacency graph

From another point of view, the affinity matrix could be also interpreted as a Gaussian weighted adjacency graph. The thresholding will control the width of the neighborhoods. This parameter chosen according to the affinity parameter plays a similar role as the parameter ϵ in case of the ϵ-neighborhood graph. A thresholding of the largest distances is equivalent to cancel edges which connect data points very distant from each other as represented in Figure 2. So it strengthens the affinity between points among the same cluster and, so, the separability between clusters.

3.2   Thresholding

However a threshold should be heuristically defined to build an automatic sparsified matrix. We define the threshold that should represents a distance adapted to any distribution of input data. To do so, we start by defining a distance D_{unif} as the distance in the case of the uniform distribution of n points in this enclosing p-th dimensional box in which the data are equidistant each other. This uniform distribution is reached when dividing the box in n smaller boxes all of the same size, each with a volume of order $D_{\max}^p/n$ where $D_{\max}$ is the maximum of the distance between two data point x_i and x_j, $\forall i, j \in \{1, .., n\}$. The corresponding edge size which defines D_{unif} is given by:

$$D_{unif} = \frac{D_{\max}}{n^{\frac{1}{p}}} \tag{4}$$

The thresholding will be function of D_{unif} for any kind of data distribution S.

4   Numerical Experiments

As numerical experiments, we first begin by testing the thresholding on geometrical examples for data set of small size. Then some tests on larger data set are investigated on image segmentation examples.

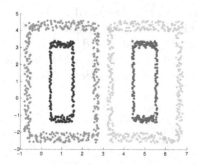

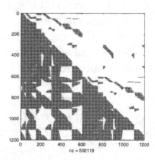

(a) Clustering result with thresholding

(b) Affinity matrix : lower triangular without threshold, upper triangular with threshold

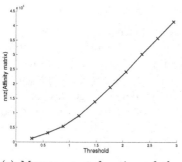

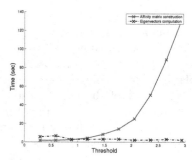

(c) Memory cost function of the threshold

(d) Timings function of the threshold

Fig. 3. Example 1: data set, sparsity of the affinity matrix, memory cost and timings

4.1   First Validations

For first validations, we consider two geometrical examples represented in Fig. 3 (a) and Fig. 4 (a) in which the clusters could not be separated by hyperplanes: the first one with four rectangles of $n = 1200$ points and the second one with a target of $n = 600$ points. The eigenvectors were provided by the reverse communication required by the Fortran library ARPACK [9].

We measure the timings in seconds, in function of the threshold, of the construction of the affinity matrix and of the computation of eigenvectors. The memory cost is evaluated in function of the threshold by the number of non-zeros elements in the affinity matrix.

We can notice on (c) sub-figure that we gain a lot of memory when we decrease the threshold i.e. when we drop the connections of points at a distance larger than it. In fact, this sub-figure shows the memory space required for the storage of the affinity matrix by using a sparse structure $(i, j, value(A_{ij}))$.

We also remark on (d) sub-figure that the time to construct the affinity matrix decreases in this case. Indeed, the computation of the component A_{ij} requires to

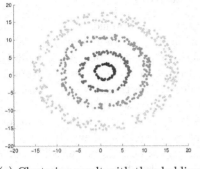

(a) Clustering result with thresholding

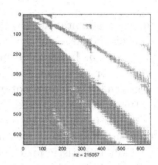

(b) Affinity matrix : lower triangular without threshold, upper triangular with threshold

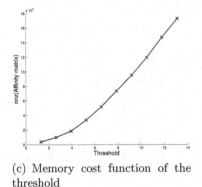

(c) Memory cost function of the threshold

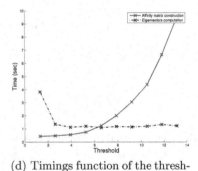

(d) Timings function of the threshold

Fig. 4. Example 2: data set, sparsity of the affinity matrix, memory cost and timings

compute an exponential (1). So because the selection of the connections we keep is done only with the distance, we don't compute the non-useful components and save a lot of floating point operations.

So we have a response to the memory consumption and timing limitations we mentioned previously. As we can see on the first validations, a thresholding strategy allows for considerable gains in terms of memory requirements and computational performance.

If we look the timing for the extraction of the eigenvectors, the time remains the same for acceptable values of the threshold. But we encounter a limit to the sparsification technique with example 2: a strong threshold could imply a very sparsified affinity matrix and an ill-conditioned matrix. In this case, the eigenvector computation becomes the most time consuming task in the sense that the algorithm from Arnoldi method does not converge.

4.2   Another Application: Image Segmentation

For image segmentation, the domain decomposition is applied geometrically on the image and also on the brightness distribution (or color levels) as shown in the figure 5. In fact, we include both 2D geometrical information and 1D brightness (or 3D color levels) information in the spectral clustering method in the sense that there does not exist some privileged directions with very different magnitudes in the distances between points along theses directions. The step between pixels and brightness (or color levels) are about the same magnitude. Thus, a new distance in the affinity measure is defined for image. In the same way, a global heuristic for the Gaussian affinity parameter is proposed in which both dimension of the problem as well as the density of points in the given 3D (or 5D for colored image) are integrated. By considering the size of the image I, the Gaussian affinity A_{ir} is defined as follows:

$$A_{ir} = \begin{cases} \exp\left(-\frac{d(I_{ij}, I_{rs})^2}{(\sigma/2)^2}\right) & \text{if } (ij) \neq (rs), \\ 0 & \text{otherwise,} \end{cases}$$

with the distance between the pixel (ij) and (rs) defined by:

$$d(I_{ij}, I_{rs}) = \sqrt{\left(\frac{i-r}{l}\right)^2 + \left(\frac{j-s}{m}\right)^2 + \left(\frac{I_{ij} - I_{rs}}{256}\right)^2} \tag{5}$$

Parallel spectral clustering was used for image segmentation [6] and we present now the first results of the sparsified parallel spectral clustering applied on image segmentation.

Computational Environment

The parallel numerical experiments were carried out on the Hyperion supercomputer[1]. Hyperion is the latest supercomputer of the CICT (Centre Interuniversitaire de Calcul de Toulouse). With its 352 bi-Intel "Nehalem" EP quad-core nodes it can develop a peak of 33TFlops. Each node has 4.5 GB memory dedicated for each of the cores and an overall of 32 GB fully available memory on the node that is shared between the cores.

3D Image Segmentation

The first example is a Mahua illustration of Benjamin Zhang Bin which presents some continuous degradation of grayscale levels. This grayscale image of 232764 data points is divided in 20 subdomains. We perform experiments with different values of the factor. The threshold is given by the product of the factor with D_{unif} defined by (4). Fig. 6 summarizes the memory consumption (black: maximum consumption on one sub-domain, red: average consumption, blue: minimum).

---

[1] http://www.calmip.cict.fr/spip/spip.php?rubrique90

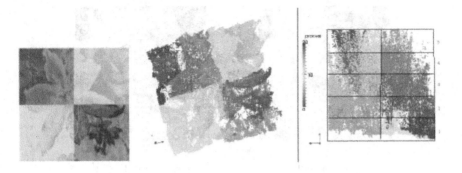

Fig. 5. Example of domain decomposition for image segmentation : geometrical decomposition on the left, brightness distribution and decomposition on the right

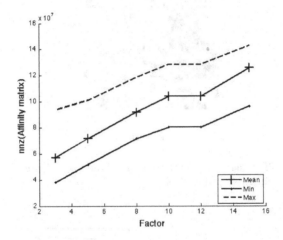

Fig. 6. Example of image segmentation in grayscale: memory cost function of the factor

As we can observe, we are able to decrease this memory consumption by a factor two on average without losing the quality of image segmentation as we can see in Fig. 7. There is no significant difference between the result with the full matrix and the one with sparsification with a factor 3.

5D Image Segmentation

The second example represents a photo of Yann Arthus Bertrand of colored fields in Vaucluse. This is a color image of 128612 points which is also divided in 20 subdomains. In Fig. 8, the memory consumption is plotted.

With this example we are able to divide by 10 this consumption when we take a factor of 1 without loss of quality as we can see in Fig. 9.

However with this example, we tried to further reduce the factor. We experiment that with the value 1, we reach a limit because with factors lower than 1 we notice a significant loss of quality in the segmentation as we can see in subfigure

  (a) original   (b) full

           (c) factor=3

Fig. 7. Example of 3D image segmentation: original data set, clustering result without thresholding and with thresholding (factor 3)

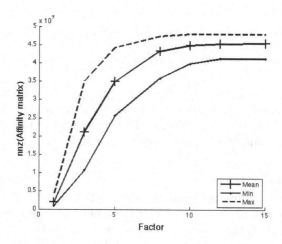

Fig. 8. Example of 5D image segmentation in colors: memory cost function of the factor

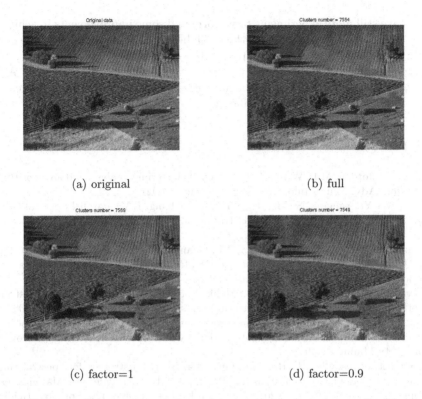

(a) original   (b) full

(c) factor=1   (d) factor=0.9

Fig. 9. Example of image segmentation: original data set, clustering result without and with thresholding (factor = 1 and factor = 0.9)

(d) of Fig. 9 that presents the results with a factor of 0.9. As D_{unif} defined by (4) represents the distance for an uniform distribution, clusters may exis if there are data points which are separated by a fraction of D_{unif}. So for a value of factor lower than 1, the thresholding could affect the clustering result.

5   Conclusion and Ongoing Works

As we mentioned in the conclusion of our work at the previous VECPAR conference [6], we have begun to study sparsification techniques in the construction of affinity matrix by dropping some components that correspond to points at a distance larger than a threshold. We validate this approach in matlab by showing that the number of non zero of the affinity matrix decreases with still some good results in terms of spectral clustering and even some gains in the time spent to compute the affinity matrix.

These results are confirmed when we use sparsification with our parallel spectral clustering solver. We are able to show that we are able to reduce significantly the size of the affinity matrix without loosing the quality of the segmentation solution.

We have still more experiments to perform and some improvements to achieve. First, we have to investigate in all the available tunings in ARPACK to be sure to use the less memory when computing the eigenvectors and eigenvalues. We will then be able to compare the timings with or without sparsification. And finally, we have to perform experiments with bigger images for which we can't have solution if we don't use sparsification.

References

1. Ng, A.Y., Jordan, M.I., Weiss, Y.: On spectral clustering: analysis and an algorithm. In: Proc. Adv. Neural Info. Processing Systems (2002)
2. Chen, W.-Y., Yangqiu, S., Bai, H., Lin, C.-J., Chang, E.Y.: Parallel Spectral Clustering in Distributed Systems. IEEE Transactions on Pattern Analysis and Machine Intelligence (2010)
3. Song, Y., Chen, W.Y., Bai, H., Lin, C.J., Chang, E.Y.: Parallel spectral clustering. In: Processing of European Conference on Machine Learning and Principles and Practice of Knowledge Discovery in Databases (2008)
4. Fowlkes, C., Belongie, S., Chung, F., Malik, J.: Spectral grouping using the Nystrom method. IEEE Transactions on Pattern Analysis and Machine Intelligence (2004)
5. Yan, D., Huang, L., Jordan, M.I.: Fast approximate spectral clustering. In: Proceedings of the 15th ACM SIGKDD International Conference on Knowledge Discovery and Data Mining (2009)
6. Mouysset, S., Noailles, J., Ruiz, D., Guivarch, R.: On a strategy for spectral clustering with parallel computation. In: Palma, J.M.L.M., Daydé, M., Marques, O., Lopes, J.C. (eds.) VECPAR 2010. LNCS, vol. 6449, pp. 408–420. Springer, Heidelberg (2011)
7. Anderson, E., Bai, Z., Bischof, C., Blackford, S., Demmel, J., Dongarra, J., Du Croz, J., Greenbaum, A., Hammarling, S., McKenney, A., et al.: LAPACK Users' guide. Society for Industrial Mathematics (1999)
8. Mouysset, S., Noailles, J., Ruiz, D.: On an interpretation of Spectral Clustering via Heat equation and Finite Elements theory. In: International Conference on Data Mining and Knowledge Engineering (2010)
9. Lehoucq, R.B., Sorensen, D.C., Yang, C.: ARPACK users' guide: solution of large-scale eigenvalue problems with implicitly restarted Arnoldi methods. SIAM (1998)

An Experimental Study
of Global and Local Search Algorithms
in Empirical Performance Tuning*

Prasanna Balaprakash, Stefan M. Wild, and Paul D. Hovland

Mathematics and Computer Science Division
Argonne National Laboratory, Argonne, IL 60439
{pbalapra,wild,hovland}@mcs.anl.gov

Abstract. The increasing complexity, heterogeneity, and rapid evolution of modern computer architectures present obstacles for achieving high performance of scientific codes on different machines. Empirical performance tuning is a viable approach to obtain high-performing code variants based on their measured performance on the target machine. In previous work, we formulated the search for the best code variant as a numerical optimization problem. Two classes of algorithms are available to tackle this problem: global and local algorithms. We present an experimental study of some global and local search algorithms on a number of problems from the recently introduced SPAPT test suite. We show that local search algorithms are particularly attractive, where finding high-preforming code variants in a short computation time is crucial.

Keywords: automatic performance tuning, search, black-box optimization.

1 Introduction

The rapid rate of innovations in computing architectures has widened the gap between the theoretical peak and the achievable performance of scientific codes [1]. Often, scientific application programmers address this issue by manually rewriting the code for the target machine, but this approach is neither scalable nor portable. Empirical performance tuning or automatic performance tuning (in short, *autotuning*) is a promising approach to address the limitations of manual tuning. This approach consists of identifying relevant code optimization techniques (such as loop unrolling, register tiling, and loop vectorization), assigning a range of parameter values using hardware expertise and application-specific knowledge, and then either enumerating or searching this parameter space to find the high performing parameter configurations for the given machine. Using

---

* This paper has been created by UChicago Argonne, LLC, Operator of Argonne National Laboratory ("Argonne"). Argonne, a U.S. Department of Energy Office of Science laboratory, is operated under Contract No. DE-AC02-06CH11357.

M. Daydé, O. Marques, and K. Nakajima (Eds.): VECPAR 2012, LNCS 7851, pp. 261–269, 2013.
© Springer-Verlag Berlin Heidelberg 2013

this approach, several researchers have achieved considerable success in tuning scientific kernels for both serial and multicore processors [1].

In large-scale empirical performance tuning, the computation time needed to enumerate all parameter configurations in a large decision space is prohibitively expensive. Hence, effective global and/or local search algorithms that examine a tiny subset of the possible configurations are required. Typically, global algorithms can be characterized by their dynamic balance between exploration of the search space and exploitation of the accumulated search history. They are theoretically guaranteed to find the globally best configuration at the expense of a long search time. In practice, however, they are run until user-defined stopping criteria are met. Examples include branch and bound, simulated annealing, genetic algorithms, and particle swarm optimization. In contrast, local search algorithms do not emphasize exploration and instead repeatedly try to move from a current configuration to a nearby improving configuration. Typically, the neighborhood of a given configuration is problem-specific and defined by the user or algorithm. These algorithms terminate when a current configuration does not have any improving neighbor and hence is locally optimal. Examples include the Nelder-Mead simplex, orthogonal search, variable neighborhood search, and trust region methods. The disadvantage of local search algorithms is that, depending on the search space and initial configuration, they can terminate with a locally optimal configuration that performs much worse than a globally optimal configuration.

Search problems in empirical performance tuning are defined by a specific combination of a kernel, an input size, a set of tunable decision parameters, a set of feasible parameter values, and a default/initial configuration of these parameters for use by search algorithms [2]. Several global and local search algorithms have been deployed for empirical performance tuning. Seymour et al. [9] performed an experimental comparison of several global (random search, a genetic algorithm, simulated annealing, particle swarm) and local (Nelder-Mead and orthogonal search) optimization algorithms. Similarly, Kisuki et al. [6] compared random search, a genetic algorithm, and simulated annealing with pyramid search and window search. In both these studies, the experimental results showed that the random search was more effective than the other algorithms tested. A reason is that in the tuning tasks considered, the number of high-performing parameter configurations is large and hence it is easy to find one of them. Moreover, we suspect that the adopted local search algorithms were less effective because they were not customized. Although Norris et al. [7] implemented the Nelder-Mead simplex method, simulated annealing, and a genetic algorithm in the empirical performance tuning framework Orio, the authors did not conduct an experimental comparison. A number of works deploy local search algorithms for empirical performance tuning. Examples include orthogonal search in AT-LAS [11], pattern search in loop optimization [8], and a modified Nelder-Mead simplex algorithm in Active Harmony [10]. However, a comparison with global search algorithms was not available. From the literature, it is not clear whether local search or global search is best suited for empirical performance tuning and, in particular, under what conditions one class may be better than another.

In this paper, we focus on a setting where the available computation time for tuning is highly limited. Our hypothesis is that appropriately modified local search algorithms can find high-performing code variants in short computation times. This is based on the rationale that the exploration component of global search algorithms is less beneficial in empirical performance-tuning problems where finding high-performing configurations in short computation time is more important than finding the optimal configuration. We conduct an experimental study of some global and local search algorithms on a number of problems from the SPAPT test suite [3]. The main contribution of the paper is empirical evidence for the effectiveness of the local search algorithms under short computation times.

2   Search Algorithms

For global search algorithms, we consider random search, a genetic algorithm, and simulated annealing. For local search algorithms, we use the Nelder-Mead simplex method and a surrogate-based search.

Random search has been shown to be effective on a number of performance-tuning tasks. The parameter configurations are sampled uniformly at random from the feasible domain $\mathcal{D}$ without replacement. At iteration k, each $x \in \mathcal{D}$ not already sampled has probability $\frac{1}{|\mathcal{D}|-k+1}$ of being selected as the point $x^{(k)}$. In the absence of other criteria, the algorithm terminates after $|\mathcal{D}|$ iterations with the global minimum.

Genetic algorithms are among the most widely used global search algorithms. These algorithms follow a common framework that consists of iteratively modifying a population of configurations by applying a set of evolutionary operations such as reproduction, recombination, and mutation. Several variants exist; the best one depends on the problem at hand and the parameters of the algorithm. We use a genetic algorithm based on [4].

Simulated annealing is inspired by the physical process of annealing. The key algorithmic component is an annealing schedule that slowly reduces the value of a temperature parameter T so that the probability of accepting a worse configuration decreases as the search progresses [5]. The mechanism of accepting worse configurations during the search helps the algorithm escape from bad local configurations encountered in the early stages of the search.

The Nelder-Mead simplex method was originally developed to solve unconstrained continuous optimization problems. It works with a simplex of $n + 1$ vertices, where n is the number of parameters. At each iteration, the simplex moves away from less promising regions of the search space using reflection, expansion, contraction, or shrink operators. We use a Nelder-Mead simplex algorithm that is customized for empirical performance tuning task; see [2] for implementation details.

Surrogate-based search is an algorithmic framework that uses inexpensive surrogates to approximate the computationally expensive objective. For our experiments, we consider a basic trust-region algorithm [12] that operates on discrete values. It starts by constructing a quadratic surrogate function by evaluating a few configurations. At each iteration, a configuration that minimizes the surrogate is evaluated, and the ratio between the true function value and the predicted surrogate value is used to monitor the quality of the surrogate. When the surrogate is accurate enough, the trust region is expanded; otherwise, the region is contracted, and a promising neighbor of the current configuration is evaluated to improve the surrogate.

3   Numerical Experiments

We evaluate the algorithms on problems from the SPAPT test suite [3], a collection of extensible and portable search problems in automatic performance tuning. These problems are implemented in an annotation-based language that can be readily processed by Orio [7]. Originally, the SPAPT problems had integer and binary parameters (scalar replacement, array copy, loop vectorization, and OpenMP) with both bound and algebraic constraints. Since the focus of our study is on bound-constrained problems with integer parameters only, we removed all algebraic constraints and binary parameters from the problems. The numerical parameters include loop unroll/jamming $\in [1,\ldots,50]$, cache tiling $\in [1, 2, 4, 8, 16, 32, 64, 128, 256, 512, 1024, 2048]$ (treated as $[1,\ldots,12]$), and register tiling $\in [1,\ldots,32]$. The number of parameters n_i ranges between 8 and 38, and the size of search space $|\mathcal{D}|$ ranges between 5.31×10^{10} and 1.24×10^{53}. Of the 18 problems in the SPAPT test suite, we use only 12. On the remaining 6 problems, since the algebraic constraints are required for the correctness of the transformation, we did not use it.

Random search (RS), the genetic algorithm (GA), simulated annealing (SA), modified Nelder-Mead simplex (mNM), and modified surrogate-based search (mSBS) were implemented and run in MATLAB version 7.9.0.529 (R2009b). We adopted the default parameter values for all the algorithms. Experiments were carried out on dedicated nodes of Fusion, a 320-node cluster at Argonne National Laboratory, comprising 2.6 GHz Intel Xeon processors with 36 GB of RAM, under the stock Linux kernel version 2.6.18 provided by RedHat.

We considered the objective value $f(x)$ at a parameter configuration x as the average computation time over 10 generated code runs. Other objective functions can be adopted, such as the median or minimum; see [3] for a discussion. For the initial configuration from which the algorithms start, we set each parameter to its lower bound. This corresponds to a code variant without any transformation. We used 100 code evaluations as the stopping criterion for each algorithm. Given a parameter configuration, a code evaluation consists of code transformation, compilation, and execution. For the size of the search space that we have, this corresponds to the evaluation of only $8.05 \times 10^{-50}\%$ ($|\mathcal{D}|=1.24 \times 10^{53}$) to 0.00000018% ($|\mathcal{D}|=5.31 \times 10^{10}$) of the total configurations.

Figure 1(a) shows a bar chart of the speedups at different time intervals. We compute "$x\%$ of max T" as follows: For each problem, max T is the maximum

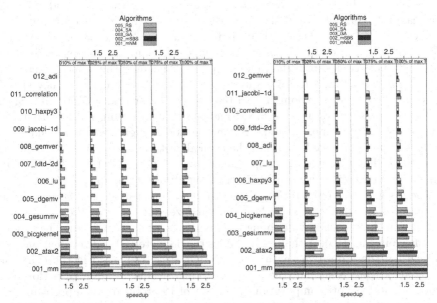

(a) Default initial configuration; default input size; 100 function evaluations

(b) Default initial configuration; large input size; 100 function evaluations

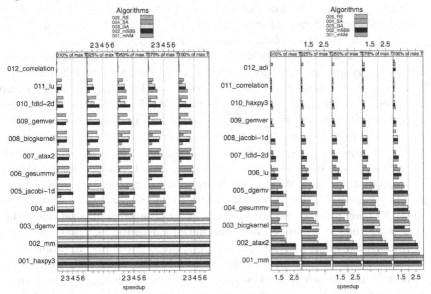

(c) Poor initial configuration; default input size; 100 function evaluations

(d) Default initial configuration; default input size; 500 function evaluations

Fig. 1. Speedups obtained by each algorithm as a function of % of the budget

elapsed time that any of the five algorithms took to complete 100 evaluations. The speedups obtained by each algorithm after 10%, 25%, 50%, and 100% of the max T is computed and shown in the figure. From the speedups obtained at these intervals, we observe that the two local search algorithms, mNM and mSBS, obtain high-quality configurations in short computation time. The main advantage here comes from the time required for the algorithms to complete 100 code evaluations. RS and GA require longer search times because they spend more time exploring the domain and tend to be slower than mNM and mSBS. The performance advantage of mNM and mSBS comes from the fact that the time per evaluation tends to be shorter once a good configuration has been found. On 9 of 12 problems, we found that the local algorithms outperformed the global algorithms. The observed speedups are between 1.15 and 3.0, respectively. On adi and correlation, we cannot detect a significant speedup.

Under the same computation budget of 100 code evaluations, we tested the behavior of the algorithms on larger input sizes (the size of the arrays and matrices in the kernels) by doubling the input size for each problem. The results are shown in Fig. 1(b). Although the times to complete 100 code evaluations are larger than those observed with smaller input sizes, the trend in the behavior of the algorithms is similar: the local search algorithms obtain high-performing code variants in short computation time. Out of 12 problems, on 8 problems the local search algorithms are better than the global search algorithms. Although mNM and mSBS find high-quality configurations in short computation time, on gessumv, given enough time GA obtains a better configuration than mNM and mSBS. On mm and correlation, we cannot detect a significant difference between the results of the global and local search algorithms.

Figure 1(c) shows the results when the starting point is set to the upper-bound values. From the exploratory studies, we found that the initial configurations with lower-bound values are reasonably good starting points and that those at the upper bounds are extremely poor. We found that mNM and SA tend to be sensitive to the starting point and obtain poor results. These algorithms also required longer search times because the parameter configurations closer to the upper bounds have longer transformation time and consequently longer compile time. Whereas SA tries to escape from the nonpromising region, mNM stagnates, spending most of the search time exploring the neighborhood of the current configuration. We found mSBS to be less sensitive than mNM or SA to the starting point because it uses randomly sampled configurations within a larger initial neighborhood to form the initial surrogate. GA uses the initial configuration only as an individual of the population in the first iteration. Since RS is independent of the starting point, it found better code variants than did mNM and SA in short computation times. The results show that the poor starting points significantly reduce the effectiveness of the local search algorithms. Out of 12, only on 6 problems did the local search algorithms, in particular, mSBS, outperform the global search algorithms. We also used the center of the hyperrectangle $\mathcal{D}$ as a starting point. The results observed are similar to those

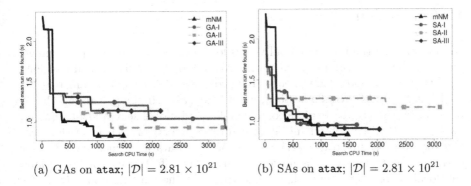

(a) GAs on `atax`; $|\mathcal{D}| = 2.81 \times 10^{21}$ (b) SAs on `atax`; $|\mathcal{D}| = 2.81 \times 10^{21}$

Fig. 2. Best objective value obtained by each algorithm as a function of search time. Each algorithm is allowed to perform 100 function evaluations. Markers are placed at every 20 evaluations.

with lower bounds as in Fig. 1(c), local algorithms being better than the global algorithms despite a slightly worse starting value than the lower bounds.

Figure 1(d) illustrates the behavior of the algorithms using a slightly larger computation budget (500 code evaluations) as the stopping criterion. The algorithms start from initial configurations in which each parameter is set to its lower-bound value. Global search algorithms benefit from a larger number of iterations. On 7 out of 12 problems local search algorithms dominate global search algorithms, but the difference in the speedups between global and local algorithms is smaller than that observed with 100 evaluations. Although local search algorithms find high-quality code variants in short times, they spend the search effort in exploring the neighborhood of a local configuration to certify local optimality.

To further test that the exploration component is the major factor affecting the performance of global search algorithms, we reduced their degree of exploration. Specifically, for GA and SA, we reduced the values of the mutation parameter μ and starting temperature parameter T, respectively. We used three GAs: GA-I (default $\mu = 0.5$), GA-II ($\mu = 0.1$), and GA-III ($\mu = 0.001$). Similarly for SA, we used SA-I (default $T = 1.0$), SA-II ($T = 0.1$), and SA-III ($T = 0.001$). Figures 2(a) and 2(b) illustrate the results of the algorithms on `atax` for 100 code evaluations. The default lower-bound configuration is used as a starting point. The results of our study show that reducing the exploration in global search algorithms is beneficial but the appropriate reduction depends on the algorithm characteristics, the problem, and the starting point. GA-I and GA-II obtain configurations with similar runtime, but the latter obtains this configuration in a shorter period of time (1200 CPU-seconds). However, an extremely small degree of exploration in GA-III leads to stagnation. In contrast, although slightly slower, SA-III obtains a better configuration than do SA-I and SA-II. Our conjecture is that given a good starting point, SA with a very low degree of exploration can be effective.

4 Conclusion

We investigated the issue of global versus local search in empirical performance tuning under short computation times. We tested illustrative global and local algorithms on bound-constrained search problems with integer parameters. We used different initial configurations, input sizes, and stopping criteria. The results show that (1) the exploration capabilities of global search algorithms are less useful; (2) given good initial configurations, local search algorithms can find high-performing code variants in short computation time; and (3) poor initial configurations can significantly reduce the effectiveness of both global and local search algorithms that are sensitive to the starting point. From the results, we conclude that when the available tuning time is severely limited, carefully customized local search algorithms are promising candidates for empirical performance-tuning problems that have integer parameters and bound constraints.

Our future work includes the following: (1) problem-specific techniques to handle binary parameters and constraints for both global and local search algorithms, (2) effective restart and multi start strategies for local search to escape from poor local configurations, (3) global algorithms that automatically adopt exploration and exploitation parameters , (3) tuning of parallel scientific codes using search algorithms, and (4) analysis of the impact of different target machines on various performance objectives.

Acknowledgments. We are grateful to the Laboratory Computing Resource Center at Argonne National Laboratory for the computing resources used in this paper.

References

1. Bailey, D., Lucas, R., Williams, S. (eds.): Performance Tuning of Scientific Applications. Chapman & Hall/CRC Computational Science (2010)
2. Balaprakash, P., Wild, S., Hovland, P.: Can search algorithms save large-scale automatic performance tuning? In: The International Conference on Computational Science (July 2011)
3. Balaprakash, P., Wild, S., Norris, B.: SPAPT: Search problems in automatic performance tuning. Procedia Computer Science 9, 1959–1968 (2012); Proceedings of the International Conference on Computational Science, ICCS 2012
4. Chipperfield, A., Fleming, P.: The MATLAB genetic algorithm toolbox. In: IEE Colloquium on Applied Control Techniques Using MATLAB (1995)
5. Kirkpatrick, S., Gelatt, C.D., Vecchi, M.P.: Optimization by simulated annealing. Science 220, 671–680 (1983)
6. Kisuki, T., Knijnenburg, P.M.W., O'Boyle, M.F.P.: Combined selection of tile sizes and unroll factors using iterative compilation. In: Proc. of the 2000 International Conference on Parallel Architectures and Compilation Techniques, Washington, DC (2000)

7. Norris, B., Hartono, A., Gropp, W.: Annotations for Productivity and Performance Portability. Computational Science, pp. 443–461. Chapman & Hall CRC Press, Taylor and Francis Group (2007)
8. Qasem, A., Kennedy, K., Mellor-Crummey, J.: Automatic tuning of whole applications using direct search and a performance-based transformation system. The Journal of Supercomputing 36(2), 183–196 (2006)
9. Seymour, K., You, H., Dongarra, J.: A comparison of search heuristics for empirical code optimization. In: Proc. of the 2008 IEEE International Conference on Cluster Computing, pp. 421–429 (2008)
10. Tiwari, A., Chen, C., Jacqueline, C., Hall, M., Hollingsworth, J.K.: A scalable auto-tuning framework for compiler optimization. In: Proc. of the 2009 IEEE International Symposium on Parallel & Distributed Processing, Washington, DC, pp. 1–12 (2009)
11. Whaley, R.C., Dongarra, J.J.: Automatically tuned linear algebra software. In: Proc. of the 1998 ACM/IEEE Conference on Supercomputing, SC 1998, Washington, DC, pp. 1–27 (1998)
12. Wild, S.M.: MNH: a derivative-free optimization algorithm using minimal norm Hessians. In: Tenth Copper Mountain Conference on Iterative Methods (April 2008), http://grandmaster.colorado.edu/~copper/2008/SCWinners/Wild.pdf

A Multi GPU Read Alignment Algorithm with Model-Based Performance Optimization

Aleksandr Drozd, Naoya Maruyama, and Satoshi Matsuoka

Tokyo Institute of Technology, 2-12-1-W8-33, Ookayama, Meguro-ku, Tokyo 152-8552
alex@smg.is.titech.ac.jp, naoya@matsulab.is.titech.ac.jp,
matsu@is.titech.ac.jp

Abstract. This paper describes a performance model for read alignment problem, one of the most computationally intensive tasks in bioinformatics. We adapted Burrows Wheeler transform based index to be used with GPUs to reduce overall memory footprint. A mathematical model of computation and communication costs was developed to find optimal memory partitioning for index and queries. Last we explored the possibility of using multiple GPUs to reduce data transfers and achieved super-linear speedup. Performance evaluation of experimental implementation supports our claims and shows more than 10fold performance gain per device.

1 Introduction

Faster and faster computing systems are developed every day to cope with ever-increasing complexity of problems that emerge in various areas of science and technology. Performance growth comes from technological advancements and mainly form architectures facilitating parallel data processing in various forms (i.e. recently GPUs). At the same time algorithms known to solve particular tasks themselves have many possibilities of improvement, taking into consideration fact that overall performance comes not just from better algorithm, but also on how it fits certain peculiarities of hardware platform and different patterns of data distribution in heterogeneous systems. GPUs and clusters of GPUs have recently become one of the main threads of supercomputing. Their computational characteristics are different from those of traditional systems and they are relatively new to software developers, which makes the above-stated issues even more important. Also while some applications have a pretty uniform data model, like those solving various matrix-based mathematical problems, in other applications data model itself is heterogeneous and its decomposition requires a profound study of balancing storage and distribution of workload parts so that we could better meet the platform characteristics and improve the overall performance.

This paper focuses on the pairwise local DNA sequence alignment problem. It is extremely computationally intensive as constant progress in sequencing technology leads to ever-increasing amounts of data to be processed. We target

M. Daydé, O. Marques, and K. Nakajima (Eds.): VECPAR 2012, LNCS 7851, pp. 270–277, 2013.

GPU-based systems that have been shown to allow for greater performance in sequence processing tasks due to their extreme parallel capacities [1].

Read alignment is basically a string matching problem and is typically done by building index of a reference and then matching queries against it. There are several types of indexes and corresponding match algorithms which were being used for alignment problem. We made a survey of existing solutions [2],[3],[4], and found that memory limitation is the performance bottleneck in all cases. Workload size for both reference sequence and query set can dramatically surpass available device memory and each index subdivision into smaller chunks to fit into memory simply doubles execution time. For example human genome contains approximately 3 billion of bases. Suffix array (array of integers giving the starting positions of suffixes of a string in lexicographical order) needs 9 bytes per base, so it will require 27 gigabytes of memory, while top modern GPUs have about 6GB. To index bigger references 64 bit integers are required and suffix array space complexity will be 17 bytes per base.

To reduce memory consumption we propose using matching algorithm based on Burrows Wheeler Transform. This algorithm is mainly used for data compression, but possibility of pattern matching using this transform was recently described[5]. Index based on BWT is more than ten times smaller than index based on suffix array. We perform an analysis of how this algorithm fits GPU characteristics and do model implementation to see if we can actually get significantly better execution time with this smaller memory footprint algorithm. This is the first contribution of this paper.

The second one is the performance model of possible memory utilization strategies. This model allowed us to find best proportions and succession of memory allocations and data transfers to maximize overall performance. We found that optimal performance is possible to achieve by using multiple GPU devices.

2   Background

In most living organisms the genetic instructions used in their development are stored in the long polymeric molecules called DNAs. To decipher this information we need to determine the order of nucleotides - the elementary building blocks of a DNA that are also called bases. This task is important for many emerging areas of science and medicine.

Modern sequencing techniques split the DNA molecule into pieces that are also called reads. Reads are processed separately to increase the sequencing throughput. Then they are aligned to the reference sequence to determine their position in the molecule. This process is called read alignment and is extremely computationally intensive, as a complete genome of such complex organisms as humans is billions of bases long, and the amount of reads data produced by sequencing machines is usually an order of magnitude bigger [6][7].

Technically read alignment is a substring matching operation: we search for a pattern of length m in reference string of length n, where $n >> m$.

Straight-forward naive approach has daunting asymptotic performance of $O(mn)$, so aligning is done by building index and than matching reads against it.

While theoretically fastest search algorithm uses suffix tree, its space complexity makes it inefficient for big references[8]. There were successful attempts to decrease memory footprint of matching algorithm or even to trade computational complexity for space consumption. In MummerGPU++ the authors replaced search algorithm based on suffix tree with one based on suffix array, which lead for another performance improvement[4].

Space complexity of suffix array is also linear, and constant multiplier under $O(n)$ is 9 bytes per symbol in case of two-bit implementation. Search complexity for suffix array is $O(m + \log n)$ where m is the length of query and n is the length of reference.

Evaluation of MummerGPU++ showed that on references over 100MB the memory limit is still taxing performance, since it leads to splitting the index into small pieces to fit into GPU memory and repeating search for each part. Search complexity does not depend (or depends very little) on index size, so splitting index in chunks increases computation time linearly. Copying index and queries to the device also takes its share of time of time. We will provide a more detailed analysis of time consumed by data transfers later on.

As the chief way to increase performance we propose using an algorithm with lesser memory footprint. Such an algorithm can be based on Burrows-Wheeler transform and some additional data structures (FM-Index) instead of suffix array. BWT was introduced in 1994 by Burrows and Wheeler[9] and was used mainly for data compression. There are some recent sequence alignment solutions using BWT, some of them are not parallel (Bowtie [10]), some are using GPUs, but for different class of alignment [11]. Also in [12] authors discuss the potential of using GPUs for exact sequence matching on single GPU.

3   BWT Based Aligner

The Burrows-Wheeler Transformation of a text T, BWT(T), is constructed as follows: The Burrows-Wheeler Matrix of T is the matrix whose rows are all distinct cyclic rotations of T$ sorted lexicographically. BWT(T) is the sequence of characters in the rightmost column of the matrix[9]. It is possible to use BWT for substring search. We adopted backward search algorithm proposed by Manzini and Ferragana [5] for GPU. Here Occ is the number of occurrences of given symbol before given position in transformed sequence. Array C contains total number of occurrences of each symbol.

BWT has a property called LF mapping: the ith occurrence of character X in the last column of the BWT matrix corresponds to the same character in original text as the ith occurrence of X in the first column. Backward search procedure (fig. 1) uses LF mapping to calculate in rounds the rows of the matrix that begin with progressively longer suffixes of the query string.

The running time of the Backward search procedure is dominated by the cost of evaluating $Occ(c, q)$. If we build a two-dimensional array OCC such that

$OCC[c][q] = Occ(c, q)$ the backward search procedure runs in $O(m)$ time and it requires $O(|\Sigma|n\log n) = O(n\log n)$ bits.

The result of the Backward_search procedure is not the position(s) of matches in the reference sequence but the range of elements in the corresponding suffix array, containing indexes of actual matches in the reference. We suggest using suffix array on a host (which usually has enough memory to store it entirely) to decipher output of Backward_search procedure in $O(1)$ time. While it is possible to resolve positions of matches using the transformed text and OCC, generating all match positions on GPU will provide unpredictable amount of results per query, i.e. each execution thread will need to use unpredictable amount of device memory, and that is unsuitable for CUDA execution model. It will also cause additional overhead for moving data from device to host. To decipher search results on the host side we simply iterate suffix array elements bound by backward search procedure output values.

We use straightforward 2bits encoding for BWT itself. To compress OCC we split the transformed text into buckets of arbitrary size. For each bucket we will store the number of occurrences of each symbol in the transformed text before the first symbol of this bucket. For example, in 64 bit implementation for buckets of 32 symbols we will need 8 bits per symbol to store compressed OCC and 8 consequent memory reads to count the number of occurrences for any symbol. It

```
i:=p,  c:=P[p],
First:=C[c]+1,  Last:=C[c+1];
while ((First <= Last)
    and (i >= 2)) do
  c:=P[i-1];
  First:=C[c]+Occ(c,First-1)+1;
  Last:=C[c]+Occ(c,Last);
  i:=i-1;
  if (Last<First)
    then return no matches
    else return <First,Last>.
```

Fig. 1. Procedure Backward_search

gives us 10 bits of index per 8 bits of reference sequence and it is possible to change this ratio by varying OCC bucket size. 64 bit suffix array need 17bytes of memory, which is 13.5 times bigger. By merely replacing suffix array with BWT we already achieved 3-4 times performance improvement for cases where the size of data is too big to fit in memory for suffix-array based software but can be processed in one pass with our approach. Fig.2a) show how increasing reference size affects performance whether index can (BWT) or can not (suffix array) fit into GPU memory. We used NVIDIA Tesla 2050 card (2.6Gb memory) on the machine with 2.67GHz 4 cores Intel Core i7 920 CPU and 12GB of RAM running under CentOS 5.4.

Experimental implementation takes reference and a set of named queries in FASTA format as input. Output is a set of positions in the reference where queries are mapped. We chose CUDA as target architecture as it is de facto standard for GPGPU programming. The algorithm was implemented in C++ for CUDA programming language.

The CUDA kernel that performs the query search is an almost straightforward implementation of procedure Backward_search, where each thread is processing its own query independently. Each thread stores results in its own preallocated global memory and accesses the reference index only by reading.

Therefore there are no race conditions and no need for synchronization. Performance profiling showed that major share of time is consumed by loading data from global memory. On references over 100mb MummerGPU++ starts to subdivide index and loses performance, while with our approach index up to several gigabytes (i.e. complete human genome) can be stored in GPU memory. For bigger reference sequence still must be subdivided. In the next chapter we present mathematical model of how memory partitioning affects performance and use it to find optimal parameters.

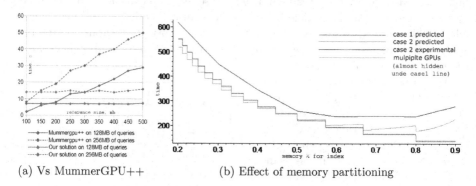

(a) Vs MummerGPU++   (b) Effect of memory partitioning

Fig. 2. Performance evaluation

4 Performance Model and Workload Balancing on Single-GPU

The theoretical complexity of matching algorithm itself is $O(q)$, where q is query length. In case of sequential execution increasing number of queries to process obviously increases execution time in the same linear manner. So we can say that the overall execution time depends linearly on the overall size of query set. We just need to keep in mind need to have query set bigger than amount of data necessary to saturate GPU parallel capacity (which is in our case approximately 10mb, much is negligibly small).

Let us call memory size S_{mem}, index size S_{idx} and query set size S_{qry}. The overall execution time consists of the computation time itself and the time spent on moving data between host and device: $T = T_{cmp} + T_{mem}$. This formula assumes the worst case scenario when there is no overlapping between computation and data transfers. Cases where such overlapping is possible will be discusses below.

Suppose we have to split the index into N_{idx} chunks of size P_{idx} each and the query set into N_{qry} chunks of P_{qry} bytes. There is an obvious correlation between N_{idx} and N_{qry}, but for the time being we shall not include it in the model to keep it simpler. We have to match each chunk of query set against each part of index, one such iteration (kernel launch) taking $C * P_{qry}$ time as complexity does not depend on index size. We have to repeat the matching procedure for each part of index and for each part of query set, which gives as execution time $T_{cmp} = C * N_{idx} * N_{qry} * P_{qry} = C * S_{qry} * N_{idx}$.

Now let's consider the communication expenses of moving index and query set parts from host to device. We have two basic options here. One option is to place one part of index on device, processing all subsets of query set one by one and then doing the same procedure for next part of index. The other option is to do the matching vice versa, i.e. matching one part of query set against all parts of index and then proceed to the next chunk of query set.

In the first case we need to copy P_{idx} bytes for each part of index, then N_{qry} times P_{qry} bytes of query subsets which equals to S_{qry} bytes and then to repeat this process N_{idx} times. Given host-to-device transfer bandwidth β communication will take $T_{mem} = \beta(P_{idx} + S_{qry}) * N_{idx} = \beta S_{idx} + \beta S_{qry} N_{idx}$ time. The overall time will be $T = C * S_{qry} * N_{idx} + \beta S_{idx} + \beta S_{qry} N_{idx} = (C + \beta)S_{qry}N_{idx} + \beta S_{idx}$.

For the second case using the same logic we get $T = C * S_{qry} * N_{idx} + \beta S_{qry} + \beta S_{idx} * N_{qry}$ overall execution time.

Let α be the share of memory occupied by index. Then each chunk of index will use αS_{mem} bytes and each chunk of queries $(1 - \alpha)S_{mem}$ bytes. We will have to split index into $N_{idx} = S_{idx}/\alpha S_{mem}$ chunks and query set into $N_{qry} = S_{qry}/(1-\alpha)S_{mem}$ chunks. Figure 2b shows how variation of α changes the overall execution time and that the first case allows for a potentially higher performance.

Actual value of C is retrieved form experiment and it depends on many parameters, like minimal required match length etc, but the asymptotic behavior will be the same. Performance of test implementation on big workloads confirms the predicted model (figure 2b).

So in the first case the overall performance increases as the index size is increased. This process continues up to the point where the memory remaining for queries is enough to run kernels with full memory saturation, which is relatively small and is not shown in figure2b.

In the second case we increase index size up until the point where communication expenses of repeating transfers of big index chunks are equal to the time spent on processing queries on extra number of index chunks. Maximal performance is better in the first case and it seems preferable from the point of view of pure GPU productivity. Moreover, it allows us to overlap communication and computation, as we can split queries without much penalty making performance even closer to ideal.

However, in this model we do not take into account the fact that results of matching of each subset of queries against each part or index need to be merged with each other. In the first case we have to store results of matching against each part of index somewhere until we process all queries and it will tax CPU-side memory/storage. This approach is completely inapplicable in a situation where queries are being streamed from some source (i.e. a sequencing machine) and we need to process each query block as it comes so we have to stay with worst case model - or we can try using multiple GPUs.

5   Multiple GPUs

Index chunk distribution among multiple GPU devices allows for smaller amount of repeatedly loaded index chunks per device. Ideally index chunks are not being moved at all. In this case theoretical performance in terms of pure GPU productivity will be even better, though not significantly, than that provided by the first approach on a single GPU device. On each device we spend $C * S_{qry} + \beta S_{qry}$ time for moving and processing all queries (once again, overlapping is possible in this case).

The process of deciphering and joining results consists of following stages. We get the ranges of suffix array elements as output of each GPU matching routine and restore actual positions of matches in reference sequence. For each device output we will have such list of positions. Then we need to merge these

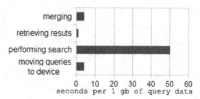

Fig. 3. Performance details

lists together and sort resulting list. It does indeed look like time consuming routine, but it obviously has $O(N_{idx})$ complexity, the same as complexity of search procedure itself. The exact multiplier depends on implementation, CPU characteristics and average number of matches for each query. However, given realistic search output, our sequential test implementation performed merging of 8 chunks of one million results in less then one second, which is definitely faster than processing corresponding amount of data on GPU (fig. 3). In previous experiments we used queries of 100 bases long, so 1 million results correspond to 100Mb of query data. In tests on both real and generated sequences multi GPU performance per device was same as for single GPU case 1. We performed benchmarking on one of the Tsubame 2.0 supercomputer nodes with 2 six-core Intel Xeon X5670 CPUs and 54GB of RAM running under SUSE Linux Enterprise Server 11 SP1 for this test. The node has three NVIDIA Tesla 2050 GPUs connected with 16 lanes of PCI Expression 2 on it. We used 100 bases long queries and set minimal match length to 40 bases. For 6GB reference sequence aligning efficiency per device was 3.55 million bases per second for single GPU and 3.7 for multi GPU implementation when all 3 devices were used. So 3 GPUs compared to single one gave us 3.11 times speed-up, i.e. 1.04 efficiency. Optimal number of devices is equal to the number of index chunks of optimal size. Increasing number of GPUs further will negatively affect the efficiency as index chunk size will be decreased.

6   Conclusion

Better software performance does not necessarily come from computational complexity of underlying algorithms. Choice of particular data structures and corresponding algorithms depends on how they meet characteristics and features of target hardware. This is particularly true for GPU devices.

This paper shows that using more compact data structures can lead to performance improvement in short read alignment problem. We refactored MummerGPU++, previous highly-efficient GPU exact-matching read alignment software by replacing suffix array with BWT and rewriting the corresponding search algorithms and get 3-4 times performance improvement. The analysis of application behavior for the case of workload size considerably exceeding device memory proves that higher performance can me achieved by intelligent strategy for data decomposition. We also showed that best performance per device for read alignment problem can be achieved by using multiple GPUs, and the optimal number of GPU devices for a particular task can be estimated from reference size.

References

1. Owens, J.D., Luebke, D., Govindaraju, N., Harris, M., Kruger, J., Lefohm, A.E., Purcell, T.J.: A survay on general-purpose computation on graphics hardware. Computer Graphics Forum 26(1), 80–113 (2007)
2. Delcher, A.L., Kasif, S., Fleischmann, R.D., Peterson, J., et al.: Alignment of whole genomes. Nucleic Acids Res. 27, 2369 (1999)
3. Schatz, M.C., Trapnell, C., Delcher, A.L., Varshney, A.: High-throughput sequence alignment using graphics processing units. BMC Bioinformatics 8, 474 (2007)
4. Gharaibeh, A., Ripeanu, M.: Size matters: Space/time tradeoffs to improve gpgpu applications performance. In: SC 2010 Proceedings of the 2010 ACM/IEEE International Conference for High Performance Computing, Networking, Storage and Analysis. IEEE Computer Society (2010)
5. Ferragina, P., Manzini, G.: Indexing compressed text. Journal of the ACM 53(4), 552–581 (2005)
6. Pop, M.: Genome assembly reborn: recent computational challenges. Briefings in Bioinformatics 10, 354 (2009)
7. Rothberg, J.M., Hinz, W., Rearick, T.M., et al.: An integrated semiconductor device enabling non-optical genome sequencing. Nature (475), 348–352 (2011)
8. Gusfield, D.: Algorithms on Strings, Trees and Sequences: Computer Science and Computational Biology. Cambridge University Press (1997)
9. Burrows, M., Wheeler, D.J.: A block-sorting lossless data compression algorithm. Technical Report 124, Digital Equipment Corporation (1994)
10. Langmead, B., Trapnell, C., Pop, M., Salzberg, S.L.: Ultrafast and memory-efficient alignment of short dna sequences to the human genome. Genome Biology 10(3), 10(25) (2009)
11. Li, R., Yu, C., Li, Y., et al.: Soap2: an improved ultrafast tool for short read alignment. Bioinformatics 15(25), 1966–1967 (2009)
12. Chen, S., Jiang, H.: An exact matching approach for high throughput sequencing based on bwt and gpus. In: 2011 IEEE 14th International Conference on Computational Science and Engineering (CSE). IEEE Computer Society (2011)

OpenMP/MPI Hybrid Parallel ILU(k) Preconditioner for FEM Based on Extended Hierarchical Interface Decomposition for Multi-core Clusters

Masae Hayashi and Kengo Nakajima

Information Technology Center, The University of Tokyo, 2-11-16 Yayoi Bunkyo-ku
Tokyo, Japan

Abstract. While ILU preconditioner is a powerful and popular preconditioning method for Krylov iterative solvers on sparse matrices derived
from finite element analysis, it have been exploerd the scalable hybrid
parallelization scheme for ILU preconditioner targetting multi/many-
core clusters. Hierarchical Interface Decompostion (HID) is a robust and
efficient parallel method for ILU preconditioner. The extended version of
HID (ExHID), our proposed method, introduces thicker level-2 connector
in order to consider fill-ins. Basing on HID and ExHID we developed hybrid parallel ILU preconditioner with fill-ins using OpenMP/MPI hybrid
parallel programing models. While inter-node parallelization is based
on HID/ExHID, we applied two different methods, multicolor based reordering and HID/ExHID to intra-node parallelization. The two implementations according to different hybrid strategy, HID(inter-node)-HID
(intra-node) and HID(inter-node)-MC(intra-node), are evaluated
through strong scaling tests and the better hybrid strategy is explored.
HID-HID generally results with better convergence and less fill-ins. On
the other hand, HID-MC could be more stable strategy than HID-HID
when increasing the number of threads per process.

1 Introduction

Domain decomposition method(DDM) is widely used parallelization method in
many finite element applications. Then distributed sparse linear systems derived
from each subdomain is considered as distributed objects[1,2] on each process.
On the other hand, preconditioining method using incomplete LU factorization
without fill-in (ILU(0)) is popular and effective preconditioner for finite element
applications. Under the parallelization based on DDM, block Jacobi-type localized preconditioner are widely used for parallel iterative solvers[3,4]. While they
provide excellent parallel performance for well-defined problems, the number of
iterations for convergence increases gradually according to the number of processors. Moreover this preconditioner decreases its robustness for ill-conditioned
problems with many processors, since it ignores the global effect of external nodes
come of inherently sequential natures of ILU preconditioner. The common remedy is to extend the overlapped elements between domains[5,6]. At the expenses

M. Daydé, O. Marques, and K. Nakajima (Eds.): VECPAR 2012, LNCS 7851, pp. 278–291, 2013.

of additional computation and communications it still only allows us to consider the global effect without updates from previous row operations. Another common remedy to parallelize ILU preconditioner is multicoloring based ordering. Multicoloring the subdomains which is assigned to processors, yeilds the parallelism from a global ordering[7]. But the archieved parallelism is limited to the color number and fiding optimal color number becomes another difficulty which relates to the performance. The Parallel Hierarchical Interface Decompostion Algorithm (PHIDAL) provides robustness and scalability for parallel ILU/IC preconditioners basing on "hierarchical interface decompostion (HID)"[8]. HID exploits a hierarchical decompostion of the graph which yields natural parallelism in the factorization process. For taking into account fill-ins, we introduced additional layers in higher level connectors defined in HID. The proposed method is called Extended HID (ExHID). We developed parallel ILU preconditioners with fill-ins (ILU(k)) basing on HID/ExHID.

To enhance our parallel ILU(k) preconditioner for multi-core environment, we applied a hybrid parallel programming model. A hybrid parallel programming model is often employed in order to archieve minimal parallelization overheads on multi-core clusters. Corse-grained parallelism is archieved through domain decompostion by message passing among nodes, while fine-grained parallelism is obtained via loop-level parallelism inside each node using compiler vased thread parallelisation techniques, such as OpenMP. HID/ExHID is applied to inter-node parallelization using message-passing interface (MPI), while two different methods are applied for intra-node parallelization using multi-threading(OpenMP). One is HID/ExHID and the other is multicolor-based reordering method. Both method are applied to distributed local data to yield the thread-level parallelism. We call the former strategy HID-HID since HID/ExHID is used for both intra-node and inter-node parallelization. And the latter is called HID-MC since HID/ExHID is used for intra-node parallelization and multicoloring is used for inter-node parallelization. Using OpenMP/MPI hybrid parallel programming model, we implemented finite element based simulations of linear elasticity problem solved by Krylov iterative solver with these hybrid parallel ILU(k) preconditioners. Developped codes are evaluated through strong scaling tests on multicore cluster called "T2K Open Supercomputer using up to 256 cores. Numerical experiments showed HID-HID generally leads to better convergence and less fill-ins. On the other hand, HID-MC could be more stable strategy than when increasing the number of threads per process.

The rest of this paper is organized as follows. Section 2 gives a overview of Hierarchical Interface Decompostion (HID) and parallel ILU preconditioning algrithm based on HID. And it descibes how fill-ins are introduced in the parallel ILU preconditioning algorithm by the extended HID. Section 3 describes tatget application based on finite element method and detailed implementations focusing on the thread level parallelization for two strategies, HID-HID and HID-MC. Section 4 contains the details on the test environment and numerical experiments followed by the conclusions section.

2   Hierarchical Interface Decompostion(HID) and Extended HID

By exploiting a static "hierarchical" decompostion of the graph, Hierarchical Interface Decompostion (HID) yeilds natural parallelism in the factorization and consider the global effect from external domain in parallel ILU/IC preconditioning process. However we cannot consider fill-ins from external domain in parallel ILU/IC preconditioning via HID. In order to consider fill-ins from external domains, we developed extended version of HID(ExHID). In this section we explain HID and our proposed method, ExHID. HID can be viewed as the methods from the angle of an ILU factorization combined with a form of nested dissection ordering in which cross points in the separators play a special role. The hierarchical decompostion starts with a partitioning of the graph with one layer of overlap. Then "stages" or "levels" are defined from thie partioning, with each level (or stage) consisting of a set of vetex grpups (small connected subgraphs). Each vertex group of a given stage is a separator for the vertex groups of a lower stage. The incomplete factorization process proceeds level by level from lowerst to higherst. Due to the separation property of the vertex groups at different levels, this process can be carried out in a highly parallel manner. These vertex groups are called connectors in definition of HID. The concept of connectors of different levels and keys are introduced for the purpose of applying this idea to general graphs as follows:

- Connectors of level-1 (C^1) Are the sets of interior points. Each set of interior points is called a sub-domain.
- A connector of level-k (C^k) (k¿1) is adjacent to k sub-domains.
- Connectors in the same level never be adjacent to each other.
- Key(u) is the set of sub-domains (connectors of level-1, C^1) connected to vertex u.

Fig.1 (left) shows the example of the partition of a 9-point gird into 4 domains. In this case, there are 4 connectors of level-1 (C^1, sub-domain), 4 connectors of level-2 (C^2) and 1 connector of level-4 (C^4). Note that different connectors of the same level are not connected directly, but are separated by connectors of higher levels. These properties provide the block structure of the coeficient matrix A through reordering the unknowns by this decompostion. By reordring the unknowns according to their level numbers, from the lowest to highest, the block strucuture would be appeared as shown in Fig.1 (right). This block strucutre leads to natural parallelism in ILU/IC decomposition or forward/backward substitution processes. Fig.2 shows pseudo code of forward substitution in ILU(0) preconditioning. The most outer loop is for levels. At the end of each level global communication is performed according to hierarchical communication table. Hierachical communication table is communication table in which export/import nodes communicated between neighbor processes are arranged by level hierarchy. This communication performed at the end of each level transfers the update information calculated in the present level to the next level. Thus HID allows us to consider the global effect of external domains in parallel ILU preconditioning, which leads to more robust parallel preconditioning than block Jacobi-type localized preconditioners.

However global effect from external doamains which can be considered via HID is confined within the case of ILU(0)/IC(0). Fig.3 shows example of domain decomposition of two dimentional 9-point grid into two domains via HID (left) and how the local data are distributed on two processes. Though it depends on the numbering assined to node A and node B, these two nodes are in the distance which can affect each other when considering 2nd level of fill-ins. Suppose node A will affect on node B as a 2nd level of fill-in, we cannot take into account the effect from node A by the given distirubted local data sets since node A and node B are in the different distributed data sets. The first remedy is simply extending overlapped elements between domains. This allows us to consider the fill-in effect on node B from node A. But the fill-in effect from node A can only be calculated without any updates even which might have been occured on node A from other nodes related to A. Thus it fundamentally results the same as blcok jacobi-type preconditioning. Then thicker layer of separators is introduced in HID. Fig.4 illustrates the level-2 connector is extended from one layer to three layers. Then node A which was in level-1 connector in Fig.3, becomes in the level 2 connector. Since the nodes are reordered according to the level from lower to higher, ILU preconditioning process proceeds level by level from lower to higher. Node A in the lowe level is always calculated prior to node B. This allows us to calculate the fill-in effect on node B from node A with updates already calculated. Extension of layer of higher level connector allows us to consider fill-ins but also leads to load inbalance in general.

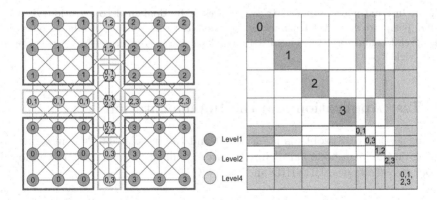

Fig. 1. Partitionoing of a 9 point grid into 4 subdomains by Hierarchical Interface Decompostion, resulted with four C^1 connectors(sub-domains), four C^2 connectors, one C^4 connectors. The numbers showed on the nodes are keys. The number of keys corresponds the level (left). Block structure is appeared in the coefficient matrix through HID reordering (right).

```
do lev= 1, LEVELtot
  do i= LEVindex(lev-1)+1, LEVindex(lev)
    SW1= WW(3*i-2,R); SW2= WW(3*i-1,R); SW3= WW(3*i  ,R)
    isL= INL(i-1)+1; ieL= INL(i)
    do j= isL, ieL
      k= IAL(j)
      X1= WW(3*k-2,R); X2= WW(3*k-1,R); X3= WW(3*k  ,R)
      SW1= SW1 - AL(9*j-8)*X1 - AL(9*j-7)*X2 - AL(9*j-6)*X3
      SW2= SW2 - AL(9*j-5)*X1 - AL(9*j-4)*X2 - AL(9*j-3)*X3
      SW3= SW3 - AL(9*j-2)*X1 - AL(9*j-1)*X2 - AL(9*j  )*X3
    enddo
    X1= SW1; X2= SW2; X3= SW3
    X2= X2 - ALU(9*i-5)*X1
    X3= X3 - ALU(9*i-2)*X1 - ALU(9*i-1)*X2
    X3= ALU(9*i  )* X3
    X2= ALU(9*i-4)*( X2 - ALU(9*i-3)*X3 )
    X1= ALU(9*i-8)*( X1 - ALU(9*i-6)*X3 - ALU(9*i-7)*X2)
    WW(3*i-2,R)= X1; WW(3*i-1,R)= X2; WW(3*i  ,R)= X3
  enddo

  call SOLVER_SEND_RECV_3_LEV(lev,.):   Communications using
                                        Hierarchical Comm. Tables.
enddo
```

Fig. 2. Forward substitution process in preconditioning. Global communication is performed at the end of each level, which allows us to caluculate the next level using updated data from previous level. This makes us parallel ILU(0) more consistent by HID than Block Jacobi-type localized method.

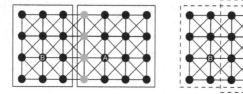

Fig. 3. Internal nodes assigned to two processes by domain decomposition via HID (left). There are two level-1 connectors (C^1, sub-domains) shown by Black nodes and one level-2 connector(C^2). The distributed local mesh is given with one overlapped layer (right).

3   Test Application and the Implementation

In this section test problems is discribed and the detailed implementations focusing on the intra-node parallelization are explained on each method employed in our two strategies, HID-HID and HID-MC.

3.1   Finite Element Based Simulations of Linearelasticity Problems

The test problem is finite element based simulations of three dimensional linearelasticity problem. Simple cube shaped analysis model is discritised by tri-linear hexahedral elements. Poisson's ratio and Young's modulus are given homogeneously for all elements and set to 0.25 and 1.0 respectively. The boundary

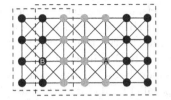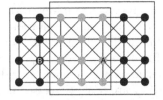

Fig. 4. The distributed local data sets when introducing thicker separator in HID (left). The level-2 connector (C^2, gray nodes) is extended to three layers. Thicker separator expand the range for global operations (right).

conditions are described in Fig.5. The Generalized Productive-type BiCG iterative solver with ILU(k) preconditioner is applied. Iterations are repeated until the norm $\|r\| / \|b\|$ is less than 10^{-8}. The code is based on the framework for parallel FEM procedures of GeoFEM[9], and GeoFEM's local data structure is applied. The local data strucutres in GeoFEM are node-based with overlapping elements[9]

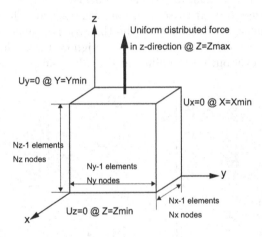

Fig. 5. Test model of simple cube geometry and the boundary conditions

3.2   HID/ExHID Ordering for Distributed Data

To yield the parallelism in the each distributed local data sets obtained by HID, we again apply HID/ExHID to the disributed data sets. We apply HID for ILU(0) and ExHID for ILU(k) to the disributed data sets on each MPI process and the resulted sub-domains (C^1) are assigned to threads and the adjacent C^k connectors to C^1s are also distributed among threads. Thus the number of sub-domains resulted by HID/ExHID corresponds to the number of threads in our implementation. And in our implementation of ExHID, extension of connector is applied in only level-2 connectors. No extensions in C^k connectors ($k > 2$).

Depending on the level of fill-ins to be considered we set the thickness. Fig.6 (right) illustrates our implementation of ExHID for the same example of 2D 9-point grid mesh. The dashed line shows the distribution of conncectors among threads. In Fig.6 (left) the decomposition by HID is illustlated for comparison. Without thicker separator node A and node B are in the same level in HID (left). Thus they are to be processed in parallel by different threads. If consider the fill-in effect on node A from node B, no updates on node B are avairable for node A. Moreover this trigger the data dependency problem between threads since one thread having node A is going to read the data on node B to calculate fill-ins on node A while another thread having node B is going to write the data on node B. By introducing thicker separator as in Fig.6 (right) put node B in level-2. Nodes in the higher level are processed after the nodes in the lower level so node B is processed after node A. This removes the data dependency and allows us to calculate the fill-in effect with update information.

However thicker separator is now applied for only level-2 connector, the same data dependency problem can happen between nodes in the higher level connectors than level 2, for example node C and node D in Fig.6 (right). Although they are in the distance which can affect when considering the 2nd level of fill-ins, they are in the same level. If these nodes are assigned to different threads, the same data dependency occurs between the threads. For avoiding such possible data dependencies in high level connectors, we ignore the fill-in effect from the node in the same level but on the different thread in our implementation.

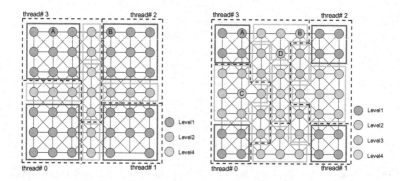

Fig. 6. The partitioning of a 9 point grid in to 4 subdomains by HID (left) and that by ExHID (right). Introducing thicker level-2 connector allows us to consider the fill-in effect on node A from node B and it also remove data dependency on it. Dashed line shows how connectors are distributed among threads.

3.3   Multicoloring Based Ordering for Distributed Data

As another parallelization method applied to distributed local data set is multicoloring, which is commonly used for parallelization of ILU factorizations. For taking into account the effect of fill-ins we apply the coloring rule which becomes strict according to level of fill-ins. For example, if we don't consider fill-ins at all, it is enough to color the nodes avoiding the adjacent nodes being in the same color. On the contrary, we apply the coloring rule so that every node are in different color from its neighbors' and "neighbors' of neighbors". Thus the number of colors increases in accordance with the level of fill-ins considered.

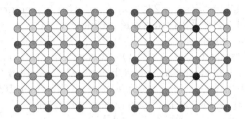

Fig. 7. Coloring rule for ILU(0) is simply to chose the color for each node avoiding the node in the same color with its adjacencies (right). For parallelization of ILU(1) each node has to be colored avoiding it in the same color with its adjacencies and the adjacencies of its adjacencies. The more colors are needed acording to the level of fill-ins to be considered.

3.4 Optimization for Memory Access

In the developed code, first touch data placement is considered. And appropriate command lines for NUMA control is applied, which is suppoted by Linux system, for efficient memory access to local memory. Minimizing memory access overhead is important for cc-NUMA architecture, such as T2K/Tokyo[10]. In order to reduce memory traffic in the system, it is important to keep the data close to the cores that runs with the data. On cc-NUMA architecture, this corresponds to making sure the pages of memory are allocated and owned by the core that works with the data contained in the page. The most common cc-NUMA page-placement algorithm is the first touch algorithm[11], in which the core first referencing a region of memory has the page holding that memory assigned to it. Very common technique in OpenMP program is to initialize data in parallel using the same loop schedule as it will be used lated in the computations.

4   Numerical Experiments

We tested two different types of hybrid parallel strategies HID-HID and HID-MC for solving the same problem discribed in section. To compare the performance for hybrid parallel method as iterative solver with ILU(k) preconditioing, we execute two numerical experiments. In both experiments we applied a strong scaling. The first test is run to compare the performances of these strategies using up to 16 nodes (256 cores) where the problem size is fixed at $3,090,903$ DOF (100^3 elements). The hybrid programming model applied is also fixed as 4x4 through this test. The second test is run to see their perfromances when the number of threads per process is incrased. The number of processes is fixed at eight and the problem size is fixed at $1,590,000$ DOF (80^3 elements).

4.1   Hardware Environment

Test environment is "T2K Open Super conputer (Todai Combined Cluster) (T2K/Tokyo), which was developed by Hitach under "T2K Open Supercomputer Alliance"[12]. T2K/Tokyo is an AMD Quad-core Opteron based combined cluster system with 952 nodes, 15,232 cores and 31 TB memory. Total peak performance is 140 TFLOPS. T2K/tokyo is an integrated system of four clusters. Number of nodes in each cluster is 512, 256, 128 and 56 respectively. Each node includes four "sockets" of AMD Quad-core Opteron processors(2.3GHz), as shown in Fig. Peack performance of each core is 9.2 GFLOPS and that of each node is 147.2 GFLOPS. Each node is connected via Myrinet-10G network. In the present work, up to 64 nodes of the system have been used. Because T2K/Tokyo is based on cache-coherent NUMA (cc-NUMA) architecture, careful design of software and data configuration is required for efficient memory access to local memory as stated in the previous section. We applied 4x4 hybrid programming model(four MPI processes x four OpemMP threads where one MPI process per one socket and four OpenMP threads per one MPI process) which is the most efficient case for this type of application on T2K.

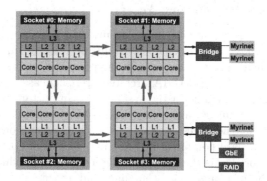

Fig. 8. The node specification of T2K/Tokyo. Each node includes four "sockets" of AMD Quad-core Opteron processors(2.3GHz). 16 cores per node.

4.2   Configuration of Hybrid Parallel Executions

We consider the fill-ins up to 2nd level of fill-ins. Thus ILU(0), ILU(1), and ILU(2) are applied to the iterative solver. In HID-HID strategy, we apply HID(3) for both ILU(1) and ILU(2). HID(3) is ExHID whose level-2 connector is extended to three layer of thickness. On the other hand, the number of colors required for parallelization of ILU(k) become larger the fill-in level k increases. In HID-MC strategy we set the number of colors for to the minimum number of colors required to yeild parallelism. The number of colors tested in the numerical experiments are eight for ILU(0), 27 for ILU(1), 64 for ILU(2).

4.3   Strong Scaling Test Up to 256 Cores

Fig.9 shows a comparison of elapsed time per iterations between HID-HID and HID-MC as increasing the number of nodes under 4x4 hybrid programming model where the problem size is fixed at $3,000,000$ DOF. For parallelization of ILU(k), we apply HID(3) while the number of colors in HID-MC is set to the minimum. Up to 256 cores(16 nodes), the similer scalabilities are observed for both strategy. Fig.10 and Fig.11 shows another comparisons on convergence and memory requirement between HID-HID and HID-MC. Fig.10 shows the iterations required for convergence and Fig.11 shows the number of fill-ins occured for ILU(1) and ILU(2). HID-HID leads to smaller values in both iterations and memory than HID-MC. Finally Fig.12 show the comparison on total solver time (elapsed time for total iterations) between HID-HID and HID-MC. Due to the larger iterations and larger number of fill-ins which is directly related to the computational the total solver time becomes larger by HID-MC than HID-HID larger iterations.

4.4   Strong Scaling Test with Different Hybrid Programming Models

Fig.13 shows solver time (elapsed time for total iterations) as the number of threads per process incrases. The number of threads is increased from one to 16 while process is fixed at eight. The problem size is here $1,590,000$ DOF. Fig.13 again shows it cost longer time by HID-MC than HID-HID. Fig.14 shows the iterations required for convergence as the number of threads per process increases.

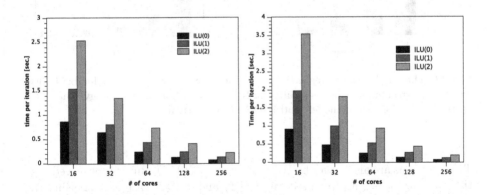

Fig. 9. Time per iteration with increase of the number of nodes under 4x4 hybrid parallele programming on the 100^3 elements problem. GPBiCG preconditioner with ILU(0), ILU(1), and ILU(2) preconditioner are applied. Intra-node parallelization for ILU(1) and ILU(2), HID(3) (ExHID adopted with thickness three) are used in HID-HID. On the other hand, the minimum number of colors, 27 colors for ILU(1) and 64 colors for ILU(2), is set in HID-MC. The result by HID-HID is shown in the left and HID-MC in the right.

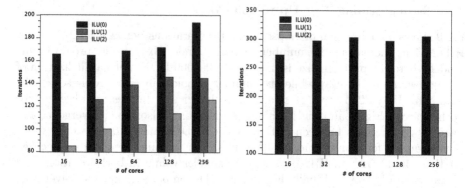

Fig. 10. Iteration with increase of the number of nodes under 4x4 hybrid parallele programming. GPBiCG preconditioner with ILU(0), ILU(1), and ILU(2) preconditioner are applied. Intra-node parallelization for ILU(1) and ILU(2), HID(3) (ExHID adopted with thickness three) are used in HID-HID. On the other hand, the minimum number of colors, 27 colors for ILU(1) and 64 colors for ILU(2), is set in HID-MC. The result by HID-HID is shown in the left and HID-MC in the right.

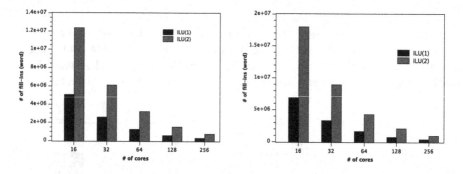

Fig. 11. The number of Fill-ins with increase of the number of nodes under 4x4 hybrid parallele programming which is resulted in ILU(1)/ILU(2) for each strategy. The result by HID-HID is shown in the left and HID-MC in the right.

The iterations are also larger by HID-MC than HID-HID. But it is observed that the more iterations are required for ILU(1) and ILU(2) case as increasing the number of therads per processe in HID-HID, while those stay almost the same for the number of threads in HID-MC. This is because our ILU(k) implementation based on ExHID can only avoid data dependency between C^1 connectors. In our implementation, we ignore the fill-in effect when the fill-in node has data dependency (i.e. the fill-in node is assigned to different thread and is in the same level) from higher level connectors, as discribed in section3.2. Such nodes to be ignored due to data dependency exist more and more in the higher level when the number of C^1 connectors increases (i.e. the number of threads increases).

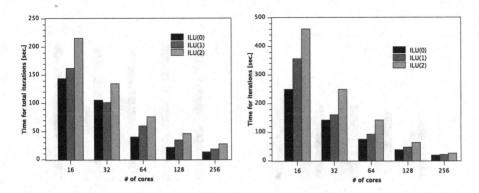

Fig. 12. Solver time (elapsed for total iterations) with increase of the number of nodes under 4x4 hybrid parallele programming. GPBiCG preconditioner with ILU(0), ILU(1), and ILU(2) preconditioner are applied. Intra-node parallelization for ILU(1) and ILU(2), HID(3) (ExHID adopted with thickness three) are used in HID-HID. On the other hand, the minimum number of colors, 27 colors for ILU(1) and 64 colors for ILU(2), is set in HID-MC. The result by HID-HID is shown in the left and HID-MC in the right.

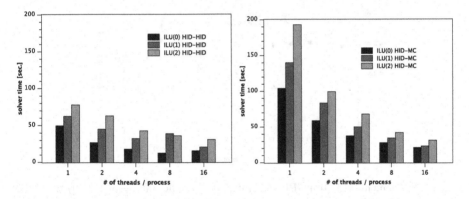

Fig. 13. Solver time (elapsed for total iterations) with increase of the number of threads per processes by GPBiCG with ILU(0), ILU(1), and ILU(2) for problem of 80^3 elements. The number MPI process is fiexed at eitht. 80^3 elements. HID-HID case is shown in left and HID-MC in right.

4.5   Number of Colors in HID-MC

For HID-MC we set the minimum number of colors needed for parallelization in previous execution. Now we increase the number of colors and compare the convergence and computation time between HID-HID and HID-MC. Fig.15 shows iterations and solver time for ILU(0)+GPBiCG executed by Hybrid 4x4 case using 16 nodes(256 cores) on small test model (80^3 elements, $1,594,323$ DOF).

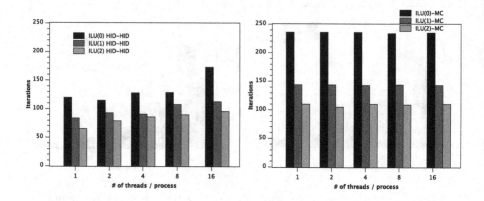

Fig. 14. Iterations with increase of the number of threads per processes by GPBiCG with ILU(0), ILU(1), and ILU(2) for problem of 80^3 elements. The number MPI process is fiexed at eitht. 80^3 elements. HID-HID case is shown in left and HID-MC in right.

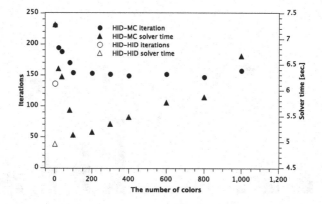

Fig. 15. 80^3 elements ILU(0)+GPBiCG, Intranode parallelization for ILU(0) HID-HID (no thicker separator), HID-MC (color number 8(minimum))

As Fig.15 displays, the iterations become shorter, while the solving time become shorter once and gradually become larger again in accordance with increase of the number of colors. This is simply because the larger number of colors leads the more frequent syncronization among threads. Thus the performance of HID-MC is directly related to the number of colors and if we can find the optimal number of colors (in this case 100 is the optimal), HID-MC can archive the close performance to HID-HID.

5   Concluding Remarks

ILU(k) preconditioner is widely used powerful preconditioner in many finite element applications and it is important to establish the hybrid parallel scheme for ILU(k) preconditioner. HID and ExHID is a robust and effective parallelization method of ILU(k) and we developed hybrid parallel scheme for ILU(k) preconditioner based on HID and ExHID. For intra-node parallelization we can have several variations of methods. In order to find more efficient strategy for hybrid parallel ILU(k), we implemented using two differetn strategies, HID-HID and HID-MC. By applying OpenMP/MPI hybrid programming model, our two implementation are evaluated on multi-core cluster using up to 256 cores. HID-HID strategy leads better convergence and fewer fill-ins than HID-MC generally. However towering many core environment more than 100, HID-MC can be more stable strategy than HID-HID. Multicoloring brings us another task how to find optimal number of colors but the flexibiliy of multicoloring which is easily applicable to the case of the large number of threads becomes advantage to HID-HID. Depending the test environment, tactical selection of hybrid strategy is important for ILU(k) preconditioner on multi/many- core clusters.

References

1. Jones, M.T., Plassmann, P.E.: Scalable iterative solution of sparse linear systems. Parallel Computing 20(5), 753–773 (1994)
2. Saad, Y., Sosonkina, M.: Distributed schur complement techniques for general sparse linear systems. SIAM Journal on Scientific Computing 21(4), 1337–1356 (2000)
3. Parallel iterative solvers of geofem with selective blocking preconditioning for nonlinear contact problems on the earth simulator (2003)
4. The Impact of Parallel Programming Models on the Linear Algebra Performance for Finite Element Simulations (2007)
5. Washio, T., Hisada, T., Watanabe, H., Tezduyar, T.E.: A robust preconditioner for fluid-structure interaction problems. Computer Methods in Applied Mechanics and Engineering 194(39-41), 4027–4047 (2005)
6. Nakajima, K.: Parallel preconditioning methods with selective fill-ins and selective overlapping for ill-conditioned problems in finite-element methods. In: Shi, Y., van Albada, G.D., Dongarra, J., Sloot, P.M.A. (eds.) ICCS 2007, Part III. LNCS, vol. 4489, pp. 1085–1092. Springer, Heidelberg (2007)
7. Saad, Y., Sosonkina, M.: Enhanced parallel multicolor preconditioning techniques for linear systems. UMSI Research Report/University of Minnesota (Minneapolis, Mn). Supercomputer Institute, vol. 99, p. 4 (1999)
8. Henon, P., Saad, Y., et al.: A parallel multistage ilu factorization based on a hierarchical graph decomposition. SIAM Journal on Scientific Computing 28(6), 2266 (2006)
9. http://geofem.tokyo.rist.or.jp/
10. Nakajima, K.: Flat mpi vs. hybrid: Evaluation of parallel programming models for preconditioned iterative solvers on "t2k open supercomputer". In: International Conference on Parallel Processing Workshops, ICPPW 2009, pp. 73–80 (2009)
11. Mattson, T., Sanders, B., Massingill, B.: Patterns for parallel programming. Addison-Wesley Professional (2004)
12. http://www.opensupercomputer.org/

Parallel Smoother Based on Block Red-Black Ordering for Multigrid Poisson Solver

Masatoshi Kawai[1], Takeshi Iwashita[2,3], Hiroshi Nakashima[2], and Osni Marques[4]

[1] Graduate School of Informatics, Kyoto University, Japan
[2] Academic Center for Computing and Media Studies, Kyoto University, Japan
[3] JST, CREST, Japan
[4] Lawrence Berkeley National Laboratory, USA

Abstract. This paper describes parallelization techniques for a multigrid solver for finite difference analysis of three-dimensional Poisson equations. We first apply our block red-black ordering for parallelization of a Gauss-Seidel (GS) smoother, whose sequentiality is often problematic in parallelization of multigrid methods. Furthermore, we introduce a new multiplicative Schwarz smoother, in which multiple GS iterations are performed in each of red-black ordered blocks. Numerical tests are conducted on a cluster of multi-processor nodes comprising four quad-core AMD Opteron processors to examine the effectiveness of these parallel smoothers. The multi-process test using 216 processes in flat-MPI model shows that the block red-black GS smoother and its multiplicative Schwarz variant achieve 1.3 and 1.8 times better performance than the conventional red-black GS smoother, respectively.

1 Introduction

Solving Poisson equation problems often plays an important role in computational science simulations. To accelerate these simulations, the development of a fast Poisson solver is demanded. This paper focuses on the finite difference method for three-dimensional Poisson equation problems. In the finite difference analysis, it is important to efficiently solve the derived linear system of equations. In this paper, the multigrid method is used as the solver of the linear system. This method is suitable for large-scale problems, because it achieves a convergence rate independent from the number of degree of freedoms [1]. In this paper, we discuss the parallelization of the multigrid solver.

It is well known that the Gauss-Seidel (GS) smoother shows good convergence for the linear system arising in the discretized Poisson equation, and is superior to other smoothers such as (weighted) Jacobi. However, the GS smoother cannot be parallelized straightforwardly due to its data-dependency. Accordingly, several GS-based smoothers have been proposed for parallelization. The hybrid smoother combining weighted Jacobi and GS smoothers is a well-known easily parallelizable smoother based on the additive Schwarz method. Another popular method is to impose red-black ordering on GS smoothing to have a convergence

M. Daydé, O. Marques, and K. Nakajima (Eds.): VECPAR 2012, LNCS 7851, pp. 292–299, 2013.

rate superior to the hybrid smoother as well as a large degree of parallelism [2]. However, when the red-black GS smoother is implemented with stride memory accesses to the data array elements, the computational time for a smoothing step becomes longer than the sequential GS smoother due to poorer cache utilization.

To remedy this problem, we first introduced the block red-black ordering to parallelize the GS smoother [3]. Next, for further improvement of the solver performance, we have proposed a parallel multiplicative Schwarz smoother based on the block red-black ordering [4]. The smoothing step of the proposed smoother is faster than that of the red-black GS smoother, while it attains a good convergence rate comparable to the sequential GS smoother. In this paper, we mainly verify the effectiveness of the block red-black GS smoother and its variant in numerical tests of multi-process implementations, the solver performance on a multi-threaded environment was reported in [4].

2 Multigrid Solver for Three-Dimensional Poisson Equation

This paper deals with a multigrid solver for the finite difference analysis of a three-dimensional Poisson equation given by

$$-\nabla^2 \phi = \rho \text{ on } \Omega, \tag{1}$$

where ρ is the given source, ϕ is the unknown spatial function, and Ω is the analyzed domain. Applying a 7-point finite difference scheme to (1), we obtain a linear system of equations to solve:

$$A^{(0)} u^0 = f^0. \tag{2}$$

In this paper, we solve (2) by means of geometrical multigrid method, in which multiple coarse grids are generated from the original grid. Using these grids, the solution process of the multigrid method is given by Alg. 1. In the algorithm, $A^{(i)}$, I_i^{i+1}, I_{i+1}^i and L denote the coefficient matrix on i-th level grid, the restriction operator from i-th level to $i+1$-th level, the prolongation operator from $i+1$-th level to i-th level, and the number of grids, respectively. The vectors u^i and $\tilde{u}^i$ are for i-th level unknowns and their approximation, respectively. In the analysis, the i-th level grid is twice as fine as the $i+1$-th level grid in each direction. We use the full-weighting restriction operator and tri-linear prolongation operator shown in, for example, [2] on the grids.

3 Parallelization of Geometric Multigrid Poisson Solver

3.1 Parallelization of Multigrid Method

The major components of a multigrid solver are a smoother and operators for restriction and prolongation. Since the restriction and prolongation operations are naturally parallelized by an usual domain decomposition, these operations

$$Smoothing \ on \ \boldsymbol{A}^{(0)} \boldsymbol{u}^0 = \boldsymbol{f}^0$$

$$\boldsymbol{f}^1 = \boldsymbol{I}_0^1 \left(\boldsymbol{f}^0 - \boldsymbol{A}^{(0)} \tilde{\boldsymbol{u}}^0 \right)$$

$$Smoothing \ on \ \boldsymbol{A}^{(1)} \boldsymbol{u}^1 = \boldsymbol{f}^1$$

$$\vdots$$

$$Solve \ \boldsymbol{A}^{(L-1)} \boldsymbol{u}^{L-1} = \boldsymbol{f}^{L-1}$$

$$\vdots$$

$$Smoothing \ on \ \boldsymbol{A}^{(1)} \boldsymbol{u}^1 = \boldsymbol{f}^1$$

$$\tilde{\boldsymbol{u}}^0 \leftarrow \tilde{\boldsymbol{u}}^0 + \boldsymbol{I}_1^0 \tilde{\boldsymbol{u}}^1$$

$$Smoothing \ on \ \boldsymbol{A}^{(0)} \boldsymbol{u}^0 = \boldsymbol{f}^0$$

Alg. 1. Procedure of L-level V-cycle multigrid method

on a grid point are mutually independent from others. On the other hand, the parallelization of the smoother is often problematic. For example, the weighted Jacobi smoother can be easily parallelized, but its convergence rate is inferior to the GS smoother. In the following subsection, we describe conventional GS-based parallel smoothers which have been used in many practical applications.

3.2 Conventional Parallel Smoothers

Hybrid Smoother. The hybrid smoother consists of the weighted Jacobi and the GS smoothers. It uses the weighted Jacobi method for boundaries of each domain-decomposed region allocated to a process/thread so that the GS smoother works on the interior region independently from those of other processes or threads. The smoother can be regarded as an additive Schwarz smoother, in which the GS method is used for the subdomain solver. The hybrid smoother is included in the popular multigrid solver library, BoomerAMG [5]. Although the hybrid smoother has the advantage of implementation easiness, it often entails a degradation in convergence. The convergence rate of the hybrid smoother depends on the number of processes/threads, and it is often worsen when that number is increased.

Red-black GS Smoother. In the 7-point finite difference scheme, each grid point has data dependence only on adjacent 6 points. Consequently, when we paint the grid points alternately by red and black, the grid points having an identical color can be updated independently of each other. That is, after the GS smoothing step is performed for red grid points in parallel, the smoothing step for black grid points is also processed in parallel. This parallel smoothing technique is called red-black GS (RB-GS) smoother. The convergence rate of the RB-GS smoother can be different from that of the sequential GS smoother in general.

However, it is known that the RB-GS smoother has good convergence when compared to the sequential one in homogeneous Poisson equation problems. Accordingly, the RB-GS smoother is the most widely-used parallel smoother for the problems.

In finite difference analyses, the RB-GS smoother is usually implemented with stride memory accesses of unknown and right-hand vector elements, because they are mapped from three-dimensional grid points with lexicographical ordering and thus red and black elements are arrayed alternately. This ordering is expedient for easy and efficient implementation of the whole of a simulation problem, and so it is for those of the restriction and prolongation. However, the efficiency of the smoother itself is degraded from the GS smoother because the stride accesses have poorer cache-line utilization causing lower cache-hit ratio and thus performance.

4 Block Red-Black Gauss-Seidel Smoother and Its Variants

In this paper, we aim to present a parallel smoother free from stride accesses for high-performance while keeping a convergence rate comparable to the conventional RB-GS smoother. For this, we first introduce block red-black ordering, which we originally proposed was for parallel ILU preconditioning [3], to parallelize a GS smoother. Next, we present a new multiplicative Schwarz smoother, being an enhanced version of the block red-black GS smoother.

4.1 Block Red-Black Ordering

Block red-black ordering is one of the parallel ordering techniques. In the ordering, the entire grid is first divided into multiple blocks. Next, red-black ordering is applied to the blocks as shown in Fig. 1, where a block of a color never has direct data-dependency on other blocks of the same color. This feature allows us to parallelize the smoothing step of the block red-black GS (BRB-GS) smoother, so that the GS smoothing is applied to all red blocks in parallel and then to all black ones also in parallel.

It is important that arbitrary ordering can be used in a block and thus the lexicographical one is used in the analysis. Consequently, this ordering makes the block-level GS smoother implemented without stride memory accesses. That is, by choosing the block size sufficiently large especially for the axis conforming to the memory address ordering, the accesses of unknown and right-hand vector elements in the GS smoothing in the block are made almost sequential. This means that a cache-line having a series of vector elements is almost fully utilized by a series of smoothing operations resulting in higher cache-hit ratio and thus more performance than the conventional (i.e., element-wise) RB-GS smoother.

As for the convergence rate, it is strongly expected that the BRB-GS's rate is sufficiently high and comparable to the RB-GS's and the sequential GS's. This expectation is based on the fact that the BRB-GS with the block size of one for

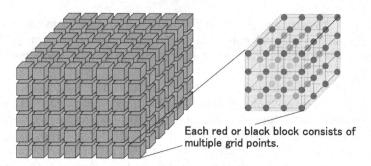

Each red or black block consists of multiple grid points.

Fig. 1. Concept of block red-black ordering

each axis is just identical to the RB-GS while extremely large block size virtually gives us the sequential GS smoothing. Though the convergence with block sizes between these two extremes above needs to be investigated with real problems, our numerical tests discussed afterward support our expectation.

Furthermore, the cache-blocking technique for smoothing and other components, namely the restriction and the prolongation operations, shown in a context of the sequential multigrid solver in [6], can be easily applied to the BRB-GS. To use the technique in the BRB-GS, we only need to set the block size to match the cache size, and to execute the restriction or the prolongation operation just after/before the smoothing step in each block. Since this technique doubles the utilization of a cache line, it should significantly improve cache-hit ratio and thus performance.

4.2 Modified Block Red-Black Gauss-Seidel Smoother

In this subsection, we introduce a modified version of the BRB-GS smoother to increase the total solver performance. In this version, we simply increase the number of GS iterations in each red/black block from 1 to $\alpha > 1$. The smoother, denoted by mBRB-GS(α) hereafter, is regarded as a multiplicative Schwarz smoother rather than a parallel GS smoother based on parallel ordering. In the following, we discuss the advantage of this multiple iterations of block smoothing.

In general, increasing the number of iterations, namely β, in a smoothing step for the *whole* grid space, leads to improved convergence. However, since grids are usually much larger than the cache size, the computational cost for one smoothing step is also increased in proportion to β. As the smoother is dominant in the multigrid solver in term of computational cost, the total computational time is also proportional to β. Consequently, increasing β rarely reduces the total computational time unless the convergence is improved by a factor of β or more.

On the other hand, increasing the number of GS steps α for a *block* in the BRB-GS is expected to have different behavior. Let t_s be the computational time required for the first smoothing step in a block. When we set the block size less than the cache size, we can expect that the computational time for the succeeding smoothing step $\tilde{t}_s$ is much less than t_s because of on-cache computation. Consequently, the computational time t_m for one smoothing step of mBRB-GS(α) is given by

$$t_m \approx t_s + (\alpha - 1)\tilde{t}_s < \alpha t_s. \tag{3}$$

From (3), even when the improvement in the convergence does not reach a factor of α, the total computational time can be reduced by increasing the number of GS steps in a block.

5 Numerical Results

5.1 Test Model and Used Parallel Computer

Numerical tests were conducted on the T2K Open Supercomputer at Kyoto University to examine the developed multigrid Poisson solver. The parallel supercomputer consists of SMP nodes, each consisting of four AMD quad-core Opteron 8356 (2.3 GHz) and 32 GB (DDR2-667) shared memory. The internal network between computational nodes, which is based on the DDR-InfiniBand technology, provides full bisection bandwidth and 8 GB/s for each node. The code was written in Fortran90 and MPI. In the present study, we only use the flat-MPI parallel programming model. It is noted that the multi-threaded implementation reported in [4] is based on OpenMP.

In the test model, the analyzed domain Ω is given by $[-0.5, 0.5]^3$ together with Dirichlet boundary condition of $\phi = 0$, and the source term is defined as

$$\rho(r) = \begin{cases} 1 & \text{if } r \leq 0.015 \\ 0 & \text{otherwise} \end{cases} \tag{4}$$

where r is the distance from the origin. To evaluate weak scalability, we fix the finest grid size per process at 128^3.

5.2 Performance Evaluation of the Multigrid Solver

Table 1 lists the computational time and the number of cycles of the multigrid solver with 1, 8, 64 and 216 cores (processes). Because the convergence of the hybrid smoother is 1.7 times slower than the sequential GS smoother, its parallel speedup is limited. On the other hand, the RB-GS and the BRB-GS smoothers attain a convergence rate comparable to that of the sequential GS smoother. However, the computational time for one multigrid cycle of the RB-GS is longer than that of the sequential GS because of the stride memory access. Consequently, only 63.7-fold (weak scaling) speedup is obtained by 216 processes compared with the sequential GS smoother. It is noted that the weak scaling speedup ratio is given by $(T_s \times P)/T$, where P is the number of processes (cores), and T_s and T are the elapsed time in sequential and parallel computations, respectively. Table 1 also indicates that the BRB-GS has advantage in

Table 1. Comparison of the computational time (s) and the number of cycles(in parenthesis) of parallel smoothers

	Number of processes			
	1	8	64	216
Seq.GS	3.55(10)		–	
Hybrid	–	9.99(17)	14.08(17)	15.08(17)
RB-GS	4.84 (9)	8.83(11)	11.39(10)	12.22(10)
BRB-GS	4.33(11)	6.22(11)	8.40(11)	9.43(11)

Table 2. Computational time (s) and the number of cycles(in parenthesis) of the multigrid solver using mBRB-GS on 216 processes

		p_o				
		1	2	3	4	5	6
	1	10.38/11	8.23/9	8.33/9	7.58/8	7.84/8	8.12/8
	2	8.10/ 9	7.42/8	6.69/7	7.21/7	7.11/7	7.58/7
p_r	3	7.54/ 8	6.75/7	6.78/7	7.05/7	7.30/7	7.59/7
	4	7.52/ 8	6.86/7	6.96/7	7.13/7	7.53/7	7.68/7
	5	7.85/ 8	7.16/7	7.45/7	7.80/7	7.92/7	8.24/7
	6	8.31/ 8	7.36/7	7.45/7	7.80/7	7.92/7	8.24/7

the computational time per cycle than the RB-GS, because of the more efficient cache utilization. Therefore the BRB-GS attains better solver performance than the conventional RB-GS smoother. The weak scaling speedup ratio of the BRB-GS reaches 81.3 by 216 processes.

Next, the performance of the mBRB-GS is examined. We carried out a preliminary test on single node using 16 treads for checking the computational time for the first and the second smoothing steps for the block, t_s and $\tilde{t}_s$. On the grid of 512^3, t_s and $\tilde{t}_s$ were measured 0.67 s and 0.11 s, respectively. The preliminary test confirms the inequality(3) because $\tilde{t}_s$ is approximately one sixth of t_s.

Table 2 shows the computational time and the number of iterations of multi-process parallel processing with 216 processes when the mBRB-GS(p_r) and the mBRB-GS(p_o) are used for the pre- and post-smoothing steps, respectively. Table 2 confirms that increasing the smoothing steps in the block leads to the improvement in the solver performance. The best result of the mBRB-GS was obtained when $(p_r, p_o)=(2, 3)$ for 216 processes. Figure 2 shows the weak scaling speedup ratio of the solver with various parallel smoothers. In the test, (p_r, p_o) of the mBRB-GS was set to (2, 3). The numerical result shows the advantage of the BRB-GS and the mBRB-GS over conventional parallel smoothers. In the multi-process parallel processing, increasing the number of GS steps in the block of the mBRB-GS improves the convergence without increasing the number of MPI communications. Consequently, the mBRB-GS achieves 1.80 times better performance than the RB-GS.

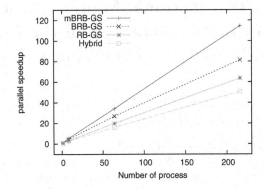

Fig. 2. Weak scaling speedup of parallel multigrid solver with various smoothers compared to a sequential solver with GS smoother

6   Conclusion

In this paper, we investigated the parallelization of a multigrid solver for three-dimensional Poisson equation problems, focusing on the parallel processing of the Gauss-Seidel (GS) smoother. First, we introduced the block red-black ordering technique to parallelize the GS smoother. In this method, the analyzed grid is divided into multiple blocks, to which the red-black ordering is applied. Numerical tests on 216 processes showed that the block red-black GS smoother can be 1.3 times faster than the conventional red-black GS smoother due to more efficient cache utilization. Next, we presented the modified version of the block red-black ordering GS smoother (mBRB-GS). In this version, we iterate GS smoothing in each block twice or more to have a like multiplicative Schwarz smoother. The smoother improves the convergence without largely increasing the computational time of one smoothing step. Consequently, the mBRB-GS can be 1.8 times faster than the RB-GS on 216 processes.

References

1. Trottenberg, U., Oosterlee, C., Achuller, A.: Multigrid. Elsevier Academic Press (2001)
2. Thoman, P.: Multigrid Methods on GPUs. VDM (2008)
3. Iwashita, T., Shimasaki, M.: Block red-black ordering: a new ordering strategy for parallelization of ICCG method. Int. J. Parallel Prog. 31, 55–75 (2003)
4. Kawai, M., Iwashita, T., Nakashima, H.: Parallel Multigrid Poisson Solver Based on Block Red-Black Ordering. In: Proc. Symposium on High Performance Computing and Computational Science, pp. 107–116 (2012) (in Japanese)
5. Henson, V., Yang, U.: BoomerAMG: A parallel algebraic multigrid solver and preconditioner. Applied Numerical Mathematics 41, 155–177 (2002)
6. Kowarschik, M., Rüde, U., Weiß, C., Karl, W.: Cache-aware multigrid methods for solving poisson's equation in two dimensions 64, 381–399 (1999)

Software Transactional Memory, OpenMP and Pthread Implementations of the Conjugate Gradients Method – A Preliminary Evaluation

Vincent Heuveline[2], Sven Janko[1], Wolfgang Karl[1], Björn Rocker[3], and Martin Schindewolf[1]

[1] Karlsruhe Institute of Technology (KIT), Chair for Computer Architecture and Parallel Processing, Haid-und-Neu-Straße 7, 76131 Karlsruhe, Germany
{sven.janko,karl,schindewolf}@kit.edu

[2] Karlsruhe Institute of Technology (KIT), Engineering Mathematics and Computing Lab (EMCL), Fritz-Erler-Str. 23, 76133 Karlsruhe, Germany
vincent.heuveline@kit.edu

[3] Robert Bosch GmbH, Corporate Sector Research and Advance Engineering, Robert-Bosch-Platz 1, 70839 Gerlingen-Schillerhöhe, Germany
bjoern.rocker@de.bosch.com

Abstract. This paper shows the runtime and cache-efficiency of parallel implementations of the Conjugate Gradients Method based on the three paradigms Software Transactional Memory (STM), OpenMP and Pthreads. While the two last named concepts are used to manage parallelization as well as synchronization, STM was designed to handle only the latter. In our work we disclose that an improved cache-efficiency does not necessarily lead to a better execution time because the execution time is dominated by the thread wait time at the barriers.

Keywords: Software Transactional Memory, OpenMP, Pthreads, Conjugate Gradients Method, Case Study.

1 Introduction and Motivation

Parallelization is state of the art in scientific computing for a long time, but also comes with the need to synchronize parallel threads of execution. Efficient synchronization is the key towards maximum performance on (shared memory) multicore architectures. Traditional synchronization primitives in OpenMP (e.g., `omp critical`) and Pthreads (e.g., locks) achieve synchronization through enforcing mutual exclusion. Threads may experience long delays when waiting for a lock to become available. In the last decade Transactional Memory (TM) has been proposed for synchronization. Instead of following the traditional pessimistic scheme of avoiding memory conflicts, TM favors an optimistic scheme that detects and resolves conflicting accesses. The goal of this strategy is to increase the scalability in regard to a high number of threads and coevally to decrease the time needed for synchronization.

M. Daydé, O. Marques, and K. Nakajima (Eds.): VECPAR 2012, LNCS 7851, pp. 300–313, 2013.
© Springer-Verlag Berlin Heidelberg 2013

In this paper, we evaluate the applicability of TM for the method of Conjugate Gradients (CG), a solver for linear systems of equations that is frequently used in many fields of application, especially in the area of structural mechanics and computational fluid dynamics.

This paper is structured as follows. Section 2 reviews related work in the area of Transactional Memory research and describes the method of CG. In Section 3 we will discuss our implementations which leads us to Section 4 where we present our results. Section 5 concludes our work and presents ideas for future work.

2   Background on Transactional Memory

Writing efficient, highly scalable and correct parallel software is a challenging task for programmers. They are in charge of the synchronization and communication of the involved threads in order to avoid memory conflicts and deadlocks. Furthermore, one should have consolidated knowledge of the mechanisms of the underlying runtime/operating system.

The idea behind TM is to simplify the process of writing parallel code by providing basic constructs for synchronization. Originally Herlihy and Moss invented TM in 1993 as an architectural extension to enable lock-free data structures [17]. The basic construct is called a transaction and guarantees to execute the comprising load and store commands with three properties: atomicity, consistency and isolation [11]. In contrast to traditional synchronization approaches that enforce mutual exclusion, transactions are executed optimistically in parallel and conflicts are detected and resolved by a TM run time system. The TM system can be implemented in hardware [21,20], software [12,13,14] or as a combination of both as hybrid TM [19,16,18]. In case of a Software Transactional Memory (STM) system a user-level library fulfills this task. All transactional accesses to shared memory are performed through this STM. Often this library comes with compiler support. Then a programmer can use a specific keyword to mark a transaction in the code. For this region of code the compiler inserts calls into the STM instead of performing accesses to shared memory directly. This approach offers the most convenience for the programmer, but also comes at some cost. The compiler makes pessimistic assumptions and, thus, may instrument more memory accesses than absolutely necessary. This phenomenon is known as over-instrumentation [22]. Further, STMs suffer from overheads due to the managing of meta data and acquiring and releasing locks [15]. In our work, we use an STM-only approach with manually instrumented memory accesses, which has the advantage that the resulting binary does not suffer from over-instrumentation through the compiler. OpenMP [6] and Pthreads APIs [7] provide the thread management for the STM.

2.1   Conjugate Gradients

The Method of **Conjugate Gradients** (CG) is a common solver in many fields of application, especially in the area of structural mechanics and computational fluid dynamics. There, finite element and volume methods (FEM/FDM) are frequently employed. Within most linearization methods linear systems have to be solved, consuming often most of the time within the solution process. If those systems are symmetric and positive definite, CG can be applied. Usually, CG is used in combination with an appropriate preconditioning depending on the problem that is solved. Within this paper, we evaluate a pure version of CG.

CG is an improvement of the methods of Steepest Descent and Conjugate Directions where the disadvantage in building the search directions disappears. By conjugation of the residuals the search directions are constructed and it is no longer needed to store the old search vectors (see [5] for a detailed explanation).

In the following, n denotes the dimension of the matrix A that is introduced in Algorithm 1. There are one matrix-vector prod-

Algorithm 1. Conjugate Gradients

1: $r_0 = b - Ax_0$, $p_0 = r_0$, A spd
2: **for** $i = 0, 1, 2, \ldots$ **do**
3:    $\alpha_i = \frac{r_i^T r_i}{p_i^T A p_i}$
4:    $x_{i+1} = x_i + \alpha_i p_i$
5:    $r_{i+1} = r_i - \alpha_i A p_i$
6:    $\beta_i = \frac{r_{i+1}^T r_{i+1}}{r_i^T r_i}$
7:    $p_{i+1} = r_{i+1} + \beta_i p_i$
8: **end for**

uct, three vector updates and two dot-products per iteration cycle. In general the matrix-vector product for computing Ap_j needs n^2 floating-point multiplications and $n^2 - n$ summations, leading to a asymptotic complexity of $O(n^2)$. The complexity for the vector updates is $O(n)$, because n multiplications and n summations for each update are needed. The inner product has also a complexity of $O(n)$. Hence the total complexity per iteration step is dominated by the matrix-vector product. If sparse matrices are used and only nonzero entries are saved the complexity decreases. Supposing a matrix having nnz nonzero entries and $nnz \ll n^2$. Now, nnz floating-point multiplications are needed and at most $nnz - 1$ summations. The total complexity is $O(nnz)$ compared to $O(n^2)$ in the dense case.

3   Implementations

In the first step we implemented the CG-algorithm as described in Section 2.1 using the C programming language and OpenMP. Then this code was transformed to a similar Pthreads variant and afterwards this version was modified using TM commands. With this approach, we were able to get results that were comparable to each other. The main calculation takes part in five for-loops, corresponding to lines 3 to 7 in Algorithm 1, each iterating n times where n still is the dimension of the underlying matrix of the algorithm.

3.1   OpenMP

In our OpenMP program the parallelization is achieved by inserting *#pragma omp for*-statements on top of each for-loop. Because a for-loop has an implicit barrier, we did not have to care about data dependencies between the several for-loops.

Listing 1.1 shows the five for-loops where most of the execution time is spent. In line 4 and 10 we make use of an OpenMP feature that is called *reduction*.

Every thread, that is part of the calculation, gets its own private copy of the variable *scp_temp*. Each thread then uses this copy for calculations inside of the loop. Afterwards an addition takes place and the variable *scp_temp* can be used as the sum of all thread-private variables. As this reduction is generated by the OpenMP compiler and hence is hidden from the programmer, this is exactly where we had to insert commands to achieve mutual exclusion when writing the Pthreads versions (with and without TM, respectively).

Listing 1.1. OpenMP parallelization

```
1   #pragma omp for private(...) schedule(static)
2   for (i=0; i<n; i++){ ... }
3   ...
4   #pragma omp for reduction(+:scp_temp) schedule(static)
5   for (i=0; i<n; i++) scp_temp += p[i]*v[i];
6   ...
7   #pragma omp for schedule(static)
8   for (i=0; i<n; i++){ ... }
9   ...
10  #pragma omp for reduction(+:scp_temp) schedule(static)
11  for (i=0; i<n; i++) scp_temp += r[i]*r[i];
12  ...
13  #pragma omp for schedule(static)
14  for (i=0; i<n; i++){ ... }
```

3.2 Pthreads

The basic idea of the OpenMP-to-Pthreads transformation was to pass the main calculation to each created thread modifying the start and end index of each for-loop. With this practice we tried to keep very close to the internal implementation of our OpenMP model. Of course, we also had to reproduce the implicit barriers in OpenMP. We achieved this by inserting explicit barriers that are implemented using the simple function shown in Listing 1.2.

3.3 Transactional Memory

The third model of the CG-algorithm was written using our Pthreads program as basis. Only a few lines in the TM-implementation differ from this code. We used the same thread creation concept and also the same barriers. We customized our code mainly in two places by inserting TM instructions to generate a transaction. With this transaction, threads will optimistically read and write the shared variable *scp_temp* concurrently. Listing 1.3 shows a TM version of the reduction that was previously mentioned in Section 3.1.

Listing 1.2. Pthreads barrier implementation

```
1   typedef struct barrier {
2     pthread_cond_t complete;
3     pthread_mutex_t mutex;
4     int count;
5     int crossing;
6   } barrier_t;
7
8   void barrier_cross(barrier_t *b) {
9     pthread_mutex_lock(&b->mutex);
10    b->crossing++;
      // one more thread through
11    if (b->crossing < b->count) {
      // if not all here, wait
12      pthread_cond_wait(&b->complete, &b->mutex);
13    } else {
14      pthread_cond_broadcast(&b->complete);
      // last thread arrived
15      b->crossing = 0;
      // Reset for next time
16    }
17    pthread_mutex_unlock(&b->mutex);
18  }
```

Listing 1.3. TM reduction

```
1   for (i = thread->start; i < thread->end; i++) {
2     scp_temp_private += p[i]*v[i]; }
3   START(thread->id, RW);
4     scp_temp_private += (double)LOAD_DOUBLE(&scp_temp);
5     STORE_DOUBLE(&scp_temp, scp_temp_private);
6   COMMIT;
```

4   Numerical Experiments

4.1   Hardware and Software Environment

All experiments were run on two computers C1 and C2 which are described in detail in Table 1. As compiler, *gcc-4.4* was invoked with options *-O3* and -g3. As Software Transactional Memory library we chose TinySTM [9,10]. TinySTM is a lightweight and efficient word-based STM implementation. Its time-based algorithm is derived from LSA and its lock-based design borrows several key elements from other word-based STMs, such as TL2.

Table 1. Experimental Setup

	Computer 1 (C1)	Computer 2 (C2)
CPU name	Intel Xeon X5670[1]	AMD Opteron 2378[2]
#Sockets	two	two
CPU frequency	2.93 GHz	2.36 GHz
RAM	12 GB	16 GB
Size of L1	32 KB	64 KB
Size of L2	256 KB	512 KB
OS	GNU/Linux (Ubuntu)	GNU/Linux (Ubuntu)
Kernel version	2.6.32-29-server	2.6.38-12-server
Architecture	x86_64	x86_64
Hyper-threading	yes	no
NUMA	yes	yes

4.2   Numerical Results

Each of our tests were run several times (>15) taking into account the exclusive computing time for the process. Afterwards we calculated the arithmetic mean of the results omitting the fastest and the slowest run. Thus, every value in the subsequent figures is an arithmetic average of at least 14 executions.

We evaluated the performance assuming a sparse matrix described by means of a CSR format. The linear system is obtained from a finite element discretization of the stationary heat equation without heat source (homogeneous case) which represents a prototype of Laplace's equation. It is equivalent to a finite differences discretization based on the 3-point-stencil. The matrix has a dimension of $5\,000\,000$ and $14\,999\,998$ nonzero entries (nnz). The residual stopping criteria for the residual is set to 10^{-13}.

Performance. As expected, with all three paradigms we could achieve significant speedups over the respective single thread execution time by increasing the number of threads from one to two, three, four and more. On Computer 1 we achieved a speedup of $S_8 = 2.72$ (OpenMP), $S_8 = 3.42$ (Pthreads) and $S_8 = 3.79$ (STM) by increasing the number of threads from one to eight. See Figure 1. The dimension of the underlying matrix was set to $5M$ in this case. Although there are clear differences in the above-named speedups, the execution time does not differ much with eight threads on C1. The good speedup with STM is also due to the high single thread overhead. A special case is 24 threads and OpenMP: the calculation takes slightly longer than with the single threaded concept. A model that describes the effects of the scheduling on the run time of the application explains the peak with 24 threads. Christmann et al. developed this model when they where researching the impact of oversubscription on the application throughput [8]. The scheduling algorithm must be fair (each process

---

[1] Registered Trademark by Intel Corporation.
[2] Registered Trademark by AMD.

gets a fair share of time), balance the load across cores (or hardware threads) and pins a process to a processing element as long as possible. Our case meets all of these assumptions. The explanation for the peak in execution time is that a fully loaded node (with 24 OpenMP threads) competes with some background process for computing resources. Eventually, after a long stall time, one of the OpenMP threads gets migrated leading to a prolonged overall execution time. Later experiments verified that a later Linux kernel (with version number 3.0.0-23-server) that enables a fair scheduling of groups instead of processes does not show this behavior anymore.

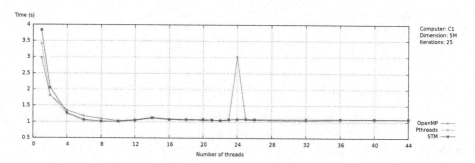

Fig. 1. Runtime analysis of the CG method (OpenMP, Pthreads, STM)

Another finding of our research is that the Pthread-program (and also the TM-program) is in the majority of cases slightly slower than the OpenMP-variation. We see mainly two causes therefor: a) more cache misses (see Section *Cache-Efficiency Analysis*) and b) more time is spent at the barriers. We will discuss the second argument in more detail now. We measured the time that the threads had to wait at each barrier in the Pthreads-program on C2. For two threads it took 7-15% of the overall execution time to wait at the barriers. Four threads waited about 25%, six threads about 43% and eight threads even about 70% of the execution time. What we discovered with this analysis is, that the time at the barriers increases rapidly if there are pairs of threads that have the same Hardware-Thread-ID. That means these threads cannot be executed in parallel because they are mapped to the same hardware entity and hence have to run one after the other. Those pairs appear even if the number of threads is less than the number of possible hardware threads in the system, which is an important insight. Apparently this is nothing the software developer is able to control.

PARSEC Barrier Tests. Another test concerning the barriers was the comparison of two slightly varying Pthreads programs. On the one hand, we used the constructs for the barriers as described in Listing 1.2, on the other hand, the PARSEC barriers were tested [2]. When using the PARSEC barriers, one can choose between two modes: 1) spinning ON and 2) spinning OFF. The results (executed on C2) are shown in Figure 2.

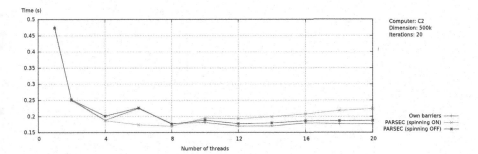

Fig. 2. PARSEC barrier comparison

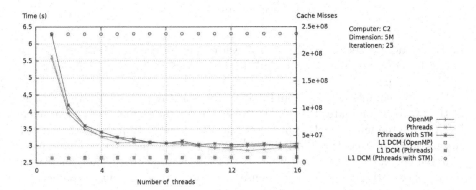

Fig. 3. Level 1 data cache misses

In general, using the PARSEC barriers did not bring strong advantages over the simple implementation which we used earlier. On the contrary, it was even slower for most configurations. Only for four to eight threads, if the spinning option was set to *ON*, it resulted in a faster runtime. As shown in Figure 2, the execution time increases for more than eight threads. That is exactly as we expected. In this example, spinning does not make any sense for a higher number of threads.

Cache-Efficiency Analysis. In order to understand the differences in runtime we also studied the cache behavior in detail. Our main focus was on the data cache, because the instruction cache analysis did not reveal noticeable results. The following designations apply to C2. As one can see in Figure 3, the data cache misses of the first level cache (L1 DCM) do not change with an increasing number of threads[3], whereas the L2 DCMs increase at the same time (see Figure 4). This holds as long as the number of threads is less or equal the number of possible hardware threads (here 8) in the system. Beyond this point the L2 DCMs are not increasing anymore. From Figure 4 we educe that there is no direct correlation of the L2 DCMs and the execution time of the program. Rising

---

[3] The DCMs of OpenMP are hidden behind the DCMs of Pthreads.

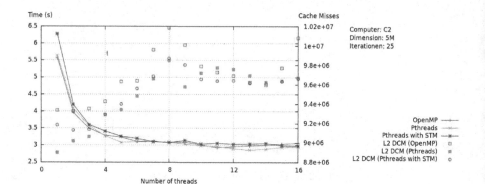

Fig. 4. Level 2 data cache misses

L2 DCMs do not necessarily bring a slower execution time and on the contrary, falling L2 DCMs do not always result in a faster execution time. This holds for all three programs.

If we now compare Figure 4 and 2, it becomes apparent that the waiting time at the barriers dominates the execution time of the programs. As one can see in Listing 1.2, the main function of the barrier construct is to pause a thread at a specific point of execution until all other threads reach the barrier. That means, that the last thread significantly increases the execution time. Thus, increasing the number of threads only makes sense, if the time that is spent at the barriers is improved, too.

4.3   Experiments with Matrices from Structural Engineering

In this section we will add additional experiments with two more matrices to provide a richer evaluation of the implemented CG variants. In order to complement the findings from the previous section, we also distinguish two more implementation variants that differ in the implementation of the reduction. The two reductions in CG are each implemented in two ways: *Fast*, and *Slow*. *Fast* uses a thread-local variable to accumulate the results over a private part of the vector that is assigned to this specific thread. Then, a single update adds the thread-local variable to the shared memory one that is guarded by a critical section or transaction. Thus, contention between threads only arises from the update of the shared memory variable. The *Fast* version of CG updates one shared memory location per thread and reduction pattern. Thus, the number of executed transactions equals the number of threads times the number of reductions per iteration. This is the reduction pattern that has also been used for the previous experiments presented in this paper. The *Slow* version updates the shared memory location in one transaction or critical section and does not use thread-local variables. Because each reduction only updates one shared memory location, OpenMP atomic is a perfect fit because it maps to a processor instruction that assures the atomicity of the update (if the processor supports

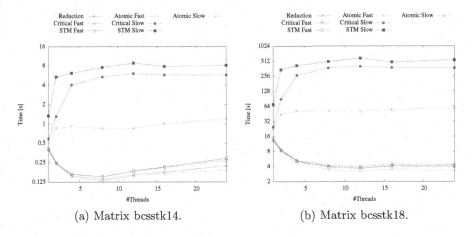

(a) Matrix bcsstk14.   (b) Matrix bcsstk18.

Fig. 5. Run times with OpenMP and all variants of synchronization mechanisms

atomics). The *Fast* version, again, uses thread-local variables whereas the *Slow* version does not. This atomicity is limited to one memory location and can not be extended. Thus, the *Atomic Fast* uses the thread-local variables to update the shared memory locations and the *Atomic Slow* updates the shared memory location for each new value. Since each value must be updated with a separate atomic instruction there is no need to distinguish between long and short sections. These self-made reductions are complemented by the OpenMP reduction, denoted as *Reduction*, that the programmer specifies through using a `#pragma omp for reduction(+:var) schedule(static)`.

The two additional matrices are taken from the matrix market[4]. This assures that other researchers may compare their results with ours. The first matrix is called bcsstk14, has a dimension of 1806 with 32630 entries. The matrix has a Frobenius norm of $6.5 * 10^{10}$ and an estimated condition number of $1.3 * 10^{10}$. The matrix is used for static analysis in structural engineering and models the roof of the Omni Coliseum in Atlanta. The second matrix, called bcsstk18, has a dimension of 11948 with 80519 entries, a Frobenius norm of $2.4 * 10^{11}$ and an estimated condition number of 65. Both matrices are from the set BCSSTRUC2 of Prof Mac Will, Georgia Institute of Technology. As experimental setup, we use again C1 with OpenMP parallelization only this time running Linux kernel version number 3.0.0-23-server that enables a fair scheduling of groups instead of processes and, thus, does not show the peak in the run time with 24 threads (cf. to Section 4.2).

Figure 5 highlights the run time and shows that the implementation strategy of the reduction is more important than the choice of the synchronization mechanisms for this reduction. Clearly all *Slow* variants perform worse than their single-threaded counter parts. This is due to the contention on the shared variables that are updated in each loop iteration. The *Fast* variants show a far better scalability due to a reduction in time with an increasing thread number.

---

[4] http://math.nist.gov/MatrixMarket

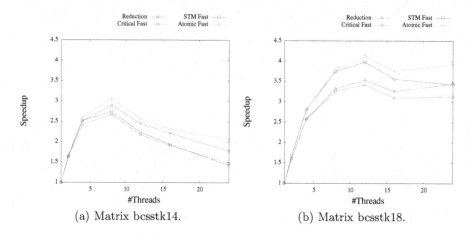

Fig. 6. Speedup of the *Fast* synchronization variants over the respective single thread performance

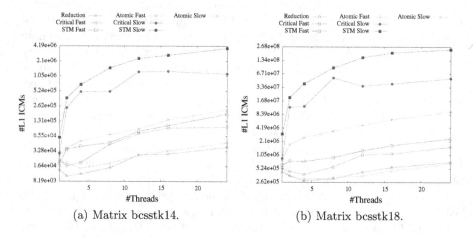

Fig. 7. L1 instruction cache misses with different synchronization variants

In Figure 6 the speedup over the respective single thread performance of the *Fast* variants. *Atomic Fast* achieves the highest speedup for bcsstk14 with 8 and for bcsstk18 with 12 threads. For bcsstk18 *STM Fast* also performs almost as good as *Atomic Fast*.

Figure 7 shows the L1 instruction misses for both matrices. The *Slow* variants have a significant higher number of instruction cache misses than the *Fast* variants. The interesting observation is that for bcsstk14 *Atomic Slow* is almost as good as *STM Fast*. This shows the large overhead in terms of instructions that is associated with using an STM system. The *Reduction* and *Atomic Fast* utilize the instruction cache the most efficiently.

For the L1 data cache misses, shown in Figure 8, the trend is similar as with the L1 instruction cache miss but the gap between STM and the other mechanisms

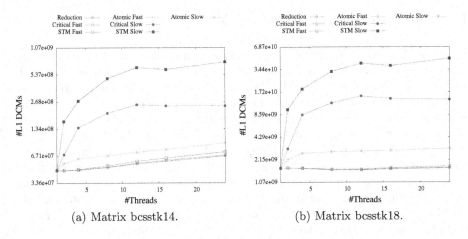

Fig. 8. L1 data cache misses of all different synchronization variants

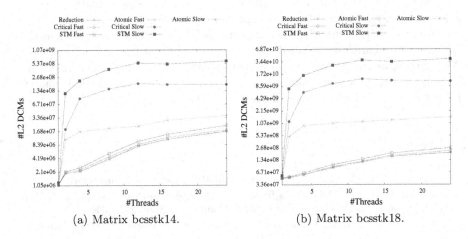

Fig. 9. L2 data cache misses with OpenMP and all variants of synchronization mechanisms

is not as big when it comes to the *Fast* variants. The *Slow* versions again have significantly more misses due to the contention on the shared variable.

Figure 9 highlights the L2 data cache misses of both matrices and implementation variants across thread counts. Again *STM Fast* is slightly worse than the other *Fast* variants but still significantly better than the *Slow* variants. For matrix bcsstk18 the gap between *STM Fast* and the rest seems smaller which may be due to the larger size of the matrix. Calculating the L2 data cache miss rate according to $\frac{L2\ data\ cache\ misses}{L2\ data\ cache\ accesses}$ yields the following results. For the *Fast* variants there is an almost linear increase in the L2 data cache miss rate with the number of threads whereas the *Slow* variants follow a logarithmic curve which results in a rate of more than 80 %. For *Fast* the rate stays well below 20 % for bcsstk18 and 30 % for bcsstk14.

5   Conclusion and Future Work

In our work we compared three similar implementations of the Conjugate Gradients Method. One that uses OpenMP, one that uses Pthreads without TM and one that uses Pthreads with TM constructs. The results showed that it is very important to reduce the waiting time at the barriers in order to improve execution time of these programs. Complementary experiments reveal that the choice for implementing the reduction is even more important than the choice of the synchronization primitive. Using thread-local variables for implementing the reduction is indispensable for a well-performing implementation. Further, these experiments with two additional matrices, lent from the static analysis in structural engineering, confirm the findings of the previous experiments regarding the cache efficiency of STM. In most cases, OpenMP is the fastest approach on both machines. This is the case because STM suffers from significantly more L1 cache misses compared to a pure OpenMP or Pthread implementation. In terms of performance, OpenMP is the first choice if the CG algorithm is used as done in this paper. As future work, the above-mentioned programs should be compared to other formulations of the Conjugate Gradients Method, such as the pipelined CG-algorithm described in [1] in order to benefit from advantages with TM. Further, we want to research the influence of using a NUMA machine (e.g., through employing a first touch policy for memory pages) on the performance of the different implementations of the CG method.

References

1. Strzodka, R., Göddeke, D.: Pipelined Mixed Precision Algorithms on FPGAs for Fast and Accurate PDE Solvers from Low Precision Components. In: IEEE Proceedings on Field-Programmable Custom Computing Machines (2006)
2. Bienia, C.: Benchmarking Modern Multiprocessors. Princeton University (January 2011)
3. Bolz, J., Farmer, I., Grinspun, E., Schröder, P.: Sparse matrix solvers on the GPU: conjugate gradients and multigrid. ACM Transactions on Graphics 22, 917–924 (2003)
4. Goodnight, N., Lewin, G., Luebke, D., Skadron, K.: A multigrid solver for boundary-value problems using programmable graphics hardware. In: Eurographics/SIGGRAPH Workshop on Graphics Hardware, pp. 102–111 (2003)
5. Saad, Y.: Iterative Methods for Sparse Linear Systems (2003)
6. OpenMP Architecture Review Board: OpenMP Application Program Interface. Version 3.1 (July 2011), http://www.openmp.org/mp-documents/OpenMP3.1.pdf
7. Butenhof, D.: Programming with POSIX threads. Addison-Wesley Longman Publishing Co., Inc. (1997)
8. Christmann, C., Hebisch, E., Weisbecker, A.: Oversubscription of Computational Resources on Multicore Desktop Systems. In: Pankratius, V., Philippsen, M. (eds.) MSEPT 2012. LNCS, vol. 7303, pp. 18–29. Springer, Heidelberg (2012)
9. Felber, P., Fetzer, C., Marlier, P., Riegel, T.: Time-Based Software Transactional Memory (2010)

10. Felber, P., Fetzer, C., Riegel, T.: Dynamic performance tuning of word-based software transactional memory. In: Proceedings of the 13th ACM SIGPLAN Symposium on Principles and Practice of Parallel Programming (2008)
11. Larus, J., Rajwar, R.: Transactional Memory. Synthesis Lectures on Computer Architecture. Morgan & Claypool Publishers (2007)
12. Lev, Y., Luchangco, V., Marathe, V., Moir, M., Nussbaum, D., Olszewski, M.: Anatomy of a Scalable Software Transactional Memory. In: Workshop on Transactional Computing TRANSACT 2009 (February 2009)
13. Saha, B., Adl-Tabatabai, A., Hudson, R., Minh, C., Hertzberg, B.: McRT-STM: a high performance software transactional memory system for a multi-core runtime. In: PPoPP 2006: Proceedings of the Eleventh ACM SIGPLAN Symposium on Principles and Practice of Parallel Programming, pp. 187–197 (2006) ISBN 1-59593-189-9
14. Dice, D., Shalev, O., Shavit, N.: Transactional Locking II. In: Dolev, S. (ed.) DISC 2006. LNCS, vol. 4167, pp. 194–208. Springer, Heidelberg (2006)
15. Cascaval, C., Blundell, C., Michael, M., Cain, H.W., Wu, P., Chiras, S., Chatterjee, S.: Software Transactional Memory: Why is it Only a Research Toy? Queue 6(5), 46–58 (2008) ISSN 1542-7730
16. Lev, Y., Moir, M., Nussbaum, D.: PhTM: Phased Transactional Memory. In: TRANSACT 2007: 2nd Workshop on Transactional Computing (August 2007)
17. Herlihy, M., Moss, E.: Transactional memory: architectural support for lock-free data structures. SIGARCH Comput. Archit. News 21(2), 289–300 (1993) ISSN 0163-5964
18. Christie, D., Chung, J., Diestelhorst, S., Hohmuth, M., Pohlack, M., Fetzer, C., Nowack, M., Riegel, T., Felber, P., Marlier, P., Rivière, E.: Evaluation of AMD's advanced synchronization facility within a complete transactional memory stack. In: EuroSys 2010: Proceedings of the 5th European Conference on Computer Systems, pp. 27–40 (2010) ISBN 978-1-60558-577-2
19. Damron, P., Fedorova, A., Lev, Y., Luchangco, V., Moir, M., Nussbaum, D.: Hybrid transactional memory. In: ASPLOS-XII: Proceedings of the 12th International Conference on Architectural Support for Programming Languages and Operating Systems, pp. 336–346 (2006) ISBN 1-59593-451-0
20. Yen, L., Bobba, J., Marty, M., Moore, K., Volos, H., Hill, M., Swift, M., Wood, D.: LogTM-SE: Decoupling Hardware Transactional Memory from Caches. In: IEEE 13th International Symposium on High Performance Computer Architecture (HPCA), pp. 261–272 (February 2007) ISBN 1-4244-0804-0
21. Hammond, L., Wong, V., Chen, M., Carlstrom, B., Davis, J., Hertzberg, B., Prabhu, M., Wijaya, H., Kozyrakis, C., Olukotun, K.: Transactional Memory Coherence and Consistency. In: Proceedings of the 31st Annual International Symposium on Computer Architecture, p. 102. IEEE Computer Society (June 2004)
22. Yoo, R., Ni, Y., Welc, A., Saha, B., Adl-Tabatabai, A., Lee, H.: Kicking the tires of software transactional memory: why the going gets tough. In: SPAA 2008: Proceedings of the Twentieth Annual Symposium on Parallelism in Algorithms and Architectures, pp. 265–274 (2008) ISBN 978-1-59593-973-9

A Smart Tuning Strategy for Restart Frequency of GMRES(*m*) with Hierarchical Cache Sizes

Takahiro Katagiri[1], Pierre-Yves Aquilanti[2,3], and Serge Petiton[4]

[1] Information Technology Center, The University of Tokyo,
2-11-16 Yayoi, Bunkyo-ku, Tokyo 113-8658, Japan
katagiri@cc.u-tokyo.ac.jp
[2] LIFL, Université Lille 1 Science et Technologie, Cite Scientifique, Bâtiment M3,
59655 Villeneuve d'Ascq Cedex, France
[3] A*STAR Computational Resource Centre,
1 Fusionopolis Way, #17-01 Connexis, Singapore 138632
aquilantip@acrc.a-star.edu.sg
[4] LIFL, Université Lille 1 Science et Technologie, Cite Scientifique, Bâtiment M3,
59655 Villeneuve d'Ascq Cedex, France
Serge.Petiton@lifl.fr

Abstract. In this paper, we propose a smart tuning strategy that uses the cache size hierarchy of current multicore architectures. Both increase and decrease auto-tuning (AT) strategies for the restart frequency of GMRES(*m*) (Generalized Minimum Residual) are evaluated with the proposed hierarchical cache sizes. This evaluation, using one node of the T2K Open Supercomputer (Univ. Tokyo), demonstrates that the proposed strategies are very efficient compared to previous strategies without hierarchical cache sizes. We test both strategies with 22 matrices from the University of Florida Sparse Matrix Collection. As a result, we find an average speedup of 1.13× (maximum 2.06×) using an increase strategy (an implementation of Xabclib), and an average speedup of 4.25× (maximum 15.1×) with a decrease strategy (Aquilanti's) using the proposed method.

Keywords: Auto-tuning, GMRES(*m*), Dynamic Restart Frequency Adjustment, Xabclib.

1 Introduction

Current computer architectures have complex structures, with multicore systems commonly utilizing non-uniform memory access and hierarchical caches. In terms of cache organization, several multiple caches are independent of cores, but one cache is shared across multiple cores. Thus, tuning the performance of software is becoming increasingly difficult. To solve this problem, auto-tuning (AT) technology is frequently used by non-experts to establish high performance computing on current architectures.

A wide range of problems is expressed through a linear system. Hence, solving sparse linear systems, such as the following, is a crucial task for scientific computing:

M. Daydé, O. Marques, and K. Nakajima (Eds.): VECPAR 2012, LNCS 7851, pp. 314–328, 2013.
© Springer-Verlag Berlin Heidelberg 2013

$$Ax = b. \tag{1}$$

When the operator A is sparse, it is common to use iterative solvers. The Generalized Minimum Residual (GMRES) algorithm [1] is considered to be powerful and can be applied to a wide range of cases. For the iterative approximation of the solution vector, the Krylov subspace is used to determine the direction in which the solution of the linear system lies, such that:

$$x_1 \in x_0 + K^m(A, r_0), \tag{2}$$

where $K^m(A, r_0) \equiv span\{r_0, Ar_0, \cdots, A^{m-1}r_0\}$ is the Krylov subspace of dimension m, x_0 is the initial guess, x_1 is the estimated vector in the first iteration, and r_0 is the initial residual.

As GMRES iterates, its computing power and memory requirements are likely to increase when the dimension of the Krylov subspace is large. As memory is limited in practice, it is common to restart GMRES after m iterations. This variant is known as the restarted GMRES [1]. The parameter m controls the restart; hence, we call this parameter the "restart frequency." It has been demonstrated that m is a critical argument, driving not only memory consumption but also the execution time required for the solver to converge. Determining m is thus a very important issue, affecting not only high-performance libraries but also research topics in AT.

1.1   Categories of AT for the Restart Frequency of GMRES(m)

As the restart frequency of GMRES(m) is very crucial for performance, several AT strategies have been proposed. In this section, we categorize the strategies as follows.

- **Increase Strategy**
The increase strategy is defined as follows. In the first phase, the restart frequency is assigned a small number, say $m = 2$. In the next phase, the frequency is increased using run-time information. Previous strategies in this category include that of Sosonkina et al. [2] and the strategy implemented in Xabclib [3]. Obviously, this strategy is good for easy problems that require only a small number of restarts to converge.
- **Decrease Strategy**
The decrease strategy is the opposite of the increase strategy. In the first phase, the frequency is assigned a maximum size. In the next phase, the frequency is decreased using run-time information. The major strategy in this category is that of Baker et al. [4]. Obviously, this strategy is good for difficult problems, which require a large number of restarts. One of the drawbacks of this strategy concerns the difficulty of finding the optimal maximum size for m.
- **Hybrid Strategy**
This strategy is a hybrid of the increase and decrease strategies. The frequency is dynamically increased or decreased according to run-time information. As the

hybrid strategy needs an initial restart frequency, we can define two subcategories according to whether the initial frequency increases or decreases.

The strategy proposed by Habu *et al.* [5] starts from a small initial frequency; hence, this is categorized as an increasing hybrid. On the other hand, the strategy proposed by Aquilanti *et al.* [6] starts from a maximum size, hence this is a decreasing type. As for the previous strategies, it depends on the convergence as to which type is better suited to the problem.

● **Other Considerations: The Target**

Although the strategies shown above have general properties, the target of their evaluations is limited to GMRES(m). There are a few implementations and preliminary evaluations for the adaptation of these strategies to other algorithms. The strategy used in Xabclib [3] has been extended to the restarted Lanczos and explicit restarted Arnoldi problems. The codes have been released to the public as a free (GNU licensed) library.

1.2   Originality of This Paper

With respect to the above categorization, we summarize the originality of this research as follows:

● **Showing Effectiveness of AT with Hierarchical Caches on Multicore Architectures**

In general, the parameter search space in restart parameter AT is huge; hence, we need some heuristics to avoid using a brute force search and to obtain reasonable parameter settings. One of the candidates for finding reasonable parameter settings for AT is to use hardware parameters. We use an AT strategy with cache information to demonstrate the effectiveness of this approach. The key to our strategy is that the cache information helps to restrain the search of m.

Aquilanti *et al.* [6] first proposed the hybrid method of decreasing type to utilize hierarchical caches for AT of the GMRES(m) algorithm. We re-evaluate this strategy from the following two viewpoints: (1) Supposing "real" multicore architectures with three kinds of cache, i.e., two levels of independent cache and one level of shared cache; (2) Evaluating the AT effect using an increase strategy.

● **Evaluating AT Strategies with Different Methods**

AT methods vary amongst the different strategies. The principal method takes the current and past (one) residual vectors, and calculates the angle between them to find a stagnation point [4,5,6]. Xabclib, however, uses a method based on multiple norms from past residuals, which are taken as "sampling" points. It then uses the ratio among these sampling points, called the MM (Maximum Minimal) ratio, to find stagnation. Our research includes a comparison between the angle calculation method and the MM ratio method.

1.3   Organization of the Paper

This paper is organized as follows. In Section 2, we explain the GMRES(*m*) method and the AT strategy for the restart frequency. Section 3 describes the proposed AT strategies (increase and decrease) for hierarchical caches. We use the strategies of Aquilanti *et al.* [6] and Xabclib [3] as examples of the two categories. In Section 4, we evaluate the proposed strategies on a multicore architecture. We use one node of the T2K Open supercomputer (Univ. Tokyo), which uses the AMD Quad Core Opteron. Finally, we summarize the findings of this paper in Section 5.

2   GMRES(*m*) and AT Strategy for the Restart Frequency

2.1   The GMRES(*m*) Algorithm

The GMRES(*m*) algorithm for this paper is based on [1]. The algorithm is shown in Fig. 1.

<1> Compute $r_0 = b - Ax_0$; $\beta = \|r_0\|_2$; $v_1 = r_0 / \beta$;

<2> if ($\|r_0\|_2$ .le. ε) then goto <16>.

<3>   Let the $(m+1) \times m$ matrix be $\overline{H}_m = \{h_{ij}\}_{1 \leq i \leq m+1, 1 \leq j \leq m}$. Set $\overline{H}_m = 0$.

<4>   *do* $j = 1, m$

<5>   Compute $\omega_j = Av_j$

<6>   *do* $i = 1, j$

<7> $h_{ij} = (\omega_j, v_j)$

<8> $\omega_j = \omega_j - h_{ij}v_j$

<9>   *enddo*

<10> $h_{i+1,j} = \|\omega_j\|_2$. If $h_{i+1,j} = 0$ then Set $m = j$; goto <12>

<11> $v_{j+1} = \omega_j / h_{j+1,j}$

<12>   *enddo*

<13>   Let the $\{v_1, v_2, ..., v_m\}$ be V_m.

<14>   Compute y_m to minimize $\|\beta e_1 - \overline{H}_m y\|_2$; $x_m = x_0 + V_m y_m$;

<15> $x_0 = x_m$; goto <1>;

<16> continue

Fig. 1. GMRES(m) algorithm

2.2   AT Strategies for the Restart Frequency of GMRES(m)

To perform AT on the restart parameter m in Fig. 1, we append the following to line
<1>:

<1-1> r_{prev} is set to previous residual vector. If this is the first iteration, then

   set $r_{prev} = r_0$.

<1-2> Line <1> in Fig.1.

<1-3> $r_{cul} = r_0$; $m = f_{AT}(r_{cul}, r_{prev})$;

The function $f_{AT}(r_{cul}, r_{prev})$ forms the AT strategy.

We first take the Xabclib strategy, which is categorized as an increase strategy. The
definition of $f_{AT}(r_{cul}, r_{prev})$ is shown in Fig. 2.

<0> SAMP = $|r_{cul}|_2$

<1>CALL **OpenATI_DAFRT**
        (5, SAMP, IRT, IATPARAM, RATPARAM, INFO)

<2> if (IRT .EQ. 1) then

<3>   MOLD = M

<4>   M = M + IATPARAM(5)

<5>   if (M .GT. MSIZE) then

<6>   M = MSIZE

<7>   endif

<8> endif

<9> return M

Fig. 2. Restart Frequency AT for Xabclib (Increase Strategy). In this example, the
OpenATI_DAFRT finds stagnation using the norms of the past five residual vectors. The "5"
is an AT parameter. If stagnation is found, the frequency is increased by IATPARAM(5),
which is set to 5 as a default value. The increase value "5" is also a tunable parameter.

Note that the stagnation state is found via the API of **OpenATI_DAFRT**, which is
provided by OpenATLib [3]. In this example, it requires the past five residual norms.
The number of past residual norms is a tunable parameter. The default
implementation of Xabclib is to use five points.

To demonstrate a decrease strategy, we consider Aquilanti's method (Fig. 3). There
are also tunable parameters in this strategy: maximum frequency (TRestart%m_max),
default frequency (TRestart%m_def), and maximum count of decrease cycles
(TRestart%m_count_max).

```
<0> resid = |r_cul|_2 ; presid = |r_prev|_2 ;
<1> max_cr = cos(8.*PI/180.); min_cr = cos(80.*PI/180.);
<2> cr = resid / presid    !! get the angle
<3> if (cr .gt. max_cr) then   !! normal cycling
<4>    M = TRestart%m_max
<5> else
<6>    if ( (cr .lt. min_cr) .or. (TRestart%m_count_max
              .lt. TRestart%m_count) ) then    !! enter an aug cycle
<7>       if (TRestart%m_aug .eq. 0) then
<8>          TRestart%m_aug = 1; TRestart%m_floor = 1;
<9>       else    !! or continue it
<10>         TRestart%m_floor = TRestart%m_floor + 1
<11>      endif
<12>      TRestart%m_count = 0;
<13>      M = TRestart%m_floor * TRestart%m_max
<14>   else
<15>      if (M - TRestart%m_incr .ge. TRestart%m_min) then
<16>         M = M - TRestart%m_incr
<17>      else
<18>         M = TRestart%m_max
<19>      endif
<20>      if ((TRestart%m_aug .eq. 1) .and. (TRestart%m_count .le.
              TRestart%m_count_max) ) then    !! if in aug cycle
<21>         if (TRestart%m_def * TRestart%m_floor
                 .lt. TRestart%m_max) then
<22>            M = TRestart%m_def * TRestart%m_floor
<23>            TRestart%m_count = TRestart%m_count + 1
<24>         else
<25>            M = TRestart%m_def;   TRestart%m_aug = 0;
<26>         endif
<27>      endif
<28>   endif
<29> endif
<30> return M
```

Fig. 3. Restart Frequency AT for Aquilanti's (Decrease) Strategy. The maximum frequency (TRestart%m_max), default frequency (TRestart%m_def), and maximum count of decrease cycles (TRestart%m_count_max) are tunable parameters in this strategy.

3   A Smart Tuning Strategy with Hierarchical Cache Sizes

3.1   Using the Vector Size of Caches to Better Estimate m

Most of the computational cost of GMRES(m) is due to the orthogonalization process in lines <4>–<12> in Fig. 1. The computational complexity of orthogonalization depends on the restart frequency m, i.e., $O(nm^2)$, where n is the dimension of matrix A.

Orthogonalization is performed in a space of dimension $n \times m$. The orthogonalization process can be parallelized by threads based on the rows of the space. (See parallel Classical Gram–Schmidt or Modified Gram–Schmidt procedures.) With respect to parallel implementation with threads, we can estimate a better value m^* with a double-precision computation using the following formula:

$$m^* = Memory\ Size / 8n / The\ number\ of\ threads, \qquad (3)$$

where the Memory Size corresponds to the sizes of the L1 cache (Independent), L2 cache (Independent), and L3 cache (Shared), etc. The memory size of the sparse matrix A is not considered in this model. It will, however, give a good estimate for the orthogonalization complexity, as all computations are performed with vectors that must be orthogonalized. In addition, if m is large, the most demanding process of GMRES(m) will be the orthogonalization. With this in mind, we consider this to be a reasonable estimation for m^*.

3.2   Principle of AT Using Hierarchical Caches

We use one socket of the AMD Opteron (Barcelona) to explain AT with hierarchical caches. The AMD Opteron has three types of cache: L1 cache (Independent, 64 KB), L2 cache (Independent, 512 KB), and L3 cache (Shared between four cores, 2 MB). Taking into account the real configuration of the caches, the idea to improve the AT strategy is summarized as follows:

- Use cache information to set maximum values of m for the AT. In the AMD Opteron case, the following hierarchy is formed (although this is not necessarily limited to the specific configuration):

  - M_MAX_{L1} : maximum size of m on the L1 cache.
  - M_MAX_{L2} : maximum size of m on the L2 cache.
  - M_MAX_{L3} : maximum size of m on the L3 cache.
  - M_MAX_{MM} : maximum size of m on the main memory.

3.3   AT for an Increase Strategy with Cache Hierarchy

Fig. 4 shows the proposed AT method with cache hierarchy for the Xabclib strategy.

```
<3-1>   if (MLEVEL .eq. 1) then
<3-2>       M = M_MAX_{L1}; MLEVEL = 2;
<3-3>       else if (MLEVEL .eq. 2) then
<3-4>           M = M_MAX_{L2}; MLEVEL = 3;
<3-5>       else if (MLEVEL .eq. 3) then
<3-6>           M = M_MAX_{L3}; MLEVEL = 4;
<3-7>       else
<3-8>           M = M + IATPARAM(5)
<3-9>       endif
<3-10>  if (M .GT. MSIZE) then
<3-11>      M = M_MAX_{MM}
<3-12>  endif
```

Fig. 4. Increase strategy (Xabclib) with cache hierarchy. Additions to line <3> in Fig. 2 are shown. Set MLEVEL = 1 before the main loop of GMRES(*m*) in Fig. 1.

3.4   AT for a Decrease Strategy with Cache Hierarchy

Fig. 5 shows the proposed AT method with cache hierarchy for Aquilanti's strategy.

```
<3'> m = TRestart%m_maxs(mlevel)
<4'> if (TRestart%mlevel .le. 3) TRestart%mlevel = TRestart%mlevel + 1
     else TRestart%mlevel = 1
<13'> m = TRestart%m_floor * TRestart%m_maxs(TRestart%mlevel)
<18'> m = TRestart%m_maxs(TRestart%mlevel)
<22'> if (TRestart%m_maxs(TRestart%mlevel) * TRestart%m_floor .lt.
         TRestart%m_max) then
<23'>    m = TRestart%m_maxs(TRestart%mlevel) * TRestart%m_floor
<26'> m = TRestart%m_maxes(TRestart%mlevel); TRestart%m_aug = 0;
```

Fig. 5. Decrease strategy (Aquilanti's) with cache hierarchy. The modifications to lines in Fig. 3 are shown. Set TRestart%mlevel = 1 before the main loop of GMRES(*m*) in Fig. 1.

3.5   Implementation Variants of the Decrease Strategy

There are some variants to the cache hierarchy strategy for Aquilanti's method. Roughly speaking, there are three ways to adapt the cache sizes: (1) using the maximum value; (2) using the minimum (default) value; or (3) a mixture of both.

We take option (3)—the maximum value for the cache hierarchy in the current level (see lines <3'> and <18'> in Fig. 5) and the default values for current level (see line <26'> in Fig. 5). The reason for this approach is to prevent setting too small a default size with respect to the execution on the previous level.

At the end, we add the following parameters to the original strategy:

- m_max = M_MAX_{MM}
- TRestart%m_maxs(1) = M_MAX_{L1}
- TRestart%m_maxs(2) = M_MAX_{L2}
- TRestart%m_maxs(3) = M_MAX_{L3}
- TRestart%m_maxs(4) = 200 (this is m_max, which is the maximum frequency parameter in Aquilanti's original strategy)

4   Numerical Experiments

4.1   Computer Environment

We used the T2K Open Supercomputer, which is a HITACHI HA8000 installed at the Information Technology Center, University of Tokyo. Each node contains four sockets of the AMD Opteron 8356 (Quad core, 2.3 GHz). The L1 cache is 64 KB/core, the L2 cache is 512 KB/core, and the L3 cache is 2 MB/4 cores. The memory on each node is 32 GB with DDR2-667 MHz. The theoretical peak is 147.2 GFLOPS/node. The inter-node connection comprises four lines of the Myri-10G with a full bisection connection. This attains 5 GB/s in both directions. We used the Intel Fortran Compiler Professional Version 11.0 with options "-O3 -m64 -openmp – mcmodel=medium."

4.2   Experimental Conditions

We used a pre-release version of Xabclib ver.1.0 [3] for the GMRES(m) implementation of both strategies. The GMRES(m) subroutine on Xabclib is **OpenATI_GMRES**. ILU(0) was chosen as a preconditioner. The convergence tolerance was set to 1.0e-08, and the time tolerance was set to 600 s. If an iteration had not converged after 600 s, the routine was forcibly stopped. This is the fundamental function of Xabclib.

We formed a solution vector x whose elements were set to 1. The right-hand-side (RHS) vector was then generated by Ax. The initial guess x_0 was set to 0.

For Aquilanti's original strategy (Fig. 3), we required a number of default parameters, which were set as follows in this experiment.

- m_def = 20
- m_min = 3
- m_max = 200
- m_incr = 3
- m_count_max = 5

4.3   Test Matrices

We used 22 non-symmetric, real matrices from the University of Florida Sparse Matrix Collection (referred to hereafter as the UF collection) [7]. The test matrices are shown in Table 1. Equation (3) was used to calculate the cached m^* size.

According to Table 1, the cached m^* sizes for L2 and L3 are the same. This is because the size of the shared L3 per core is the same as the L2 (512 KB) when we use 16 cores on the AMD Opteron. We set the cached sizes of m^* to M_MAX_{L1}, M_MAX_{L2}, and M_MAX_{L3}, respectively.

Table 1. Test matrices from the UF collection and cached m^* sizes on the AMD Quad Core Opteron. N is the dimension of the matrix, and NNZ is the number of non-zero elements.

Name	N	NNZ	L1 Cached m^* Size	L2 Cached m^* Size	L3 Cached m^* Size (16 cores)
chipcool0	20082	281150	6.5	52.2	52.2
chem_master1	40401	201201	3.2	26.0	26.0
torso1	116158	8516500	1.1	9.0	9.0
torso2	115967	1033473	1.1	9.0	9.0
torso3	259156	4429042	0.5	4.0	4.0
memplus	17758	126150	7.4	59.0	59.0
ex19	12005	259879	10.9	87.3	87.3
poisson3Da	13514	352762	9.7	77.6	77.6
poisson3Db	85623	2374949	1.5	12.2	12.2
airfoil_2d	14214	259688	9.2	73.8	73.8
viscoplastic2	32769	381326	2.0	32.0	32.0
xenon1	48600	1181120	1.3	21.6	21.6
xenon2	157464	3866688	0.4	6.7	6.7
wang3	26064	177168	2.5	40.2	40.2
wang4	26068	177196	2.5	40.2	40.2
ecl32	51993	380415	1.3	20.2	20.2
sme3Da	12504	874887	5.2	83.9	83.9
sme3Db	29067	2081063	2.3	36.1	36.1
sme3Dc	42930	3148656	1.5	24.4	24.4
epb1	14734	95053	4.4	71.2	71.2
epb2	25228	175027	2.6	41.6	41.6
epb3	84617	463625	0.8	12.4	12.4

4.4   Results and Discussion

Effect on the Increase Strategy (Xabclib)

Fig. 6 shows the effect of AT with hierarchical cache sizes for the increase strategy (Xabclib) on the T2K (16 threads).

According to Fig. 6, the performance for several matrices is improved. The average speedup of execution time compared to the original is 1.13×. In particular, the airfoil_2d matrix from the UF collection establishes a speedup of more than 2×. We examined the evolution of the restart frequency for this case. Fig. 7 illustrates the change in m using the original Xabclib method and our proposed cache size AT strategy for airfoil_2d.

According to Fig. 7, the frequency of m is increased dramatically, from 9 to 73, after the first restart process. This is because airfoil_2d is not a large matrix; the 73 vectors are all cached in L2. Hence, after the first restart, our strategy can suddenly increase m up to 73. This causes faster convergence than in the original strategy.

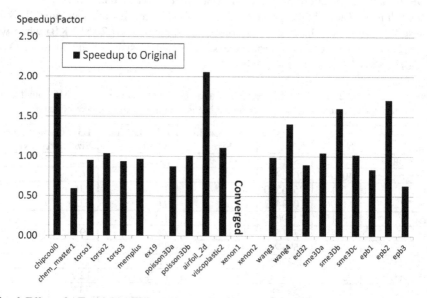

Fig. 6. Effect of AT with hierarchical cache sizes for the increase strategy (Xabclib) on the T2K (16 threads). Execution time of the original is normalized to 1. Speedup factor greater than 1 implies faster convergence than the original. The ex19 and xenon2 matrices do not converge in the ILU(0) preconditioner, whereas xenon1 converges if we use the cache size strategy.

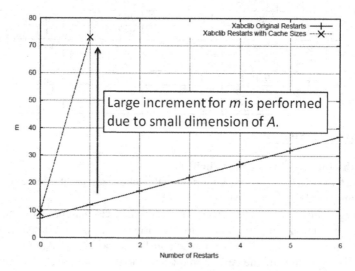

Fig. 7. Change in restart frequency using the Xabclib strategy in the airfoil_2d

Effect on the Decrease Strategy (Aquilanti's)

Fig. 8 shows the effect of AT with hierarchical cache sizes for the decrease strategy (Aquilanti's) on the T2K (16 threads).

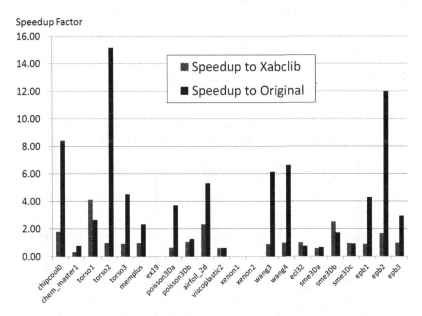

Fig. 8. Effect of AT with hierarchical cache sizes for the decrease strategy (Aquilanti's) on the T2K (16 threads). Execution time of the original (Aquilanti's) is normalized to 1. Speedup factors greater than 1 implies faster convergence than the original. Speedup compared to the original Xabclib increase strategy is also shown. The ex19, xenon1, and xenon2 matrices did not converge in the ILU(0) preconditioner.

According to Fig. 8, the performance for several matrices is strongly improved. The average speedup factor is 4.25×. In addition, the average speedup compared to the original Xabclib strategy is 1.29×, which is considerable. Therefore, this implies that the crucial effect is due to our cache size strategy.

The torso2 matrix, in particular, established a speedup of more than 15× the original decrease strategy, and torso1 achieved a 4× speedup compared to the original Xabclib increase strategy. We examined the evolution of the restart frequency for both of these matrices, and have plotted these in Fig. 9 in order to help explain these phenomena.

According to Fig. 9 (a), the maximum frequency in the proposed strategy is limited, as it still uses the cached *m* sizes for L1 and L2. The torso2 matrix is large, hence the cached *m* for L2 is only 9, whereas the original strategy uses *m* = 200, the default maximum. In addition to this, torso2 needs a very small value of *m* to converge. This fact leads to the enormous speedup compared to the original.

In contrast, torso1 is a difficult problem in that it requires almost maximum frequency, i.e., 200, to converge. In the first phase, the cache size strategy is trying to find better values for m within L1, L2, and L3; however, this does not give convergence because m is very small i.e., 9, in this case. After searching all cache sizes, the strategy retains the default maximum size, i.e., 200. On the other hand, the Xabclib strategy is to increase m step-by-step until it reaches 200. This causes very slow convergence compared to Aquilanti's strategy.

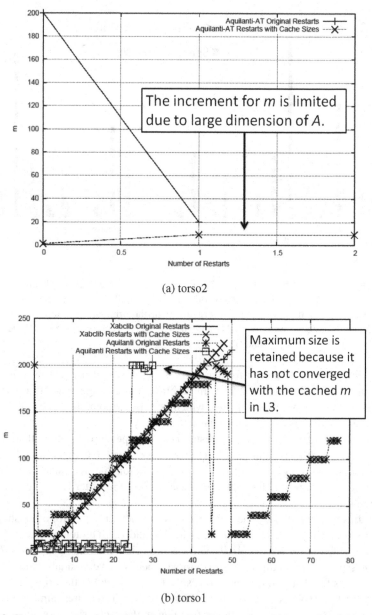

(a) torso2

(b) torso1

Fig. 9. Change in restart frequencies for Aquilanti's strategy on (a) torso2 and (b) torso1

From the above discussions concerning the statistically significant gains in speedup, we can conclude that our proposed cache size strategy is crucial to AT for both increase and decrease strategies.

It is difficult to implement a good decrease strategy without cache information; this is because it is generally difficult to set a better maximum restart frequency. Using the cache size strategy, it is clear that no additional cost is needed to set the maximum value. As a result, the decrease strategy is faster than the increase strategy in terms of average speedup.

5   Conclusion

In this paper, we proposed a smart tuning strategy using hierarchical cache sizes for a current multicore architecture. We have proposed an auto-tuning (AT) strategy for the restart frequency of both increase and decrease GMRES(*m*) methods.

As a result of performance tuning using one node of the T2K Open Supercomputer composed of an AMD Quad Core Opteron (16 cores), the proposed AT strategies were found to be very efficient compared to the original strategies without hierarchical cache sizes.

We evaluated the proposed strategies on 22 matrices from the University of Florida Sparse Matrix Collection. The results showed an average speedup of 1.13× for the increase method (an implementation of Xabclib) and an average speedup of 4.25× for the decrease method (Aquilanti's) using the proposed strategy.

One of the drawbacks to the traditional decrease strategy is the difficulty in determining the optimal maximum restart frequency—if too large a value is specified, the algorithm takes a long time; however, if too small a value is specified, it may not converge at all. According to the results of our numerical experiment, we found that the performance of the decrease strategy was improved by a factor of 15. This was caused by the selection of appropriate values for the maximum restart frequency based on cache size information.

In addition, we found that the decrease strategy is better than the increase strategy, in terms of the average speedup, if the maximum frequency is set appropriately. The hierarchical cache information is a crucial factor in setting an appropriate maximum frequency.

As the L2 and L3 cache sizes are the same when we use 16 cores for the AMD Quad Core Opteron, the evaluation of several multicore architectures is important future work. The proposed strategy does not permit "multiple" re-use of the cache information. Constructing such a strategy is also vital in future research.

Acknowledgments. This work is supported by the FP3C "Framework and Programming for Post Petascale Computing" Project, funded by Strategic International Research Cooperative Program, JST, Japan, and Agence National de la Recherche (ANR), France. The authors would like to thank members of the Xabclib project for the use of a pre-released version of Xabclib ver.1.0.

References

1. Saad, Y., Schults, M.: GMRES: A Generalized Minimal Residual Algorithm For Solving Nonsymmetric Linear Systems. SIAM J. Sci. Stat. Comput. 7(3), 856–869 (1986)
2. Sosonkina, M., Watson, L., Kapania, R.: A New Adaptive GMRES Algorithm for Achieving High Accuracy. Technical Report, Computer Science, Virginia Polytechnic Institute and State University (1996)
3. Xabclib Project, http://www.abc-lib.org/Xabclib/index.html
4. Baker, A., Jessup, E., Kolev, Tz.: A Simple Strategy for Varying The Restart Parameter in GMRES(m). Journal of Computational and Applied Mathematics 230(2), 751–761 (2009)
5. Habu, M., Nodera, T.: GMRES(m) Method with Changing The Restart Cycle Dynamically. Trans. IPS Japan 43(6), 1795–1803 (2002) (in Japanese)
6. Aquilanti, P.-Y., Petiton, S., Calandra, H.: Parallel Auto-tuned GMRES Method to Solve Complex Non-Hermitian Linear System. In: Proceedings of iWAPT 2010 (2010)
7. The University of Florida Sparse Matrix Collection, http://www.cise.ufl.edu/research/sparse/matrices/

Adaptive Off-Line Tuning
for Optimized Composition of Components
for Heterogeneous Many-Core Systems

Lu Li, Usman Dastgeer, and Christoph Kessler

PELAB, IDA, Linköping University
S-581 83 Linköping, Sweden
{lu.li,usman.dastgeer,christoph.kessler}@liu.se

Abstract. In recent years heterogeneous multi-core systems have been given much attention. However, performance optimization on these platforms remains a big challenge. Optimizations performed by compilers are often limited due to lack of dynamic information and run time environment, which makes applications often not performance portable. One current approach is to provide multiple implementations for the same interface that could be used interchangeably depending on the call context, and expose the composition choices to a compiler, deployment-time composition tool and/or run-time system. Using off-line machine-learning techniques allows to improve the precision and reduce the run-time overhead of run-time composition and leads to an improvement of performance portability. In this work we extend the run-time composition mechanism in the PEPPHER composition tool by off-line composition and present an adaptive machine learning algorithm for generating compact and efficient dispatch data structures with low training time. As dispatch data structure we propose an adaptive decision tree structure, which implies an adaptive training algorithm that allows to control the trade-off between training time, dispatch precision and run-time dispatch overhead.

We have evaluated our optimization strategy with simple kernels (matrix-multiplication and sorting) as well as applications from RODINIA benchmark on two GPU-based heterogeneous systems. On average, the precision for composition choices reaches 83.6 percent with approximately 34 minutes off-line training time.

Keywords: Autotuning, Heterogeneous architecture, GPU.

1 Introduction

Recently GPU-based heterogeneous multi-core system have been given much attention, because GPUs have shown remarkable performance advantage over CPUs for suitable computations with sufficiently large problem size. However, effective utilization of those systems often requires much programming effort

M. Daydé, O. Marques, and K. Nakajima (Eds.): VECPAR 2012, LNCS 7851, pp. 329–345, 2013.
© Springer-Verlag Berlin Heidelberg 2013

(programmability problem), and moreover, we often observe a performance decrease when porting the code to a new platform without re-optimization (performance portability problem).

For building performance portable applications, one solution is to provide multiple implementation variants of the same functionality that may execute on different platforms, internally use different programming models, different algorithms and/or different compilation settings, or encapsulate library calls or accelerator-specific code. The execution time of such variants will generally depend on the resources available for execution (e.g., cores or accelerator) and other call context properties such as problem sizes, but also on tunable parameters of the implementation variants themselves such as buffer sizes or tiling factors.

The PEPPHER [6,4] component model provides a XML-based metadata language that allows to specify descriptors that externally annotate PEPPHER components and interfaces. A component is an annotated software module adhering to an PEPPHER interface for which multiple implementation variants may be available. Beyond the traditional functional interface properties such as parameter types and direction, component metadata of an implementation variant includes the implemented interface (functionality), dependences on other PEPPHER components or third-party software packages, compilation commands, tunable parameters, platform and resource requirements, and possibly also statically provided performance models that allow to predict average-case execution time as a function of values taken from a call context instance. Hence, PEPPHER allows to delay and expose the selection decisions to later stages (e.g., at runtime) when more information about the invocation context and resource availability (e.g. from the run-time environment) is available. In this way, the selection of an implementation variant for an interface function call is completely automatized and not hardcoded in the application, allowing for automated re-optimizing of the selection mechanism when porting a PEPPHER application to a new platform.

In order to better utilize different kinds of processing units by appropriate automatic selection, a reasonably good performance model for predicting the fastest implementation variant for a given context instance is required. The two trends for building such performance models are towards an analytical model and an empirical model. It is normally considered that modern computer systems (including heterogeneous ones) are too complex for a reasonably good analytical performance model, thus empirical models constructed from measurements of test code on the target system have become more practical nowadays. Machine-learning techniques have shown potential for building such empirical performance models. In essence, machine learning constructs from results of example runs a surrogate function that approximates an unknown selection function for a (new) target architecture.

Empirical automated performance tuning (or autotuning for short) of best-variant selection by measurements and learning can be performed on-line or off-line. On-line learning is done at runtime, after first instrumented invocations of components have been executed with random selection decisions, and represents the selection function in an internal data structure, such as a hash table as applied in StarPU [3].

On-line machine learning performs selection decisions purely based on recorded performance history data and thus does not require any additional performance modeling information by the component provider, but can not offer good prediction results until enough representative example measurements are collected, and incurs additional runtime overhead for that. Off-line tuning can ease the problem by actively invoking those representative training examples manually or automatically; however, the number of training examples generated with a straightforward strided scanning of context property values (e.g., problem sizes) grows very large if suitable precision of performance prediction and best-variant selection shall be achieved.

In this work we suggest a new approach to off-line tuning with a novel adaptive generation of training data and representation of the constructed selection (dispatch) function. In our approach, the training time can be reduced remarkably while a reasonable prediction precision can still be achieved. It can also be integrated with compile time tools such as composition tools, thus enhance static composition by better precision. Furthermore, it can be integrated with run-time systems such as StarPU by dynamically exposing only the best implementations of the different kinds of processing units to reduce run-time selection overhead.

The remainder of this paper is organized as follows: Section 2 introduces the PEPPHER component model and composition tool. In section 3 we discuss our adaptive offline tuning approach in detail. In section 4 we show and discuss experimental results. Section 5 lists related work; section 6 concludes and discusses future work.

2   PEPPHER Components and Composition

A *PEPPHER component* is an annotated software module that implements a specific functionality declared in a *PEPPHER interface*. A PEPPHER interface is defined by an *interface descriptor*, an XML document that specifies the name, parameter types and the access types (read, write or both) of a function to be implemented, and in addition specifies which performance metrics (e.g. average case execution time) the prediction functions of component implementations must provide. Interfaces can be generic in static entities such as element types or code; genericity is resolved statically by expansion, as with C++ templates.

Applications for PEPPHER are currently assumed to be written in C/C++. Several component variants may implement the same functionality (as defined by a PEPPHER interface), e.g. by different algorithms or for different execution platforms. These implementation variants can exist already as part of some standard library (e.g. CUBLAS components for CUDA) or can be provided by the programmer. The PEPPHER framework provides support for implementation repository to manage evolution of implementation variants to increase the re-use potential in the long run. Also, more component implementation variants may be generated automatically from a common source module, e.g. by special compiler transformations or by instantiating or binding tunable parameters. These variants differ by their resource requirements and performance behavior, and thereby become alternative choices for composition whenever the (interface) function is called.

In order to prepare and guide variant selection, component implementations need to expose their relevant properties explicitly to the composition tool. Each PEPPHER component implementation variant thus provides its own *component descriptor*, an XML document that contains information (meta-data) about properties such as the provided and required interface(s), source files, compilation commands and resource requirements, tunable parameters, further constraints on composition, and a reference to a performance prediction function.

The `main` module of a PEPPHER application is also annotated by its own XML descriptor, which states e.g. the target execution platform and the overall optimization goal.

The PEPPHER framework automatically keeps track of the different implementation variants for the identified components, technically by storing their descriptors in repositories that can be explored by the composition tool. The composition tool reads the metadata of interfaces and components used in the application and generates, for each call to a PEPPHER interface, the necessary code for pre-selecting (dispatching) a suitable implementation variant and creating a task for the PEPPHER runtime system that will execute that call. Composition points of PEPPHER components are restricted to calls on general-purpose execution units only. Consequently, all component implementations using hardware accelerators such as GPUs must be wrapped in CPU code containing a platform-specific call to the accelerator.

Component invocations result in *tasks* that are managed by the PEPPHER run-time system and executed non-preemptively. PEPPHER components and tasks are *stateless*. However, the parameter data that they operate on do have state. For this reason, parameters passed in and out of PEPPHER components may be wrapped in special portable, generic, STL-like *container* data structures such as `Vector` and `Matrix` with platform-specific implementations that internally keep track of, e.g., in which memory modules of the target system which parts of the data are currently located or mirrored (*smart containers*). The container state becomes part of the call context information as it is relevant for performance prediction.

Composition tool Composition is the selection of a specific implementation variant (i.e., callee) for a call to component-provided functionality and the allocation of resources for its execution. Composition is made *context-aware* for performance optimization if it depends on the current *call context*, which consists of selected input parameter properties (such as size) and currently available resources (such as cores or accelerators). The context parameters to be considered and optionally their *ranges* (e.g., minimum and maximum value) are declared in the PEPPHER interface descriptor. We refer to this considered subset of a call context instance's parameter and resource values shortly as a *context instance*, which is thus a tuple of concrete values for context properties that might influence callee selection. Hence, composition maps context instances to implementation variants [12].

Composition can be done either statically or dynamically. *Static composition* constructs off-line a *dispatch function* that is evaluated at runtime for a context instance to return a function pointer to the expected best implementation variant [12]. *Dynamic composition* generates code that delegates the actual composition to a context-aware runtime system that records performance history and construct a dispatch mechanism on-line to be used and updated as the application proceeds.

Composition can even be done in multiple stages: First, static composition can narrow the set of candidates for the best implementation variant per context instance to a few ones that are registered with the context-aware runtime system that takes the final choice among these at runtime.

Dynamic composition is the default composition mechanism in PEPPHER. In the special case where sufficient meta-data for performance prediction is available for all selectable component variants, composition can be prepared completely statically and co-optimized with resource allocation and scheduling, thus bypassing the runtime system; see e.g. [11,12].

The *PEPPHER composition tool* [6] deploys the components and builds an executable PEPPHER application. It recursively explores all interfaces and components that (may) occur in the given PEPPHER application by browsing the interfaces and components repository.

The composition tool processes the set of interfaces (descriptors) bottom-up in reverse order of their components' *required interfaces* relation (lifted to the interface level) [12]. For each interface (descriptor) and its component implementations, the composition tool performs the following tasks:

1. It reads the descriptors and internally represents the metadata of all component implementations that match the target platform, expands generic interfaces and components, and generates platform-specific header files from the interface descriptor.
2. It looks up prediction data from the performance data repository or runs microbenchmarking code on the target platform, as specified in the components' performance meta-data.
3. It generates composition code in the form of *stubs* (proxy or wrapper functions) that will perform context-aware composition at runtime. If sufficient performance prediction metadata is available, it constructs performance data and dispatch tables for static composition by evaluating the performance prediction functions for selected context scenarios [11,12], which could be compacted by machine learning techniques [5]. Otherwise, the generated composition code contains calls to the PEPPHER run-time system to delegate variant selection to runtime, where the runtime system can access its recorded performance history to guide variant selection, in addition to other criteria such as operand data locality.

4. It calls the native compilers, as specified for each component, to produce a binary of every patched component source.

Finally, it links the application's main program and its compiled components together with the generated and compiled stubs, the PEPPHER library and the PEPPHER runtime system to obtain an executable program.

3   Adaptive Off-Line Tuning

3.1   Motivation

Consider a typical example where a component's implementation variants for execution on different kinds of processors show performance advantages for different variants with respect to different input sizes, as shown in Figure 1. In a subrange of call context instance values (here, of the number of array elements to sort) where one implementation variant runs fastest among all implementations variants we call that implementation variant the *winner* for that range of input sizes.

We can map a n-dimensional range to a n-dimensional space. A specific context instance can also be considered as a point in a n-dimensional space. Some points or hyperplanes divide winning ranges of different implementations, we call those the transition points or hyperplanes. Ideally if all those points or hyperplanes can be found effectively, we can construct a compact representation which requires small overhead for both store and look-up, and it will provide 100 percent precision of winner prediction.

One may argue that the characteristics shown in Figure 1 may not apply for other problems. In this paper, we test three other benchmark applications, and these applications surprisingly conform to the characteristics of Figure 1, which shows an interesting property: The winning range for each implementation variant is convex, i.e., if two points on a one-dimensional space have the same winner, then it wins on all points between these. Our pruning strategy in this paper is based on this convexity assumption: for n-dimensional space, if all vertices of a space have the same winner, then it wins on all points in the space. Based on this assumption, we construct an algorithm and data structure to approximate and represent these transition points.

3.2   Hybrid Static/Dynamic Composition with Off-Line Training

Unlike static composition, dynamic composition can be guided by access to the run-time context for each invocation, and thus owns prerequisites for better selection precision at the cost of some run-time overheads. The hope is that the time saved by invoking the fastest implementation variant is larger than the overhead of the dynamic selection process, and thus portable performance is increased.

Dynamic composition with on-line training by the runtime system shows some disadvantages: it requires a certain number of representative executions before

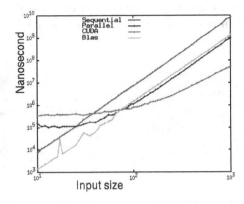

Fig. 1. Performance for matrix-matrix multiplication variants

it can offer acceptable selection precision for dynamic composition; however, it is often not guaranteed that those representative executions will happen during a sufficiently long period of time. As an alternative, we consider off-line training and dynamic composition. In off-line training, measuring performance for every possible runtime context instance (which would offer perfect selection and precise representation of this information) is often not feasible, thus a dynamic composer is forced to make predictions based on a limited set of training examples.

The space $\mathcal{C} = I_1 \times ...I_D$ of context instances for a component with D attributes in the context instances is spanned by the D context attribute axes with considered (user-specified or default) finite intervals I_i of discrete values, for $i = 1, ..., D$. A continuous subinterval of an I_i is called a *range*, and any cross product of such subintervals on the D axes is called a *subspace* of $\mathcal{C}$. Hence, subspaces are "rectangular", i.e., subspace borders are orthogonal to the axes of $\mathcal{C}$.

In an experimental version of our composition tool, we offer a precision-controllable offline-trainer and dynamic composer based on ranges, i.e. it tries to automatically approximate the (usually, non-rectangular and possibly non-convex) subsets in $\mathcal{C}$ where one particular implementation variant performs better than all the others, by a set of subspaces.

Our idea is to find sufficiently precise approximations by adaptively recursive splitting of subspaces by splitting the intervals I_i, $i = 1, ..., D$. Hence, subspaces are organized in a hierarchical way (following the subspace inclusion relation) and represented by a 2^D-ary tree (cf. binary space partitioning trees and quadtrees/octrees etc.).

Our algorithm for off-line measurement starts from a trivial tree $T_{\mathcal{C}}$ that has just one node, the root (corresponding to the whole $\mathcal{C}$), which is linked to its 2^D corner points (here, the 2^D outer corners of $\mathcal{C}$) that are stored in a separate table of recorded performance measurements. The implementation variants of the component under examination are run with each of the corresponding 2^D context instances, possibly multiple times for averaging, using a context instance generator provided with the metadata of the component; a variant whose execution exceeds a timeout for a context instance are aborted and not considered further for that context instance. Now we know the winning implementation variant for each corner point and store it in the performance table, too, and T_C is properly initialized.

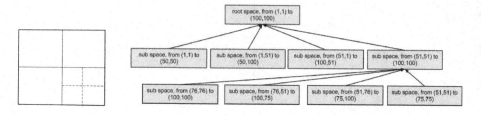

Fig. 2. Cutting a space recursively into subspaces, and the resulting dispatch tree

Consider any leaf node v in the current tree T_t representing a subspace $S_v = R_1^v \times \ldots \times R_D^v$. If the same specific implementation variant runs fastest on all context instances corresponding to the 2^D corners of S_v, we stop further exploration of that subspace and will always select that implementation whenever a context instance at run-time falls within that subspace. Otherwise, the subspace S_v may be refined further. Accordingly, the tree is extended by creating new children below v which correspond to the newly created subspaces of S_v.

By iteratively splitting the ranges in FIFO order, we generate an adaptive tree structure to represent the performance data and selection choices, which we call *dispatch tree*.

The user can specify a *maximum depth* (training depth) for this iterative refinement of the dispatch tree, which implies an upper limit on the runtime lookup time, and also a maximum tree size (number of nodes) beyond which any further refinement is cut off. Third, the user may specify a timeout for overall training time, after which the dispatch tree is considered final.

Run-time lookup searches through the dispatch tree starting from the root and descending into subspace nodes according to the current runtime context instance. If the search ends at a *closed leaf*, i.e., a leaf node with equal winners on all corners of its subspace, the winning implementation variant can be looked up in the node. If the search ends in an *open leaf* with different winners on its borders (e.g., due to reaching the specified cut-off depth), we perform an approximation within that range by choosing the implementation that runs fastest on the subspace corner with the shortest Euclidean distance from the run-time context instance.

The deeper the algorithm explores the tree, the better precision the dynamic composer can offer for the composition choice; however, it requires more off-line training time and more runtime lookup overhead as well. We give the option to let the user decide the trade-off between training time and precision by setting the cut-off depth, size and time in the component interface descriptor.

3.3 Example for Hybrid Composition with Adaptive Off-Line Training

Let us consider a matrix-matrix multiplication example with two implementation variants, the well-known sequential version and a parallel version parallelized by pthreads with a fixed number of 4 threads. In the off-line training phase, performance data is measured by one execution per context instance; at execution time of the composed code with dynamic selection, performance is averaged over 10 runs per context instance.

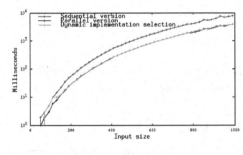

Fig. 3. Execution time for hybrid composition with a 41-node lookup tree determined by the adaptive refinement training algorithm with cut-off depth 3. — The hardware we use is a multi-core system with 16 CPUs, where each CPU is an Intel(R) Xeon(R) CPU E5520 running at 2.27GHz with 8192 KB cache. The operating system is Linux 3.0-ARCH and the compiler is gcc 4.6.1.

As the resources (here, number of threads for OpenMP) was fixed, a context instance is just a triple consisting of the three problem sizes that define the operand matrix dimensions. The training space of context instances was chosen as $[1 : 1000, 1 : 1000, 1 : 1000]$, i.e., comprising 10^9 possible context instances (input sizes). As tree data structure we used an octree with simultaneous refinement of subspaces along all three dimensions. The cut-off depth for the tree was set to 3. With these settings, the off-line training time (i.e., for the tree construction including the measurements on the target system) takes 228 seconds and the constructed tree has 41 nodes, where the adaptive tree refinement is done mostly for subspaces with smaller problem sizes. By comparing the composed code at runtime with the actually fastest component for each context measured for square test matrices (see Figure 3), we find that the tree lookup yields a dynamic selection precision of 92%. From Figure 3 we can also see that the overhead for performing dynamic selection is rather negligible. For some context instance the dynamically selected implementation variant runs even faster than the same one without dynamic selection; such anomalies are mostly due to the operating system's interruptions during measuring; in principle, the composed code should always run slightly slower than the best individual component, due to run-time lookup overhead.

3.4   Selection

At initialization time, we read the dispatch tree from the file generated in the training phase, and add a translation table that maps each implementation variant's symbolic name to its function address.

The wrapper function generated by the composition tool selects the relevant parameter values from the call context and uses them to look up the dispatch tree and thereby the right function address, which is filled in the descriptor for the task to be submitted to the runtime system. For open leaves, it chooses the winner of the corner that has the shortest Euclidean distance from the actual context instance.

4   Experimental Results

Platform. We use two GPU based heterogeneous systems called Fermi and Cora. A brief description of the two platforms is shown in Table 1.

Table 1. Platform description

Machine name	CPU cores	CPU type	GPUs	GPU type	OS	Compiler
Fermi	16	Intel(R) Xeon(R) CPU E5520 @ 2.27GHz	2	two Tesla M2050	3.2.1-2-ARCH	gcc 4.6.2 and nvcc V0.2.1221
Cora	16	Intel(R) Xeon(R) CPU X5550 @ 2.67GHz	3	two nVidia Tesla C2050 and one Tesla C1060	RHEL 5.6	gcc 4.1.2 and nvcc V0.2.1221

Benchmark. For the evaluation we have chosen 4 benchmark problems: matrix-matrix multiplication, sorting, and two RODINIA benchmarks: path finder and backpropagation. A detailed description is shown in Table 2.

Table 2. Benchmark test settings

Benchmark	feature modeling	Range	space size	Implementation variants
Matrix-matrix multiplication	row size, column size of first matrix; column size of second matrix	(1, 1, 1) to (3000, 3000, 3000)	2.7E+10	Sequential implementation, CUDA implementation, Blas implementation, Pthread implementation
Sorting	array size; discretization of array values distribution (sampled number of inversions)	(1,0) to (100000,10)	1000000	bubble sort, insertion sort, merge sort, quick sort, CUDA thrust sort (only on Fermi)
Path finder	row; column	(1,1) to (10000,20000)	200000000	OMP implementation, CUDA implementation
Back propagation	array size	(1000) to (100000)	99000	OMP implementation, CUDA implementation

Methodology. We first train each benchmark problem with training depth from 0 to 4. If the training time exceeds 3 hours then we terminate the training process. Each benchmark is trained twice, with one version which prunes closed space in the tree representation and another which performs no pruning at all.

The test points are chosen evenly from the training space so that every subspace in the dispatch tree is used for performance prediction.

Table 3. Test results for 4 benchmarks on Fermi (td: Training depth; tt: Training time; ato: average time overhead on dynamic selection; nn: Number of nodes generated in the tree representation)

td	pruning closed space				no pruning for closed space			
Matrix-matrix multiplication on Fermi, 343 test points								
	tt (s)	Precision (%)	ato (μs)	nn	tt (s)	Precision (%)	ato (μs)	nn
0	85	51	17	1	88	50	15.9	1
1	755	48	21	9	762	48	20.4	9
2	6118	62	23	73	6252	62	23	73
Sorting on Fermi, 110 test points								
	pruning closed space				no pruning for closed space			
0	233	36	4	1	233	36	3.6	1
1	1035	61	4.9	5	1035	64	4.9	5
2	2485	80	5.5	17	4071	80	5.6	21
Back propagation on Fermi, 20 test points								
	pruning closed space				no pruning for closed space			
0	7	55	9	1	6	55	9	1
1	7	80	11	3	6	80	10	3
2	8	90	13.6	5	8	90	11.8	7
3	7	95	12	7	13	95	12.5	15
4	8	100	13.1	9	18	100	14	31
Path finder on Fermi, 200 test points								
	pruning closed space				no pruning for closed space			
0	36	59	12.6	1	29	59	12.7	1
1	161	77	16.5	5	122	77	14.5	5
2	371	86	16.5	17	497	86	15.8	21
3	609	95	16.8	45	1992	95	20.9	85

Experimental Results on Two Machines. The test results for 4 benchmarks on Fermi are shown in Table 3. In particular, for backpropagation, the performance behavior for different training depths on Fermi are shown in Figure 4. The results for the 4 benchmarks on Cora are shown in Tables 4.

Discussion. The sorting, pathfinder and backpropagation benchmarks have shown a good result. The result for the matrix-matrix multiplication benchmark is a little disappointing, because it has a relatively large training space. Most subspaces in its tree are open ones and for the points near their corners the Euclidean distance criterion can give a better approximation while in the large central area of these subspaces, the precision can not be guaranteed. Since we train on a large space, which means large input sizes, a single training execution may take a long time; for this reason, training depths larger than 3 become not practical and not considered in this benchmark testing.

From the test results we can see that in most cases the precision of prediction of the winner implementations increases with the depth of the dispatch tree. This is expected because, as open subspaces can be partly closed by exploring deeper levels, the precision increases. This trade-off is exposed to the users.

We also can see that for a relatively short training time, we get a reasonable prediction precision in total which means pruning closed subspaces works and the assumption that we can treat all points in a closed subspace equally holds for those benchmarks. Another evidence for the assumption is the comparison between two version of test results, one which perform closed subspace pruning

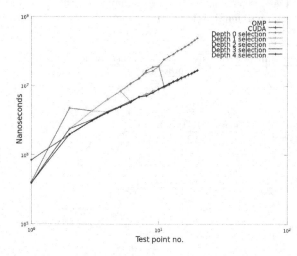

Fig. 4. Performance with maximum depths 0 to 4 for the backpropagation benchmark on Fermi

and one which does not. We get almost the same results from the two set of tests on all benchmarks we use, thus it is safe not to explore closed space in the training phase.

The time overhead for run-time selection is acceptable, on the level of microseconds. Since we only explore a shallow depth of a dispatch tree, the number of nodes generated is small, too, so the memory overhead is acceptable as well.

As for the relation between precision and performance, we can illustrate it in Figure 4 for backpropagation. Comparing constant invocation of the OpenMP implementation variant with dynamic selection among all available variants, we see that for subspaces where the OpenMP variant wins, the performance of all variants only differs by a few microseconds; for the subspace where OpenMP does not win, we gain performance. The performance gained might be remarkable if some variant scaling badly is constantly invoked. From the figures we can also see that wrong decisions for points within open subspaces often happen near transition points between different winners, and often the performance difference of implementation variants at points near transition points is low, thus a wrong decision does not yield a performance penalty as large as in other points in the subspace.

In general, our approach can pick the best implementation variant for most of the cases for the different platforms.

We observed an anomaly for exploration of subspace in matrix-matrix multiplication on Fermi. When the depth increases from 0 to 1, with more training time, the precision drops. One possible explanation is that when splitting some space where the winner on one of the corners is shared by a minority of the other corners, the Euclidean distance criterion will cause a majority of points to be predicted wrongly, which with the coarser dispatch tree are predicted correctly. Continuing to refine that subspace may make the precision increase again; however, continuing the exploration for matrix-matrix multiplication on such large space is so time-consuming that we have to postpone further investigation of this problem to future work.

Table 4. Test results for 4 benchmarks on Cora

	Matrix-matrix multiplication on Cora, 343 test points							
td	pruning closed space				no pruning for closed space			
	tt (s)	Precision (%)	ato (μs)	nn	tt (s)	Precision (%)	ato (μs)	nn
0	67	48	17.6	1	63	48	18.3	1
1	634	49	22.5	9	621	49	22.2	9
2	5115	67	26.4	73	5009	68	26.2	73
	Sorting on Cora, 110 test points							
	pruning closed space				no pruning for closed space			
0	162	34	5.3	1	159	35	5.5	1
1	714	62	7.1	5	710	62	8.5	5
2	1747	80	8.6	17	2809	78	8.6	21
	Back propagation on Cora, 20 test points							
	pruning closed space				no pruning for closed space			
0	3	55	11.9	1	3	60	13	1
1	4	85	12.6	3	4	90	14.7	3
2	4	95	16.1	5	5	95	14	7
3	4	100	13.4	7	7	95	15.2	15
	Path finder on Cora, 200 test points							
	pruning closed space				no pruning for closed space			
0	21	39	12.5	1	21	39	12.1	1
1	97	67	14.5	5	92	67	16.1	5
2	219	82	15.6	17	400	82	16.2	21
3	367	95	15.9	45	1511	95	18.1	85

5 Related Work

Techniques for automated performance tuning have been considered extensively in previous work; they are applied e.g. in generators of optimized domain-specific libraries (such as basic linear algebra [25,14,21], reduction [26], sorting [15,21] or signal transforms [8,20,17,7]), iterative compilation frameworks (e.g. [16]), or for the optimized composition of general program units [11,13,2,1,24], e.g. the components in our case.

Automated performance tuning usually involves three fundamental preparatory tasks: (1) search through the space of context property values, (2) generation of training data and measurements on the target system, (3) learning a decision function / rule (e.g. for best variant selection, decomposition, or settings for tunable parameters), or alternatively (3a) learning a predictor for performance and then (3b) decide / optimize based on that predictor among the remaining options. In our approach, these three tasks are tightly coupled to limit the amount of measurement time and representation size required, while most other approaches decouple at least two of these tasks.

Search, measurements and learning can each be performed off-line (i.e., at deployment time or compile time) or on-line (i.e., at run time), or as a combination of both. In our approach, all tasks are done off-line at component deployment time, while all are performed at runtime in the StarPU runtime system by continuously recording measurements from the running program and using these data for future decisions [2].

Kessler and Löwe [11] propose a methodology for optimized composition of grey-box components. The component provider offers additional knowledge such as time functions for performance prediction, which might include data obtained from microbenchmarking, measuring, direct prediction or hybrid prediction. Predictions are made for a regularly sampled (dense) space of context instances, including composition of prediction functions for recursive components in a dynamic programming algorithm. Based on those predictions, a dispatch table and dynamic selection code are generated and injected into the components for runtime selection. The dispatch tables can be a-posteriori compressed using various machine learning techniques such as decision tree, decision graph, Bayesian classifier and SVM, where the decision tree was empirially found to be most effective [5]. In contrast, our current work does the compression a-priori, thus avoiding excessive prediction or measurements.

PetaBricks [1] provides a framework with language and compiler support for exposing implementation variant choices. It also contains an off-line autotuning system which starts to test with a small input size and doubles the size of the input on each later iteration. They assume that optimal choices for smaller subproblems are independent of the larger problem, so they construct new composition candidates incrementally from small input sizes to larger ones. The algorithmic choices are made off-line in the output of the compiler.

Elastic computing [24] lets the programmer transparently utilize the heterogeneous computing resources by providing a library of elastic functions. The autotuner trains itself from measurements (which are not further specified) and then uses a linear regression model for predicting performance of untested input values.

Grewe and O'Boyle [9] suggested an approach for statically choosing the best mapping between tasks and unit types (CPU, GPU). Static features such as numbers of float operations, are extracted from a set of programs, and and scheduling decisions are fed to a SVM classifer. Then at compile time, the decision for distribution of work load on different kinds of processors is made.

ABCLibScript[10] is a directive system that provides autotuning functionality on numerical computations within the FIBER framework. The choice of performance-related parameters, such as unrolling depth and block length, is specified for training execution. A performance model is also specified by the users, and generated together with training results. At run-time, best code regions are selected.

Danylenko et al. [5] compares 4 different machine-learning approaches, Decision Trees, Decision Diagrams, Naive Bayes and SVM on sorting benchmark in the field of context-aware composition for a-posteriori compression of the dispatch function. Results show Decision Diagram performs better in scalability, and almost the same in prediction accuracy and decision overhead comparing with other 3 approaches.

Singer and Veloso [19] applied a back-propagation neural netork for performance prediction in the field of signal processing. Results show that choices of different combination of features affect remarkably the prediction precisions.

[22] presents an unsupervised learning approach (fuzzy clustering algorithm) for a machine learning based compiler. Significant reduction in the training cost is achieved by grouping training programs into clusters using 'ratio of assembly

instructions to the total program instructions' as a feature vector. After clustering, they carried training executions on one (randomly selected) representative from each cluster, recording the best execution configuration for each of the selected programs. This is an alternative approach to reduce training time.

In [23], Wang and O'Boyle developed two predictor functions (data-sensitive and data-insensitive) to predict the best OpenMP execution configuration (number of OpenMP threads, scheduling policy) for an OpenMP program on a given architecture. They use two machine learning algorithms (Artificial Neural Network and Support Vector Machine) and train them using code, data and runtime features extracted via source to source instrumentation.

Our approach can be considered as an adaptive variant of decision tree learning. Decision tree learning, often based on C4.5 [18] or similar tools, is also used in many other approaches, e.g. in [20,21,26,7]. A direct comparison of our learning algorithm with C4.5 and other learning methods is planned for future work.

6 Conclusions and Future Work

We have developed an adaptive off-line training algorithm and dispatch tree representation that allows to pick the best implementation variants for most of the cases on different GPU-based heterogeneous machines, hence it improves performance portability. Our method allows to reduce training time and enables the user to trade off prediction precision, runtime overhead and training time.

Our approach for pruning closed space is based on the assumption that, if corners of a space show a common winner, all points in the space would have the same winner, which holds in most of our benchmark applications. The assumption needs to be further investigated with more applications, and refined prediction methods for open spaces should be developed. Note that, in cases where the user knows that the assumption does not hold, a better accuracy could then be enforced by also refining closed space within the given depth limit, at the expense of a larger dispatch tree and longer training time.

Further improvements of our method are possible and will be considered in future work. For instance, timeouts for individual measurements (applicable on CPUs) and aborting variants under measurement that exceed the current winner of a training point can save more training time. Also, the user may accept a tolerance such that even suboptimal variants not slower than the winner by that tolerance could also be considered winners in order to close spaces earlier.

Acknowledgments. This work was partly funded by EU FP7 project PEPPHER (www.peppher.eu) and by SeRC. We also thank the Scientific Computing group at the University of Vienna, Austria, for letting us use their GPU server *Cora* on which some of the measurements reported in this paper were taken.

References

1. Ansel, J., Chan, C.P., Wong, Y.L., Olszewski, M., Zhao, Q., Edelman, A., Amarasinghe, S.P.: PetaBricks: A language and compiler for algorithmic choice. In: Proceedings of the 2009 ACM SIGPLAN Conference on Programming Language Design and Implementation, PLDI 2009, pp. 38–49. ACM (2009)
2. Augonnet, C., Thibault, S., Namyst, R.: Automatic calibration of performance models on heterogeneous multicore architectures. In: Lin, H.-X., Alexander, M., Forsell, M., Knüpfer, A., Prodan, R., Sousa, L., Streit, A. (eds.) Euro-Par 2009. LNCS, vol. 6043, pp. 56–65. Springer, Heidelberg (2010)
3. Augonnet, C., Thibault, S., Namyst, R., Wacrenier, P.-A.: StarPU: A Unified Platform for Task Scheduling on Heterogeneous Multicore Architectures. Concurrency and Computation: Practice and Experience, Special Issue: Euro-Par 2009 23, 187–198 (2011)
4. Benkner, S., Pllana, S., Träff, J.L., Tsigas, P., Dolinsky, U., Augonnet, C., Bachmayer, B., Kessler, C., Moloney, D., Osipov, V.: PEPPHER: Efficient and productive usage of hybrid computing systems. IEEE Micro 31(5), 28–41 (2011)
5. Danylenko, A., Kessler, C., Löwe, W.: Comparing machine learning approaches for context-aware composition. In: Apel, S., Jackson, E. (eds.) SC 2011. LNCS, vol. 6708, pp. 18–33. Springer, Heidelberg (2011)
6. Dastgeer, U., Li, L., Kessler, C.: Performance-aware dynamic composition of applications for heterogeneous multicore systems with the PEPPHER composition tool. In: Proc. 16th Int. Workshop on Compilers for Parallel Computers (CPC 2012), Padova, Italy (January 2012)
7. de Mesmay, F., Voronenko, Y., Püschel, M.: Offline library adaptation using automatically generated heuristics. In: Int. Parallel and Distr. Processing Symp. (IPDPS 2010), pp. 1–10 (2010)
8. Frigo, M., Johnsson, S.G.: Fftw: An adaptive software architecture for the FFT. In: Proc. IEEE Int. Conf. on Acoustics, Speech, and Signal Processing, vol. 3, pp. 1381–1384 (May 1998)
9. Grewe, D., O'Boyle, M.F.P.: A static task partitioning approach for heterogeneous systems using openCL. In: Knoop, J. (ed.) CC 2011. LNCS, vol. 6601, pp. 286–305. Springer, Heidelberg (2011)
10. Katagiri, T., Kise, K., Honda, H., Yuba, T.: Abclibscript: a directive to support specification of an auto-tuning facility for numerical software. Parallel Computing 32(1), 92–112 (2006)
11. Kessler, C.W., Löwe, W.: A framework for performance-aware composition of explicitly parallel components. In: Parallel Computing: Architectures, Algorithms and Applications (ParCo 2007). Advances in Parallel Computing, vol. 15, pp. 227–234. IOS Press (2007)
12. Kessler, C.W., Löwe, W.: Optimized composition of performance-aware parallel components. In: Proc. 15th Int. Workshop on Compilers for Parallel Computers (CPC 2010) (July 2010)
13. Kessler, C.W., Löwe, W.: Optimized composition of performance-aware parallel components. Concurrency and Computation: Practice and Experience 24(5), 481–498 (2012); Published online in Wiley Online Library, doi: 10.1002/cpe.1844 (September 2011)
14. Li, X., Garzarán, M.J.: Optimizing matrix multiplication with a classifier learning system. In: Ayguadé, E., Baumgartner, G., Ramanujam, J., Sadayappan, P. (eds.) LCPC 2005. LNCS, vol. 4339, pp. 121–135. Springer, Heidelberg (2006)

15. Li, X., Garzarán, M.J., Padua, D.: A dynamically tuned sorting library. In: Proc. ACM Symp. on Code Generation and Optimization (CGO 2004), pp. 111–124 (2004)

16. Park, E., Kulkarni, S., Cavazos, J.: An evaluation of different modeling techniques for iterative compilation. In: Proc. Int. Conf. on Compilers, Architectures and Synthesis for Embedded Systems (CASES 2011) (October 2011)

17. Püschel, M., Moura, J.M.F., Johnson, J.R., Padua, D., Veloso, M.M., Singer, B.W., Xiong, J., Franchetti, F., Gacic, A., Voronenko, Y., Chen, K., Johnson, R.W., Rizzolo, N.: Spiral: Code generation for DSP transforms. Proceedings of the IEEE 93(2) (February 2005)

18. Ross Quinlan, J.: C4.5: programs for machine learning. Morgan Kaufmann Publishers Inc., San Francisco (1993)

19. Singer, B., Veloso, M.: Learning to predict performance from formula modeling and training data. In: Proc. 17th Int. Conf. on Machine Learning, pp. 887–894 (2000)

20. Singer, B., Veloso, M.: Learning to construct fast signal processing implementations. Journal of Machine Learning Research 3, 887–919 (2002)

21. Thomas, N., Tanase, G., Tkachyshyn, O., Perdue, J., Amato, N.M., Rauchwerger, L.: A framework for adaptive algorithm selection in STAPL. In: Proceedings of the ACM SIGPLAN Symposium on Principles and Practice of Parallel Programming (PPoPP), pp. 277–288. ACM (2005)

22. Thomson, J., O'Boyle, M., Fursin, G., Franke, B.: Reducing training time in a one-shot machine learning-based compiler. In: Gao, G.R., Pollock, L.L., Cavazos, J., Li, X. (eds.) LCPC 2009. LNCS, vol. 5898, pp. 399–407. Springer, Heidelberg (2010)

23. Wang, Z., O'Boyle, M.F.P.: Mapping parallelism to multi-cores: a machine learning based approach. SIGPLAN Not. 44(4), 75–84 (2009)

24. Wernsing, J.R., Stitt, G.: Elastic computing: A framework for transparent, portable, and adaptive multi-core heterogeneous computing. In: Proceedings of the ACM SIGPLAN/SIGBED 2010 Conference on Languages, Compilers, and Tools for Embedded Systems (LCTES), pp. 115–124. ACM (2010)

25. Whaley, R.C., Petitet, A., Dongarra, J.: Automated empirical optimizations of software and the ATLAS project. Parallel Computing 27(1-2), 3–35 (2001)

26. Yu, H., Rauchwerger, L.: An adaptive algorithm selection framework for reduction parallelization. IEEE Trans. on Par. and Distr. Syst. 17(10), 1084–1096 (2006)

A Domain-Specific Compiler
for Linear Algebra Operations

Diego Fabregat-Traver and Paolo Bientinesi

AICES, RWTH Aachen, Germany
{fabregat,pauldj}@aices.rwth-aachen.de

Abstract. We present a prototypical linear algebra compiler that automatically exploits domain-specific knowledge to generate high-performance algorithms. The input to the compiler is a target equation together with knowledge of both the structure of the problem and the properties of the operands. The output is a variety of high-performance algorithms, and the corresponding source code, to solve the target equation. Our approach consists in the decomposition of the input equation into a sequence of library-supported kernels. Since in general such a decomposition is not unique, our compiler returns not one but a number of algorithms. The potential of the compiler is shown by means of its application to a challenging equation arising within the *genome-wide association study*. As a result, the compiler produces multiple "best" algorithms that outperform the best existing libraries.

1 Introduction

In the past 30 years, the development of linear algebra libraries has been tremendously successful, resulting in a variety of reliable and efficient computational kernels. Unfortunately these kernels are limited by a rigid interface that does not allow users to pass knowledge specific to the target problem. If available, such knowledge may lead to domain-specific algorithms that attain higher performance than any traditional library [1]. The difficulty does not lay so much in creating flexible interfaces, but in developing algorithms capable of taking advantage of the extra information.

In this paper, we present preliminary work on a linear algebra compiler, written in Mathematica, that automatically exploits application-specific knowledge to generate high-performance algorithms. The compiler takes as input a target equation and information on the structure and properties of the operands, and returns as output algorithms that exploit the given information. In the same way that a traditional compiler breaks the program into assembly instructions directly supported by the processor, attempting different types of optimization, our linear algebra compiler breaks a target operation down to library-supported kernels, and generates not one but a family of viable algorithms. The decomposition process undergone by our compiler closely replicates the thinking process of a human expert.

M. Daydé, O. Marques, and K. Nakajima (Eds.): VECPAR 2012, LNCS 7851, pp. 346–361, 2013.
© Springer-Verlag Berlin Heidelberg 2013

We show the potential of the compiler by means of a challenging operation arising in computational biology: the *genome-wide association study* (GWAS), an ubiquitous tool in the fields of genomics and medical genetics [2,3,4]. As part of GWAS, one has to solve the following equation

$$\begin{cases} b_{ij} := (X_i^T M_j^{-1} X_i)^{-1} X_i^T M_j^{-1} y_j & \text{with } 1 \le i \le m \\ M_j := h_j \Phi + (1 - h_j)I & \text{and } 1 \le j \le t, \end{cases} \tag{1}$$

where X_i, M_j, and y_j are known quantities, and b_{ij} is sought after. The size and properties of the operands are as follows: $b_{ij} \in \mathcal{R}^p$, $X_i \in \mathcal{R}^{n \times p}$ is full rank, $M_j \in \mathcal{R}^{n \times n}$ is symmetric positive definite (SPD), $y_j \in \mathcal{R}^n$, $\Phi \in \mathcal{R}^{n \times n}$, and $h_j \in \mathcal{R}$; $10^3 \le n \le 10^4$, $1 \le p \le 20$, $10^6 \le m \le 10^7$, and t is either 1 or of the order of 10^5.

At the core of GWAS lays a linear regression analysis with non-independent outcomes, carried out through the solution of a two-dimensional sequence of the Generalized Least-Squares problem (GLS)

$$b := (X^T M^{-1} X)^{-1} X^T M^{-1} y. \tag{2}$$

While GLS may be directly solved, for instance, by MATLAB, or may be reduced to a form accepted by LAPACK [5], none of these solutions can exploit the specific structure pertaining to GWAS. The nature of the problem, a sequence of correlated GLSs, allows multiple ways to reuse computation. Also, different sizes of the input operands demand different algorithms to attain high performance in all possible scenarios. The application of our compiler to GWAS, Eq. 1, results in the automatic generation of dozens of algorithms, many of which outperform the current state of the art by a factor of four or more.

The paper is organized as follows. Related work is briefly described in Section 2. Sections 3 and 4 uncover the principles and mechanisms upon which the compiler is built. In Section 5 we carefully detail the automatic generation of multiple algorithms, and outline the code generation process. In Section 6 we report on the performance of the generated algorithms through numerical experiments. We draw conclusions in Section 7.

2 Related Work

A number of research projects concentrate their efforts on domain-specific languages and compilers. Among them, the SPIRAL project [6] and the Tensor Contraction Engine (TCE) [7], focused on signal processing transforms and tensor contractions, respectively. As described throughout this paper, the main difference between our approach and SPIRAL is the inference of properties. Centered on general dense linear algebra operations, one of the goals of the FLAME project is the systematic generation of algorithms. The FLAME methodology, based on the partitioning of the operands and the automatic identification of loop-invariants [8,9], has been successfully applied to a number of operations, originating hundreds of high-performance algorithms.

The approach described in this paper is orthogonal to FLAME. No partitioning of the operands takes place. Instead, the main idea is the mapping of operations onto high-performance kernels from available libraries, such as BLAS [10] and LAPACK.

3   The Compiler Principles

In this section we expose the human thinking process behind the generation of algorithms for a broad range of linear algebra equations. As an example, we derive an algorithm for the solution of the GLS problem, Eq. 2, as it would be done by an expert. Together with the derivation, we describe the rationale for every step of the algorithm. The exposed rationale highlights the key ideas on top of which we founded the design of our compiler.

Given Eq. 2, the **first concern is the inverse operator** applied to the expression $X^T M^{-1} X$. Since X is not square, the inverse cannot be distributed over the product and the expression needs to be processed first. The attention falls then on M^{-1}. The inversion of a matrix is costly and not recommended for numerical reasons; therefore, since M is a general matrix, we **factor** it. Given the structure of M (SPD), we choose a Cholesky factorization, resulting in

$$LL^T = M$$
$$b := (X^T(LL^T)^{-1}X)^{-1}X^T(LL^T)^{-1}y, \tag{3}$$

where L is square and lower triangular. As L is square, the inverse may now be distributed over the product LL^T, yielding $L^{-T}L^{-1}$. Next, we process $X^T L^{-T} L^{-1} X$; we observe that the quantity $L^{-1}X$ **appears multiple times**, and may be computed and reused to **save computation**:

$$W := L^{-1}X$$
$$b := (W^T W)^{-1}W^T L^{-1}y. \tag{4}$$

At this point, since W is not square and the inverse cannot be distributed, there are two **alternatives**: 1) multiply out $W^T W$; or 2) factor W, for instance through a QR factorization. In this example, we choose the former:

$$S := W^T W$$
$$b := S^{-1}W^T L^{-1}y. \tag{5}$$

One can prove that S is SPD, suggesting yet another factorization. We choose a Cholesky factorization and distribute the inverse over the product:

$$GG^T = S$$
$$b := G^{-T}G^{-1}W^T L^{-1}y. \tag{6}$$

Now that all the remaining inverses are applied to triangular matrices, we are left with a series of products to compute the final result. Since all operands are matrices except the vector y, we compute Eq. 6 from right to left to **minimize the number of flops**. The final algorithm is shown in Alg. 1, together with the names of the corresponding BLAS and LAPACK building blocks.

Algorithm 1. Solution of the GLS problem as derived by a human expert	
1 $LL^T = M$	(POTRF)
2 $W := L^{-1}X$	(TRSM)
3 $S := W^T W$	(SYRK)
4 $GG^T = S$	(POTRF)
5 $y := L^{-1}y$	(TRSV)
6 $b := W^T y$	(GEMV)
7 $b := G^{-1}b$	(TRSV)
8 $b := G^{-T}b$	(TRSV)

Three ideas stand out as the guiding principles for the thinking process:

- The first concern is to deal, whenever it is not applied to diagonal or triangular matrices, with the inverse operator. Two scenarios may arise: a) it is applied to a single operand, A^{-1}. In this case the operand is factored with a suitable factorization according to its structure; b) the inverse is applied to an expression. This case is handled by either computing the expression and reducing it to the first case, or factoring one of the matrices and analyzing the resulting scenario.
- When decomposing the equation, we give priority to a) common segments, i.e., common subexpressions, and b) segments that minimize the number of flops; this way we reduce the amount of computation performed.
- If multiple alternatives leading to viable algorithms arise, we explore all of them.

4 Compiler Overview

Our compiler follows the above guiding principles to closely replicate the thinking process of a human expert. To support the application of these principles, the compiler incorporates a number of modules ranging from basic matrix algebra support to analysis of dependencies, including the identification of building blocks offered by available libraries. In the following, we describe the core modules.

Matrix Algebra. The compiler is written using Mathematica from scratch. We implement our own operators: addition (plus), negation (minus), multiplication (times), inversion (inv), and transposition (trans). Together with the operators, we define their precedence and properties, as commutativity, to support matrices as well as vectors and scalars. We also define a set of rewrite rules according to matrix algebra properties to freely manipulate expressions and simplify them, allowing the compiler to work on multiple equivalent representations.

Inference of Properties. In this module we define the set of supported matrix properties. As of now: identity, diagonal, triangular, symmetric, symmetric positive definite, and orthogonal. On top of these properties, we build an inference engine that, given the properties of the operands, is able to infer properties of complex expressions. This module is extensible and facilitates incorporating additional properties.

Building Blocks Interface. This module contains an extensive list of patterns associated with the desired building blocks onto which the algorithms will be mapped. It also contains the corresponding cost functions to be used to construct the cost analysis of the generated algorithms. As with the properties module, if a new library is to be used, the list of accepted building blocks can be easily extended.

Analysis of Dependencies. When considering a sequence of problems, as in GWAS, this module analyzes the dependencies among operations and between operations and the dimensions of the sequence. Through this analysis, the compiler rearranges the operations in the algorithm, reducing redundant computations.

Code Generation. In addition to the automatic generation of algorithms, the compiler includes a module to translate such algorithms into code. So far, we support the generation of MATLAB code for one instance as well as sequences of problems.

To complete the overview of our compiler, we provide a high-level description of the compiler's *reasoning*. The main idea is to build a tree in which the root node contains the initial target equation; each edge is labeled with a building block; and each node contains intermediate equations yet to be mapped. The compiler progresses in a breadth-first fashion until all leaf nodes contain an expression directly mapped onto a building block.

While processing a node's equation, the search space is constrained according to the following criteria:

1. if the expression contains an inverse applied to a single (non-diagonal, non-triangular) matrix, for instance M^{-1}, then the compiler identifies a set of viable factorizations for M based on its properties and structure;
2. if the expression contains an inverse applied to a sub-expression, for instance $(W^T W)^{-1}$, then the compiler identifies both viable factorizations for the operands in the sub-expression (e.g., $QR = W$), and segments of the sub-expression that are directly mapped onto a building block (e.g., $S := W^T W$);
3. if the expression contains no inverse to process (as in $G^{-T} G^{-1} W^T L^{-1} y$, with G and L triangular), then the compiler identifies segments with a mapping onto a building block.

When inspecting expressions for segments, the compiler gives priority to common segments and segments that minimize the number of flops.

All three cases may yield multiple building blocks. For each building block — either a factorization or a segment— both a new edge and a new children node are created. The edge is labeled with the corresponding building block, and the node contains the new resulting expression. For instance, the analysis of Eq. 4 creates the following sub-tree:

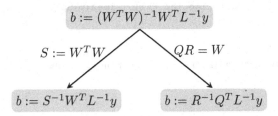

In addition, thanks to the *Inference of properties* module, for each building block, properties of the output operands are inferred from those of the input operands.

Each path from the root node to a leaf represents one algorithm to solve the target equation. By assembling the building blocks attached to each edge in the path, the compiler returns a collection of algorithms, one per leaf.

Our compiler has been successfully applied to equations such as pseudo-inverses, least-squares-like problems, and the automatic differentiation of BLAS and LAPACK operations. Of special interest are the scenarios in which sequences of such problems arise; for instance, the study case presented in this paper, genome-wide association studies, which consist of a two-dimensional sequence of correlated GLS problems.

The compiler is still in its early stages and the code is not yet available for a general release. However, we include along the paper details on the input and output of the system, as well as screenshots of the actual working prototype.

5   Compiler-Generated Algorithms

We detail now the application to GWAS of the process described above. Box 1 includes the input to the compiler: the target equation along with domain-specific knowledge arising from GWAS, e.g, operands' shape and properties. As a result, dozens of algorithms are automatically generated; we report on three selected ones.

5.1   Algorithm 1

To ease the reader, we describe the process towards the generation of an algorithm similar to Alg. 1. The starting point is Eq. 1. Since X is not square, the inverse operator applied to $X^T(h\Phi + (1 - h)I)^{-1}X$ cannot be distributed over the product; thus, the inner-most inverse is $(h\Phi + (1 - h)I)^{-1}$. The inverse is applied to an expression, which is inspected for viable factorizations and segments. Among the identified alternatives are a) the factorization of the operand Φ according to its properties, and b) the computation of the expression $h\Phi + (1 - h)I$.

```
equation = {
  equal[b,
    times[
      inv[times[
          trans[X],
          inv[plus[ times[h, Phi], times[plus[1, minus[h]], id] ]],
          X]
      ],
      trans[X],
      inv[plus[ times[h, Phi], times[plus[1, minus[h]], id] ]],
      y
    ]
  ]
};

operandProperties = {
  {X,   {''Input'',  ''Matrix'',  ''FullRank''} },
  {y,   {''Input'',  ''Vector'' } },
  {Phi, {''Input'',  ''Matrix'',  ''Symmetric''} },
  {h,   {''Input'',  ''Scalar'' } },
  {b,   {''Output'', ''Vector'' } }
};

expressionProperties = {
  inv[plus[ times[h, Phi], times[plus[1, minus[h]], id] ]], ''SPD'' };

sizeAssumptions = { rows[X] > cols[X] };
```

Box 1. Mathematica input to the compiler

Here we concentrate on the second case. The segment $h\Phi + (1 - h)I$ is matched as the SCAL-ADD building block (scaling and addition of matrices); the operation is made explicit and replaced:

$$M := h\Phi + (1 - h)I$$
$$b := (X^T M^{-1} X)^{-1} X^T M^{-1} y. \tag{7}$$

Now, the inner-most inverse is applied to a single operand, M, and the compiler decides to factor it using multiple alternatives: Cholesky ($LL^T = M$), QR ($QR = M$), eigendecomposition ($ZWZ^T = M$), and SVD ($U\Sigma V^T = M$). All the alternatives are explored; we focus now on the Cholesky factorization (POTRF routine from LAPACK):

$$LL^T = M$$
$$b := (X^T L^{-T} L^{-1} X)^{-1} X^T L^{-T} L^{-1} y. \tag{8}$$

After M is factored and replaced by LL^T, the inference engine propagates a number of properties to L based on the properties of M and the factorization applied. Concretely, L is square, triangular and full-rank.

Next, since L is triangular, the inner-most inverse to be processed in Eq. 8 is $(X^T L^{-T} L^{-1} X)^{-1}$. In this case two routes are explored: either factor X (L is triangular and does not need further factorization), or map a segment of the expression onto a building block. We consider this second alternative. The compiler identifies the solution of a triangular system (TRSM routine from BLAS) as a common segment appearing three times in Eq. 8, makes it explicit, and replaces it:

$$W := L^{-1} X$$
$$b := (W^T W)^{-1} W^T L^{-1} y. \qquad (9)$$

Since L is square and full-rank, and X is also full-rank, W inherits the shape of X and is labelled as full-rank. As W is not square, the inverse cannot be distributed over the product yet. Therefore, the compiler faces again two alternatives: either factoring W or multiplying $W^T W$. We proceed describing the latter scenario while the former is analyzed in Sec. 5.2. $W^T W$ is identified as a building block (SYRK routine of BLAS), and made explicit:

$$S := W^T W$$
$$b := S^{-1} W^T L^{-1} y. \qquad (10)$$

The inference engine plays an important role deducing properties of S. During the previous steps, the engine has inferred that W is full-rank and rows[W] > cols[W]; therefore the following rule states that W is SPD.[1]

```
isSPDQ[ times[ trans[ A_?isFullRankQ ], A_ ] /; rows[A] > cols[A]
   := True;
```

This knowledge is now used to determine possible factorizations for S. We concentrate on the Cholesky factorization:

$$GG^T = S$$
$$b := G^{-T} G^{-1} W^T L^{-1} y. \qquad (11)$$

In Eq. 11, all inverses are applied to triangular matrices; therefore, no more treatment of inverses is needed. The compiler proceeds with the final decomposition of the remaining series of products. Since at every step the inference engine keeps track of the properties of the operands in the original equation as well as the intermediate temporary quantities, it knows that every operand in Eq. 11 are matrices except for the vector y. This knowledge is used to give matrix-vector products priority over matrix-matrix products, and Eq. 11 is decomposed

---

[1] In Mathematica notation, the symbols _, _?, and /; indicate a pattern, a constrained pattern, and a condition, respectively. The rule reads: the matrix $A^T A$ is SPD if A is full rank and has more rows than columns.

accordingly. In case the compiler cannot find applicable heuristics to lead the decomposition, it explores the multiple viable mappings onto building blocks. The resulting algorithm, and the corresponding output from Mathematica, are assembled in Alg. 2, CHOL-GWAS.

	Algorithm 2. CHOL-GWAS	
1	$M := h\Phi + (1-h)I$	(SCAL-ADD)
2	$LL^T = M$	(POTRF)
3	$W := L^{-1}X$	(TRSM)
4	$S := W^T W$	(SYRK)
5	$GG^T = S$	(POTRF)
6	$y := L^{-1}y$	(TRSV)
7	$b := W^T y$	(GEMV)
8	$b := G^{-1}b$	(TRSV)
9	$b := G^{-T}b$	(TRSV)

```
tmp1 = - (h id) + 1 id + h Phi
L2 L2ᵀ = tmp1
tmp5 = Xᵀ L2⁻ᵀ
tmp10 = tmp5 tmp5ᵀ
L3 L3ᵀ = tmp10
tmp23 = L2⁻¹ y
tmp31 = tmp5 tmp23
tmp40 = L3⁻¹ tmp31
tmp55 = L3⁻ᵀ tmp40
b = tmp55
```

5.2 Algorithm 2

In this subsection we display the capability of the compiler to analyze alternative paths, leading to multiple viable algorithms. At the same time, we expose more examples of algebraic manipulation carried out by the compiler. The presented algorithm results from the alternative path arising in Eq. 10, the factorization of W. Since W is a full-rank column panel, the compiler analyzes the scenario where W is factored using a QR factorization (GEQRF routine in LAPACK):

$$QR := W$$
$$b := ((QR)^T QR)^{-1}(QR)^T L^{-1}y. \tag{12}$$

At this point, the compiler exploits the capabilities of the *Matrix algebra* module to perform a series of simplifications:

$$b := ((QR)^T QR)^{-1}(QR)^T L^{-1}y;$$
$$b := (R^T Q^T QR)^{-1} R^T Q^T L^{-1}y;$$
$$b := (R^T R)^{-1} R^T Q^T L^{-1}y;$$
$$b := R^{-1}R^{-T}R^T Q^T L^{-1}y;$$
$$b := R^{-1}Q^T L^{-1}y. \tag{13}$$

First, it distributes the transpose operator over the product. Then, it applies the rule

```
times[ trans[ q_?isOrthonormalQ, q_ ] -> id,
```

included as part of the knowledge-base of the module. The rule states that the product $Q^T Q$, when Q is orthogonal with normalized columns, may be rewritten (->) as the identity matrix. Next, since R is square, the inverse is distributed over the product. More mathematical knowledge allows the compiler to rewrite the product $R^{-T} R^T$ as the identity.

In Eq. 13, the compiler does not need to process any more inverses; hence, the last step is to decompose the remaining computation into a sequence of products. Once more, y is the only non-matrix operand. Accordingly, the compiler decomposes the equation from right to left. The final algorithm is put together in Alg. 3, QR-GWAS.

Algorithm 3. QR–GWAS		
1	$M := h\Phi + (1 - h)I$	(SCAL-ADD)
2	$LL^T = M$	(POTRF)
3	$W := L^{-1}X$	(TRSM)
4	$QR = W$	(GEQRF)
5	$y := L^{-1}y$	(TRSV)
6	$b := Q^T y$	(GEMV)
7	$b := R^{-1}b$	(TRSV)

```
tmp1 = - (h id) + 1 id + h Phi
L2 L2ᵀ = tmp1
tmp5 = Xᵀ L2⁻ᵀ
Q10 R10 = tmp5ᵀ
tmp16 = L2⁻¹ y
tmp21 = Q10ᵀ tmp16
tmp29 = R10⁻¹ tmp21
b = tmp29
```

5.3 Algorithm 3

This third algorithm exploits further knowledge from GWAS, concretely the structure of M, in a manner that may be overlooked even by human experts.

Again, the starting point is Eq. 1. The inner-most inverse is $(h\Phi + (1-h)I)^{-1}$. Instead of multiplying out the expression within the inverse operator, we now describe the alternative path also explored by the compiler: factoring one of the matrices in the expression. We concentrate in the case where an eigendecomposition of Φ (SYEVD or SYEVR from LAPACK) is chosen:

$$ZWZ^T = \Phi$$
$$b := (X^T(hZWZ^T + (1-h)I)^{-1}X)^{-1}$$
$$X^T(hZWZ^T + (1-h)I)^{-1}y \tag{14}$$

where Z is a square, orthogonal matrix with normalized columns, and W is a square, diagonal matrix.

In this scenario, the *Matrix algebra* module is essential; it allows the compiler to work with alternative representations of Eq. 14. We already illustrated an example where the product $Q^T Q$, Q orthonormal, is replaced with the identity matrix. The freedom gained when defining its own operators, allows the compiler to perform also the opposite transformation:

```
id -> times[ Q, trans[ Q ] ];
id -> times[ trans[ Q ], Q ];
```

To apply these rules, the compiler inspects the expression $hZWZ^T + (1-h)I$ for orthonormal matrices: Z is found to be orthonormal and used instead of Q in the right-hand side of the previous rules. The resulting expression is

$$b := (X^T(hZWZ^T + (1-h)ZZ^T)^{-1}X)^{-1}$$
$$X^T(hZWZ^T + (1-h)ZZ^T)^{-1}y. \tag{15}$$

The algebraic manipulation capabilities of the compiler lead to the derivation of further multiple equivalent representations of Eq. 15. We recall that, although we focus on a concrete branch of the derivation, the compiler analyzes the many alternatives. In the branch under study, the quantities Z and Z^T are grouped on the left- and right-hand sides of the inverse, respectively:

$$(X^T(Z(hW + (1-h)I)Z^T)^{-1}X)^{-1};$$

then, since both Z and $hW + (1-h)I$ are square, the inverse is distributed:

$$(X^T(Z^{-T}(hW + (1-h)I)^{-1}Z^{-1})X)^{-1};$$

finally, by means of the rules:

```
inv[ q_?isOrthonormalQ ] -> trans[ q ];
inv[ trans[ q_?isOrthonormalQ ] ] -> q;
```

which state that the inverse of an orthonormal matrix is its transpose, the expression becomes:

$$(X^TZ(hW + (1-h)I)^{-1}Z^TX)^{-1}.$$

The resulting equation is

$$b := (X^TZ(hW + (1-h)I)^{-1}Z^TX)^{-1}$$
$$X^TZ(hW + (1-h)I)^{-1}Z^Ty. \tag{16}$$

The inner-most inverse in Eq. 16 is applied to a diagonal object (W is diagonal and h a scalar). No more factorizations are needed, $hW + (1-h)I$ is identified as a SCAL-ADD building block, and exposed:

$$D := hW + (1-h)I$$
$$b := (X^TZD^{-1}Z^TX)^{-1}X^TZD^{-1}Z^Ty. \tag{17}$$

D is a diagonal matrix; hence only the inverse applied to $X^TZD^{-1}Z^TX$ remains to be processed. Among the alternative steps, we consider the mapping of the common segment X^TZ, that appears three times, onto the GEMM building block (matrix-matrix product):

$$K := X^TZ$$
$$b := (KD^{-1}K^T)^{-1}KD^{-1}Z^Ty. \tag{18}$$

From this point on, the compiler proceeds as shown for the previous examples, and obtains, among others, Alg. 4, EIG-GWAS.

Algorithm 4. EIG-GWAS	
1	$ZWZ^T = \Phi$	(SYEVX)
2	$D := hW + (1-h)I$	(ADD-SCAL)
3	$K := X^T Z$	(GEMM)
4	$V := KD^{-1}$	(SCAL)
5	$S := VK^T$	(GEMM)
6	$QR = S$	(GEQRF)
7	$y := Z^T y$	(GEMV)
8	$b := Vy$	(GEMV)
9	$b := Q^T b$	(GEMV)
10	$b := R^{-1} b$	(TRSV)

```
Z1 W1 Z1^T == Phi
tmp2 == - (h id) + 1 id + h W1
tmp7 == X^T Z1
tmp13 == tmp7 tmp2^-1
tmp20 == tmp13 tmp7^T
Q31 R31 == tmp20
tmp36 == Z1^T y
tmp51 == tmp13 tmp36
tmp66 == Q31^T tmp51
tmp76 == R31^-1 tmp66
b == tmp76
```

At first sight, Alg. 4 might seem to be a suboptimal approach. However, as we show in Sec. 6, it is representative of a family of algorithms that play a crucial role when solving a certain sequence of GLS problems within GWAS.

5.4 Cost Analysis

We have illustrated how our compiler, closely replicating the reasoning of a human expert, automatically generates algorithms for the solution of a single GLS problem. As shown in Eq. 1, in practice one has to solve one-dimensional ($t = 1$) or two-dimensional ($t \approx 10^5$) sequences of such problems. In this context we have developed a module that performs a loop dependence analysis to identify loop-independent operations and reduce redundant computations. For space reasons, we do not further describe the module, and limit to the automatically generated cost analysis.

The list of patterns for the identification of building blocks included in the *Building blocks interface* module also incorporates the corresponding computational cost associated to the operations. Given a generated algorithm, the compiler composes the cost of the algorithm by combining the number of floating point operations performed by the individual building blocks, taking into account the loops over the problem dimensions.

Table 1 includes the cost of the three presented algorithms, which attained the lowest complexities for one- and two-dimensional sequences. While QR-GWAS and CHOL-GWAS share the same cost for both types of sequences, suggesting a very similar behavior in practice, the cost of EIG-GWAS differs in both cases. For the one-dimensional sequence the cost of EIG-GWAS is not only greater in theory, the practical constants associated to its terms increase the gap. On the contrary, for the two-dimensional sequence, the cost of EIG-GWAS is lower than the cost of the other two. This analysis suggests that QR-GWAS and CHOL-GWAS are better suited for the one-dimensional case, while EIG-GWAS is better suited for the two-dimensional one. In Sec. 6 we confirm these predictions through experimental results.

Table 1. Computational cost for the three algorithms selected by the compiler

Scenario	QR-GWAS	CHOL-GWAS	EIG-GWAS
One instance	$O(n^3)$	$O(n^3)$	$O(n^3)$
1D sequence	$O(n^3 + mpn^2)$	$O(n^3 + mpn^2)$	$O(n^3 + mpn^2 + mp^2n)$
2D sequence	$O(tn^3 + mtpn^2)$	$O(tn^3 + mtpn^2)$	$O(n^3 + mpn^2 + mtp^2n)$

5.5   Code Generation

The translation from algorithms to code is not a straightforward task; in fact, when manually performed, it is tedious and error prone. To overcome this difficulty, we incorporate in our compiler a module for the automatic generation of code. As of now, we support MATLAB; an extension to Fortran, a much more challenging target language, is planned. We provide here a short overview of this module.

Given an algorithm as derived by the compiler, the code generator builds an *abstract syntax tree* (AST) mirroring the structure of the algorithm. Then, for each node in the AST, the module generates the corresponding code statements. Specifically, for the nodes corresponding to *for* loops, the module not only generates a **for** statement but also the specific statements to extract subparts of the operands according to their dimensionality; as for the nodes representing the building blocks, the generator must map the operation to the specific MATLAB routine or matrix expression. As an example of automatically generated code, the MATLAB routine corresponding to the aforementioned EIG-GWAS algorithm for a two-dimensional sequence is illustrated in Fig. 1.

```
function [b] = GWAS_9_2(X, y, Phi, h, sn, sp, nXs, nys)
    b = zeros(sp, nXs * nys);
    [Z1, W1] = eig( Phi );
    for j = 1:nys
        y_j = y(:, j);
        h_j = h(j);
        T_1 = - (h_j * eye(sn)) + 1 * eye(sn) + h_j * W1;
        T_5 = Z1' * y_j;
        for i = 1:nXs
            X_i = X(:, sp*(i-1)+1 : sp*i);
            T_2 = X_i' * Z1;
            T_3 = T_2 / T_1;
            T_4 = T_3 * T_2';
            [Q31, R31] = qr( T_4 );
            T_6 = T_3 * T_5;
            T_7 = Q31' * T_6;
            T_8 = R31 \ T_7;
            b(:, i + (j-1)*nXs) = T_8;
        end
    end
end
```

Fig. 1. MATLAB code corresponding to EIG-GWAS

6   Performance Experiments

We turn now the attention to numerical results. In the experiments, we compare the algorithms automatically generated by our compiler with LAPACK and GenABEL [11], a widely used package for GWAS-like problems. For details on GenABEL's algorithm for GWAS, GWFGLS, we refer the reader to [12]. We present results for the two most representative scenarios in GWAS: one-dimensional ($t = 1$), and two-dimensional ($t > 1$) sequences of GLS problems.

The experiments were performed on an 12-core Intel Xeon X5675 processor running at 3.06 GHz, with 96GB of memory. The algorithms were implemented in C, and linked to the multi-threaded GotoBLAS and the reference LAPACK libraries. The experiments were executed using 12 threads.

We first study the scenario $t = 1$. We compare the performance of QR-GWAS and CHOL-GWAS, with GenABEL's GWFGLS, and GELS-GWAS, based on LAPACK's GELS routine. The results are displayed in Fig. 2. As expected, QR-GWAS and CHOL-GWAS attain the same performance and overlap. Most interestingly, our algorithms clearly outperform GELS-GWAS and GWFGLS, obtaining speedups of 4 and 8, respectively.

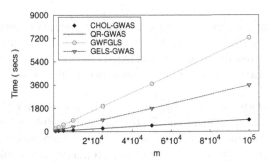

Fig. 2. Timings for a one-dimensional sequence of GLS problems within GWAS. Problem sizes: $n = 10,000$, $p = 4$, $t = 1$. The improvement in the performance of our algorithms is due to a careful exploitation of both the properties of the operands and the sequence of GLS problems.

Next, we present an even more interesting result. The current approach of all state-of-the-art libraries to the case $t > 1$ is to repeat the experiment t times with the same algorithm used for $t = 1$. On the contrary, our compiler generates the algorithm EIG-GWAS, which particularly suits such scenario. As Fig. 3 illustrates, EIG-GWAS outperforms the best algorithm for the case $t = 1$, CHOL-GWAS, by a factor of 4, and therefore outperforms GELS-GWAS and GWFGLS by a factor of 16 and 32 respectively.

The results remark two significant facts: 1) the exploitation of domain-specific knowledge may lead to improvements in state-of-the-art algorithms; and 2) the library user may benefit from the existence of multiple algorithms, each matching a given scenario better than the others. In the case of GWAS our compiler

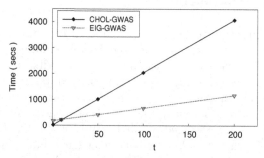

Fig. 3. Timings for a two-dimensional sequence of GLS problems within GWAS. Problem sizes: $n = 5,000$, $p = 4$, $m = 10^6$. CHOL-GWAS is best suited for the scenario $t = 1$, while EIG-GWAS is best suited for the scenario $t \gg 1$.

achieves both, enabling computational biologists to target larger experiments while reducing the execution time.

7   Conclusions

We presented a linear algebra compiler that automatically exploits domain-specific knowledge to generate high-performance algorithms. Our linear algebra compiler mimics the reasoning of a human expert to, similar to a traditional compiler, decompose a target equation into a sequence of library-supported building blocks.

The compiler builds on a number of modules to support the replication of human reasoning. Among them, the *Matrix algebra* module, which enables the compiler to freely manipulate and simplify algebraic expressions, and the *Properties inference* module, which is able to infer properties of complex expressions from the properties of the operands.

The potential of the compiler is shown by means of its application to the challenging *genome-wide association study* equation. Several of the dozens of algorithms produced by our compiler, when compared to state-of-the-art ones, obtain n-fold speedups.

As future work we plan an extension to the *Code generation* module to support Fortran. Also, the asymptotic operation count is only a preliminary approach to estimate the performance of the generated algorithms. There is the need for a more robust metric to suggest a "best" algorithm for a given scenario.

Acknowledgements. The authors gratefully acknowledge the support received from the Deutsche Forschungsgemeinschaft (German Research Association) through grant GSC 111.

References

1. Bientinesi, P., Eijkhout, V., Kim, K., Kurtz, J., van de Geijn, R.: Sparse direct factorizations through unassembled hyper-matrices. Computer Methods in Applied Mechanics and Engineering 199, 430–438 (2010)
2. Lauc, G., et al.: Genomics Meets Glycomics–The First GWAS Study of Human N-Glycome Identifies HNF1α as a Master Regulator of Plasma Protein Fucosylation. PLoS Genetics 6(12), e1001256 (2010)
3. Levy, D., et al.: Genome-wide association study of blood pressure and hypertension. Nature Genetics 41(6), 677–687 (2009)
4. Speliotes, E.K., et al.: Association analyses of 249,796 individuals reveal 18 new loci associated with body mass index. Nature Genetics 42(11), 937–948 (2010)
5. Anderson, E., Bai, Z., Bischof, C., Blackford, S., Demmel, J., Dongarra, J., Du Croz, J., Greenbaum, A., Hammarling, S., McKenney, A., Sorensen, D.: LAPACK Users' Guide, 3rd edn. Society for Industrial and Applied Mathematics, Philadelphia (1999)
6. Püschel, M., Moura, J.M.F., Johnson, J., Padua, D., Veloso, M., Singer, B., Xiong, J., Franchetti, F., Gacic, A., Voronenko, Y., Chen, K., Johnson, R.W., Rizzolo, N.: SPIRAL: Code generation for DSP transforms. Proceedings of the IEEE, Special Issue on "Program Generation, Optimization, and Adaptation" 93(2), 232–275 (2005)
7. Baumgartner, G., Auer, A., Bernholdt, D.E., Bibireata, A., Choppella, V., Cociorva, D., Gao, X., Harrison, R.J., Hirata, S., Krishnamoorthy, S., Krishnan, S., Chung Lam, C., Lu, Q., Nooijen, M., Pitzer, R.M., Ramanujam, J., Sadayappan, P., Sibiryakov, A., Bernholdt, D.E., Bibireata, A., Cociorva, D., Gao, X., Krishnamoorthy, S., Krishnan, S.: Synthesis of high-performance parallel programs for a class of ab initio quantum chemistry models. Proceedings of the IEEE (2005)
8. Fabregat-Traver, D., Bientinesi, P.: Knowledge-based automatic generation of partitioned matrix expressions. In: Gerdt, V.P., Koepf, W., Mayr, E.W., Vorozhtsov, E.V. (eds.) CASC 2011. LNCS, vol. 6885, pp. 144–157. Springer, Heidelberg (2011)
9. Fabregat-Traver, D., Bientinesi, P.: Automatic generation of loop-invariants for matrix operations. In: Computational Science and its Applications, International Conference, pp. 82–92. IEEE Computer Society, Los Alamitos (2011)
10. Dongarra, J., Croz, J.D., Hammarling, S., Duff, I.S.: A set of level 3 basic linear algebra subprograms. ACM Trans. Math. Softw. 16(1), 1–17 (1990)
11. Aulchenko, Y.S., Ripke, S., Isaacs, A., van Duijn, C.M.: Genabel: an R library for genome-wide association analysis. Bioinformatics 23(10), 1294–1296 (2007)
12. Fabregat-Traver, D., Aulchenko, Y.S., Bientinesi, P.: Fast and scalable algorithms for genome studies. Technical report, Aachen Institute for Advanced Study in Computational Engineering Science (2012),
http://www.aices.rwth-aachen.de:8080/aices/preprint/documents/AICES-2012-05-01.pdf

Designing Linear Algebra Algorithms
by Transformation: Mechanizing the Expert Developer

Bryan Marker[1], Jack Poulson[2], Don Batory[1], and Robert van de Geijn[1]

[1] Dept. of Computer Science
The Univ. of Texas at Austin
{bamarker,batory,rvdg}@cs.utexas.edu
[2] Institute for Computational Engineering and Sciences
The Univ. of Texas at Austin
poulson@cs.utexas.edu

Abstract. To implement dense linear algebra algorithms for distributed-memory computers, an expert applies knowledge of the domain, the target architecture, and how to parallelize common operations. This is often a rote process that becomes tedious for a large collection of algorithms. We have developed a way to encode this expert knowledge such that it can be applied by a system to generate mechanically the same (and sometimes better) highly-optimized code that an expert creates by hand. This paper illustrates how we have encoded a subset of this knowledge and how our system applies it and searches a space of generated implementations automatically.

1 Introduction

Parallelizing and optimizing dense linear algebra (DLA) algorithms for distributed-memory machines is typically accomplished by domain experts very familiar with both linear algebra and the oddities of the target machine. When a DLA expert has no experience with distributed-memory code and wants to implement an algorithm, (s)he must live with an existing library, learn a lot about that architecture, or find an experienced developer. This is inefficient and, as we argue, unnecessary because the work of an expert is very mechanical and systematic, and therefore automatable.

We use pipe-and-filter graphs and graph transformations to codify the fundamental algorithms and distributed-memory expertise used in Elemental [20], a domain-specific language with functionality similar to ScaLAPACK [8] and PLAPACK [26]. Doing so enables us to automate the activities of experts: selecting algorithms, composing algorithms, and applying optimizations. In this paper, *we show how expert-tuned, high-performance parallel code for a handful of prototypical examples for distributed-memory architectures can be mechanically produced by a tool.*

We call this approach Design by Transformation (DxT) [23,19], pronounced "dext". We explain the basic ideas behind DxT that were developed while studying distributed-memory examples. Further, we describe how we automatically generate code using DxT and show performance results. Lastly, we explain our future ambitions for DxT.

M. Daydé, O. Marques, and K. Nakajima (Eds.): VECPAR 2012, LNCS 7851, pp. 362–378, 2013.

Partition $A \rightarrow \left(\begin{array}{c|c} A_{TL} & \star \\ \hline A_{BL} & A_{BR} \end{array} \right)$ where A_{TL} is 0×0

while $m(A_{TL}) < m(A)$ do

Repartition $\left(\begin{array}{c|c} A_{TL} & \star \\ \hline A_{BL} & A_{BR} \end{array} \right) \rightarrow \left(\begin{array}{c|c|c} A_{00} & \star & \star \\ \hline A_{10} & A_{11} & \star \\ \hline A_{20} & A_{21} & A_{22} \end{array} \right)$ where A_{11} is $b \times b$

Variant 1	Variant 2	Variant 3
$A_{10} := A_{10}\mathrm{tril}(A_{00})^{-T}$	$A_{11} := A_{11} - \mathrm{tril}(A_{10}A_{10}^{T})$	$A_{11} := \mathrm{chol}(A_{11})$
$A_{11} := A_{11} - \mathrm{tril}(A_{10}A_{10}^{T})$	$A_{11} := \mathrm{chol}(A_{11})$	$A_{21} := A_{21}\mathrm{tril}(A_{11})^{-T}$
$A_{11} := \mathrm{chol}(A_{11})$	$A_{21} := A_{21} - A_{20}A_{10}^{T}$	$A_{22} := A_{22} - \mathrm{tril}(A_{21}A_{21}^{T})$
	$A_{21} := A_{21}\mathrm{tril}(A_{11})^{-T}$	

Continue with $\left(\begin{array}{c|c} A_{TL} & A_{TR} \\ \hline A_{BL} & A_{BR} \end{array} \right) \leftarrow \left(\begin{array}{c|c|c} A_{00} & \star & \star \\ \hline A_{10} & A_{11} & \star \\ \hline A_{20} & A_{21} & A_{22} \end{array} \right)$

endwhile

Fig. 1. Blocked algorithms for computing the Cholesky factorization. $m(A)$ stands for the number of rows of A and $\mathrm{tril}(A)$ indicates the lower triangular part of A. The '$\star$' symbol denotes entries that are not referenced.

2 What an Expert Does

To appreciate our work, let us examine what steps an expert follows in order to produce *by hand* a highly-optimized, parallel implementation of a dense matrix operation in Elemental. We choose Cholesky factorization, an operation that is simple yet prototypical of this class of operations, targeting a cluster architecture as a vehicle to illustrate expert activities. Section 3 explains how we automate this process.

From Specification to Algorithm to Sequential Code. Over the last decade, the FLAME project has developed a repeatable process by which loop-based families of algorithms for dense matrix operations, such as those in Basic Linear Algebra Subprograms (the BLAS [10,9,16]) and LAPACK [1], can be systematically derived [14]. FLAME uses formal derivation [5] and yields a number of algorithmic variants for each operation so the best for a given situation can be chosen[1]. In Figure 1, we show the three known blocked algorithmic variants that result when applied to Cholesky factorization[2]. Blocked algorithms cast most computation in terms of matrix-matrix operations (level-3 BLAS [10]), which can attain high performance on cache-based architectures. Unblocked algorithms can be obtained by setting the block size $b = 1$. Henceforth, we use Variant 3 of Cholesky factorization as our running example.

The FLAME project has produced a library, `libflame` [27], with functionality comparable to that of the widely-used LAPACK library [1]. The algorithms encoded in FLAME were systematically derived and then represented in code using an API, FLAME/C [4], that allows the code to closely resemble the algorithm of Figure 1. Code

---

[1] This expert task of deriving algorithms has been mechanized [3].

[2] *chol* calculates the Cholesky factor of the input, and *tril* returns the lower-triangular portion.

```Chol(Lower, A11);```	```A11_Star_Star = A11;```
	 ```LocalChol(Lower, A11_Star_Star);```
	 ```A11 = A11_Star_Star;```

```Trsm(Right, Lower, Transpose, NonUnit,```
 ```   (T)1, A11, A21);```
	```A21_VC_Star = A21;```
	 ```A11_Star_Star = A11;```
	 ```LocalTrsm```
	 ``` (Right, Lower, Transpose, NonUnit,```
	 ``` (T)1, A11_Star_Star, A21_VC_Star);```
	 ```A21 = A21_VC_Star;```

```TriangularRankK(Lower, Transpose,```
 ```   (T)-1, A21, A21,```
 ```   (T)1, A22);```
	```A21_MC_Star = A21;```
	 ```A21_MR_Star = A21;```
	 ```LocalTriangularRankK```
	 ``` (Lower, Transpose,```
	 ``` (T)-1, A21_MC_Star, A21_MR_Star,```
	 ``` (T)1, A22);```

(a) Original code.	(b) Inline routines.

```A11_Star_Star = A11;```
 ```LocalChol(Lower, A11_Star_Star);```
 ```A11 = A11_Star_Star;```
 	
 ```A21_VC_Star = A21;```
 ```A11_Star_Star = A11;```
 ```LocalTrsm```
 ```   (Right, Lower, Transpose, NonUnit,```
 ```   (T)1, A11_Star_Star, A21_VC_Star);```
 ```\\ A21 = A21_VC_Star;```
 ```A21_MC_Star = A21_VC_Star;```
 ```A21 = A21_MC_Star;```
 	
 ```\\ A21_MC_Star = A21;```
 ```A21_VC_Star = A21;```
 ```A21_MC_Star = A21_VC_Star;```
 ```\\ A21_MR_Star = A21;```
 ```A21_VC_Star = A21;```
 ```A21_MR_Star = A21_VC_Star;```
 ```LocalTriangularRankK```
 ```   (Lower, Transpose,```
 ```   (T)-1, A21_MC_Star, A21_MR_Star,```
 ```   (T)1, A22);```
	```A11_Star_Star = A11;```
	 ```LocalChol(Lower, A11_Star_Star);```
	 ```A11 = A11_Star_Star;```
	
	 ```A21_VC_Star = A21;```
	
	 ```LocalTrsm```
	 ``` (Right, Lower, Transpose, NonUnit,```
	 ``` (T)1, A11_Star_Star, A21_VC_Star);```
	 ```A21_MC_Star = A21_VC_Star;```
	 ```A21 = A21_MC_Star;```
	
	 ```A21_MR_Star = A21_VC_Star;```
	 ```LocalTriangularRankK```
	 ``` (Lower, Transpose,```
	 ``` (T)-1, A21_MC_Star, A21_MR_Star,```
	 ``` (T)1, A22);```

(c) Inline communication.	(d) Remove redundant communication.

Fig. 2. Sequence of optimizations of the loop-body in Variant 3 of Cholesky.

in the style of the Elemental library [20] follows this approach; the code closely resembles the algorithm.

Elemental. If one were to code the algorithm in Figure 1 directly in Elemental code, the loop body would look like Figure 2(a). To see the hidden parallelism for these operations, we must review Elemental [20]. Elemental is a new dense linear algebra library for distributed-memory architectures that uses a 2-dimensional, cyclic distribution of data with blocksize of 1 over a 2-dimensional grid of processors. Specifically, it views the p processes as an $r \times c = p$ grid, and the data is stored, by default, in a distribution that cyclically wraps the rows and columns of the matrix around the process grid (denoted $[M_C, M_R])^3$: element (i, j) of a matrix is stored on process $(i\%r, j\%c)$.

Besides this 2-dimensional distribution, Elemental supports other data distributions and ways to switch between them. This allows a programmer to parallelize an algorithm and its sub-operations in many ways. Elemental is implemented in C++, and matrices are stored in classes that know about distributions[4]. Switching between distributions in the code is accomplished by overloading the '=' operator in the matrix classes, meaning that the '=' operator hides specifics about the communication required to switch between distributions. Behind '=' is code to re-format the data into buffers and call MPI collective communication routines for combinations of distributions.

We use Elemental because it provides a domain-specific language in which we can start with a sequential algorithm and apply expertise about distributed-memory systems to parallelize and optimize an implementation. Two key insights an expert uses that must be codified are the management of redistributions (an operation that represents pure overhead due to the communication involved) and the parallelization of sub-operations. An expert trades more communication (which increases overhead) for more parallelism (which allows useful computation to complete sooner). We explain these considerations below using the Cholesky example, but we remind readers they are representative of a large class of operations in this domain.

Elemental's code lends itself well to identifying and codifying insights about these components because all common operations are abstracted and layered to be modular. The Elemental library uses these common operations across codes. For example there are a finite number of data redistribution functions that are used repeatedly, hidden behind the '=' operator. That code includes the MPI layer, data storage information, etc. The local (sequential BLAS and LAPACK-like) functions are called using the familiar APIs and are wrapped to work with Elemental's matrix class. Elemental's distributed BLAS and LAPACK functionality is built on top of these layers. On top of that layer is Elemental's solver functionality. Lastly, user applications are built on top of the Elemental library. *All of this layering and modularity makes mechanizing expert selections of algorithms and optimizations easier because the inherent structure of the domain is exposed.* Further, this results in common patterns of function calls. An expert knows the optimizations to apply to these repeated patterns across codes for different applications. From the expert's perspective, this layering and separation of concerns improves productivity and makes the code easier to port [18]. See [20] for more details on Elemental.

How an Expert Optimizes for Distributed-Memory Architectures. With this background on Elemental, we now give a high-level explanation of how an expert takes a sequential algorithm and optimizes it for a distributed-memory architecture. Doing so motivates the codification of domain expertise. Consider the sequential, Variant 3, lower-triangular Cholesky algorithm of Figure 1. It can achieve very good performance

---

[3] We provide the Elemental notation for distributions so the graphs below can be understood, but we will not fully explain what the notation means. Please see [20] for more information.

[4] In the Elemental coding convention, variables are named by a submatrix and the name of the data distribution is appended.

on sequential machines, but it is only implicitly parallel if routines Chol, Trsm, and TriangularRankK are parallelized.

To optimize this code, an expert focuses on the loop body, which we show again in Figure 2(a), and inlines the choices of parallelized implementations of its three operations to yield Figure 2(b):

- A_{11} is distributed among processes ($[M_C, M_R]$) and Chol(Lower, A11) represents a small part of all computation. Thus, a convenient way to perform this operation is to bring all data to all processes ($[\star, \star]$), and to then perform the operation redundantly. A11_Star_Star = A11 performs the allgather that duplicates data to all processes. LocalChol(...) then locally performs the factorization on the processes, and A11 = A11_Star_Star places the updated data back in A11 (with no communication).
- Next, consider the update $A_{21} := A_{21}\text{tril}(A_{11})^{-T}$. If one partitions A_{21} into rows as $A_{21} = \begin{pmatrix} a_{21,0}^T \\ a_{21,1}^T \\ \vdots \end{pmatrix}$ and redistributes A_{21} so that rows are assigned to processes in a cyclic order ($[V_C, \star]$), then the processes can compute

$$\begin{pmatrix} a_{21,k}^T \\ a_{21,k+p}^T \\ \vdots \end{pmatrix} := \begin{pmatrix} a_{21,k}^T \\ a_{21,k+p}^T \\ \vdots \end{pmatrix} \text{tril}(A_{11}^{-T})$$

locally in parallel *if* A_{11} is duplicated on all nodes ($[\star, \star]$). A21_VC_Star = A21 redistributes A_{21}. A11_Star_Star = A11 duplicates, again, A_{11}. The local computation is performed by LocalTrsm(...). The data is placed back in A21 by A21 = A21_VC_Star.
- Similarly, the call to TriangularRankK(...) is parallelized by redistributions of data A21_MC_Star = A21 and A21_MR_Star = A21 ($[M_C, \star]$ and $[M_R, \star]$, respectively) followed by a local computation.

Details of what the distributions are and how exactly they are accomplished are not crucial to our discussion [20]. The resultant code provides a hint as to why optimizations are needed: the statements A11 = A11_Star_Star and A11_Star_Star = A11 can be replaced by the more efficient single A11 = A11_Star_Star, which eliminates unnecessary communication.

Further, an Elemental expert knows that redistributions like A21_MC_Star = A21 can be composed from two or more redistributions via intermediate distributions. Some choices of possible substitutions are exposed in Figure 2(c). In most instances, this simply inlines intermediate distributions that were previously hidden. For example replacing A21 = A21_VC_Star by

A21_MC_Star = A21_VC_Star; A21 = A21_MC_Star.

An expert knows this is inefficient as data is distributed from one distribution to another and back to the original distribution instead of redistributing only as necessary. However, the astute reader may notice redundant redistributions which, when removed, yield the code in Figure 2(d) (e.g., A21_MC_Star = A21_VC_Star is removed).

Summary. An expert performs (consciously or subconsciously) the previously described steps to optimize the code of Figure 2(a). Machine-specific details influence how updates in the loop body are parallelized and which operations are expensive and can/need to be optimized. We show in the next section how to mechanize these steps by codifying knowledge about distributed-memory computing and related optimizations using graph transformations. The keys are (1) layer algorithms, and (2) explicitly codify implementation knowledge about (i.e. options for) algorithms, layers, communication, and target architectures – details that were inlined in this section.

3   Toward a Mechanical Expert

The previous section showed, step-by-step, the process a domain expert uses to parallelize and optimize a sequential algorithm. The process is not only systematic but also applies to a broad class of operations in the domain of dense matrix computations. In this section, we discuss how DxT mechanizes this process.

Using Graphs to Model Algorithms and Code. The classic (and arguably greatest to date) example of automated software development is relational query optimization (RQO) [24,25]. A query evaluation program (QEP) is represented by a relational algebra expression. A query optimizer rewrites this expression, using relational algebra identities, to an equivalent expression (program) that has better performance. The optimized expression is then translated to code, thereby synthesizing an efficient QEP implementation. The keys to RQO are (a) representing the design of QEPs as relational algebra expressions and (b) optimizing these expressions to produce efficient programs.

We follow the same paradigm but in a data-flow graph setting using graph transformations. The starting point for our optimization is the loop-body in Figure 2(a) or, equivalently, Figure 1, which is represented as the data-flow graph of Figure 3. The inputs to this graph are the submatrices of A and the results of the loop body (outputs of the graph) are the updated submatrices of A. The boxes

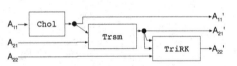

Fig. 3. Cholesky variant 3 algorithm loop body in data-flow graph. This uses abstractions (solid boxes) for component operations; implementations must be chosen.

represent the update operations of the loop body (e.g. Chol, TriangularRankK, Trsm). These operations, also called abstractions, have no implementation details. Just as in the starting algorithm, abstraction boxes only have precondition and postconditions to specify their functionality (we omit these details here).

Abstractions are implemented by algorithms; the pairing of operations with their algorithms form algebraic identities (a.k.a. refinements) of the domain. Algorithms can reference lower-level operations, which have their own implementing algorithms, and this recurses. This codifies the layering of software libraries. There could potentially be many refinements for an operation, each depending on, for example, architecture-specific details, the method of parallelization, or numerical stability characteristics. An expert explores such options or instinctively chooses one out of experience.

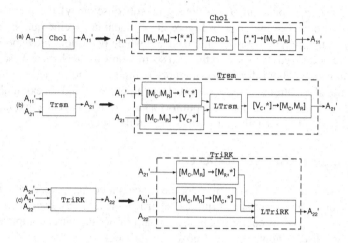

Fig. 4. Sample refinements to implement some operations for distributed-memory. The boxes with → are redistribution operations in Elemental (i.e. "=" operators) from one distribution, represented on the left of the arrow, to another, represent on the right of the arrow. The other solid boxes represent local computation.

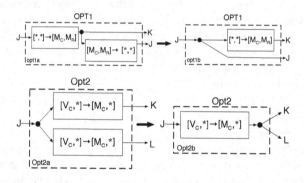

Fig. 5. Sample optimizing graph transformations to remove unnecessary redistribution

In the Cholesky algorithm, an expert replaces abstractions with implementation details as shown in Figure 2(b), which are represented as graph transformations in Figure 4. That figure only shows the best refinement for each abstraction in this algorithm; keep in mind there are others. The boxes represent either redistribution operations (i.e. the "=" operator in the code) or a sequential LAPACK or BLAS function call. Some of the redistribution boxes are abstractions that can be further refined with implementation options like an expert changing code from Figure 2(b) to Figure 2(c). These refinements are also represented as graph transformations.

Optimizations are identities of the form $exp_1 = exp_2$, which allows us to replace one expression (DLA subprogram) exp_1 with another, often more efficient, expression exp_2. An expert developer recognizes the inefficient redistribution code patterns in Figure 2(c) and optimizes the code to end with Figure 2(d). We can represent these optimizations as graph transformations such as those shown in Figure 5. The top is equivalent to the optimization needed to replace A11 = A11_Star_Star and A11_Star_Star = A11 with just A11 = A11_Star_Star and the bottom removes the redundant A21_MC_Star = A21_VC_Star in the Cholesky example.

Using a Graph Representation. By refining the operations of the Cholesky algorithm in Figure 2(a), the top layer of code is flattened to expose redistribution in Figure 2(b). These redistributions are another layer of operations that can be refined in various ways. By refining some of them as in Figure 2(c), we can break through this layer to expose inefficient redistributions that can be removed to create the optimized implementation in Figure 2(d). All of these steps can be encoded as graph transformations that can be applied to more algorithms than just this Cholesky example.

By exploring the space of equivalent graphs (implementations) for a given DLA application and selecting the graph with the best performance characteristics, an efficient DLA program is synthesized. Source is produced by translating the optimized graph to code. These refinements and optimizations are the same as those experts know well and currently apply repeatedly by hand; they can be thought of as transformations that incrementally elaborate DLA programs. One goal of our work is to enable experts to identify and encode these transformations so they can be mechanically applied. Just as functions are modularized for re-use, these transformations can be modularized for re-use, hence the name 'Design by Transformation'.

Given a portfolio of basic, local (sequential) operations and redistribution primitives, cost functions for each primitive, and a target sequence of DLA operations (e.g. as given in Figure 2(a)), a mechanical system employs transformations that an expert would apply by hand. Doing so produces all implementations that have merit (meaning they are best by some measure for some subset of operands) and a mechanism by which to choose from these implementations (e.g. cost functions for implementations).

Searching the Space of Implementations. To an observer, an expert implementing an algorithm appears to follow instincts to select refinements and optimizations. In fact, though, (s)he explores possibilities and assesses cost (implicitly or explicitly). How do we enable a mechanical system to choose the best implementation using "instincts"? We do not (*yet*). Instead of encoding the heuristics and instincts of an expert, we currently use a breadth-first (or brute-force) search that works well for all of the algorithms we have studied thus far. By iteratively applying all possible transformations to an input algorithm's graph, our method generates a search space of all implementations, both good and bad. For all examples we describe in this paper, DxTer takes at most four hours to generate the search space; Cholesky takes less than 5 seconds. By associating a cost with every implementation, the best in the search space can, in principle, be picked out analytically. Thus, our prototype system employs run-time cost estimates for redistribution and computation operations in Elemental to find the best-performing codes. We want the system to see that the code of Figure 2(d) is better than the code of Figure 2(a) by summing operation costs and determining which takes less time to

Table 1. Representative first-order approximations for the cost of operations

Operation	Cost
LocalChol $(n \times n)$	$\gamma n^3/3$
LocalTrsm (Right, Lower, $n \times n$, $m \times n$)	γmnn
All_Star_Star = All $(m \times n)$	$\alpha \lceil \log_2 p \rceil + \beta \frac{p-1}{p} mn$
A21_MC_Star = A21_VC_Star $(m \times n)$	$\alpha \lceil \log_2 c \rceil + \beta \frac{c-1}{c} \frac{m}{r} n$

execute. It should then choose Figure 2(d) out of all implementations in the search space, just as an expert would.

Finding the optimal implementations by cost estimates requires information about the machine such as communication costs, computation speed, and the number of processors. Further, the problem size affects which algorithm is optimal; different parallelization schemes yield varying performance based on the matrix size. We consider a range of problem sizes and find implementations that are optimal for some subset of that range, and use cost functions to then choose which implementation to employ when at run time the problem size is known. Cross-over points between the best implementations occur often. To identify these in our system, we need more accurate models of computation and communication. Fortunately, a mechanical system does not care how complicated the expressions become, which we hope to investigate in future research. An expert rarely attempts this degree of optimization and accuracy since it requires careful analysis that is too error-prone and time consuming to perform by hand. Automation overcomes this hurdle.[5]

For DLA we have reasonable cost estimates. First-order approximations for sequential operations can be given in terms of the number of floating point operations that are performed as a function of the size of operands. For example the matrix multiplication $C = AB$ where C, A, and B are $m \times n$, $m \times k$ and $k \times n$, respectively, takes time (costs) $\gamma 2mkn$ where γ is the time for a floating point operation. The cost of every computation kernel can be approximated by the operation count multiplied by γ.[6]

The data redistributions found in Elemental are implemented using MPI collective communication routines. Lower-bound costs of the common algorithms under idealized models of communication are known [7] in terms of coefficients α and β, which capture the latency and cost per item transferred, respectively. For example redistributing an $n \times n$ block of A_{11} as in line All_Star_Star = All on p processes requires an all-gather operation, which has a lower bound cost of approximately $\alpha \log_2(p) + \beta \frac{p-1}{p} n^2$.

Sample cost functions from our Cholesky example are in Table 1. They are a subset of those necessary to enable the prototype we describe in the next section. They only include higher-order terms and are first-order approximations meant to distinguish good (lower-cost) implementations of an algorithm from others that the system generates.

---

[5] Readers may note that this is exactly the RQO paradigm (described above) applied to DLA implementations.

[6] A second-order approximation would take algorithm performance variation into account, but for now we stick to first-order approximations since this is generally good enough for an expert implementing algorithms by hand.

It turns out these estimates are good enough for the examples we have studied so far, but we expect to improve them to find the best code for more complicated algorithms. For example, we have encountered situations where collective communications are suboptimally implemented on a specific architecture while some other architectures provide hardware support for such redistributions.

4 Experimental Results

We developed a prototype system to test the power of DxT and the cost functions described in the previous section. We call this prototype DxTer [17], pronounced "dexter". We now describe our initial findings.

The results shown in this section were taken from the Lonestar cluster at the Texas Advanced Computing Center. We used 20 nodes, each with 2 Intel Xeon hexa-core processors running at 3.33 GHz.[7] The combined theoretical peak performance of all 240 cores is 3200 GFLOPS. For each problem, we tested a range of algorithmic block sizes and a set of process grid configurations and show the best results. Two-thirds of peak performance is shown at the top of the graphs.

Cholesky Variant 3. We encoded the most useful refinements for a handful of common operations (e.g. BLAS functions) as well as Elemental redistributions to enable DxTer to implement the Cholesky example. [8] From Figure 3, DxTer is able to mechanically produce, without human intervention, hundreds of loop body implementations including all versions in Figure 2. Each of these is Elemental code for Cholesky Variant 3. This allows DxTer to explore the space of implementations. Additionally, we encoded more complicated and subtle expert opti-

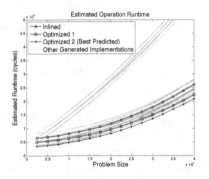

Fig. 6. Cholesky Variant 3 estimated runtime in processor cycles on 240 cores

mizations that are out of the scope of this paper. With these DxTer generates an even better implementation than that of Figure 2(d). This superior implementation is identical to that coded by the expert developer of Elemental.

In its current incarnation, DxTer applies all possible graph re-writes to enumerate the entire search space of implementations. It then uses symbolic cost functions like those those described in Section 3 to choose the best of all the mechanically generated implementations (this is a breadth-first search). Figure 6 shows the cost estimates for the most interesting generated implementations across a range of problem sizes (we omit

[7] We used versions 11.1 of the Intel compiler, 1.6 of MVAPICH2, 1.8.0 of ScaLAPACK, and 1.30 of the GotoBLAS.

[8] There are currently 88 refinement transformations and 382 optimization transformations (most of which are exactly the same except the distributions to which they apply). The transformations presented above are included and are indicative of the complexity of others.

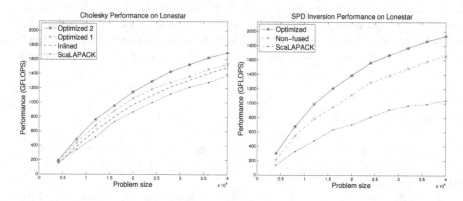

Fig. 7. Cholesky Variant 3 (left) and SPD Inversion (right) implementation performance. Two-thirds of peak performance is at the top of the graphs.

clearly sub-optimal choices). To make the choice of which is best, we fixed the machine-specific parameters that appear in the cost functions. We take the process grid to be 16×15. γ, a measure of machine speed, is set to be 1, and the other machine parameters are set as reasonable multiples of γ. We then determined the Cholesky implementation in Elemental has a lower cost, i.e. run-time, than any of the hundreds of automatically generated implementations. In Figure 6, this implementation's cost estimate is at the bottom of the graph, labeled "Optimized 2."

In Figure 7 (left), we show the performance results of the code of Figure 2(c) (labeled "Inlined"), the code of Figure 2(d) (labeled "Optimized 1"), and the further optimized code (labeled "Optimized 2"). DxTer automatically generated all of these implementations. We leave out the performance of the original code as it is similar to that of the inlined code. Notice that if a domain expert only implemented the algorithm directly and did not optimize considering the machine, the inlined code performance would be what (s)he would see. It shows what happens when one calls high-level operations with hidden inefficiencies. The difference between "Optimized 1" and "Optimized 2" shows the performance gained when complex optimizations are understood and applied. It is clear that expert optimizations are necessary for good performance.

Additional Operations. DxTer was designed to be applied to most, if not all, of the operations supported by `libflame`, the initial development used Cholesky as the driving example. Once this worked, DxTer was applied to other operations to examine how easily the methodology can be applied to new algorithms and extended with new knowledge and reusing knowledge that is already encoded.

Our first experiment was to apply DxTer to a specific algorithm for triangular solve with multiple right-hand sides (`Trsm`) that casts the computation in the loop-body in terms of operations that are very similar to those in the loop-body of Cholesky factorization Variant 3. As expected getting hundreds of implementations from DxTer took very little work on our part because existing transformations were re-used. DxTer's costs models point to the same implementation as the hand-tuned as the best out of the hundreds that were automatically generated.

Next, we tested DxTer's implementation of Cholesky Variant 2. It requires a different flavor of parallelization since the bulk of computation is in the Gemm operation, which requires local computations to be summed (reduced) across processes. With this refinement encoded, DxTer was again able to produce the same optimized implementation code that an expert created. Adding such a transformation to DxTer was an easy change, and it will henceforth be explored for any algorithm with a Gemm operation.

DxTer was also applied to triangular matrix multiplication and triangular matrix inversion. For the former, the expert made an implementation error that *produced wrong results*; DxTer generated correct code. For the inversion operation, DxTer generated a version that had slightly better theoretical numerical properties (with equivalent performance). DxTer had a transformation encoded to use Trsm with a triangular matrix before inversion instead of Trmm after inversion. The expert developer applied this transformation to other algorithms, so it was encoded into DxTer, but he forgot to apply it here. Elemental has been updated with both of the differences DxTer discovered.

Complex Operations. Our greatest triumphs to date came when we applied DxTer to two much more complex operations, $A := L^{-1}AL^{-H}$, which is important in reducing a generalized Hermitian symmetric positive-definite eigenvalue problem to a standard problem, and a fused symmetric, positive-definite (SPD) matrix inversion algorithm.

The parallelization of $A := L^{-1}AL^{-H}$, or two-sided triangular solve, is discussed extensively in [21]. The loop body of one of five algorithmic variants is shown in Figure 8. This algorithm is significantly more complex than those we described previously, but it is similar in style to them and many other DLA algorithms. In addition to the BLAS refinements already encoded, refinements for Axpy had to be added as well as some additional parallelization schemes for TriangularRankK. DxTer will now explore them with any algorithm that uses them. Lastly, the unblocked operation $L^{-1}AL^{-H}$, specific to this algorithm, was added. After these additions, DxTer was able to generate *tens of thousands* of implementations. The existing optimizations described

$$A_{10} := L_{11}^{-1}A_{10}$$
$$A_{20} := A_{20} - L_{21}A_{10}$$
$$A_{11} := L_{11}^{-1}A_{11}L_{11}^{-H}$$
$$Y_{21} := L_{21}A_{11}$$
$$A_{21} := A_{21}L_{11}^{-H}$$
$$A_{21} := W_{21} = A_{21} - \tfrac{1}{2}Y_{21}$$
$$A_{22} := A_{22} -$$
$$\qquad (L_{21}A_{21}^{H} + A_{21}L_{21}^{H})$$
$$A_{21} := A_{21} - \tfrac{1}{2}Y_{21}$$

Fig. 8. Loop body of $A :=$ $L^{-1}AL^{-H}$

above enabled much of this variety; no new optimizations were needed. DxTer's cost models were used to automatically choose a "best" implementation from those generated. The chosen version was slightly better than the optimized version implemented by the expert developer of Elemental. He had forgotten to apply an optimization that was used in other algorithms. Our tool had no such excuses and found the superior implementation.

Next, we applied DxTer to a fused-loop algorithm for SPD matrix inversion. We encoded variant 2 of the algorithm described in [6], the loop body of which is shown in Figure 9. With no additional transformations, DxTer generated *hundreds of thousands* of implementations and chose the same implementation as the expert developed. Figure 7 (right) shows the resulting performance. The "Optimized" version is the implementation generated by DxTer. The "Non-fused" version is the implementation that uses the optimized Cholesky factorization, triangular matrix inversion, and triangular matrix-matrix multiply operations, described above, in succession (also generated by DxTer). This prohibits some optimizations to reduce communication allowed by the fused-loop algorithm. ScaLAPACK

$$A_{11} := Chol(A_{11})$$
$$A_{01} := A_{01}A_{11}^{-1}$$
$$A_{00} := A_{00} + A_{01}A_{01}^T$$
$$A_{12} := A_{11}^{-T}A_{12}$$
$$A_{02} := A_{02} - A_{01}A_{12}$$
$$A_{22} := A_{22} - A_{12}^TA_{12}$$
$$A_{01} := A_{01}A_{11}^{-T}$$
$$A_{12} := -A_{11}^{-1}A_{12}$$
$$A_{11} := A_{11}^{-1}$$
$$A_{11} := A_{11}A_{11}^T$$

Fig. 9. Loop body of SPD matrix inversion

uses a non-fused version of the algorithm. Performance is better here than for Cholesky because the communication is better ammortized over more computation.

To recap these examples demonstrate that it is possible to generate high-performance DLA code mechanically. Indeed, the original motivation for a tool like DxTer was to simplify the burden of experts. We believe DxT is a practical basis to do this.

5 Related Work

Our paper takes a giant step forward for a vision that has been part of the FLAME project since its inception. In the first dissertation that resulted from the project [13], "The Big Picture" was expressed that already captured the idea of encoding algorithms and expert knowledge and mechanically transforming it into code. There, too, optimized parallel implementations were the goal. At the time, the PLAPACK library [26] played the role of a domain specific language much like the Elemental library does in this paper. Many implementations were generated by a system coded in Mathematica, and performance estimates were generated from annotations with cost functions of the algorithms. The present work benefits from an extra ten years of insights during which dozens of papers were published that slowly filled in the blanks of knowledge that now enable the current, more sophisticated, approach based graph transformations.

DxT is similar in goal to the SPIRAL project [22], which primarily focuses on the domain of Digital Signal Processing (DSP). SPIRAL automatically generates high-performance kernels for target architectures. It starts with a mathematical description of the operation in a DSL and applies rewrite transformations to recursively replace operations with implementation code. It uses learning techniques and performance testing to explore the space of implementations. DxT is aimed at higher-level operations, built on lower-level, architecture-specific functions like the BLAS, which allows us to use relatively accurate cost models instead of online-search. We can envision building on SPIRAL-generated kernels, though, instead of the hand-tuned components we use today. We would need the kernels as well as cost estimates for this to work.

Autotuning is often viewed as a way to automatically improve performance [28]. DxT is different in that it generates the search space from a high-level understanding of how algorithms can be transformed. Also, we generate parameterized cost estimates which then guide us to the best implementation(s). We can envision adding autotuning to this approach in order to then choose the best parameters like, for example, the algorithmic block size.

The Broadway compiler [15] had a similar goal as ours to encode expert knowledge to enable optimized code generation. Library function annotations enable Broadway to choose the best implementation at a call site. Broadway is not able to optimize by replacing inlined code with a better implementation, though. Further, Broadway does not search the space of implementations, which is necessary to avoid local minima.

Finally, The Tensor Contraction Engine (TCE) [2] aims to generate high-performance code for a tensor contraction expressed at a high-level using a DSL. It does so by applying transformations to reduce computational complexity, space complexity, communication cost, and then data access. The transformations and cost models of TCE are very similar in spirit and goal to DxT. TCE is geared specifically to tensor contractions while DxT is more general (i.e. not just DLA algorithms [23]). We aim to make transformations easy to understand and encode, so DxT can be applied to other domains.

6   Future Work

Our larger goal is to automatically generate libraries of algorithms for DLA by encoding knowledge about operations and many target architectures. A system would then transform this knowledge into optimized algorithms based on cost estimates, automatically generating families of implementations for more than just distributed-memory. Our prototype shows promising results on distributed-memory targets for operations that are indicative of most operations found in the domain of DLA. Obviously, there remain many additional problems before we can reach this goal, such as the topics below.

Adding Knowledge. We have not yet included all possible transformations in our prototype system. Instead, we have incrementally added those needed by algorithms as we target them for experiments. For example similar to the more advanced optimizations an expert applies for a particular target architecture, there are likely other target-specific optimizations that should be added to the system. Also, the cost functions that were used were first-order approximations for the true cost of the various operations. Better parameterized costs estimates can eventually be incorporated to predict machine-specific performance oddities.

Other Target Architectures. We chose first to test with distributed-memory algorithms for three reasons: (1) We knew a large number of possible algorithms would result; (2) We suspected first-order approximations for the cost of operations would suffice for large matrix sizes; and (3) A considerable penalty would observed if a clearly wrong optimization was chosen, so the benefits of optimization would be clearly visible in experiments. Another important application of DxT would target optimization of sequential and multithreaded dense linear algebra libraries . There, communication would show up in the form of copying of data into contiguous buffers for performance reasons and computation would be performed by so-called inner kernels [11,12,28]. While in principle this is similar to what we have described in this paper, we suspect that in practice the cost functions need to be more accurate and sophisticated. With knowledge encoded to optimize for different architectures, DLA software could be optimized at all layers of code. We will pursue this in future research.

Pruning the Search Space. For now our breadth-first approach to search is sufficient. It only takes hours for even the most complex operations. When optimizing at all architectural levels, though, the search space will become substantially larger, and this cost will become prohibitive. We must study how to prune the space to limit an explosion of choices and we must study more advanced searching techniques. This is an active area of research and will be covered in more detail in a follow-up paper.

7 Conclusion

Using Design by Transformation, we have demonstrated that it is possible to mechanize the actions of an expert dense linear algebra developer to parallelize an algorithm for a distributed-memory target. We presented multiple non-trivial case studies that showed we could reproduce automatically what experts today produce manually. One of the more complicated examples clearly indicates as DLA design problems become more complex, a mechanized expert can produce even better code than manual designs. DLA codes targeting distributed-memory architectures and the related optimizations have similar structure to the examples we have explored. Therefore, we believe our prototype's successes thus far indicate potential for success for a large body of DLA algorithms for distributed-memory computers and even other targets.

The key to DxT is exposing the inherent structure of the DLA domain – this is accomplished by capturing the fundamental operations of the domain using layered designs. Further, we codify fundamental algorithms that implement the operations and optimizations that naturally arise in this domain. Given this structure, we explained that the manual process that a DLA expert uses to design efficient algorithms is so systematic that we could mechanize these tasks. We presented a tool that accomplished this goal. Further, we explained why we believe DxT is not limited to distributed-memory and how it can be used to optimize code beyond what is currently possible by hand. As such we expect this paper to be the first of many to explore the topics described above.

Acknowledgements. Marker was sponsored by a fellowship from Sandia National Laboratories and an NSF Graduate Research Fellowship under grant DGE-1110007. Poulson was sponsored by a fellowship from the Institute of Computational Engineering and Sciences. Batory is supported by the NSF's Science of Design Project CCF 0724979. This research was also partially sponsored by NSF grants OCI-0850750 and CCF-0917167 as well as by a grant from Microsoft. This research used resources at the Texas Advanced Computing Center. *Any opinions, findings and conclusions or recommendations expressed in this material are those of the author(s) and do not necessarily reflect the views of the National Science Foundation (NSF).*

References

1. Anderson, E., et al.: LAPACK Users' Guide. SIAM, Philadelphia (1992)
2. Auer, A., et al.: Automatic code generation for many-body electronic structure methods: The tensor contraction engine. Molecular Physics (2005)
3. Bientinesi, P.: Mechanical Derivation and Systematic Analysis of Correct Linear Algebra Algorithms. PhD thesis, Department of Computer Sciences, The University of Texas, 2006. Technical Report TR-06-46 (September 2006)
4. Bientinesi, P., et al.: Representing linear algebra algorithms in code: The FLAME application programming interfaces. ACM Trans. Math. Soft. 31(1), 27–59 (2005)
5. Bientinesi, P., et al.: Families of algorithms related to the inversion of a symmetric positive definite matrix. ACM Trans. Math. Soft. 35(1) (2008)
6. Bientinesi, P., et al.: Families of algorithms related to the inversion of a symmetric positive definite matrix. ACM Trans. Math. Softw. 35(1), 1–22 (2008)
7. Chan, E., et al.: Collective communication: theory, practice, and experience. Concurrency and Computation: Practice and Experience 19(13), 1749–1783 (2007)
8. Choi, J., et al.: Scalapack: A scalable linear algebra library for distributed memory concurrent computers. In: Proceedings of the Fourth Symposium on the Frontiers of Massively Parallel Computation, pp. 120–127. IEEE Comput. Soc. Press (1992)
9. Dongarra, J.J., et al.: An extended set of FORTRAN basic linear algebra subprograms. ACM Trans. Math. Soft. 14(1), 1–17 (1988)
10. Dongarra, J.J., et al.: A set of level 3 basic linear algebra subprograms. ACM Trans. Math. Soft. 16(1), 1–17 (1990)
11. Goto, K., van de Geijn, R.: High-performance implementation of the level-3 BLAS. ACM Trans. Math. Softw. 35(1), 1–14 (2008)
12. Goto, K., van de Geijn, R.A.: Anatomy of high-performance matrix multiplication. ACM Trans. Math. Softw. 34(3), 1–25 (2008)
13. Gunnels, J.A.: A Systematic Approach to the Design and Analysis of Parallel Dense Linear Algebra Algorithms. PhD thesis, Department of Computer Sciences. The University of Texas (December 2001)
14. Gunnels, J.A., et al.: Flame: Formal linear algebra methods environment. ACM Trans. Math. Soft. 27(4), 422–455 (2001)
15. Guyer, S., Lin, C.: Broadway: A compiler for exploiting the domain-specific semantics of software libraries. Proceedings of the IEEE 93, 342–357 (2005); Special issues on program generation, optimization, and adaptation
16. Lawson, C.L., et al.: Basic linear algebra subprograms for Fortran usage. ACM Trans. Math. Soft. 5(3), 308–323 (1979)
17. Marker, B., et al.: Dxter: An automated software generation prototype for dense linear algebra. In: Preparation
18. Marker, B., et al.: Programming many-core architectures - a case study: Dense matrix computations on the intel scc processor. Concurrency and Computation: Practice and Experience (to appear)
19. Marker, B., et al.: Designing linear algebra algorithms by transformation: Mechanizing the expert developer. In: PPoPP 2012: Proceedings of the Seventeenth ACM SIGPLAN Symposium on Principles and Practices of Parallel Programming, 2 pages (2012) (to appear)
20. Poulson, J., et al.: Elemental: A new framework for distributed memory dense matrix computations. ACM Transactions on Mathematical Software (accepted)
21. Poulson, J., et al.: Parallel algorithms for reducing the generalized hermitian-definite eigenvalue problem. ACM Transactions on Mathematical Software (submitted)

22. Püschel, M., et al.: SPIRAL: Code generation for DSP transforms. Proceedings of the IEEE, Special Issue on "Program Generation, Optimization, and Adaptation" 93(2), 232–275 (2005)
23. Riche, T.L.: et al. Software architecture design by transformation. Computer Science report TR-11-19, Univ. of Texas at Austin (2011)
24. Selinger, P.G., et al.: Access Path Selection in a Relational Database Management System. In: ACM SIGMOD (1979)
25. Ullman, J.D., et al.: Database Systems: The Complete Book, 1st edn. Prentice Hall PTR, Upper Saddle River (2001)
26. van de Geijn, R.A.: Using PLAPACK: Parallel Linear Algebra Package. The MIT Press (1997)
27. Van Zee, F.G.: Libflame: The Complete Reference (2009), http://www.lulu.com
28. Whaley, R.C., Dongarra, J.J.: Automatically tuned linear algebra software. In: Proceedings of SC (1998)

Accelerating the Reorthogonalization of Singular Vectors with a Multi-core Processor

Hiroki Toyokawa[1,2], Hiroyuki Ishigami[2], Kinji Kimura[2], Masami Takata[3], and Yoshimasa Nakamura[2]

[1] CyberAgent, Inc., Shibuya Markcity West 1-12-1 Dogenzaka, Shibuya-ku, Tokyo 150-0043, Japan
[2] Graduate School of Informatics, Kyoto University, Yoshida-honmachi, Sakyo-ku, Kyoto 606-8501, Japan
[3] Academic Group of Information and Computer Sciences, Nara Women's University, Kitauoyanishi-machi, Nara-city, Nara 630-8506, Japan

Abstract. The dLV twisted factorization is an algorithm to compute singular vectors for given singular values fast and in parallel. However the orthogonality of the computed singular vectors may be worse if a matrix has clustered singular values. In order to improve the orthogonality, reorthogonalization by, for example, the modified Gram-Schmidt algorithm should be done. The problem is that this process takes a longer time. In this paper an algorithm to accelerate the reorthogonalization of singular vectors with a multi-core processor is devised.

1 Introduction

Singular value decomposition (SVD) is one of most important matrix operations [1][2]. SVD is applied for data analysis, signal processing and has many other applications to engineering. Therefore SVD has been well studied for many years and some effective SVD algorithms are devised. The QR decomposition, the bisection algorithm, the divide and conquer algorithm (D&C), the MR^3 algorithm, and the I-SVD algorithm are known[1][9][8][12][13][17].

Some of those algorithms can be divided to computations of 2 phases. The first is the computation of singular values. The second is the computation of singular vectors of the computed singular values. Examples of such algorithms include the bisection algorithm and the I-SVD algorithm.

We can select an algorithm for computing singular vectors for given singular values. The dLV twisted factorization [6][7] is one of the possible choices and can compute singular vectors fast. However if a matrix has clustered singular values, the orthogonality of singular vectors computed by the dLV twisted factorization can be worse. In order to improve the orthogonality, we have to use other algorithms to compute singular vectors of such singular values. The inverse iteration algorithm with the modified Gram-Schmidt reorthogonalization can be used for this purpose, and the orthogonality becomes better indeed. The algorithm is known to be slower because the computational complexity is larger than the

M. Daydé, O. Marques, and K. Nakajima (Eds.): VECPAR 2012, LNCS 7851, pp. 379–390, 2013.

dLV twisted factorization. Furthermore in parallel computation with a multi-core/many-core processor, it is often that few computation cores are busy for computation but the others are idle [14][15]. Such a phenomenon occurs because the ordinary implementation of the modified Gram-Schmidt reorthogonalization is not parallelized and the task scheduling is not well-done.

In this paper, a new solution for faster computation of the SVD is devised. The main idea is as follows: we introduce a faster reorthogonalization algorithm which utilize all of the computation cores. For more efficient computation, We estimate the computation time to compute the singular vectors, and we arrange these tasks by the greedy algorithm, which is used for two-dimensional bin packing problem. The selection and arrangement of all of the tasks and algorithms for the computation are decided by auto-tuning technique. Finally we do the numerical experiments and examine the usefulness of the new reorthogonalization algorithm.

2   Algorithms of SVD Which Computes Singular Values and Singular Vectors Individually

There are some algorithms for SVD which consists of 2 phases of computation. The procedure of such algorithms is described as follows:

1. Compute singular values.
2. Compute all singular vectors of the computed singular values.

The bisection algorithm and the I-SVD algorihm can be classified to the group of algorithms. In the first phase, algoirthms which compute singular values are used. In the bisection algorithm, we prepare the Golub-Kahan matrix of the given matrix if the given matrix is bidiagonal. Then we can compute singular values by the use of the subroutine of the bisection algorithm which is implemented as the xSTEBZ in LAPACK. In the I-SVD algorithm, the mdLVs algorithm is used for this purpose.

In the second phase, algorithms which compute singular vectors of designated singular values are used. Examples of such algorithms include the inverse iteration and the dLV twisted factorization.

In this paper, we assume that given matrices are bidiagonal, and we use the bisection algorithm to obtain singular values.

2.1   The Hybrid Algorithm for Computing Singular Vectors

The dLV twisted decomposition can compute singular vectors of the given singular values. However it is known that the orthogonality of the computed singular vectors may be ill if the singular values are in the clusters. In order to improve the orthogonality, we have to compute singular vectors of singular values in the clusters in a different way.

The inverse iteration algorithm with the modified Gram-Schmidt reorthogonalization can also compute singular vectors. The orthogonality of the computed singular vectors is better than those of the singular vectors computed by the dLV twisted decomposition. But the reorthogonalization is slower than the dLV twisted decomposition. In this paper, we call this algorithm *the mGS algorithm*.

Thus we can compute all the singular vectors in the hybrid algorithm as follows (Fig. 1):

1. Compute singular values and some singular vectors.
2. Find clusters of close singular values.
3. Compute singular vectors corresponding to the close singular values with the mGS algorithm.
4. Compute singular vectors of the other singular values by the dLV twisted factorization.

In the first phase, all of the singular values are computed with the bisection algorithm. In the second phase, all of the computed singular values are checked whether there are any singular values which are very close to the next singular value, namely, whether the distance of the neighboring singular values is sufficiently small. In the program used for the experiments described in Sect. 4, a pair of singular values (σ_i, σ_{i+1}) $(\sigma_i < \sigma_{i+1})$ are regarded as close if the pair satisfies either condition as follows:

$$\frac{\sigma_{i+1}^2 - \sigma_i^2}{\|M\|_1} < \text{ortol}_1,$$

$$\frac{\sigma_{i+1} - \sigma_i}{\|M\|_1} < \text{ortol}_2,$$

where σ_i is the i-th singular value, M is the given bidiagonal matrix, $\text{ortol}_1 = 10^{-8}$, and $\text{ortol}_2 = 10^{-3}$. Note that singular values are nonnegative real numbers. The groups of such singular values are called *clusters*. The singular vectors which are corresponding to the clustered singular values are computed in the first phase are discarded since the orthogonality may be ill.

In the third phase, the singular vectors corresponding to the clustered singular values are computed with the mGS algorithm.

In the fourth phase, singular vectors which are not in the clusters are computed by the dLV twisted factorization.

This idea of the hybrid algorithm can be applied to the I-SVD algorithm. The detail is described in [15].

2.2   Problems of the Performance

The computation time of the mGS algorithm is $O(k^2 N)$, where k is the size of clusters $(1 \leq k \leq N)$. Thus if the k is larger, the whole computation time can be much longer. As a preliminary experiment, we measure the computation time to achieve SVD by the method described in Sect. 2.1. Here we prepare matrices of 2 types, M_1 and M_2. The characteristics of them is described in Table 1 and Fig. 2. This experiment is done on the computer whose features are described in Table 3.

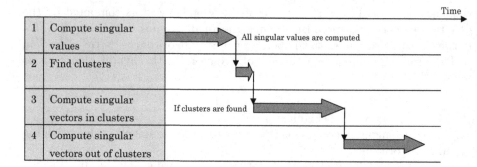

Fig. 1. The main flow of the hybrid algorithm

Table 1. The 3 types of matrices

Matrix type	M_1	M_2
Bidiagonal elements	Random	1
Subdiagonal elements	Random	$1, 10^{-6}$
Distribution of singular values	Some are clustered	Clustered

$$M_2 = \begin{pmatrix} \hat{W} & 10^{-6} & & & \\ & \hat{W} & 10^{-6} & & \\ & & \ddots & \ddots & \\ & & & \hat{W} & 10^{-6} \\ & & & & \hat{W} \end{pmatrix}, \hat{W} = \begin{pmatrix} 1 & 1 & & & \\ & 1 & 1 & & \\ & & \ddots & \ddots & \\ & & & 1 & 1 \\ & & & & 1 \end{pmatrix}$$

$\hat{W}: \hat{N} \times \hat{N}$ matrix, $\hat{N} = 17$.

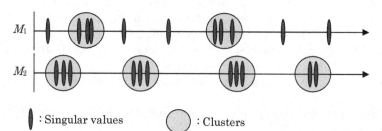

Fig. 2. The distribution of singular values

Table 2. The computation time to achieve SVD

Matrix size N	M_1	M_2
3400	2.6	4.3
6800	9.9	29.7
10200	24.2	130.3
in second: [s]		

Table 2 shows the computation time of several matrix of size N. This shows that it takes much longer time to achieve SVD of M_2 compared to M_1.

Fig. 3 shows the number of the working CPU cores during the computation of the M_2 ($N = 6800$). According to the graph, only 1 core works at the latter time and it lengthens the total computation time, while the other cores are idle. It is obviously inefficient.

The M_2 contains 17 large clusters. The computer which is used for the numerical experiment has 8 computation cores. Therefore 16 (= 8 × 2) clusters are computed by each cores, but the left 1 cluster is computed by 1 core after computing the 16 clusters (Fig. 4).

In this article, we accelerate the total computation of SVD by improving the above mentioned inefficiency.

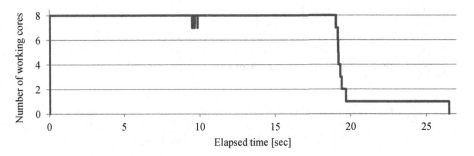

Fig. 3. The changes of the number of working cores

Table 3. The features of computers

CPU	Intel Xeon CPU X5570 2.93GHz
	(2 processors × 4 cores)
Memory	32GBytes
OS	Fedora Linux 17 (x86 64bit)
Compiler	icpc, ifort 12.1 (-O3 option)
Libraries	Intel Math Kernel Library 10.3

3 Accelerating the Reorthogonalization of Singular Vectors

In order to accelerate the reorthogonalization, we first introduce another inverse iteration method in Sect. 3.1. In Sect. 3.2, a hybrid algorithm to utilize the alternative inverse iteration method as well as the mGS algorithm is explained.

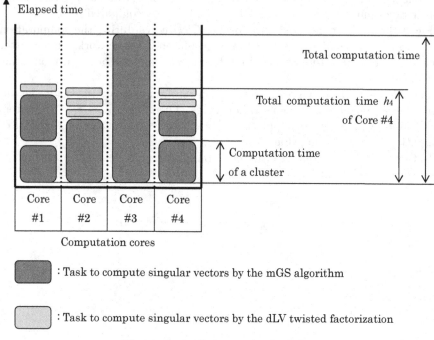

Fig. 4. An inefficient task scheduling

3.1 Inverse Iteration Method with the Compact WY Orthogonalization

The algorithm in this section is based on the idea of the use of the block House-holder orthogonalization in terms of the compact WY representation[16]. The calculation mainly consists of the product of matrices and vectors and rank-1 update operation. Thus these operations can be done with the level-2 BLAS. These computations can be parallelized using the BLAS for parallel computation such as Intel Math Kernel Library, GotoBLAS 2, and so on. The detail of the algorithm is explained in [3]. In this paper, this algorithm is called *the cWY algorithm*.

During the execution of this algorithm, all of the computation cores are used for the computation. Consequently the computation finishes faster than the mGS algorithm, providing that sufficient cores are used for the computation.

3.2 Utilization of the 2 Algorithms

Using the cWY algorithm, we can accelerate the reorthogonalization by using all of the cores. However the product of *CPU time* and *the number of CPU cores* is apt to be larger than that of the inverse iteration and the modified Gram-Schmidt reorthogonalization. Therefore we should not use the cWY algorithm blindly. In fact, we should apply the mGS algorithm to small clusters, and apply

Table 4. The features of the two algorithms

Algorithm	mGS	cWY
Required computation cores for 1 cluster	1 core	All cores
Computation time	Longer	Shorter

mGS: The inverse iteration and the modified Gram-Schmidt reorthogonalization
cWY: The cWY algorithm

the cWY algorithm to large clusters, which takes much longer time. Table 4 shows the features of the 2 algorithms.

To utilize the 2 algorithms in a well-defined hybrid form, we have to keep in mind the 2 points as follows:

1. The algorithm which is applied to a cluster should be selected properly.
2. Computation for all the clusters should be scheduled to each other properly.

These points are described in the next paragraph.

Selection of the Algorithm. Judging from Table 4, the main strategy for selecting algorithm is that: for clusters of small number of singular values, the modified Gram-Schmidt reorthogonalization should be selected. For large clusters, the cWY algorithm should be selected. For more proper selection, we should estimate the computation time for each cluster using the 2 algorithms, and select the faster one.

However it is difficult to determine the obvious conditions and thresholds to judge which algorithm is faster before the computation starts.

It is known that the computational complexity of mGS algorithm is $O(k^2 N)$, and that of the cWY algorithm is $O(k^2 N)$, except for coefficients. First we collect enough sample data of $\{N, k, t_{a,\mathrm{mGS}}, t_{a,\mathrm{cWY}}\}$, where $t_{a,\mathrm{mGS}}$ is the actual time to compute singular vectors by the mGS algorithm, and $t_{a,\mathrm{cWY}}$ is the actual time to compute singular vectors by cWY algorithm. Then we can obtain the estimated computation times $t_{e,\mathrm{mGS}}$ and $t_{e,\mathrm{cWY}}$ of the given $\{N, k\}$ by using the least-squares method, where $t_{e,\mathrm{mGS}}$ is the estimated computation time of the modified Gram-Schmidt algorithm and $t_{e,\mathrm{cWY}}$ is the estimated computation time of the cWY algorithm, respectively.

Assuming that enough data of $\{N, k, t_{a,\mathrm{cWY}}\}$ are given, the procedure to estimate the computation time $t_{e,\mathrm{cWY}}$ of the cWY algorithm is as follows:

1. Fix k and regard $t_{e,\mathrm{cWY}}$ as a linear function of N, and estimate its value at the given N by the least-squares method by using $\{t_{a,\mathrm{cWY}}\}$.
2. Repeat Step 1. with several values $k = \{k_i\}$, and estimate $\{t_{e,\mathrm{cWY}}\}_{k=k_i}$.
3. Regard $\{t_{e,\mathrm{cWY}}\}_{k=k_i}$ as a sample data at the given N, and estimate $t_{e,\mathrm{cWY}}$ for given $\{N, k\}$ by the least-squares method.

The computation time $t_{e,\mathrm{mGS}}$ can also be estimated by the same way.

Scheduling the Tasks. For more acceleration, the computation should also be scheduled well. Scheduling tasks of computing singular vectors in clusters can be regarded as the two-dimensional bin packing problem. By scheduling, the total computation time should be shorter, but we should not consume time for the scheduling itself. Therefore we adopt the greedy algorithm, which can make so good answer and does not cost so longer time.

Assuming that computation time of all of the tasks are estimated by the technique described above, the procedure of the greedy algorithm which is used in this paper is as follows:

1. Compute the current total computation time $\{h_i\}$ of each cores, where h_i means the total computation time of tasks which are assigned to core $\#i$ (see Fig. 4).
2. Assign a task to the core whose total computation time h_i is minimum.
3. Go to Step 1 until all of the tasks are assigned.

3.3   The New Algorithm

Based on the techniques above, the procedure of the new algorithm can be described as follows:

1. Only once: Measure the actual computation time for some combinations of parameters $\{N, k, t_a\}$ and collect the measured sample data.
2. Compute all the singular values by the mdLVs algorithm.
3. Find clusters of close singular values.
4. Select algorithms and schedule the tasks.
   (a) Estimate the computation time of all the tasks.
   (b) Arrange the tasks by the estimated computation time in descending order.
   (c) Set $m \leftarrow 0$.
   (d) Estimate the computation time $\left\{t_{e,\mathrm{mGS}}^{(i)}\right\}_{i=m+1}^{M}$, $\left\{t_{e,\mathrm{cWY}}^{(i)}\right\}_{i=1}^{m}$ of the clusters, and set the estimated computation time
   $$\left\{t_e^{(i)}\right\}_{i=1}^{M} = \begin{cases} t_{e,\mathrm{mGS}}^{(i)} & (m < i \leq M) \\ t_{e,\mathrm{cWY}}^{(i)} & (1 \leq i \leq m) \end{cases}.$$
   (e) Arrange the tasks to minimize the total computation time by the greedy algorithm.
   (f) Check whether the total estimated computation time is shorter than that of previous trial. If the time is not shorter, we adopt the previous selection of algorithms and task schedule, and go to Step 5
   (g) Set $m \leftarrow m + 1$ and go to Step 4d.
5. Compute all the singular vectors in clusters by the selected algorithms.
6. Compute other singular vectors by the dLV twisted factorization.

The Step 1 should be done only once. The results of the step can be stored in a file, and we can load the results from the file and can reuse them. After the procedure, the computation become more efficient and the total computation time will be shorter (Fig. 5). The Step 1 should be only once. Once this step is done, we do not have to do it again. Therefore this step should be done before the computation, for example, on installing the library or at night. The Step 1 and Step 4 are added to the original procedure described in Sect. 2.1.

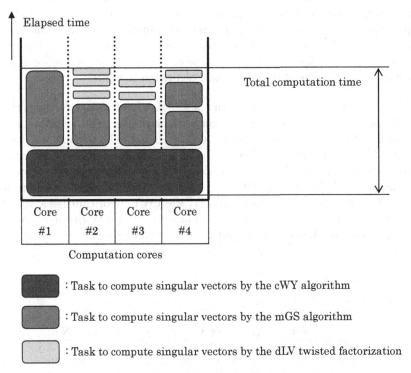

: Task to compute singular vectors by the cWY algorithm

: Task to compute singular vectors by the mGS algorithm

: Task to compute singular vectors by the dLV twisted factorization

Fig. 5. An efficient task scheduling

4 Numerical Experiments

In order to examine the effect of the new algorithm, we give numerical experiments. In the experiments, we see the length of shortened time by the new algorithm. The computers described in Table 3 are used for the experiments. For the experiments, we use the 2 types of matrices $\{M_1, M_2\}$, which are described in Sect. 2.2. We use several size N of these matrices.

In the experiments, we use the GotoBLAS2 for matrix operations. This library is called parallelly and individually by working threads, which execute their own tasks. During the computation of the mGS algorithm, the operation of the library in 1 task should be done by only 1 core. In order to force the policy, we use `omp_set_num_threads` function to adjust the number of threads which are used for matrix operations.

The results of the experiments are shown in Table 5 for M_1, and in Table 6 for M_2. The experiments are done under the condition that the number of cores allocated for the experiments is 1, 2, 4, and 8.

Result of M_1. According to Table 5, any effect in acceleration of the new algorithm is not observed. That is because the new algorithm accelerates only the computation for clusters. Table 7 shows the distribution of the size of the clusters of M_1 ($N = 20400$). According to the table, the size of most of the clusters are less than 10. Only 1 cluster is larger, but the cluster size is not so large compared to that of M_2. Thus the effect of the new algorithm is so little.

Result of M_2. According to Table 6, the new algorithm accelerates the orthogonalization if N is large enough and 8 cores are used. The most computation for M_2 consists of the reorthogonalization because there are large clusters. The size of clusters $k = N/17$. Furthermore at the end of the computation, the computation of the remaining 1 large core is done on 1 core while other cores are idle. Thus the effect of the new algorithm can be observed if enough cores are used for the cWY algorithm. In this computer, the cWY algorithm is slower than the mGS algorithm if 2 cores are used. Thus the cWY algorithm is not used when 2 cores are used, and the computation time is not shorten by the proposed algorithm.

Table 5. The computation time of M_1

Matrix size N	Maximum cluster size max $\{k\}$	1 core Org.	2 cores Org.	New	4 cores Org.	New	8 cores Org.	New
3400	52	16	8	9	5	5	3	3
6800	95	64	33	33	18	18	10	10
10200	169	143	73	75	42	42	24	24
13600	237	253	131	135	75	74	45	45
17000	296	396	207	210	119	118	73	72
20400	369	573	306	307	177	176	110	110

Org.: The original algorithm, New: The new algorithm
in second : [s]

Table 6. The computation time of M_2

Matrix size N	Maximum cluster size max $\{k\}$	1 core Org.	2 cores Org.	New	4 cores Org.	New	8 cores Org.	New
3400	200	16	8	10	6	5	4	4
6800	400	111	58	59	41	42	28	30
10200	600	390	201	203	139	143	126	130
13600	800	912	472	474	328	324	308	304
17000	1000	1725	899	903	639	633	598	585
20400	1200	2901	1509	1509	1059	1065	1037	1020

Org.: The original algorithm, New: The new algorithm
in second : [s]

Table 7. The distribution of the size of clusters ($N = 20400$, Matrix: M_1)

Cluster size k	Count
2	65
3	6
4	5
5	1
6	1
10	1
14	1
25	1
369	1

5 Conclusion

In this paper, a new method to shorten the time for reorthogonalization of singular vectors and then the time for SVD. If the singular values of a given matrix are heavily clustered, the reorthogonalization process is apt to take much longer time in the previous works[14][15]. The new algorithm in this paper can shorten this process by using 1) the cWY algorithm, 2) selecting algorithm to reorthogonalization, and scheduling the tasks. The new algorithm especially works well, providing that there are some large clusters and the computation for the large cluster is done on few cores while other tasks completes. The effect of the new algorithm is checked by some numerical experiments.

As a future work, the predominance in performance and accuracy of the new algorithm should be examined using more types of matrices. Test matrices which have designated singular values or whose conditional number is large can be generated[10][11]. The new algorithm is desired to be compared to existing algorithms implemented in LAPACK.

References

1. Demmel, J.W.: Applied Numerical Linear Algebra. SIAM, Philadelphia (1997)
2. Golub, G., Van Loan, C.: Matrix Computation, 3rd edn. John Hopkins Univ. Press, Baltimore (1996)
3. Ishigami, H., Kimura, K., Nakamura, Y.: Implementation and performance evaluation of inverse iteration with new reorthogonalization algorithm. In: Proceedings of International Conference on Parallel and Distributed Processing Techniques and Applications (PDPTA 2011), vol. II, pp. 775–780 (2011)
4. Iwasaki, M., Nakamura, Y.: Accurate computation of singular values in terms of shifted integrable schemes. Japan Journal of Industrial and Applied Mathematics 23, 239–259 (2006)
5. Iwasaki, M., Nakamura, Y.: Positivity of dLV and mdLVs algorithms for computing singular values. Electronic Transactions on Numerical Analysis 38, 184–201 (2011)
6. Iwasaki, M., Sakano, S., Nakamura, Y.: Accurate twisted factorization of real symmetric tridiagonal matrices and its application to singular value decomposition. Transactions of the Japan Society for Industrial and Applied Mathematics 15, 461–481 (2005) (in Japanese)

7. Konda, T., Takata, M., Iwasaki, M., Nakamura, Y.: A new singular value decomposition algorithm suited to parallelization and preliminary results. In: Proceedings of IASTED International Conference on Advances in Computer Science and Technology (ACST 2006), pp. 79–85 (2006)

8. Nakamura, Y.: Functionality of Integrable System. Kyoritsu Publishing, Tokyo (2006) (in Japanese)

9. Parlett, B.N., Dhillon, I.S.: Relatively robust representations of symmetric tridiagonals. Linear Algebra Appl. 309, 121–151 (2000)

10. Takata, M., Kimura, K., Iwasaki, M., Nakamura, Y.: Algorithms for generating bidiagonal test matrices. In: Proceedings of International Conference on Parallel and Distributed Processing Techniques and Applications (PDPTA 2007), pp. 732–738 (2007)

11. Takata, M., Kimura, K., Nakamura, Y.: Generating algorithms for matrices with large condition number to evaluate singular value decomposition. In: Proceedings of International Conference on Parallel and Distributed Processing Techniques and Applications (PDPTA 2010), pp. 619–625 (2010)

12. Takata, M., Kimura, K., Iwasaki, M., Nakamura, Y.: Performance of a new scheme for bidiagonal singular value decomposition of large scale. In: Proceedings of IASTED International Conference on Parallel and Distributed Computing and Networks (PDCN 2006), pp. 304–309 (2006)

13. Takata, M., Kimura, K., Iwasaki, M., Nakamura, Y.: Implementation of library for high speed singular value decomposition. Journal of Information Processing Society of Japan 47 SIG7(ACS 14), 91–104 (2006)

14. Toyokawa, H., Kimura, K., Takata, M., Nakamura, Y.: On parallelism of the I-SVD algorithm with a multi-core processor. JSIAM Letters 1, 48–51 (2009)

15. Toyokawa, H., Kimura, K., Takata, M., Nakamura, Y.: On parallelization of the I-SVD algorithm and its evaluation for clustered singular values. In: Proceedings of International Conference on Parallel and Distributed Processing Techniques and Applications (PDPTA 2009), pp. 711–717 (2009)

16. Yamamoto, Y., Hirota, Y.: A parallel algorithm for incremental orthogonalization based on the compact WY representation. JSIAM Letters 3, 89–92 (2011)

17. I-SVD Library, http://www-is.amp.i.kyoto-u.ac.jp/lab/isvd/download/

Auto-tuning the Matrix Powers Kernel with SEJITS

Jeffrey Morlan, Shoaib Kamil, and Armando Fox

Computer Science Division, University of California at Berkeley
Berkeley, CA 94720, USA
{jmorlan,skamil,fox}@cs.berkeley.edu

Abstract. The matrix powers kernel, used in communication-avoiding Krylov subspace methods, requires runtime auto-tuning for best performance. We demonstrate how the SEJITS (Selective Embedded Just-In-Time Specialization) approach can be used to deliver a high-performance and performance-portable implementation of the matrix powers kernel to application authors, while separating their high-level concerns from those of auto-tuner implementers involving low-level optimizations. The benefits of delivering this kernel in the form of a specializer, rather than a traditional library, are discussed. Performance of the matrix powers kernel specializer is evaluated in the context of a communication-avoiding conjugate gradient (CA-CG) solver, which compares favorably to traditional CG.

1 Introduction

Krylov subspace methods (KSMs) are iterative algorithms in linear algebra used to solve linear systems (given matrix A and vector b, solve $Ax = b$ for x) or to find eigenvalues and eigenvectors (given A, solve $Ax = \lambda x$ for λ and x) when the matrix is large and sparse, making direct solvers impractical. The solution vectors these methods produce in the first i iterations lie in the vector space spanned by the vectors $\{x_0, Ax_0, \ldots, A^i x_0\}$ for some starting vector x_0; this kind of space is called a Krylov subspace.

Conventionally, KSMs access the matrix A with one or more sparse matrix-vector multiplications (SpMVs) per iteration. Since an SpMV must read a matrix entry from memory for every two useful floating-point operations, making it a highly memory-bound operation, Demmel et al. have proposed *communication-avoiding* algorithms that improve performance by trading redundant computation for memory traffic [1]. In communication-avoiding KSMs, SpMV is replaced by the *matrix powers kernel*, which computes $Ax, A^2 x, \ldots, A^k x$ (or some equivalent basis that spans the same vector space) for matrix A, vector x, and a small constant k. Once the computation has been performed, the next k steps of the solver can proceed without further memory accesses to A by combining vectors from this set. Thus, memory traffic can be reduced – by up to a factor of k in the best case – but obtaining the best performance requires substantial tuning.

M. Daydé, O. Marques, and K. Nakajima (Eds.): VECPAR 2012, LNCS 7851, pp. 391–403, 2013.
© Springer-Verlag Berlin Heidelberg 2013

A difficulty in auto-tuning the matrix powers kernel is that optimal code depends on not only the machine architecture but also on the specific problem instance (namely, the placement of nonzeros in the matrix A), so runtime auto-tuning is necessary to get high performance and performance portability. Moreover, it is desirable to separate concerns of the application writers using KSM solvers and programmers implementing auto-tuning. To do this, we take advantage of SEJITS (Selective Embedded Just-In-Time Specialization) [2], a programming methodology for maximizing separation of concerns between programmers working in specific problem domains and programmers who know how to write efficient low-level code for the kinds of computation used in these domains. The idea is to enable domain experts to express their applications in code without needing to deal with low-level optimizations. This is accomplished by defining a domain-specific language (DSL) or API, embedded in a high-level general purpose language such as Python. The efficiency expert, with the help of a SEJITS framework such as Asp ("Asp is SEJITS for Python") [3], writes a *specializer* to compile efficient implementations for the domain-specific code.

This can be seen as a generalization of the common practice of writing a library in a low-level language with bindings allowing it to be called from a high-level language. The ability to generate code at runtime makes specializers applicable in cases where a library would not be able to provide sufficient flexibility and performance. This could be because the computation itself is too general for a library: an example of this is the domain of stencil computations [4]. It would not be possible to compile implementations of all possible stencils up front, but a specializer can lower a stencil function, given as code in its Python-embedded DSL, down to C++, making it capable of generating optimized code for arbitrary stencils.

Although the matrix powers kernel's computation is not application specific, since SEJITS can generate and compile code at runtime, this approach to delivering auto-tuning avoids the combinatorial explosion of code variants implied by the large space of possible optimizations. Application writers can get both high performance and performance portability without having to be concerned with the low-level optimizations making them possible. SEJITS also allows the tuning logic to be written in the high-level language, making it easier to write and maintain while still keeping it well separated from applications. These benefits would be difficult to obtain if the kernel were delivered as a conventional library.

This paper describes a specializer for the matrix powers kernel, which was built on the Asp framework. Section 2 describes the specializer and the various optimizations implemented in it, and Section 3 contains performance results. Finally, discussion of the benefits that SEJITS brings to this kernel is in Section 4.

2   Implementation

The overall structure of the specializer and code using it is shown in Figure 1. The application first calls the tuner, passing in the sparse matrix A and the

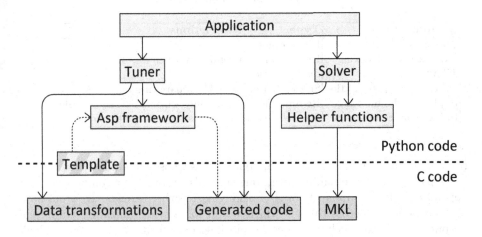

Fig. 1. Overview of specializer and related code. Solid arrows indicate calls, dotted arrows indicate processing of code.

constant k to be used. The tuner attempts to produce an optimized plan for computing matrix powers, comprising both the code to do this computation and its input data derived from the matrix. To do this, the tuner iterates over feasible ranges of the optimization parameters described in Section 2. It calls into static C code to do the necessary transformations on the matrix data, and uses the Asp infrastructure to generate the specialized computational code from a template and compile it. Each candidate plan is benchmarked by running it in a loop until more than a half second has elapsed to get an accurate measurement of its execution time.

An object representing the fastest plan found is returned back to the application, which can then use it in a KSM solver. The solver invokes a method on the object to execute the matrix powers kernel; for other linear algebra operations that KSMs need, the specializer module also provides helper functions which are simply wrappers around Intel Math Kernel Library (MKL) [5] BLAS operations.

2.1 Optimizations

The optimizations of the matrix powers kernel, summarized in Table 1, fall into two major categories: those that reduce memory traffic by storing data more efficiently, and those that re-order computation to parallelize it or make better use of cache. The latter must obey the constraint that if matrix entry A_{ij} is nonzero, then for each level $e : 0 \leq e < k$, component j of $A^e x$ must be computed before it can be used in computing component i of $A^{e+1} x$. Because of this, their effectiveness is highly dependent upon the structure of the matrix's nonzero entries, making runtime auto-tuning necessary for best performance.

The first optimization is to allow for parallelism by *thread blocking*. The matrix rows are partitioned among a number of threads, with each thread being

Table 1. Summary of optimizations

Optimization	Type	Restrictions
Thread blocking	Re-ordering	
Explicit cache blocking	Re-ordering	Useful only when $k > 1$
Tiling	Size reduction	
Symmetric representation	Size reduction	$A = A^T$; square tiles only
Implicit cache blocking	Re-ordering	Useful only when $k > 1$; square tiles only
Index array compression	Size reduction	Block must be sufficiently small

responsible for computing the vector components whose indices are that of rows in its partition. In general, however, one thread's set of $A^{e+1}x$ components will depend on a few components of $A^e x$ belonging to other threads. To avoid having dependencies across threads which would force synchronization after each output vector and preclude the cache blocking described later, a thread block contains not only the rows in its partition, but also any additional rows needed for redundantly computing other threads' components as shown in Figure 2.

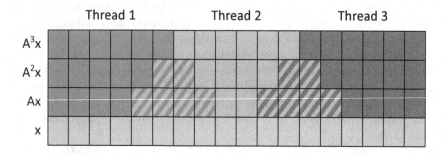

Fig. 2. Thread blocking of an 18x18 tridiagonal matrix. Colors indicate which thread computes each component; the striped components are redundantly computed by two different threads.

Minimizing the redundant computation means choosing a good partitioning, i.e. one with few dependencies between thread blocks. One way would be to make a graph with one vertex per matrix row, add an edge between vertex i and vertex j whenever $A_{ij} \neq 0$, and partition this graph. For a more accurate model of communication volume, what is implemented is to build a hypergraph in which each vertex has a net containing all vertices that have a dependency on it in k steps [6] (that is, net i contains vertex j if $(A^T + I)^k_{ij} \neq 0$, ignoring cancellation), and partition the hypergraph with the PaToH [7] library. Because this can be very time-consuming for larger k, the current tuner only partitions a $k = 1$ hypergraph regardless of the actual value of k.

If a thread block is too large to fit in the processor's cache, then without further division it would be read from RAM k times. We can improve this with *cache blocking*: divide each thread block into sufficiently small cache blocks, and compute the entries for all k vectors in one cache block before moving on to the next. This way, the contents of the thread block are only read once for all k vectors. *Explicit cache blocking* is done in an identical manner to thread blocking; each thread block is simply subdivided recursively until each piece is small enough. An alternative is implicit cache blocking, which is done later on.

Nonzero entries in a sparse matrix are often close together, and this can be taken advantage of by *tiling*. By default, blocks are stored in compressed sparse row (CSR) format, which consists of three arrays: an array of nonzero values (one floating point number for each nonzero), an array of column indices corresponding to those values (one index for each nonzero), and an array indicating where each row begins and ends (one index for each row plus one more, since the end of one row is the start of the next). Tiling a block modifies this to store some fixed-size tile instead of individual values; a tile is stored if any of its individual values are nonzero (Figure 3). This results in a larger values array to hold the extra zeros, but smaller column index and row pointer arrays, which can often make for a net decrease in size. When either dimension of the tile size is even, it also becomes possible to use SIMD instructions to do two multiply-adds at a time, which is implemented via compiler intrinsics.

In many applications, the matrix is *symmetric*, meaning that $A_{ij} = A_{ji}$ for all i, j. When this is the case, the leading square of every block is symmetric as well, since the columns are permuted into the same order as the rows. All entries below the main diagonal can be omitted without losing any information, as they are merely reflections of entries above the diagonal; this optimization can reduce the memory size of a block by almost half. However, it alters the structure of the computation: computing component i now requires going through not only the entries in row i, but also any entry with a column index of i. This adds additional dependencies between components for the purposes of implicit cache blocking, possibly reducing its efficacy.

Unlike the partitioning of the matrix A into thread blocks or of a thread block into explicit cache blocks, where each block is internally stored as a separate

$$\begin{bmatrix} 4 & 5 & & \\ 6 & 7 & & \\ & & & 8 \\ & & 9 & \end{bmatrix}$$

values: 4 5 6 7 8 9
colidx: 0 1 0 1 3 2
rowptr: 0 2 4 5 6

$$\begin{bmatrix} 4 & 5 & & \\ 6 & 7 & & \\ & & 0 & 8 \\ & & 9 & 0 \end{bmatrix}$$

values: 4 5 6 7 0 8 9 0
colidx: 0 1
rowptr: 0 1 2

Fig. 3. A 4x4 block and its representation in compressed sparse row format, before and after 2x2 tiling

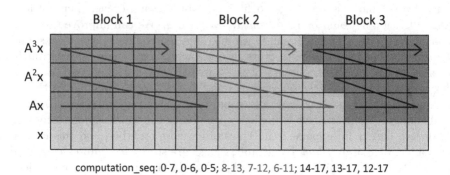

computation_seq: 0-7, 0-6, 0-5; 8-13, 7-12, 6-11; 14-17, 13-17, 12-17

Fig. 4. Implicit cache blocking. Arrows represent the order of computation within each block; the left block is done first, then the middle block, then the right.

matrix, in *implicit cache blocking* (Figure 4) the partitioning into cache blocks is not reflected in the internal data structures in this way. Instead, an array is created that lists the indices of components that need to be computed at each level of each cache block; this array determines the sequence to perform the computation in, which would otherwise simply be one level after another. Since this array will often contain long sequences of increasing integers, it may be stanza-encoded. There is no need for any redundant computation: while creating the array, keep track of what level each entry will have been computed up to at the current point, and just omit any redundancy.

Finally, arrays of indices in each block can be compressed from 32-bit to 16-bit if the block is sufficiently small. If a block is implicitly cache blocked and has fewer than 2^{16} rows, the computation sequence can be compressed. If any block has no more than 2^{16} columns, its `colidx` array can be compressed. With fewer than 2^{16} nonzero tiles, the `rowptr` array can be compressed as well.

The tuner logic currently works as follows: for each possible number of threads (specified in the configuration), both explicit and implicit cache blocking are attempted. For explicit blocking, the tuner iterates over a range of maximum block sizes (5M bytes down to 250K, dividing by 2 each time) recursively bisecting each thread block until all cache blocks are below the maximum size. For implicit blocking, it iterates over a range of the number of implicit blocks per thread (1 to 256, multiplying by 2 each time). With both types of cache blocking, the tile size of each explicit cache block or thread block is chosen to give the smallest memory footprint. Symmetric representation and index compression are used if possible, but implicit blocking is tried both with and without symmetric representation.

3   Results

To test the specializer in a realistic context, we have implemented in Python a communication-avoiding variant of the conjugate gradient (CG) method, a

Krylov method for solving symmetric positive definite linear systems. The basic structure of communication-avoiding CG is shown in Algorithm 1. It produces a sequence of solution approximation vectors x_i and residual vectors $r_i = b - Ax_i$, using a three-term recurrence which relates x_{i-1}, x_i, x_{i+1}, and r_i (details of this are described in [8]). On each iteration, it applies the matrix powers kernel to the current residual vector r_{ki+1}; the power vectors, along with vectors from the previous iteration, form a basis B from which all vectors produced in this iteration will be a linear combination. The recurrence relation is used to compute a matrix D giving the iteration's output vectors in terms of B columns, with dot products computed using the Gram matrix $G = B^T B$. Finally, the output vectors are made explicit by multiplying B and D. Constructions of the Gram matrix and the final output vectors are done by calling the specializer's BLAS wrappers.

Algorithm 1. CA-CG algorithm outline

1: $x_0 \leftarrow 0$
2: $x_1 \leftarrow$ initial guess
3: $r_0 \leftarrow 0$
4: $r_1 \leftarrow b - Ax_1$
5: **for** $i = 0, 1, \ldots$ **do**
6:    Use matrix powers kernel to compute $[Ar_{ki+1}, \ldots, A^k r_{ki+1}]$
7:    $B \leftarrow [x_{ki}, x_{ki+1}, r_{ki-i+2}, \ldots, r_{ki+1}, Ar_{ki+1}, \ldots, A^k r_{ki+1}]$
8:    $G \leftarrow B^T B$
9:    Compute matrix D of output vectors in terms of B
10:    $[x_{ki+i}, x_{ki+i+1}, r_{ki+2}, \ldots, r_{ki+i+1}] \leftarrow BD$
11: **end for**

To demonstrate performance portability, the CA-CG solver was tested on three different multi-core machines: an Intel Xeon (Figure 5), another Intel Xeon with a large number of cores (Figure 6), and an AMD Opteron (Figure 7). The five test matrices are from the University of Florida Sparse Matrix Collection [9] and were chosen for being positive definite and so compatible with CG, being reasonably well-conditioned, and having the kind of locality in their structures that makes it possible to avoid communication. A matrix labeled 149K/10.6M has 149 thousand rows and 10.6 million nonzero elements. The solver is generally several times faster than SciPy's serial implementation of conventional CG, the baseline performance a high-level language application writer could obtain without using any additional libraries. When k is allowed to be greater than 1, it often beats MKL's parallel CG implementation as well (geometric mean is 159% faster for matrix powers alone and 35% faster altogether).

For the $k > 1$ case, the time given is the time per iteration divided by k, since one iteration is mathematically equivalent to k iterations of conventional CG. Although CA-CG is more susceptible to accumulating error in the x vectors, for every matrix tested, if CA-CG did converge to a given tolerance then it did so in nearly the expected number of iterations. The dark part of each bar shows

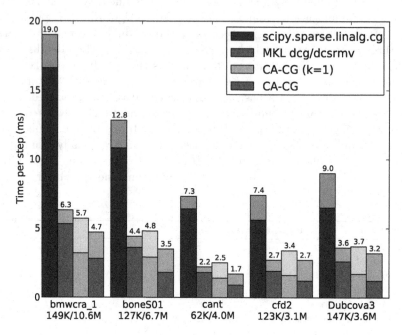

Fig. 5. CG solver performance on 2-socket Intel Xeon X5550 (8 cores, 2.67GHz)

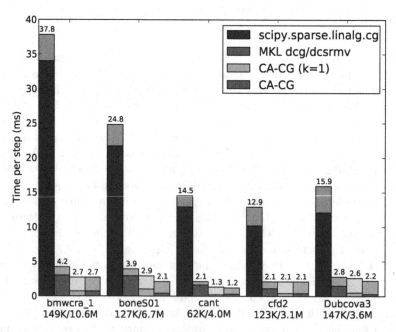

Fig. 6. CG solver performance on 4-socket Intel Xeon X7560 (32 cores, 2.27GHz)

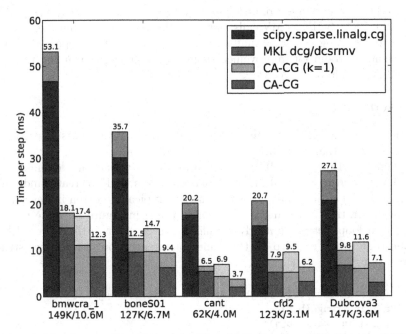

Fig. 7. CG solver performance on 2-socket AMD Opteron 2356 (8 cores, 2.3GHz)

Table 2. CG solver timing data for Figures 5–7

Platform	Solver	Time per step (ms; sparse/dense)				
		bmwcra_1	boneS01	cant	cfd2	Dubcova3
	SciPy	16.6/2.4	10.8/2.0	6.4/0.9	5.6/1.8	6.5/2.5
2-socket	MKL	5.3/1.0	3.6/0.8	1.8/0.4	1.9/0.8	2.6/1.0
Intel Xeon X5550	CA-CG (k=1)	3.2/2.5	2.9/1.9	1.4/1.1	1.6/1.8	1.7/2.0
(8 cores, 2.67GHz)	CA-CG (best)	2.8/1.9	1.8/1.7	0.9/0.8	1.2/1.5	1.2/2.0
		k=2	k=2	k=3	k=2	k=2
	SciPy	34.0/3.8	21.7/3.1	12.9/1.6	10.2/2.7	12.1/3.8
4-socket	MKL	3.0/1.2	2.9/1.0	1.6/0.5	1.1/1.0	1.5/1.3
Intel Xeon X7560	CA-CG (k=1)	0.7/2.0	1.0/1.9	0.2/1.1	0.4/1.7	0.5/2.1
(32 cores, 2.27GHz)	CA-CG (best)	0.7/2.0	0.4/1.7	0.3/0.9	0.4/1.7	0.3/1.9
		k=1	k=2	k=2	k=1	k=2
	SciPy	46.7/6.4	30.1/5.6	17.6/2.6	15.3/5.4	20.7/6.4
2-socket	MKL	14.9/3.2	9.6/2.9	5.4/1.1	5.2/2.7	6.6/3.2
AMD Opteron 2356	CA-CG (k=1)	11.1/6.3	9.7/5.0	4.3/2.6	5.1/4.4	5.9/5.7
(8 cores, 2.3GHz)	CA-CG (best)	8.6/3.7	6.2/3.2	2.0/1.7	3.1/3.1	2.9/4.2
		k=2	k=2	k=3	k=3	k=2

time spent on matrix powers while the light part shows time in the remainder of the solver. This does not include time spent in tuning before calling the solver, which is on the order of a few minutes for each matrix and value of k, or typically about 4000–10000 SciPy SpMV calls; however, this cost can be amortized across multiple solves using the same matrix, as tuning need not be repeated. There is also plenty of room for improvement regarding reducing the tuning time, as discussed in section 6.

4   Discussion

From the experience of developing this specializer, several benefits of writing a specializer rather than a traditional library are observable.

One benefit is that the SEJITS framework provides a ready-made templating system for generating code. SEJITS templates are less work to create, and often cleaner, than the ad-hoc code generation scripts typically written in developing auto-tuned libraries. An example of template use is in Figure 8, where normal and unrolled loops integrate nearly seamlessly, in contrast to the more confusing code that would exist to do the same code generation using direct string concatenation.

```
    for (jb = A->browptr[ib]; jb < A->browptr[ib+1]; ++jb) {
%       for i in xrange(b_m):
%           for j in xrange(b_n):
            y[ib*${b_m} + ${i}] += A->bvalues[jb*${b_m*b_n} + ${i*b_n + j}]
                                * x[A->bcolidx[jb]*${b_n} + ${j}];
%           endfor
%       endfor
    }
```

Fig. 8. Template code for computing one row (having index `ib`) of the matrix-vector multiplication $y = Ax$. `b_m` and `b_n` are the tile height and width, respectively. In Asp's template language, lines beginning with % are template directives, and ${} substitutes the value of an expression.

Another benefit of writing a specializer is that it allows the auto-tuning logic to be written in the high-level language. Not only does this make it easier to write but it also makes it more extensible; if someone wishes to plug in a more advanced auto-tuner, this can be done without having to modify and re-install the specializer.

Finally, being able to generate and compile code at runtime means the combinatorial explosion of all possible code variants does not cause exponential growth in the size of the specializer. Each combination of parameters for basis, tile size, symmetric representation, implicit cache blocking and index compression requires its own compiled code variant to work efficiently. The set of all possible combinations already numbers in the hundreds, which would make for a large library; adding more features and optimizations could render the library approach unworkable.

5 Related Work

The idea of using multiple variants with different optimizations is a cornerstone of auto-tuning. Auto-tuning was first applied to dense matrix computations in the PHiPAC library (Portable High Performance ANSI C) [10]. Using parameterized code generation scripts written in C, PHiPAC generated variants of generalized matrix multiply (GEMM) with a number of optimizations plus a search engine, to, at install time, determine the best GEMM routine for the particular machine. The technology has since been broadly disseminated in the ATLAS package (`math-atlas.sourceforge.net`). Auto-tuning libraries include OSKI (sparse linear algebra) [11], SPIRAL (Fast Fourier Transforms) [12], and stencils [13,14], in each case showing large performance improvements over non-autotuned implementations. With the exception of SPIRAL and Pochoir, all of these code generators use ad-hoc Perl or C with simple string replacement, unlike the template and tree manipulation systems provided by SEJITS.

The OSKI (Optimized Sparse Kernel Interface) library [11] precompiles 144 variants of each supported operation based on install-time hardware benchmarks and includes logic to select the best variant at runtime, but applications using OSKI must still intermingle tuning code (hinting, data structure preparation, etc.) with the code that performs the calls to do the actual computations.

ABCLibScript [15] is a tool to create auto-tuned libraries from files which, like SEJITS templates, contain efficiency-level code combined with scripting directives to control code variant generation. It is geared towards specific kinds of optimizations and tuning searches, whereas SEJITS tries to provide a more general framework suitable for any domain.

6 Future Work

There are several ways this specializer might be improved or extended. Variations on the matrix powers kernel required by more sophisticated solvers could be added, such as preconditioning, or simultaneous computation of powers of A and A^T as in BiCG. More optimizations could be added based on the extensive existing knowledge of optimizing sparse matrix-vector multiplication. The tuner could be made more advanced, by using a performance model or machine learning, in order to effectively cover a larger search space of possible optimizations without taking excessively long as the current brute-force approach would; note that this would not require changes to the underlying C code.

Currently, the hypergraph partitioning used is the most time-consuming part of the tuning. However, such partitioning is most beneficial when the matrix is highly non-symmetric. One simple optimization would be to use non-hypergraph partitioning for symmetric matrices, such as the matrices used with the CA-CG solver; this could also be extended to other matrices that are not highly non-symmetric. In addition, the search could be implemented using more intelligent mechanisms such as hill climbing or gradient ascent. Such search strategies would also be amenable to cases when the user wants a tuning decision within a specified time bound, in which case hill climbing or gradient ascent could be used for a few iterations until the maximum bound is reached.

Tuning decisions could potentially be reused for matrices with the same structure; such matrices are commonly used in finite element computations, where the actual values in the matrix may change, but the elements appear in the same locations across modeling problems.

7 Conclusion

Though originally motivated by domains where a library is unsuitable due to the generality of the desired computational kernel, the SEJITS methodology also proves useful for domains where generality comes not from the kernel itself but from the need to tune it for performance. Although the matrix powers kernel could plausibly be written as a library, as a specializer it demonstrates how writing auto-tuners as specializers has benefits for both efficiency-level programmers and for productivity-level programmers who wish to extend the tuning logic.

Acknowledgements. Thanks to Erin Carson and Nicholas Knight for providing the initial version of the code for the matrix powers kernel. Thanks also to Mark Hoemmen, Marghoob Mohiyuddin, and James Demmel for feedback and suggestions on this paper.

This work was performed at the UC Berkeley Parallel Computing Laboratory (Par Lab), supported by DARPA (contract #FA8750-10-1-0191) and by the Universal Parallel Computing Research Centers (UPCRC) awards from Microsoft Corp. (Award #024263) and Intel Corp. (Award #024894), with matching funds from the UC Discovery Grant (#DIG07-10227) and additional support from Par Lab affiliates National Instruments, NEC, Nokia, NVIDIA, Oracle, and Samsung.

References

1. Mohiyuddin, M., Hoemmen, M., Demmel, J., Yelick, K.: Minimizing communication in sparse matrix solvers. In: Supercomputing 2009, Portland, OR (November 2009)
2. Catanzaro, B., Kamil, S., Lee, Y., Asanović, K., Demmel, J., Keutzer, K., Shalf, J., Yelick, K., Fox, A.: SEJITS: Getting productivity and performance with selective embedded JIT specialization. In: Workshop on Programming Models for Emerging Architectures, PMEA 2009, Raleigh, NC (October 2009)
3. Kamil, S.: Asp: A SEJITS implementation for Python, https://github.com/shoaibkamil/asp/wiki
4. Kamil, S., Coetzee, D., Fox, A.: Bringing parallel performance to Python with domain-specific selective embedded just-in-time specialization. In: Proceedings of the 10th Python in Science Conference, SciPy 2011, Austin, TX (2011)
5. Intel: Math Kernel Library, http://software.intel.com/en-us/articles/intel-mkl/

6. Carson, E., Demmel, J., Knight, N.: Hypergraph partitioning for computing matrix powers (October 2010), http://www.cs.berkeley.edu/~knight/cdk_CSC11_abstract.pdf
7. Catalyürek, Ü.V.: Partitioning Tools for Hypergraph, http://bmi.osu.edu/~umit/software.html
8. Hoemmen, M.: Communication-avoiding Krylov subspace methods. PhD thesis, EECS Department, University of California, Berkeley (April 2010)
9. Davis, T., Hu, Y.: The University of Florida sparse matrix collection, http://www.cise.ufl.edu/research/sparse/matrices
10. Bilmes, J., Asanović, K., Chin, C.W., Demmel, J.: Optimizing matrix multiply using PHiPAC: a Portable, High-Performance, ANSI C coding methodology. In: Proceedings of International Conference on Supercomputing, Vienna, Austria (July 1997)
11. Vuduc, R., Demmel, J.W., Yelick, K.A.: OSKI: A library of automatically tuned sparse matrix kernels. Journal of Physics Conference Series 16(i), 521–530 (2005)
12. Püschel, M., Moura, J.M.F., Johnson, J., Padua, D., Veloso, M., Singer, B., Xiong, J., Franchetti, F., Gacic, A., Voronenko, Y., Chen, K., Johnson, R.W., Rizzolo, N.: SPIRAL: Code generation for DSP transforms. Proceedings of the IEEE, Special Issue on "Program Generation, Optimization, and Adaptation" 93(2), 232–275 (2005)
13. Kamil, S., Chan, C., Oliker, L., Shalf, J., Williams, S.: An auto-tuning framework for parallel multicore stencil computations. In: IPDPS 2010, pp. 1–12 (2010)
14. Tang, Y., Chowdhury, R.A., Kuszmaul, B.C., Luk, C.K., Leiserson, C.E.: The Pochoir stencil compiler. In: Proceedings of the 23rd ACM Symposium on Parallelism in Algorithms and Architectures, SPAA 2011, pp. 117–128. ACM, New York (2011)
15. Katagiri, T., Kise, K., Honda, H., Yuba, T.: ABCLibScript: a directive to support specification of an auto-tuning facility for numerical software. Parallel Comput. 32(1), 92–112 (2006)

Auto-tuning of Numerical Programs
by Block Multi-color Ordering Code Generation
and Job-Level Parallel Execution

Tatsuya Abe[1] and Mitsuhisa Sato[2,1]

[1] Advanced Institute for Computational Science, RIKEN, Hyogo, Japan
abet@riken.jp
[2] Center for Computational Sciences, University of Tsukuba, Ibaraki, Japan
msato@cs.tsukuba.ac.jp

Abstract. Multi-color ordering is a parallel ordering that allows programs to
be parallelized by application to sequentially executed parts of the programs.
While multi-color ordering parallelizes sequentially executed parts with data dependences and increases the number of parts executed in parallel, improved performance by multi-color ordering is sensitive to differences in the architectures
and systems on which the programs are executed. This sensitivity requires us to
tune the numbers of colors; i.e., modify programs for each architecture and system. In this work, we develop a code generator based on multi-color ordering and
automatically tune the number of colors using a job-level parallel scripting language Xcrypt. Furthermore, we support block multi-color ordering that avoids the
disadvantage of stride accesses in the original multi-color ordering, and evaluate
and clarify the effectiveness of block multi-color ordering.

Keywords: Block multi-color ordering, source-to-source code generation, job-
level parallel execution, scripting parallel language.

1 Introduction

Parallelization is a method applied to programs that allows us to parallelize sequentially executed parts of the program. Parallel orderings are parallelizations that extend
the parts of a program executed in parallel by changing the order of computation in the
program [1]. Why do the parts of a program executed in parallel increase when changing the order of computation in the program? To answer this, consider the following
program as an example.

```
x(:) = 0
x(1) = 1
do i = 2, 100
    x(i) = x(i-1)
end do
```

M. Daydé, O. Marques, and K. Nakajima (Eds.): VECPAR 2012, LNCS 7851, pp. 404–419, 2013.

In the above program, x(i) is initially 0, and then becomes 1 for any i. In this program, computation order is significant. If x(3) were computed in advance, x(3) would remain 0, whereas x(i) would be 1 for all i except 3. This is because x(i) depends on x(i-1), and this data dependence obviously prevents the loop in the program from being parallelized.

In the above program, the computation order cannot be changed since the semantics of the program relies on the order. However, this is not always the case. There are some programs whose order of computation can be semantically changed. For example, recall the Gauss-Seidel method for solving systems of linear equations.

The Gauss-Seidel method is an iterative method. Roughly speaking, the point of the Gauss-Seidel method is to update values immediately, i.e., to use values at the n-th time step to calculate the n-th time step:

$$u_i^n = \frac{1}{a_{ii}}(b_i - \sum_{j=1}^{i-1} a_{ij}u_j^n - \sum_{j=i+1}^{n} a_{ij}u_j^{n-1})$$

where u_i^n denotes the i-th column of solution candidates at the n-th step and a_{ij} denotes the element in the i-th row and j-th column of the coefficient matrix. This calculation is different from that in the Jacobi method where values at the n-th step are not only used in the n-th step, but also in the $(n + 1)$-th step:

$$u_i^n = \frac{1}{a_{ii}}(b_i - \sum_{j=1}^{i-1} a_{ij}u_j^{n-1} - \sum_{j=i+1}^{n} a_{ij}u_j^{n-1}) \ .$$

While both the Gauss-Seidel and Jacobi methods are used to solve systems of linear equations, the values of u at the n-th step in these two methods are different. Nevertheless, this is not actually a problem. The aim in solving a system of linear equations is to find a solution of the system of linear equations. Both the Gauss-Seidel and Jacobi methods give us solution candidates for each system of linear equations. These solution candidates can be checked by being substituted into the system of linear equations. In this sense, differences in intermediate values along the way do not constitute a problem.

The Gauss-Seidel method contains data dependences. If we ignore these data dependences by changing the ordering, that is, we forcibly parallelize sequentially executed parts of a program, then intermediate values along the way would be different from those in the original program. Therefore, we must choose a legal ordering that leads to legal solutions.

In this paper, we propose automatic parallelization by using a parallel ordering called *multi-color ordering*. Furthermore, our method supports *block multi-color ordering* that avoids the disadvantage of the original multi-color ordering, namely, stride accesses, which result in degraded performance. Our method relies on code generation using block multi-color ordering, and job-level parallel execution. The proposed method provides programmers with an environment in which to modify and tune programs by block multi-color ordering.

It should be noted that multi-color ordering changes the algorithm, i.e., the semantics of the program. This distinguishes the work in this paper from studies on compilers.

Compilers do not change the semantics of programs although they may raise numerical errors by exchanging the order of operations in programs as a result of optimization[1]. This work is beyond the scope of compilers, and targets what compilers should not do. It is up to the programmers to apply the proposed method to their programs, and in this respect, great care should be taken.

We formalize the parallelization of block multi-color ordering, thus allowing it to be implemented by computers. We would also like to handle a large number of programs instantiated with parameters simultaneously. Using existing tools like Bash for this seems to be adequate. However, this is not the case. Given that anyone can execute a program in a parallel or distributed computational environment, higher portability is required for tools on computers. In reality, parallel and distributed computational environments often have batch job systems to manage and control the execution of jobs, and keep the execution of one job separate from that of another. These batch job systems usually require text files containing *job scripts*. In other words, when executing a large number of programs simultaneously in a parallel or distributed computational environment, we need to create a large number of job scripts. It is tedious to do this using existing tools like Bash.

In this work we propose using a scripting language domain specific to job-level parallel execution, namely Xcrypt [2]. Xcrypt absorbs differences between computational environments and is suitable for parameter sweeps since it was originally designed for executing and controlling such programs. Since we have developed a code generator for block multi-color ordering, i.e., applying block multi-color ordering to programs is semi-automatic, we can easily automate the execution of programs with block multi-color ordering in parallel and distributed computational environments using Xcrypt. In this paper we introduce the method by means of an example.

Outline. In Sect. 2 we introduce multi-color ordering in detail. In Sect. 3 we explain block multi-color ordering, which overcomes the disadvantage of multi-color ordering as described in Sect. 2. In Sect. 4 we give an overview of how to automatically tune programs to which block multi-color ordering has been applied. In Sect. 5 we explain the code generator based on block multi-color ordering. In Sect. 6 we introduce the job-level parallel scripting language Xcrypt, which is used to execute block multi-color ordered programs automatically, while in Sect. 7 we evaluate our method through experimental results. In Sect. 8 we conclude this work with reference to future work.

2   Multi-color Ordering

Multi-color ordering is a parallel ordering that parallelizes programs with data dependences by exchanging the order of computation of columns of matrices [1]. For the purpose of understanding parallelization, consider the differences in the settings of three-dimensional real space $\mathbb{R}^3$ for the following program:

---

[1] For example, the Intel Composer XE compiler has a compile option -par-report to show reports on parallelization, while version 12.1 has the option -guide to provide detailed hints. However, the compiler does not give any hints that change the semantics of the program.

```fortran
integer x_num, y_num, z_num
real(8), allocatable :: u(:,:,:), f(:,:,:), dgn(:,:,:)
real(8), allocatable :: axp(:,:,:), axm(:,:,:)
real(8), allocatable :: ayp(:,:,:), aym(:,:,:)
real(8), allocatable :: azp(:,:,:), azm(:,:,:)
integer i, j, k
...
do k = 2, z_num+1
   do j = 2, y_num+1
      do i = 2, x_num+1
         u(i, j, k) = (f(i, j, k)-&
                 axp(i, j, k)*u(i+1, j, k)-&
                 axm(i, j, k)*u(i-1, j, k)-&
                 ayp(i, j, k)*u(i, j+1, k)-&
                 aym(i, j, k)*u(i, j-1, k)-&
                 azp(i, j, k)*u(i, j, k+1)-&
                 azm(i, j, k)*u(i, j, k-1)&
                 )/dgn(i, j, k)
      end do
   end do
end do
```

Fig. 1. Sample code in Fortran 90

$$u^n(i, j, k) = (f(i, j, k)$$
$$- a_{x+}(i, j, k)u^{n-1}(i + 1, j, k) - a_{x-}(i, j, k)u^n(i - 1, j, k)$$
$$- a_{y+}(i, j, k)u^{n-1}(i, j + 1, k) - a_{y-}(i, j, k)u^n(i, j - 1, k)$$
$$- a_{z+}(i, j, k)u^{n-1}(i, j, k + 1) - a_{z-}(i, j, k)u^n(i, j, k - 1)$$
$$) / dgn(i, j, k)$$

where $dgn(i, j, k)$, $f(i, j, k)$, $a_{x+}(i, j, k)$, $a_{x-}(i, j, k)$, $a_{y+}(i, j, k)$, $a_{y-}(i, j, k)$, $a_{z+}(i, j, k)$, and $a_{z-}(i, j, k)$ are various functions from the three-dimensional space to the set of real numbers $\mathbb{R}$. $u^n(i, j, k)$ denotes the value of (i, j, k) at the n-th time step. The equation shows that $u^{n-1}(i + 1, j, k)$, $u^{n-1}(i, j + 1, k)$, $u^{n-1}(i, j, k + 1)$, $u^n(i - 1, j, k)$, $u^n(i, j - 1, k)$, and $u^n(i, j, k - 1)$ are used to obtain $u^n(i, j, k)$. This is done by iterating loops controlled by the third (z-axis), second (y-axis), and first (x-axis) arguments in order, that is, the equation is coded as in Fig. 1 where f, axp, axm, ayp, aym, azp, azm, and dgn denote f, a_{x+}, a_{x-}, a_{y+}, a_{y-}, a_{z+}, a_{z-}, and dgn, respectively. In the program, no loop can be parallelized, i.e., iterations in loops cannot be executed in parallel since each pair of iterations in the loops has data dependences.

Multi-color ordering forces such loops to be parallelized, i.e., some iterations in the loops are executed in parallel. It appears that parallelization changes the semantics of

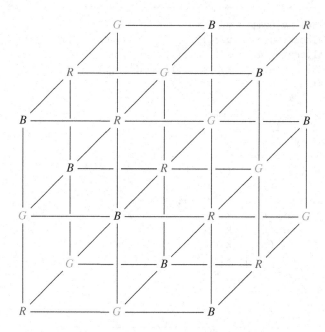

Fig. 2. Multi-color ordering of lattice points

the program. This is true. However, multi-color ordering does not force all iterations in the loops to be executed in parallel. Adjacent pairs on the three-dimensional lattice are not executed in parallel. An example with the number of colors set to 3, is illustrated in Fig. 2 where the lattice points executed in parallel are colored in the same color for simplicity. Here, R, G, and B are colored in red, green, and blue, respectively.

The program is one for stencil computation. Any lattice point immediately affects adjacent lattice points. Adjacent pairs should not be considered for parallel execution. Multi-color ordering executes computations at distinct lattice points. Actually, the semantics of a program to which multi-color ordering has been applied is different from that of the original program. Therefore, we cannot apply multi-color ordering to all programs and should take care in applying this ordering to programs.

Iterative methods for solving systems of linear equations are typical examples of programs to which we can apply multi-color ordering. These methods involve a sequence of initial values that is iteratively updated to a new sequence, and finally becomes a solution. Since a sequence is often proved to be a legal solution by substituting the sequence in the linear equations and analyzing the residual errors, intermediate values on the way do not matter. A change in ordering affects only the convergence rate. Therefore, we can apply multi-color ordering to such iterative methods.

We conclude this section by showing the effect of applying multi-color ordering to a program. In general, there are certain sufficient conditions that prevent adjacent lattice points from being colored in the same color. To enable us to consider these sufficient conditions in this section, we consider an equation in one-dimensional space. The following program:

```
do i = 2, x_num+1
    u(i) = (f(i)-axp(i)*u(i+1)-axm(i)*u(i-1))/dgn(i)
end do
```

is translated to

```
mulco_color_num = 3
do mulco_color = 0, mulco_color_num-1
    do i = 2+mulco_color, x_num+1, mulco_color_num
        u(i) = (f(i)-axp(i)*u(i+1)-axm(i)*u(i-1))/dgn(i)
    end do
end do
```

where mulco_color_num denotes the number of colors. Here, it is assumed to be 3. The mulco_color is a loop variable over $0, \ldots,$ mulco_color_num. When mulco_color is 0, computations at the red lattice points are executed in the loop. Similarly, when it is 1, computations at the green points are executed, and when mulco_color is 2, computations at the blue points are executed. In other words, multi-color ordering is a parallelization that executes independent computations (on non-adjacent lattice points) in parallel.

3   Block Multi-color Ordering

Multi-color ordering parallelizes sequentially executed parts with data dependences and increases the number of parts executed in parallel. However, the naïve implementation of multi-color ordering described in the previous section has the disadvantage of stride accesses, that is, jumps in accessing elements in arrays, thereby decreasing the performance of programs.

To avoid this disadvantage, block multi-color ordering, which is called *block red-black ordering* when two colors are used, has been proposed [3]. A deterioration in performance is caused by stride accesses to elements in arrays. Block multi-color ordering prevents iterations in loops from becoming too fine-grained and keeps loop iterations coarse-grained. Whereas multi-color ordering colors lattice points, block multi-color ordering colors sets of lattice points called blocks. While computations in blocks are executed sequentially, blocks are processed in parallel. When the block size is 1, block multi-color ordering is simply multi-color ordering. Conversely, the larger the grain size, the more sequential a program with block multi-color ordering becomes.

Block sizes for block multi-color ordering are customizable. The following are examples of blocks with sizes 1 and 2:

$$\text{——} R \text{——} G \text{——} B \text{——} R \text{——} G \text{——} B \text{——} R \text{——}$$

$$\text{——} R \text{——} R \text{——} G \text{——} G \text{——} B \text{——} B \text{——} R \text{——}$$

where lattice points R, G, and B are colored in red, green, and blue, respectively. Thus, block multi-color ordering decreases the number of stride accesses in programs and enables us to benefit from parallel computational environments. Extensions of the applicable scope of block multi-color ordering are therefore studied [4].

4   Toward Auto-tuning

Block multi-color ordering is one of the parallelizations that can be applied to programs. As described in Sect. 3, we can customize block sizes and the number of colors. However, block multi-color ordering requires us to modify the program source code according to the block size and number of colors. Moreover, the performance of programs to which multi-color ordering has been applied is known to be sensitive to the number of colors [5].

In this work we relieve programmers of modifying their source code in order to apply block multi-color ordering to their programs. Concretely, we develop a source code generator for programs written in Fortran 90. The code generator analyzes the Fortran 90 source code both lexically and syntactically, and returns a program with block multi-color ordering.

Furthermore, we provide programmers with an environment for automatically tuning their programs with block multi-color ordering. Since the code generator enables us to apply block multi-color ordering to programs automatically, it is sufficient to set the block size and number of colors for automatic tuning of our programs. Then, we use *job-level parallel execution*. Job-level parallel execution is the coarsest form of parallel execution, and does not require any modifications to the programs for execution in parallel. Job-level parallel execution is suitable for tuning programs with parameter sweeps. We make it possible to apply block multi-color ordering to programs automatically by using techniques for code generation and job-level parallel execution.

5   A Code Generator for Block Multi-Color Ordering

We have developed a code generator based on block MULti Color Ordering (Mulco). Mulco takes Fortran 90 source code and returns source code that includes block multi-color ordering. Mulco supports programs with computations in n-dimensional space ($n = 1, 2, 3$). Mulco also takes as parameters, the number of colors and the x-axis block size. However, this version of Mulco limits the y- and z-axes block sizes to 1. This is because it is sufficient to allow only the x-axis block size to be customizable for the purpose of removing stride accesses, which is the main disadvantage of multi-color ordering.

We reuse the sample code for the three-dimensional space given in Section 2 with the only difference being that the outermost loop has an annotation !$bmulco(3) as shown in Fig. 3. Mulco applies block multi-color ordering to a program when it finds !$bmulco($n$) in the source code of the program, where n denotes the number of dimensions. Since n is 3 in the sample code, Mulco applies block multi-color ordering in a three-dimensional space to the program. Mulco translates the sample code into that given in Fig. 4, in which we have manually added extra carriage returns owing to

```
!$bmulco(3)
do k = 2, z_num+1
  do j = 2, y_num+1
    do i = 2, x_num+1
      u(i, j, k) = (f(i, j, k)-&
              axp(i, j, k)*u(i+1, j, k)-&
              axm(i, j, k)*u(i-1, j, k)-&
              ayp(i, j, k)*u(i, j+1, k)-&
              aym(i, j, k)*u(i, j-1, k)-&
              azp(i, j, k)*u(i, j, k+1)-&
              azm(i, j, k)*u(i, j, k-1)&
              )/dgn(i, j, k)
    end do
  end do
end do
```

Fig. 3. Sample Fortran 90 code with an annotation

the limitations on page size for the paper. Actually, Mulco adds the minimal carriage returns compatible with the Fortran 90 grammar.

Variables mulco_color_num and mulco_block_size_n ($n = 1, 2, 3$) denote the number of colors and the sizes of blocks (1: x-axis, 2: y-axis, and 3: z-axis), respectively. Although mulco_block_size_2 and mulco_block_size_3 give the sizes of the y- and z-axes blocks, the current version of Mulco does not support them, i.e., they are fixed at 1. The number of colors and the size of the x-axis blocks are given to Mulco as arguments. Variables mulco_block_num_2 and mulco_block_num_3 denote the number of y- and z-axes blocks, respectively. Since we fix the sizes of the y- and z-axes blocks to 1, the numbers of y- and z-axes blocks correspond with the numbers of y- and z-axes lattice points, respectively. Variable mulco_block_num_1 denotes the minimum number of x-axis blocks containing all colors, i.e., the quotient of the number of x-axis blocks and the number of colors.

Variable mulco_color is the loop variable that represents the color using values 1 or 2, since mulco_color_num is 2. As mentioned in Sect. 2, when the number of colors is 2, the ordering is called a block red-black ordering. In the language of block red-black ordering, all blocks in red (or black) are processed before any blocks in black (or red, resp.).

Directives !$omp parallel and !$omp do are OpenMP directives [6]. The directive !$omp do dictates that iterations in the loop should be executed in parallel. Mulco generates source code in which the outermost loop (z-axis) in a space is parallelized. It should be possible to parallelize a loop of colors when the number of colors is greater than the number of threads the computer can handle. However, we have not achieved anything significant in our experimental setting discussed in Sect. 7. Therefore, we refrain from referring to this in the paper.

```
!$bmulcoed(3)
mulco_color_num = 2
mulco_block_size_3 = 1
mulco_block_size_2 = 1
mulco_block_size_1 = 16
mulco_block_num_3 = ((z_num+1-2)+1)/mulco_block_size_3
mulco_block_num_2 = ((y_num+1-2)+1)/mulco_block_size_2
mulco_block_num_1 = &
((x_num+1-2)+1)/(mulco_color_num*mulco_block_size_1)
do mulco_color = 0, mulco_color_num-1
!$omp parallel
!$omp do
do mulco_block_3 = 0, mulco_block_num_3-1
do k = 2+mulco_block_size_3*mulco_block_3, &
       2+mulco_block_size_3*(mulco_block_3+1)-1
   do mulco_block_2 = 0, mulco_block_num_2-1
      mulco_color_remainder = &
      mod((mulco_color+mulco_block_3+mulco_block_2),&
          mulco_color_num)
   do j = 2+mulco_block_size_2*mulco_block_2, &
          2+mulco_block_size_2*(mulco_block_2+1)-1
      do mulco_block_1 = 0, mulco_block_num_1-1
      do i = 2+mulco_block_size_1*(mulco_color_remainder+&
             mulco_block_1*mulco_color_num), &
             2+mulco_block_size_1*(mulco_color_remainder+&
             mulco_block_1*mulco_color_num+1)-1
         u(i, j, k) = ( f(i, j, k)-&
             axp(i, j, k) * u(i+1, j, k)-&
             axm(i, j, k) * u(i-1, j, k)-&
             ayp(i, j, k) * u(i, j+1, k)-&
             aym(i, j, k) * u(i, j-1, k)-&
             azp(i, j, k) * u(i, j, k+1)-&
             azm(i, j, k) * u(i, j, k-1) &
             )/dgn(i, j, k)
      end do
   end do
   end do
   end do
end do
end do
!$omp end do
!$omp end parallel
end do
```

Fig. 4. Translation of the Fortran 90 code based on block multi-color ordering

Variables i, j, and k are loop variables ranging from the initial to terminal points of the blocks. Note that the initial point of *i* is offset by mulco_color_remainder, the sum of mulco_color, mulco_block_3, and mulco_block_2 modulo mulco_color_num. This is the point of block multi-color ordering that prevents any adjacent pair of lattice points from being colored in the same color.

As an aside, Mulco can also generate code with a multi-color ordering as given in Sect. 2 when finding the annotation !$mulco(1).

6 A Scripting Language for Job-Level Parallel Execution

In this section we introduce the scripting language for job-level parallel execution, Xcrypt, developed by Hiraishi et al. [2]. In high performance computing, a program is usually executed as a job through a batch job system to prevent other jobs from interfering with that job. It is necessary to create a text file called a job script in order to submit a job to a batch job system. In addition, the format of the job script depends on the particular batch job system, making it difficult or tedious to submit, and moreover, to control a large number of jobs.

Xcrypt is a domain specific language for controlling jobs. In Xcrypt there are different layers for system administrators, module developers, and end users allowing administrators to configure Xcrypt systems, developers to provide useful modules for Xcrypt users, and end users to use Xcrypt. With this mechanism, end users can control their jobs without considering differences in systems.

Xcrypt is implemented almost as a superset[2] of the Perl programming language, with most of the new functionality implemented as functions or modules in Perl. This decreases the cost of learning a new scripting language that differs from existing scripting languages.

The example Xcrypt script shown in Fig. 5 is used to highlight the syntax of Xcrypt. The statement use base qw (limit core) is a Perl statement that declares super classes similar to the notion in object oriented programming languages. Module core is a required module in Xcrypt, while module limit limits the number of jobs submitted at any one time. Xcrypt has special methods before and after that are executed before and after executing a job, respectively. Module limit increments and decrements a semaphore in the method before. These methods are used as hooks for a job, allowing the number of submitted jobs to be controlled.

Module data_extractor provides a way to extract data from text files. Various methods of data_extractor are used to obtain the elapsed time of execution later in this script.

Function limit::initialize(1) sets the value of the semaphore to control the number of submitted jobs.

---

[2] Xcrypt has certain additional reserved keywords apart from those in Perl and includes name spaces. However, Xcrypt supports the complete syntax of Perl, i.e., an interpreter for Xcrypt can execute any script written in Perl.

```
use base qw (limit core);
use data_extractor;

limit::initialize(1);

foreach $j (1..9) {
    $c = 2**$j;
    foreach $i (0..(9-$j)) {
        $x = 2**$i;
        spawn {
            system("./mulco bmulco3.f90 $c $x;" .
                   "ifort -openmp bmulco3_mulco.f90;" .
                   "time ./a.out");
        }_after_in_job_{
            my $self = shift;
            $fh0 = data_extractor->new($self->JS_stderr);
            $fh0->extract_line_rn('real');
            $fh0->extract_column_nn('end');
            my @output = $fh0->execute();
            open ($fh1, '>>', 'result.dat') or die $!;
            print $fh1 "$c $x 1 1 $output[0]\n";
        };
    }
}
sync;
```

Fig. 5. Example script in Xcrypt

Variables c and x range over the colors and the x-axis block sizes, respectively. Anything executed on computation nodes is written in a block statement spawn[3]. Since the statement system executes the string given as an argument as a command, the command is executed not at hand but on a computation node. Although a job script is required for communicating with a batch job system as previously mentioned, Xcrypt automatically creates an appropriate job script from the descriptions in the spawn block. Program mulco refers to Mulco as explained in the previous section. Mulco takes the source code written in Fortran 90, the number of colors, and the x-, y-, and z-axes block sizes as arguments. We use the Intel Composer XE compiler version 12.1 with option -openmp to use OpenMP. Mulco generates a Fortran 90 file with the name (*the file name of the source code*)_mulco.f90. Xcrypt also includes a block called _after_in_job_. Everything about extracting data in this block is asynchronously done after the main processing of the job. It is implemented by Xcrypt's special method after as described before. It is useful to describe this in the terminology of object oriented languages.

---

[3] Strictly speaking, spawn and the following _after_in_job_ are not block statements, but functions in Perl. Perl and various other modern scripting languages have such mechanisms as syntax extensions.

Functions new, extract_line_rn, and extract_column_nn are methods in the module data_extractor as described. Function sync waits for all threads generated by spawn to complete.

7 Experimental Results

In this section we present experimental results for the elapsed time in executing numerical programs on one computation node[4]. The computer used in the experiments has the following specifications:

CPU:	Intel Xeon X5650 2.67 GHz
Cores:	12 (6 cores × 2)
Memory:	12 GB
OS:	CentOS 6.2
Compiler:	Intel Composer XE 12.1.2

In addition, we used -openmp as a compiler option as described in the previous section, since the programs are parallelized using OpenMP. The program given in Fig. 3 was used in these experiments. It makes use of nine arrays of real(8). Since the memory size is 12 GB, the number of elements in a single array cannot be greater than $(12 \cdot 10^9)/(9 \cdot 2^3)$. Thus, we fix the size of a space at $512 * 512 * 512$, since $(12 \cdot 10^9)/(9 \cdot 2^3 \cdot (2^9)^3) \approx 1.24$ holds. Through the experiments we set out to investigate the relationship between the number of colors and the size of blocks. We set the number of colors to be greater than 1. Since the one-dimensional size of an array is 512, the maximum size of a block side is 256.

First, we give the results for varying numbers of colors and block sizes in Table 1. Some values in Table 1 are given as n/a owing to the limitation on the number of colors and block sizes as explained above.

Next, we investigated the effect of varying the number of colors. Figure 6 illustrates the results with the block size fixed at 1, i.e., non-block multi color ordering. As described in Sect. 4 we can see that the performance of the program is sensitive to the number of colors in block multi-color ordering, although not to the extent of being called fragile. In general, there is an optimal threshold for the number of colors for parallel execution. If the number of colors is less than the threshold, we cannot benefit from parallel environments. If the number of colors is greater than the threshold, the overhead of parallel execution contributes the greater part of the elapsed times. However, we did not obtain any results such as these. In our experiments, two colors was almost always the best. Even with a lesser number of colors, parallel execution of the inner loop may contribute to improved performance. Sensitivity to the number of colors has been mentioned with respect to multi-color ordering. In fact, this can be seen in

---

[4] In parallel and distributed computational environments the outermost loop in a program should be parallelized not at thread-level, but at process-level, e.g., by using certain MPI libraries and not OpenMP. This conforms to our proposal in Sect. 6 for using Xcrypt to execute programs in parallel and distributed computational environments. We assume that this is done manually. Mulco should be developed as generating a program in which the outermost loop is parallelized not by OpenMP, but by MPI.

Table 1. Elapsed times (colors versus block size)

Time (sec.)	Colors								
Block size	2	4	8	16	32	64	128	256	512
2^0	3.516	4.304	6.505	6.716	6.716	6.920	6.105	6.107	7.119
2^1	3.504	4.518	5.316	5.116	5.116	4.715	4.718	4.919	n/a
2^2	3.503	4.118	4.517	4.517	4.504	4.133	4.117	n/a	n/a
2^3	3.326	3.703	3.923	3.917	3.717	3.726	n/a	n/a	n/a
2^4	3.317	3.704	3.514	3.512	3.306	n/a	n/a	n/a	n/a
2^5	3.103	3.318	3.516	3.305	n/a	n/a	n/a	n/a	n/a
2^6	3.117	3.115	3.304	n/a	n/a	n/a	n/a	n/a	n/a
2^7	3.103	3.118	n/a	n/a	n/a	n/a	n/a	n/a	n/a
2^8	2.903	n/a	n/a	n/a	n/a	n/a	n/a	n/a	n/a
2^9	3.092	3.092	3.092	2.892	3.090	2.891	3.091	2.891	2.893
2^{10}	3.091	2.891	2.889	2.893	2.890	2.892	2.891	2.891	n/a
2^{11}	2.905	2.891	2.886	2.890	2.889	2.891	2.892	n/a	n/a
2^{12}	2.896	2.891	2.893	2.892	2.891	2.894	n/a	n/a	n/a
2^{13}	2.891	2.891	2.891	2.889	2.891	n/a	n/a	n/a	n/a
2^{14}	2.892	2.892	2.890	2.891	n/a	n/a	n/a	n/a	n/a
2^{15}	2.890	2.886	2.888	n/a	n/a	n/a	n/a	n/a	n/a
2^{16}	2.891	2.893	n/a	n/a	n/a	n/a	n/a	n/a	n/a
2^{17}	2.891	n/a	n/a	n/a	n/a	n/a	n/a	n/a	n/a
2^{18}	2.891	2.890	2.891	2.891	2.910	2.893	3.291	3.692	4.490
2^{19}	2.891	2.890	2.890	2.893	3.093	3.092	3.701	4.291	n/a
2^{20}	2.890	2.892	2.894	3.090	3.093	3.691	4.292	n/a	n/a
2^{21}	2.891	2.892	3.091	3.091	3.692	4.292	n/a	n/a	n/a
2^{22}	2.891	2.888	3.291	3.691	4.274	n/a	n/a	n/a	n/a
2^{23}	2.888	3.290	3.692	4.292	n/a	n/a	n/a	n/a	n/a
2^{24}	3.093	3.489	4.293	n/a	n/a	n/a	n/a	n/a	n/a
2^{25}	3.491	4.292	n/a	n/a	n/a	n/a	n/a	n/a	n/a
2^{26}	4.292	n/a	n/a	n/a	n/a	n/a	n/a	n/a	n/a

the histogram in Table 1 with only a single color. However, it seems to be unnecessary to adjust the number of colors when the block size is large. We have not been able to obtain any explicit results thus far.

Finally, we fixed the number of colors at 2, and obtained results for varying block sizes as shown in Fig. 7. The best elapsed time is less than those for block size 1. That is, the bar graph in Fig. 7 confirms that block multi-color ordering clearly contributes to better performance of this program than non-block multi-color ordering. The best time is found when the block size is 2^{23}. Since the number of lattice points of an object is 2^{27}, the number of iterations is $2^{27}/2^{23} = 16$. Since the number 16 is the upper bound of more than the number of computational cores 12, it can be considered to be the best to coarse-grainedness, an essence of block multi color ordering.

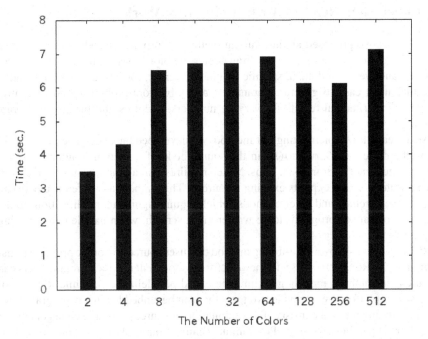

Fig. 6. Elapsed times (colors)

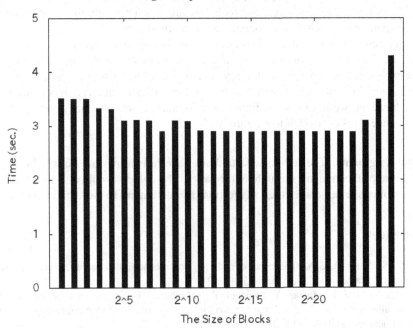

Fig. 7. Elapsed times (block sizes)

8   Conclusion, Related Work, and Future Work

In this paper we proposed an auto-tuning method incorporating code generation and job-level parallel execution. The resulting code generator is based on block multi-color ordering and we use a domain specific language, Xcrypt, for job-level parallel execution, making it easy to generate parameters and jobs from programs, and to control the jobs. Experimental results for varying numbers of colors and blocks sizes were presented.

As a means for implementing our method, we developed the code generator Mulco. Using the directive !$bmulco($n$) in the source code of a Fortran program instructs Mulco to generate Fortran code. Thus, Mulco requires a small modification to the original Fortran code and exploits existing resources. This is the main difference between the proposed method and other methods, which require significant modifications to the original program, or programs to be written from scratch, when parallelizing existing programs.

ROSE is a well-known auto-tuning method that uses source-to-source program transformation [7]. ROSE provides a framework for development in which programmers can execute their methods including optimizations and parallelizations. While ROSE provides such a platform, we provide two tools that can be embedded in any program flow. Our auto-tuning method consists of a combination of source-to-source code generation and job-level parallel execution. Programmers can use our tools as components in their program flow when and wherever they choose. Hitherto we have mostly used the so-called Unix tools as typified by GNU tools. We are developing our tools with specific focus on parallel and distributed computational environments.

The current version of Mulco does not support changes in the sizes of y- and z-axis blocks, that is, Mulco in this version cannot divide a research object into blocks of the shape except cuboid[5]. A support for any y- and z-axis blocks remains a future work. From the results of the experiments in this work we found that larger block sizes are better; however, we have not yet found the maximum block size that gives optimum performance. This is also left to a future work.

Acknowledgments. The authors wish to thank Takeshi Iwashita and Masatoshi Kawai for their helpful comments throughout this work. The authors would also like to thank the anonymous reviewers for their suggestions and comments to the submitted draft.

References

1. Duff, I.S., Meurant, G.A.: The effect of ordering on preconditioned conjugate gradient. BIT Numerical Mathematics 29(4), 635–657 (1989)
2. Hiraishi, T., Abe, T., Miyake, Y., Iwashita, T., Nakashima, H.: Xcrypt: Flexible and intuitive job-parallel script language. In: The 8th Symposium on Advanced Computing Systems and Infrastructures, pp. 183–191 (2010) (in Japanese)
3. Iwashita, T., Shimasaki, M.: Block red-black ordering: A new ordering strategy for parallelization of ICCG method. International Journal of Parallel Programming 31(1), 55–75 (2003)

---

[5] This is a suggestion by Takeshi Iwashita at the workshop.

4. Kawai, M., Iwashita, T., Nakashima, H.: Parallel multigrid Poisson solver based on block red-black ordering. In: High Performance Computing Symposium, Information Processing Society of Japan, pp. 107–116 (2012) (in Japanese)
5. Doi, S., Washio, T.: Ordering strategies and related techniques to overcome the trade-off between parallelism and convergence in incomplete factorizations. Parallel Computing 25(13-14), 1995–2014 (1999)
6. Chandra, R., Menon, R., Dagum, L., Kohr, D., Maydan, D., McDonald, J., Holzmann, G.J.: Parallel Programming in OpenMP. Morgan Kaufmann (2000)
7. Quinlan, D.J.: Rose: Compiler support for object-oriented frameworks. Parallel Processing Letters 10(2), 215–226 (2000)

Automatic Parameter Optimization for Edit Distance Algorithm on GPU

Ayumu Tomiyama and Reiji Suda

The University of Tokyo

Abstract. In this research, we parallelized the dynamic programming algorithm of calculating edit distance for GPU, and evaluated the performance. In GPU computing, access to the device memory is likely to be one of the primal bottleneck due to its high latency, and this effect gets noticeable especially when sufficient number of active threads cannot be secured because of the lack of parallelism or overuse of GPU resources. Then, we constructed a model that approximates the relations between the values of parameters and the execution time considering latency hiding, and by using this model, we devised a method of automatic tuning of parallelization parameters in order to attain high performance stably even when the problem size is relatively small.

1 Introduction

The problem of analyzing the similarities of a given string with patterns and finding their optimum alignment is called approximate string matching. It is one of the important problem in information science applied not only to text retrieval but also to a variety of fields including computational biology. The computation of approximate string matching, however, includes some computationally expensive steps. In this research, we worked on acceleration of the algorithm for calculating Levenshtein edit distance by using graphics processing units (GPUs) and made an evaluation of its performance.

GPUs, as the name suggests, were originally created as a hardware for image processing. They have quite a lot of computing units, which produce high peak performance at a relatively-low cost. Then, the idea to utilize the huge computing power of GPU for calculation other than image processing was born under the name of GPGPU (General Purpose computing on GPUs,) and today this technique has a wide variety of applications. Although GPGPU programming became easily accessible thanks to the improvement of GPU architectures and appearance of some useful developing environments including CUDA (Compute Unified Device Architecture) [7] provided by NVIDIA, it is generally still difficult to exploit the full performance of GPUs. In order to utilize all of the computing units in GPUs, it is important to select highly parallelized algorithm and to create a large number of threads. The cost of memory access is also a major factor determining overall performance. To reduce this cost, it is important in terms of latency hiding to increase the number of simultaneously executed threads by

M. Daydé, O. Marques, and K. Nakajima (Eds.): VECPAR 2012, LNCS 7851, pp. 420–434, 2013.

setting a limitation on the size of used resource per thread and increasing the total number of threads itself. Moreover, efficient utilization of low-capacity but fast shared memory is also a common practice. These optimization can not be achieved only by the selection of parallel algorithms, but it is also indispensable to adjust values of parameters related to the parallelization depending on the capacity of used GPUs.

As for approximate string matching on GPU, there are several researches on implementation of parallel versions of Smith-Waterman algorithm for GPUs. The Smith-Waterman algorithm closely resembles the edit distance algorithm in the dependence relationship between computed elements, so almost the same parallelization and optimization scheme can be applied.

The origin of the implementation of the Smith-Waterman algorithm for GPUs dates back to the age before the CUDA became common. Liu et al. [3] implemented the Smith-Waterman algorithm on GPU by mapping the algorithm on rendering pipeline with using OpenGL API. They parallelized the algorithm by making use of the fact that the cells on the same anti diagonal in the computation region can be simultaneously calculated. After the appearance of CUDA, several researches on the parallel Smith-Waterman algorithm on GPU are conducted. Manavski et al. [5] and Munekawa et al. [6] implemented the algorithm by using CUDA. These implementations are similar in that alignments of multiple combination of strings are calculated in parallel, but different in that each thread computes the whole alignment (inter-task parallelization) in the approach of Manavski et al, while it is covered by whole threads in a thread block (intra-task parallelization) in the approach of Munekawa et al. Liu et al. [4] not only improved the performance but also eased the restriction caused by overuse of memory on GPU by covering both of the inter-task and inner-task parallelization. Ling et al. [2] also resolved the restriction on the length of the strings arisen from the limitation of available resources of GPUs by adopting a divide and conquer approach. Dohi et al. [1] improved the performance by using some technique including the divide and conquer approach and an idea of reducing the cost of thread synchronization.

Also after those, several implementations appear and show pretty good performance as compared with the ones for CPU accelerated by some heuristic methods. Most of them put emphasis on the performance when there is a lot of queries, which enables simultaneous calculation of matchings of multiple combinations of strings and patterns, and ensures sufficient number of threads. On the other hand, they seem to be giving little consideration on the case when the number of active threads is limited because of the lack of queries relative to the capacity of the used GPUs. When sufficient number of active threads is not secured, the performance drops beause of failure in latency hiding or load balancing.

In this article, we propose a parallel algorithm of calculating edit distance of a pair of strings on GPU. We analyzed the relationship between computation time and the values of parameters associated with parallelization: the block size parameters, and the number of active threads. As a result, we confirmed that the latency of the device memory and the hiding of it have the strongest effect on the performance other than the simple number of arithmetic and memory reference instructions. Then, considering them, we constructed an estimation model of execution time, and applied it to our system to automatically select the optimum block size.

The rest of this article is constituted as follows. First we introduce the parallel edit distance algorithm with brief explanation of GPU architecture in Chapter 2. Then, in Chapter 3, we present our estimation model of computation time. Then we show the experimental results in Chapter 4, and finally we conclude this article in Chapter 5.

2   Edit Distance Algorithm and Parallelization

First, we introduce the definition of edit distance and the serial version of the dynamic programming algorithm for calculating it. Then we propose our parallel algorithm after showing the features of CUDA GPUs. Note that in this research we assume the use of GPUs supporting CUDA whose compute capability is 1.2 or 1.3.

2.1   Dynamic Programming Algorithm

Edit distance of two strings is defined as the minimum number of editing operations required for transforming one string into the other. There are three types of operations available: insertion, deletion, and substitution of a character. For exapmle, edit distance of strings "change" and "hunger" is three, for the former string can be transformed into the latter one by the following three operations: deletion of 'c', substitution of 'u' for 'a', and insertion of 'r'.

In this paper, we let $d(str1, str2)$ denote edit distance of the two strings $str1, str2$. Also, $|str|$ denotes the length of string str, $str[i]$ the i-th character of str $(0 \le i < |str|)$, $str[i..j]$ the substring from the i-th character to the j-th character of str $(0 \le i \le j < |str|)$.

Edit distance $d(str1, str2)$ can be calculated by using dynamic programming. Specifically, it can be obtained from edit distances $d(str1[0..i], str2[0..j])$ of prefixes of the strings. Considering how to match the last characters of $str1[0..i]$ and $str2[0..j]$ by using one or none of the three editing operations, the edit distance $d(str1[0..i], str2[0..j])$ can be calculated according to the following formula:

$$d(str1[0..i], str2[0..j]) = min(d(str1[0..i-1], str2[0..j]) + 1,$$
$$d(str1[0..i], str2[0..j-1]) + 1,$$
$$d(str1[0..i-1], str2[0..j-1]) + c(str1[i], str2[j]))$$
$$(1)$$

where $c(str1[i], str2[j])$ is 0 if $str1[i]$ equals to $str2[j]$, and otherwise this value is 1.

Therefore, the edit distance $d(str1, str2)$ can be calculated by completing a table of edit distances $d(str1[0..i], str2[0..j])$ of prefixes of $str1$ and $str2$ like

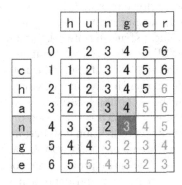

Fig. 1. Table of edit differences of suffixes

one shown in figure 1. First, the values of leftmost cells and topmost ones in the table are trivially obtained. On the other hand, the value of each inner cell can be obtained from those of the immediate left, upper left, and upper cells of its own by using the formula (1), so they can be sequentially calculated from the upper left corner to the lower right corner. The amount of calculation is $O(|str1||str2|)$.

The calculation has a certain level of parallelism, and there is also flexibility in what order to calculate values of cells.

2.2   GPU Architecture

GPU mainly consists of several multiprocessors (MPs) and a device memory. A MP is a group of eight simple processors called scalar processors (SPs). SPs in the same MP simultaneously execute the same instruction on different data like SIMD instructions. GPU provides great performance by making its numerous SPs carry the same operation in parallel.

The device memory is accessible from CPU and all MPs in the GPU. Generally, GPU program first copies data from host memory on CPU to the device memory on GPU. Then, it runs parallel codes, called CUDA kernels, which make each MP read data from the device memory, perform calculation, and write the results back to the memory. Finally it transfers the results to the host memory. Besides the device memory, each MP has its own low-latency memory called shared memory. Via shared memory, each SP can access computational results of other SPs in the same MP very fast.

In order to make GPU execute tasks, we have to give it a group of threads called grid. A grid consists of numerous groups of threads called thread blocks. The relationship of grid, thread block, and thread corresponds to that of GPU, MP, and SP. Each thread block is assigned to a MP, and each thread in the thread block is assigned to a SP in the MP. Each SP (or MP) concurrently executes several threads (or thread blocks) by frequently switching the thread (or thread block) to be executed.

In order to exploit the performance of GPU, there is a lot of things to be considered. In this research, we put emphasis on mainly two points.

One point is the hierarchical structure of tasks and the hardware. What SPs can do is restricted in that SPs in the same MP can simultaneously execute only the same instruction, so the sequence of executed instructions should be made as nearly equal as possible among threads in the same thread block by avoiding conditional branching. On the other hand, threads in the same thread block have advantage in that they can communicate with each other very fast by utilizing the shared memory and fast synchronization instruction, whereas communication among threads in different thread blocks is not supported. Therefore, it is an important matter how to split the entire processing into grids, into thread blocks, and into threads considering regularity and dependency of calculation.

The other point is the latency hiding. In GPU computing, access to the device memory is one of the primal bottleneck because of its high latency. Generally, it is hidden by executing instructions of other threads during the latency. Therefore, it is important to secure sufficient number of active threads in order to fill the latency. The number of active threads depends not only on the number of total threads but also on the amount of resources each thread block uses such as shared memory and registers.

2.3 Parallelization and Blocking

We parallelized the dynamic programming algorithm for GPU, and in order to reduce grid execution and access to the device memory, we brought in blocked algorithm.

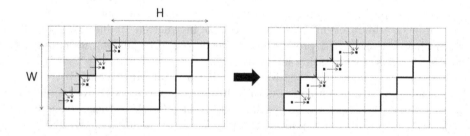

Fig. 2. Shape of a block and calculational procedure

Figure 2 shows the shape of a block by which the computational region of the dynamic programming algorithm is divided in this parallel algorithm, and how to calculate values of cells in the block in parallel. As explained in Section 2.1, the value of each cell can be directly calculated if those of the immediate left, upper left and upper cells are available. Therefore, if all the values of gray cells in Figure 2 are available, values of the leftmost cells in all the W rows of the parallelogram block can be simultaneously calculated. Once their values are calculated, then by using them, values of all their immediate right cells can be simultaneously calculated. In the same way, values of W diagonally aligned cells are calculated in parallel, and it takes H steps to complete the processing of all the cells in the block, in sequence from left cells to right ones.

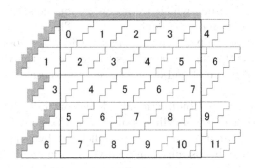

Fig. 3. Blocking

Figure 3 shows how to divide the computational region into the parallelogram blocks and what order to process the blocks. The blocks can be processed in ascending order of the number, and all blocks assigned the same number can be processed in parallel without inter-block communication. In parallelization of this algorithm for GPU, a grid is assigned to process all the parallelogram blocks tagged the same number, and in the grid, each thread block is assigned to calculate values of cells in a block region. In a block region, just W values of cells which are in different rows each other can be calculated in parallel at any step, so the processing of a block region can be evenly parallelized by assigning each thread calculation of values of all cells in a row in the block.

To be more precise, behavior of a thread block is expressed in Figure 4. First, threads load necessary data, values of gray part in the left block of Figure 4 and

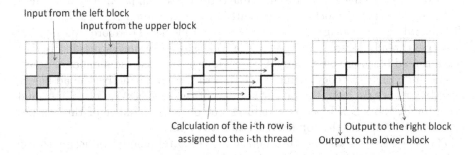

Fig. 4. Processing of a thread block

characters in the strings, from the device memory and store them on shared memory. The values of gray cells are just calculated by the previous grid execution. Then, each thread starts calculation of values of the assigned cells. W threads in the thread block act in synchronization, and process diagonally aligned cells in parallel, receiving necessary data, which was just calculated in the previous step by the neighbor threads, via shared memory. Finally, after all the cells are processed, they store the results which are necessary for the subsequent calculation in the next grid execution on the device memory. The necessary region includes only boundary part of the parallelogram block, which is colored gray in the left block of Figure 4, so the thread block has to store only the values of gray part in the right block of Figure 4 and do not have to do those of the whole inside of the block.

Furthermore, we adopted bidirectional calculation. Edit distance can also be calculated by putting together edit distance of prefixes of the two strings and that of suffixes. They can be calculated in parallel, and by using this property, the process of calculating edit distance can be divided into two half-size processings and post-processing whose amount of calculation is at most proportional to the length of the strings. By this division, the total amount of calculation slightly increases by the post-processing. Instead, the parallelism doubles, so the number of thread blocks per grid, and naturally that of threads, also doubles. In GPU computing, it is important to increase the number of threads in terms of latency hidind and load balancing, and as a whole it mostly results in an improvement in performance.

3   Optimization of Block Size

It is important to configure appropriate values of block size parameters, W and H in Figure 2, because they determine various quantities related to performance of the parallel algorithm for GPU.

First, block size parameters determine the number of thread blocks in each grid, which is important in terms of load balancing. At the same time, they indirectly influence the number of active threads per MP by determining the size of resources, such as shared memory and registers, needed by a thread block. In grid execution, each MP executes multiple thread blocks in parallel by frequently switching the thread block to be executed. Here, the resources of the MP are distributed to the concurrently executed thread blocks. Therefore, the number of active thread blocks is limited by the total size of resources divided by the size a thread block uses. Basically it is desirable to increase the number of active threads in terms of latency hiding. Note that the number of active thread blocks per MP can not be more than eight because of the specification of CUDA, and accordingly, too small block size reduces not only the number of threads per thread block but also that of active threads per MP.

The block size parameters also determine the total amount of access to the device memory and other additional calculation arisen from parallelization, including grid executions. Generally, selecting small block size increases such cost.

Besides, it is also important to appropriately configure the proportion of W to H in order to reduce the cost.

Therefore, the block size should be carefully chosen considering trade-offs among these factors. In this research, we constructed a model to estimate the grid execution time from the parameters, and made the system to choose the block size which minimize the total computation time estimated on the model.

Considering GPU architecture, there are not so many options of appropriate values of block size parameters, so we adopted full search: estimating the total computation time for all of the options, and actually calculating edit distance on the block size which minimize the estimated execution time.

In the following sections, we introduce a model for estimating the grid execution time, and then how to estimate values of parameters in the model.

3.1   Model of Grid Execution Time

In order to choose optimum block size parameters, we constructed a model to estimate the grid execution time from block size parameters W and H, and the maximum number B_{MP} of thread blocks per MP, which is obtained from the total number B_{total} of thread blocks and the number N_{MP} of MPs as follows

$$B_{MP} = \left\lceil \frac{B_{total}}{N_{MP}} \right\rceil .$$

First, we calculated the maximum number B_{act} of active blocks. It is limited by the following two factors. One is the used resource size. In our algorithm, the number of used registers is not so large as to limit the number of active threads, but the shared memory usage may do. A thread block uses $(W + H)$ of character size area for storing the compared substring, and $(W + H)$ of integer size area for sharing values of the cells among the threads. Then, the number of active blocks is not more than the quotient of the total size of shared memory per MP and the used amount described above. The other factor is the specification of CUDA. The maximum number of active threads per MP is 1024, so B_{act} must not be more than the quotient of 1024 and the number of thread per thread block. In addition, the number B_{act} of active blocks itself is also limited not to exceed 8. Consequently, B_{act} is obtained as the maximum number such that all the above conditions are satisfied.

Then, we introduce our model to estimate the grid execution time from W, H, and B_{act}. In this parallel algorithm, each thread block first loads necessary data from the device memory all at once. Then, after synchronizing all threads in the block, threads start calculating values of the cells assigned to themselves with synchronization and communication with the neighbor threads through the shared memory. Finally, they write the results back on the device memory. The amount of main calculation per thread block is approximately proportional to the number of cells in the block region, that is to say, the product of W and H. On the other hand, the size of the data read from and written on the device memory has linear relationship to W and H. Besides, there are trivial processing whose amount is at most linear to W or H. Therefore, as the most fundamental approximation, the total amount of calculation of the grid per MP can be expressed as $B_{MP}(a_0WH+a_1W+a_2H+a_3)$, where each a_i is a constant. This approximation, however, may not be accurate depending on the degree of latency hiding.

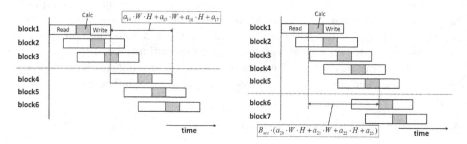

Fig. 5. Latency hiding

Figure 5 represents a simplified model of the flow of instruction execution in a thread block of the parallel edit distance algorithm. The left figure corresponds to the case when B_{act} is three and the latency of the device memory access is not fully hidden, while the right one corresponds to the case when B_{act} is five and the latency is fully hidden. In this algorithm, threads in the same thread blocks are synchronized before and after the access to the device memory. Therefore, during the memory access of one thread, only threads in different active thread blocks can contribute to latency hiding by executing the main task of calculating edit distance. The latency is completely hidden when the total latancy of one thread is shorter than the occupation time of computing units by threads in all active blocks except one.

Based on this model, the grid execution time, if the total number of blocks B_{MP} is less than B_{act}, can be simply approximated by the expression

$$T_{BS}(W, H, B_{MP}) = B_{MP} \cdot (a_{00}WH + a_{01}W + a_{02}H + a_{03}) + (a_{04}WH + a_{05}W + a_{06}H + a_{07}).$$

The first term of the expression represents the main computation time of all the threads, while the second term represents the latency remained unhidden.

On the other hand, it depends on whether the latency is fully hidden or not how much extra time is needed for grid execution when the number of thread blocks in the grid increases by B_{act}. When the latency is not fully hidden, as described in the left one of Figure 5, a new thread block should wait for the completion of an old one without executable instruction in the MP, so the extra time can be approximated by the turn around time of execution of one thread block, the expression of which is

$$T_{DL}(W, H, B_{act}) = (a_{14}WH + a_{15}W + a_{16}H + a_{17}).$$

When the latency is fully hidden, some sort of calculation is always executed on the MP even during the latency of memory access, so the extra time can be approximated by the total amount of calculation of B_{act} thread blocks expressed as

$$T_{DH}(W, H, B_{act}) = B_{act} \cdot (a_{20}WH + a_{21}W + a_{22}H + a_{23}).$$

By using these functions, the grid execution time in each condition can be approximated by the expression

$$T_{TL}(W, H, B_{MP}, B_{act}) = T_{BS}(W, H, B_{MP}\%B_{act}) + T_{DL}(W, H, B_{act}) \cdot \left\lfloor \frac{B_{MP}}{B_{act}} \right\rfloor$$

if the latency is not fully hidden, and otherwise, it is approximated by

$$T_{TH}(W, H, B_{MP}, B_{act}) = T_{BS}(W, H, B_{MP}\%B_{act}) + T_{DH}(W, H, B_{act}) \cdot \left\lfloor \frac{B_{MP}}{B_{act}} \right\rfloor.$$

Whether the latency is fully hidden or not can be determined by comparing the function values: fully hidden when the value of $T_{DL}(W, H, B_{act})$ is smaller than that of $T_{DH}(W, H, B_{act})$, and otherwise, not fully hidden. Therefore, the execution time is also approximately expressed as

$$max(T_{TH}(W, H, B_{MP}, B_{act}), T_{TL}(W, H, B_{MP}, B_{act})).$$

In this way, the problem of estimating grid execution time comes down to that of estimating parameters a_{ij} in the functions T_{BS}, T_{DL}, and T_{DH}. Note that, in practice, we extended the forms of the functions T_{DL} and T_{DH} into

$$T_{DL}(W, H, B_{MP}) = B_{MP} \cdot (a_{10}WH + a_{11}W + a_{12}H + a_{13}) + \\ (a_{14}WH + a_{15}W + a_{16}H + a_{17})$$
$$T_{DH}(W, H, B_{MP}) = B_{MP} \cdot (a_{20}WH + a_{21}W + a_{22}H + a_{23}) + \\ (a_{24}WH + a_{25}W + a_{26}H + a_{27}),$$

the same in form as T_{BS}, for improving the quality and for convenience sake.

3.2 Parameter Estimation

We introduce the way of estimating values of parameters a_{ij} in the functions T_{BS}, T_{DL}, and T_{DH} from the sample data of block size parameters, the number of thread blocks, and the grid execution time.

As explained in the previous section, the grid execution time is approximated by the greater value of the two: linear combination of T_{BS} and T_{DL}, and that of T_{BS} and T_{DH}. Moreover, the three functions have linear relationships to the parameters a_{ij}. Therefore, values of the parameters a_{ij} can be approximated by using linear least-squares method if it is possible to determine in which case each performance data belongs to: the case where the latency is fully hidden or the other.

Then, we adopted iterative refinement method alternately repeating classification and parameter estimation. From the viewpoint of the model introduced in the previous section, whether latency is fully hidden or not basically depends on the ratio between the number of times loading from and storing on the device memory per thread and the total amount of calculation of all active thread blocks except one. By comparing this ratio to appropriately predetermined threshold, we can roughly judge in which class each performance data belongs, and by regarding this result as initial classification, values of the parameters can be estimated by the linear least-squares method. Once the parameter values are approximately obtained, the classification can be updated by another criterion: which of the two estimating function T_{TH} and T_{TL} each sample of performance data is near to. Repeating this cycle of parameter estimation and classification a few times, better approximation of values of parameters can be obtained.

4   Experiments and Results

In this chapter, we show some results of estimation of grid execution time, and efficiency of block size optimization based on this estimation.

Here, we used two GPUs: an old one called Quadro FX 4800, and a relatively new one called Tesla C2075, which is based on Fermi architecture. Quadro FX 4800 has 24 MPs, each of which has 16384 bytes of shared memory and can execute at most 1024 threads in parallel, while Tesla C2075 has 14 MPs and 49152 bytes of shared memory at each, and the maximum number of active threads per MP is 1536.

We measured the grid execution time varying the block size W from 32 to 512 on multiples of 32, H from 16 to $(1024 - W)$ on multiples of 16, and the total number of blocks B_{total} from 1 to 504. Then, we estimated parameter values of the model of grid execution time by the method introduced in Section 3, and compared estimated time by the model with the measured value.

The actual grid execution time and estimated one on some H from 160 to 384 at step 32 when W was 32 are shown in Figures 6 - 9. In these graphs, the horizontal axis corresponds to the number of thread blocks B_{total} while the vertical axis corresponds to the execution time on the millisecond time scale.

When W is 32, the number of active threads per MP is limited by the restriction of that of active blocks per MP, so in this example the latency of the device memory access was not fully hidden. Therefore, the execution time jumped at points where B_{total} exceeded multiples of $B_{act} \cdot N_{MP}$, which corresponds to the left case in Figure 5. Broadly speaking, the graph of estimated time reproduced that of actual time well.

However, they have some errors in detail. Figure 10 and Figure 11 show the ratio of estimated grid execution time to actual one when W was 32, with B_{total} on the horizontal axis. In both graphs, periodical waves were observed, but they are qualitatively different. As Figure 6 shows, the slow increment of time, represented by T_{BS} in the previous chapter, is almost the same independent of the range of the number of thread blocks on Quadro FX 4800, so its error is

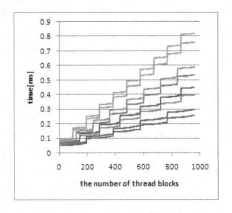

Fig. 6. Actual grid execution time
($W = 32$, Quadro FX 4800)

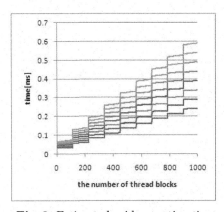

Fig. 7. Estimated grid execution time
($W = 32$, Quadro FX 4800)

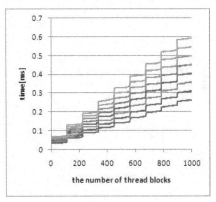

Fig. 8. Actual grid execution time
($W = 32$, Tesla C2075)

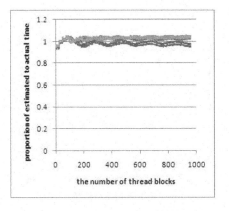

Fig. 9. Estimated grid execution time
($W = 32$, Tesla C2075)

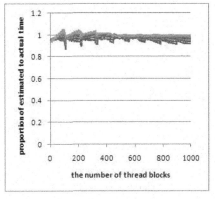

Fig. 10. Approximation ratio
($W = 32$, Quadro FX 4800)

Fig. 11. Approximation ratio
($W = 32$, Tesla C2075)

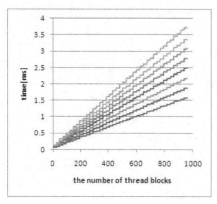

Fig. 12. Actual grid execution time ($W = 256$, Quadro FX 4800)

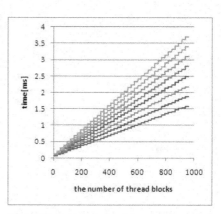

Fig. 13. Estimated grid execution time ($W = 256$, Quadro FX 4800)

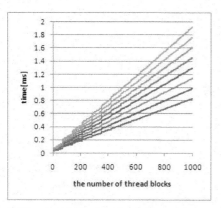

Fig. 14. Actual grid execution time ($W = 256$, Tesla C2075)

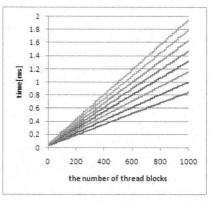

Fig. 15. Estimated grid execution time ($W = 256$, Tesla C2075)

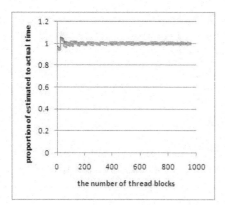

Fig. 16. Approximation ratio ($W = 256$, Quadro FX 4800)

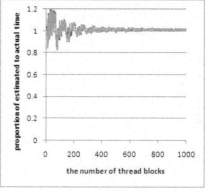

Fig. 17. Approximation ratio ($W = 256$, Tesla C2075)

reduced by refining the approximation of T_{BS}. In Figure 8, on the other hand, the slow increment when the number of thread blocks is from 1 to 104 and that when it is from 937 to 1040 is visibly different. It seems to be caused by adoption of device memory cache and development of scheduling system on Tesla C2075. Therefore, in order to improve the approximation accuracy on new GPUs, it is indispensable to revise our model itself which approximate the execution time by the sum of the slow increment T_{BS} and the rapid increment T_{DH} or T_{DL}.

Figures 12 - 15 show the actual and estimated time of grid execution respectively when W was 256. In this case, the number of active threads per MP was so large that the latency of the device memory access was fully hidden by instruction execution of other threads. Therefore, the execution time constantly increased every N_{MP} threads, which corresponds to the right case in Figure 5.

Figure 16 and Figure 17 show the ratio of estimated grid execution time to actual one when W was 256.

Just as the case when W was 32, there were some errors in T_{BS} when the number of blocks is small. In other points, however, the model accurately approximated the execution time, and the approximation error was within about 3% of the actual value except for the case with too small number of thread blocks.

The results of experiments suggests that most of the approximation error in our model was in T_{BS} and T_{DL}, that is to say, the approximation in case when the latency was not fully hidden. Our model puts emphasis simply on the number of instructions and the effect of latency hiding of the device memory access because they have the most powerful influence on the performance, but there are some other factors we did not considered. For example, it is officially informed that registers of GPU are also source of generating delays, and it is also hidden by securing sufficient number of active threads. Moreover, as mentioned above, the influence of device memory cache is not negligible on new machines. Our model cannot adapt to the two-stage bends of performance curve caused by latency hiding of both the device memory and registers, which may be a cause of the approximation error. This remains to be a future work.

5   Conclusion

In this research, we parallelized the dynamic programming algorithm for calculating edit distance on GPU. In GPU Computing, the cost of the device memory access is in many cases a primary factor to slow down the performance because of its high latency, so it is important to utilize the shared memory as a cache and to hide the latency by meantime-executed arithmetic instructions, in addition to merely reducing the number of access itself. Besides, load balancing is also important in order to exploit the full performance of the GPU's vast computing resources.

Considering these facts, we suggested a blocking algorithm, and constructed a model to estimate grid execution time for the purpose of optimizing block size. In our model, especially taking into account the latency of memory reference instructions, we expressed the relationship between the number of active threads, the block size, and the grid execution time. By selecting the block size to minimize the total computation time obtained from this model, we tried to find optimum trade-off between load balancing and the cost of memory access and other extra processing accompanied with the block splitting. Consequently, we succeeded in automatically selecting nearly optimum block size in terms of computation time.

Our model is specific to the dynamic programming algorithms, but there are several problems which have similar data dependency, the Smith-Waterman algorithm, SOR method, preprocessing of ICCG method, and so forth. Therefore, we think our model has some application range. In the edit distance algorithm, the number of active threads is determined almost exclusively by the length of strings, but other algorithms usually consume more GPU resources, and this may limit the number of active threads. In such cases, it is indispensable to consider the effect of latency hiding as done in our work. Application of the model for grid execution time to more complicated algorithms is one of the challenges for the future, as well as improvement in accuracy of the current model and its adaptation to new GPUs.

Acknowledgements. This work is partially supported by CREST project "ULP-HPC: Ultra Low-Power, High-Performance Computing via Modeling and Optimization of Next Generation HPC Technologies", JST.

References

1. Dohi, K., Benkrid, K., Ling, C., Hamada, T., Shibata, Y.: Highly efficient mapping of the Smith-Waterman algorithm on CUDA-compatible GPUs. In: ASAP, pp. 29–36 (2010)
2. Ling, C., Benkrid, K., Hamada, T.: A parameterisable and scalable Smith-Waterman algorithm implementation on CUDA-compatible GPUs. In: 2009 IEEE 7th Symposium on Application Specific Processors, pp. 94–100 (2009)
3. Liu, Y., Huang, W., Johnson, J., Vaidya, S.: GPU accelerated Smith-Waterman. In: Alexandrov, V.N., van Albada, G.D., Sloot, P.M.A., Dongarra, J. (eds.) ICCS 2006. LNCS, vol. 3994, pp. 188–195. Springer, Heidelberg (2006)
4. Liu, Y., Maskell, D.L., Schmidt, B.: CUDASW++: optimizing Smith-Waterman sequence database searches for CUDA-enabled graphics processing units. BMC Research Notes 2(1), 73 (2009)
5. Manavski, S.A., Valle, G.: CUDA compatible GPU cards as efficient hardware accelerators for Smith-Waterman sequence alignment. BMC Bioinformatics 9(suppl. 2), S10 (2008)
6. Munekawa, Y., Ino, F., Hagihara, K.: Design and implementation of the Smith-Waterman algorithm on the CUDA-compatible GPU. In: BIBE, pp. 1–6 (2008)
7. NVIDIA Corporation. NVIDIA CUDA C Programming Guide Version 4.0

Automatic Tuning of Parallel Multigrid Solvers Using OpenMP/MPI Hybrid Parallel Programming Models

Kengo Nakajima

Information Technology Center, The University of Tokyo, 2-11-16 Yayoi,
Bunko-ku, Tokyo 113-8658, Japan
nakajima@cc.u-tokyo.ac.jp

Abstract. The multigrid method with OpenMP/MPI hybrid parallel
programming model is expected to play an important role in large-scale
scientific computing on post-peta/exa-scale supercomputer systems. Because
the multigrid method includes various choices of parameters, selecting the
optimum combination of these is a critical issue. In the present work, we focus
on the selection of single-threading or multi-threading in the procedures of
parallel multigrid solvers using OpenMP/MPI parallel hybrid programming
models. We propose a simple empirical method for automatic tuning (AT) of
related parameters. The performance of the proposed method is evaluated on
the T2K Open Supercomputer (T2K/Tokyo), the Cray XE6, and the Fujitsu
FX10 using up to 8,192 cores. The proposed method for AT is effective, and the
automatically tuned code provides twice the performance of the original one.

Keywords: Multigrid, Hybrid Parallel Programming Model, Automatic Tuning.

1 Introduction

To achieve minimal parallelization overheads on multi-core clusters, a multi-level
hybrid parallel programming model is often employed. In this method, coarse-grained
parallelism is achieved through domain decomposition by message passing among
nodes, and fine-grained parallelism is obtained via loop-level parallelism inside each
node by using compiler-based thread parallelization techniques such as OpenMP.
Another often used programming model is the single-level *flat MPI* model, in which
separate single-threaded MPI processes are executed on each core.

In previous works [1,2], OpenMP/MPI hybrid parallel programming models were
implemented in 3D finite-volume simulation code for groundwater flow problems
through heterogeneous porous media using parallel iterative solvers with multigrid
preconditioning. The performance and the robustness of the developed code was
evaluated on the T2K Open Supercomputer at the University of Tokyo (T2K/Tokyo)
[3,4] using up to 8,192 cores for both weak and strong scaling computations.
Furthermore, a new strategy for solving equations at the coarsest level (coarse grid
solver) was proposed and evaluated in [2], and the new coarse grid solver improved

M. Daydé, O. Marques, and K. Nakajima (Eds.): VECPAR 2012, LNCS 7851, pp. 435–450, 2013.

the scalability of the multigrid solver dramatically. The OpenMP/MPI hybrid parallel programming model, in which one MPI process was applied to a single quad-core socket of the T2K/Tokyo with four OpenMP threads (HB 4×4) [1,2], demonstrated the best performance and robustness for large-scale ill-conditioned problems by appropriate optimization and coarse grid solvers. In [5], performance of parallel programming models for algebraic multigrid solvers in Hypre Library [6] have been evaluated on various multicore HPC platforms with more than 10^5 cores, such as IBM BlueGene/P, and Cray XT5. The *MultiCore SUPport library (MCSup)* [5] provides a framework, in which the optimization processes described in [1,2] are applied automatically. Results show that threads of an MPI process should always be kept on the same socket for optimum performance to achieve both memory locality and to minimize OS overhead for cc-NUMA architecture. This corresponds to HB 4×4 programming model in [1,2].

The concepts of OpenMP/MPI hybrid parallel programming models can be easily extended and applied to supercomputers based on heterogeneous computing nodes with accelerators/co-processors, such as GPUs and/or many-core processors by Intel Many Integrated Core Architecture. Multigrid is a scalable method for solving linear equations and for preconditioning Krylov iterative linear solvers, and it is especially suitable for large-scale problems. The multigrid method is expected to be one of the powerful tools on post-peta/exa-scale systems. It is well known that the multigrid method includes various choices of parameters. Because each of these strongly affects the accuracy, the robustness, and the performance of multigrid procedures, selection of the optimum combination of these is very critical. In OpenMP/MPI hybrid parallel programming models, the number of threads strongly affects the performance of both the computation and the communications in multigrid procedures [1].

In the present work, we focus on the selection of single-threading or multi-threading in procedures of parallel multigrid solvers using OpenMP/MPI hybrid parallel programming models. We propose a new method of automatic tuning (AT) of the parameters. The proposed method implemented in the code in [2] is evaluated by using up to 8,192 cores of the T2K/Tokyo, the Cray XE6 [7], and the Fujitsu FX10 [3]. The rest of this paper is organized as follows. In Section 2, an overview of the target hardware is provided. In Section 3, we outline the target application and give a summary of the results in [1] and [2]. In Section 4, details of our new method for automatic tuning and the results of the computations are described, while some final remarks are offered in Sections 5.

2   Hardware Environment

Table 1 and Fig. 1 summarize features of the architectures of the three target systems used in the present work. The T2K/Tokyo was developed by Hitachi under the "T2K Open Supercomputer Alliance" [4]. It is a combined cluster system with 952 nodes, 15,232 cores and 31 TB memory. The total peak performance is 140 TFLOPS. Each node includes four sockets of AMD quad-core Opteron (Barcelona) processors (2.3 GHz), as shown in Fig. 1(a). Each socket is connected through HyperTransport links, and the computing nodes are connected via a Myrinet-10G network, which has a multi-stage cross-bar network topology. In the present work, 512 nodes of the system are evaluated. Because T2K/Tokyo is based on *cache-coherent* NUMA (cc-NUMA) architecture, careful design of both the software and the data configuration is required for efficient access to local memory.

Table 1. Summary of specifications: Computing node of the target systems

	T2K/Tokyo	Cray XE6	Fujitsu FX10
Core #/Node	16	24	16
Size of Memory/node (GB)	32	32	32
Peak Performance/node (GFLOPS)	147.2	201.6	236.5
Peak Memory Bandwidth/node (GB/sec)	42.7 8×DDR2 667MHz	85.3 8×DDR3 1333MHz	
STREAM/Triad Performance/node (GB/sec) [8]	20.0	52.3	64.7
B/F Rate	0.136	0.260	0.274

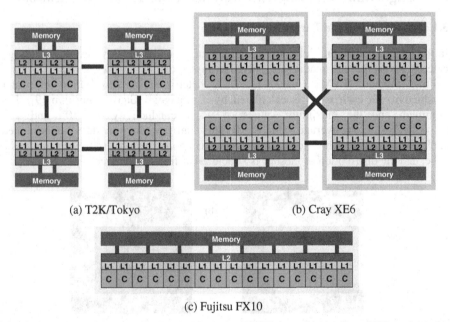

(a) T2K/Tokyo   (b) Cray XE6

(c) Fujitsu FX10

Fig. 1. Overview of a computing node of (a) the T2K/Tokyo, (b) the Cray XE6, and (c) the Fujitsu FX10 (C: core, L1/L2/L3: cache, Memory: main memory)

Each node of the Cray XE6 (Hopper) system at the National Energy Research Scientific Computing Center (NERSC), Lawrence Berkeley National Laboratory [7] includes two sockets of 12-core AMD Opteron (Magny-Cours) processors (2.1 GHz). Each socket of the Magny-Cours consists of two *dies*, each of which consists of six cores. Four dies are connected through HyperTransport links (Fig. 1(b)). The Magny-Cours has more HyperTransport links than the Barcelona processor, and the four dies are connected more tightly. An entire system consists of 6,384 nodes, 153,216 cores, and 212 TB memory. The total peak performance is 1.28 PFLOPS. The computing nodes are connected via Cray's Gemini network, which has a 3D torus network topology. In the present work, 128 nodes of the system are evaluated. Both the T2K/Tokyo and the Cray XE6 are based on cc-NUMA architecture. Each die with six cores of the Cray XE6 corresponds to a socket with four cores of the T2K/Tokyo.

The Fujitsu FX10 (Oakleaf-FX) system at the University of Tokyo [3] is Fujitsu's PRIMEHPC FX10 massively parallel supercomputer with a peak performance of 1.13 PFLOPS. The Fujitsu FX10 consists of 4,800 computing nodes of SPARC64™ IXfx processors with 16 cores (1.848 GHz). SPARC64™ IXfx incorporates many features for HPC, including a hardware barrier for high-speed synchronization of on-chip cores [3]. An entire system consists of 76,800 cores and 154 TB memory. Nodes are connected via a six-dimensional mesh/torus interconnect called "Tofu" [3]. In the present work, 128 nodes of the system are evaluated. On the SPARC64™ IXfx, each of the 16 cores can access the memory in a uniform manner (Fig. 1(c)).

3   Algorithms and Implementations of the Target Application

3.1   Overview of the Target Application

In the target application, Poisson's equations for groundwater flow problems through heterogeneous porous media are solved using a parallel *cell-centered* 3D finite-volume method (FVM) (Fig.2) [1,2]. A heterogeneous distribution of water conductivity in each mesh is calculated by a sequential Gauss algorithm [9]. The minimum and the maximum values of water conductivity are 10^{-5} and 10^{5}, respectively, and the average value is 1.0. This configuration provides ill-conditioned coefficient matrices whose condition number is approximately 10^{10}. Each mesh is a cube, and the distribution of meshes is structured as finite-difference-type voxels.

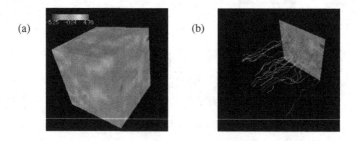

(a)   (b)

Fig. 2. Example of groundwater flow through heterogeneous porous media
(a) Distribution of water conductivity, (b) Streamlines

The conjugate gradient (CG) solver with a multigrid preconditioner (MGCG) is applied for solving the Poisson's equations [1,2]. A very simple geometric multigrid with a V-cycle, where eight children form one parent mesh in an isotropic manner for structured finite-difference-type voxels, is applied [1,2]. The *level* of the finest grid is set to 1 and the level is numbered from the finest to the coarsest grid, where the number of meshes is one in each domain (MPI process). Multigrid operations at each level are done in parallel manner, but the operations at the coarsest levels are executed on a single MPI process by gathering the information of entire processes. The total number of meshes at the coarsest level is equal to the number of MPI processes. IC(0) with additive Schwarz domain decomposition (ASDD) [1,2], because of its robustness, is adopted as the smoothing operator at each level. The 3D code is parallelized by domain decomposition using MPI for communications between partitioned domains [1,2]. In the OpenMP/MPI hybrid parallel programming model, multithreading by OpenMP is applied to each partitioned domain. The reordering of elements in each domain allows the construction of local operations without global dependency to achieve the optimum parallel performance of IC operations in multigrid processes. In the present work, Reverse Cuthill-McKee (RCM) with cyclic-multicoloring (CM-RCM) [10] is applied. The number of colors is set to 2 at each level (CM-RCM(2)) for efficiency [1,2]. The following three types of optimization procedures for cc-NUMA architectures are applied to the OpenMP/MPI hybrid parallel programming models [1,2]:

- Appropriate command lines for NUMA control, with "--cpunodebind" and "--localalloc", where memory locality is kept, and each thread can access data on the memory of each socket efficiently [1]
- First touch data placement [1,2]
- Reordering for contiguous "sequential" access to memory [1]

Furthermore, optimization of the coarse grid solver proposed in [2] is applied.

3.2   Results (Weak Scaling)

The performance of weak scaling is evaluated using between 16 and 8,192 cores of the T2K/Tokyo. The number of finite-volume meshes per each core is 262,144 ($=64^3$); therefore, the maximum total problem size is 2,147,483,648. The following three types of OpenMP/MPI hybrid parallel programming models are applied as follows, and the results are compared with those of flat MPI:

- **Hybrid 4×4 (HB 4×4):** Four OpenMP threads for each of four sockets in Fig. 2(a), four MPI processes in each node
- **Hybrid 8×2 (HB 8×2):** Eight OpenMP threads for two pairs of sockets, two MPI processes in each node
- **Hybrid 16×1 (HB 16×1):** Sixteen OpenMP threads for a single node, one MPI process in each node

In Fig. 3, (a) and (b) show the performance of the MGCG solver. An improved version of the coarse grid solver (C2) proposed in [2] is applied, where a multigrid based on the V-cycle with IC(0) smoothing is applied until convergence ($\varepsilon=10^{-12}$) at the coarsest level [2]. Both figures show the scalable features of the developed method. The number of iterations until convergence and the elapsed time for MGCG solvers at 8,192 cores are as follows:

- **Flat MPI:**   70 iterations, 35.7 sec.
- **HB 4×4:**   71 iterations, 28.4 sec.
- **HB 8×2:**   72 iterations, 32.8 sec.
- **HB 16×1:**   72 iterations, 34.4 sec.

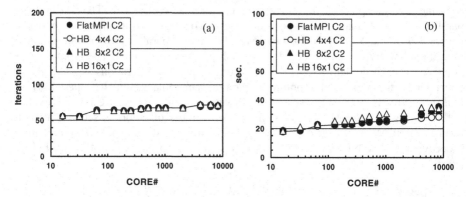

Fig. 3. Performance of MGCG solver with CM-RCM(2) on the T2K/Tokyo using up to 8,192 cores, weak scaling: 262,144 meshes/core, maximum total problem size: 2,147,483,648. (a) Number of iterations for convergence, (b) Computation time for MGCG solvers with improved coarse grid solver (C2) applied.

MGCG is a *memory-bound* process, and the performance of memory access is very critical. The performance of HB 4×4 is the best, primarily because all data for each process are guaranteed to be on the local memory of each socket, and so the most efficient memory access is possible. HB 4×4 is the best according to both the elapsed computation time and the performance in a single iteration [2]. Flat MPI is also better than the others for a small number of cores, but it consists of a larger number of MPI processes than OpenMP/MPI hybrid parallel programming models. Moreover, the problem size for the coarse grid solver is larger than that of these hybrid parallel programming models. Therefore, its performance gets worse for a larger number of cores due to the overhead of communications and coarse grid solvers.

3.3   Results (Strong Scaling) and the Optimization of Communication

The performance of strong scaling is evaluated for a fixed size of problem with 33,554,432 meshes (=512×256×256) using between 16 and 1,024 cores of the T2K/Tokyo [1]. Figure 4(a) provides the parallel performance of the T2K/Tokyo based on the performance of flat MPI with 16 cores using the original coarse grid solver in [1]. At 1,024 cores, the parallel performance is approximately 60% of the performance at 16 cores. Decreasing of the parallel performance of HB 16×1 is very significant. At 1,024 cores, HB 16×1 is slower than flat MPI, although the convergence is much better [1]. Communications between partitioned domains at each level occur in the parallel multigrid procedures. Information at each domain boundary is exchanged by using the functions of MPI for point-to-point communications. In this procedure, copies of arrays to/from sending/receiving buffers are made, as shown in Fig. 5. In the original code using OpenMP/MPI hybrid parallel programming models, this type of operation for the memory copy is parallelized by OpenMP. But the overhead of OpenMP is significant if the length of the loop is short at the coarser levels of the multigrid procedure and the number of threads is large. If the length of the loop is short, operations by a single thread might be faster than those by multi-threading.

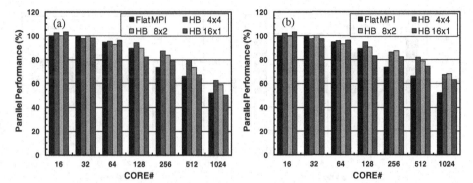

Fig. 4. Performance of MGCG solver with CM-RCM(2) on the T2K/Tokyo using up to 1,024 cores, strong scaling: 33,554,432 meshes (=512×256×256). (a) Parallel performance based on the performance of flat MPI with 16 cores, (a) Initial case, (b) Optimized case: "*LEVcri=2*" in Fig. 7 is applied for OpenMP/MPI hybrid parallel programming models [1].

```
!C
!C-- SEND
        do neib= 1, NEIBPETOT
          istart= levEXPORT_index(lev-1,neib) + 1
          iend  = levEXPORT_index(lev  ,neib)
          inum  = iend - istart + 1
!$omp parallel do private (ii)
          do k=, istart, iend
            WS(k)= X(EXPORT_ITEM(k))
          enddo
!$omp end parallel do
          call MPI_ISEND (WS(istart), inum, MPI_DOUBLE_PRECISION,      &
     &                    NEIBPE(neib), 0, SOLVER_COMM, req1(neib), ierr)
        enddo
```

Fig. 5. Point-to-point communications for information exchange at the domain boundary (sending process), copies of arrays to/from sending/receiving buffers occur

In [1], the effect of switching from multi-threading to single-threading at coarser levels of the multigrid procedure was evaluated. Figure 6(a) shows the results of HB 16×1 with 1,024 cores (64 nodes) for the strong scaling case. The "communication" part includes processes of the memory copies shown in Fig. 5. "*LEVcri=0*" is the original case, and it applies multi-threading by OpenMP at every level of the multigrid procedure. "*LEVcri=k* (*k*>0)" means applying multi-threading if the level of the grid is smaller than *k*. Therefore, single-threading is applied at every level if "*LEVcri=1*", and multi-threading is applied at only the finest grid (level=1) if

"*LEVcri=2*". Generally, "*LEVcri=2*" provides the best performance at 1,024 cores for all of HB 4×4, HB 8×2, and HB 16×1. The optimized HB 16×1 with "*LEVcri=2*" is 22% faster than that of the original case, although the effect of switching is not so clear for HB 4×4. Figure 4(b) shows the effects of this optimization with "*LEVcri=2*" for all OpenMP/MPI hybrid cases. The performance of HB 8×2 and HB 16×1 are much improved at a large number of cores, and HB 8×2 is even faster than HB 4×4 at 1,024 cores, while the performance with a fewer number of cores does not change.

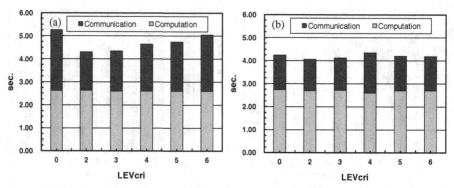

Fig. 6. Effect of switching from multi-threading to single-threading at coarse levels of the multigrid procedure in operations of memory copy for communications at domain boundaries using 1,024 cores for a strong scaling case with 33,554,432 meshes (=512×256×256), "*LEVcri=0*": applying multi-threading by OpenMP at every level of the multigrid procedure (original case), "*LEVcri=k (k>0)*": applying multi-threading if the grid level is smaller than k. (a) HB 16×1, (b) HB 4×4.

4   Automatic Tuning (AT) of Multigrid Processes

4.1   Overview

In multigrid procedures with OpenMP/MPI hybrid parallel programming models, most of the processes are parallelized by OpenMP at each level. But the overhead of OpenMP is significant if the length of the loop is short at coarser levels and the number of threads is large. If the length of the loop is short, operations by a single thread might be faster than those by multi-threading, as shown in 3.3. In 3.3, the optimum parameter ("*LEVcri=2*") is determined by comparing the results of cases with different values of *LEVcri* between 0 and 6. But this optimum parameter depends on various conditions, such as problem size, number of processors, number of threads per each MPI process, architecture of hardware, performance of computing nodes, communication performance of network, etc. Therefore, automatic tuning (AT) is helpful for the selection of the optimum parameters.

4.2   Method for Automatic Tuning (AT)

In the present work, we focus on the selection of single-threading or multi-threading in procedures of parallel multigrid solvers using OpenMP/MPI hybrid parallel programming models. The method for AT of related parameters was proposed, and was implemented to the code in [2], which was optimized for cc-NUMA architectures, such as the T2K/Tokyo and the Cray XE6. Finally, the proposed method is evaluated using the three supercomputer systems. In the present work, the following three types of parallel programming models are evaluated:

- **Hybrid 4×4/6×4 (HB 4×4/6×4):** Four MPI processes on each node. Four OpenMP threads/MPI process for the T2K/Tokyo and the Fujitsu FX10, six threads/MPI process for the Cray XE6
- **Hybrid 8×2/12×2 (HB 8×2/12×2):** Two MPI processes on each node. Eight OpenMP threads/MPI process for the T2K/Tokyo and the Fujitsu FX10, 12 threads/MPI process for the Cray XE6
- **Hybrid 16×1/24×1 (HB 16×1/24×1):** One MPI process in each node. Number of threads corresponds to the number of cores on each node (16: T2K/Tokyo, Fujitsu FX10, 24: Cray XE6)

We focus on the automatic selection of single-threading or multi-threading in the following three procedures of parallel multigrid solvers using the OpenMP/MPI hybrid parallel programming models:

(A)   Smoothing operations at each level of the V-cycle
(B)   Point-to-point communications at domain boundaries with the memory copies described in 3.3
(C)   Smoothing operations at each level of the coarse grid solver

Process (A) corresponds to smoothing operations for each MPI process at each level of the V-cycle, whereas process (C) corresponds to smoothing operations at each level of the coarse grid solver on a single MPI process [2]. The level of the multigrid for switching from multi-threading to single-threading is defined as $LEVcri_A$ for process (A), $LEVcri_B$ for process (B), and $LEVcri_C$ for process (C). The definition of $LEVcri_X$ is the same as that of $LEVcri$ in 3.3 [1]. If "$LEVcri_X=0$", multi-threading is applied at every level of the process (X). "$LEVcri_X=k$ ($k>0$)" means multi-threading is applied to the process (X) if the level of the grid is smaller than k. The *policy* for optimization is defined as a combination of these three parameters ($LEVcri_A$, $LEVcri_B$, and $LEVcri_C$). In the present work, this policy is represented by a three-digit number, where each digit corresponds to each $LEVcri_X$. For example, the policy represented by "542" means "$LEVcri_A=5$", "$LEVcri_B=4$", and "$LEVcri_C=2$". In the present work, we develop a method for AT that can automatically define the optimum policy (*i.e.*, the combination of optimum $LEVcri_X$'s) for various kinds of hardware and software conditions. $LEVcri_A$ and $LEVcri_C$ are defined by a critical loop length $LOOPcri$, which is a parameter for selection of single-threading or multi-threading, and is calculated by the simple *off-line* benchmark shown in Fig. 7. This benchmark simulates typical and costly processes in smoothing operations, such as sparse-matrix-vector products and forward/backward substitutions for IC(0) operations [1,2]. Six off-diagonal components are used because the target application is based on cell-centered structured 3D meshes with six surfaces. This off-line benchmark compares the performance of loops with single-threading and multi-threading for various loop lengths, N, and automatically introduces the critical loop length *LOOPcri*. *LOOPcri* is a function of the number of threads, the computational performance of each core and the memory bandwidth. Table 2 shows the *LOOPcri*

```
!$omp parallel do private (i,k)
   do i= 1, N
      Y(i)= D(i)*X(i)
      do k= 1, 6
         Y(i)= Y(i) + A(k,i)*X(i)
      enddo
   enddo
!$omp end parallel do

   do i= 1, N
      Y(i)= D(i)*X(i)
      do k= 1, 6
         Y(i)= Y(i) + A(k,i)*X(i)
      enddo
   enddo
```

Fig. 7. Off-line benchmark, which defines critical loop length *LOOPcri* for the selection of single-threading/multi-threading. If loop length N is larger than *LOOPcri*, multi-threading is applied.

calculated by a single node of each supercomputer system for each parallel programming model. *LOOPcri* of the FX10 is smaller because of its hardware barrier for high-speed synchronization of on-chip cores. Generally, this off-line benchmark needs to be performed just once, as long as computational environment, such as version of the compiler, does not change significantly. We just provide *LOOPcri* as one of the input parameters of the application, in which the proposed method for AT is implemented. If the loop length is larger than *LOOPcri*, multi-threading is applied. Optimization of process (B) (*i.e.*, selection of $LEVcri_B$) is done by a run-time tuning procedure. This run-time tuning procedure is embedded as one of the subroutines of the target application written in FORTRAN. This subroutine (**comm_test**) is called by the main program of the application before starting of the real computations. This subroutine (**comm_test**) compares the performance of single-threaded and multi-threaded versions of communication functions at each level, and chooses the faster one for the real computations, as shown in Fig. 8. This procedure is very convenient and reliable because it can evaluate the combined performance for both MPI communication and memory copying by the real functions used in MGCG solvers of the target application. Moreover, this run-time tuning procedure is not costly: it takes less than 0.05 sec. for all cases in the present work. Finally, Figure 9 summarizes the procedure of proposed method for AT.

Table 2. LOOPcri measured by the simple off-line benchmark in Fig. 7

	T2K/Tokyo	Cray XE6	Fujitsu FX10
HB 4×4/ 6×4	256	64	32
HB 8×2/12×2	512	128	32
HB 16×1/24×1	1,024	256	32

1. At each *level* of V-cycle, execution time $T(m, level)$ for point-to-point communications at domain boundaries (including memory copies and MPI communications) with multi-threading using OpenMP is measured.
2. At each *level* of V-cycle, execution time $T(s, level)$ for point-to-point communications at domain boundaries (including memory copies and MPI communications) with single-threading (without OpenMP) is measured.
3. Each of 1. and 2. is repeated for 50 times at each level, and average execution time, $T_{ave}(m, level)$ and $T_{ave}(s, level)$ are calculated.
4. If $T_{ave}(m, level) > T_{ave}(s, level)$, single-threading is adopted at this level of V-cycle. Otherwise, multi-threading is adopted.
5. Optimum $LEVcri_B$ is determined through these procedures.

Fig. 8. Procedure of **subroutine comm_test** for run-time tuning for optimization of process (B) (point-to-point communications at domain boundaries with the memory copies)

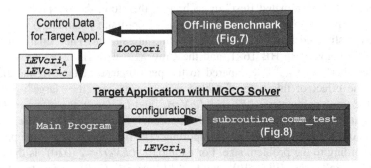

Fig. 9. Procedures for the proposed method of AT

4.3 Preliminary Results

The method for AT proposed in 4.2 is implemented with the code in [2]. CM-RCM(2) reordering [1,2] is applied at each level; therefore, the loop length at each level is half of the problem size. The following three types of problem sizes are evaluated:

- Large: 2,097,152   (=128×128×128) meshes per each node
- Medium: 524,288   (=128×64×64) meshes per each node
- Small: 65,536   (=64×32×32) meshes per each node

Tables 3, 4, and 5 show the results for the three types of problem sizes on the T2K/Tokyo (512 nodes, 8,192 cores), the Cray XE6 (128 nodes, 3,072 cores), and the Fujitsu FX10 (128 nodes, 2,048 cores), respectively. For each case, two types of codes are developed and applied.

The first code is based on the method for AT described in 4.2, and Fig.9. This code is implemented so that critical loop length $LOOPcri$ in Table 2 provides the optimum $LEVcri_A$ and $LEVcri_C$ automatically, while the run-time tuning procedure by a subroutine (comm_test) of the first code described in Fig. 8 provides the optimum $LEVcri_B$ automatically. The rows with "*AT*" in Tables 3, 4, and 5 show the results of this first code and provide the speed-up compared to that of the original case (policy="000"). The three-digit numbers in parentheses are the policy for optimization provided by AT.

In the second code, each of $LEVcri_A$, $LEVcri_B$, and $LEVcri_C$ can be explicitly specified, where two of these parameters are fixed as "0" in the present work. According to the results by the second code, the best combination of parameters can be estimated. The rows with "*Estimated Best*" in Tables 3, 4, and 5 show the results of the second code (speed-up and corresponding policy for optimization). Figures 10, 11, and 12 show the effect of the individual parameter. The performance is 1.00 for the original case (policy="000").

The effect of tuning by switching from multi-threading to single-threading is significant on the T2K/Tokyo and the Cray XE6 if the problem size per node is small and the number of threads per node is large. Generally, "AT" provides better performance than "Estimated Best" in each case; therefore, the AT procedure works well. The optimum policies (*i.e.*, combinations of optimum parameters) provided by "AT" and "Estimated Best" in Tables 3, 4, and 5 are similar. The "small" size cases for the T2K/Tokyo with HB 16×1, and the Cray XE6 with HB 24×1 provide 1.75–1.96 times speed-up by AT, compared to the performance of the original cases. In contrast, the effect of this type of tuning is very small for the "large" size cases. Among the three parameters ($LEVcri_A$, $LEVcri_B$, and $LEVcri_C$), the effect of $LEVcri_B$ is the most critical in the "small" and "medium" size cases. Therefore, accurate estimation of the optimum $LEVcri_B$ is important. The optimum value of the parameter varies according to the problem size. For example, "$LEVcri_B=2$ (020)" is the best for "medium" size cases (Fig. 10), while "$LEVcri_B=1$ (010)" provides the best performance for "small" size cases (Fig. 11). The effect of tuning by switching from multi-threading to single-threading is very small on the Fujitsu FX10 (Table 5 and Fig. 12). This is because of its hardware barrier for high-speed synchronization of on-chip cores. Multi-threading provides higher efficiency even for short loops on the Fujitsu FX10, as shown in Table 2. Finally, Table 6 compares the performance of the three types of OpenMP/MPI hybrid parallel programming models on three supercomputers for "small" size cases, where the effect of the proposed AT procedure is the most significant. Generally, HB 4×4/6×4 provides the best performance for original, except Fujitsu FX10. But HB 8×2 for T2K/Tokyo and Fujitsu FX10 optimized by the proposed AT procedure is even faster than optimized HB 4×4.

Table 3. Results on the T2K/Tokyo with 512 nodes (8,192 cores): Speed-up compared to the original case (policy="000") and policy for optimization (three-digit number in parentheses)

		Large	Medium	Small
HB 4×4	AT	1.02 (522)	1.04 (422)	1.08 (322)
	Estimated Best	1.01 (522)	1.03 (613)	1.14 (612)
HB 8×2	AT	1.01 (522)	1.10 (412)	1.40 (311)
	Estimated Best	1.02 (532)	1.08 (412)	1.33 (512)
HB 16×1	AT	1.03 (421)	1.22 (421)	1.96 (311)
	Estimated Best	1.02 (633)	1.10 (521)	1.89 (511)

Table 4. Results on the Cray XE6 with 128 nodes (3,072 cores): Speed-up compared to the original case (policy="000") and policy for optimization (three-digit number in parentheses)

		Large	Medium	Small
HB 6×4	AT	1.02 (522)	1.09 (422)	1.49 (322)
	Estimated Best	1.02 (642)	1.07 (432)	1.40 (612)
HB 12×2	AT	1.01 (632)	1.11 (521)	1.57 (311)
	Estimated Best	1.01 (641)	1.08 (621)	1.68 (421)
HB 24×1	AT	1.05 (531)	1.18 (531)	1.75 (321)
	Estimated Best	1.05 (641)	1.18 (541)	1.63 (421)

Table 5. Results on the Fujitsu FX10 with 128 nodes (2,048 cores): Speed-up compared to the original case (policy="000") and policy for optimization (three-digit number in parentheses)

		Large	Medium	Small
HB 4×4	AT	1.01 (532)	1.01 (532)	1.06 (422)
	Estimated Best	1.00 (053)	1.01 (023)	1.06 (622)
HB 8×2	AT	1.00 (532)	1.02 (532)	1.06 (422)
	Estimated Best	1.00 (042)	1.01 (542)	1.04 (522)
HB 16×1	AT	1.00 (042)	1.01 (642)	1.06 (422)
	Estimated Best	1.00 (052)	1.01 (030)	1.08 (532)

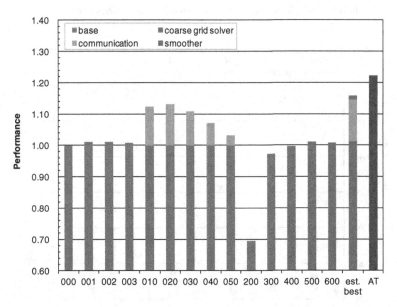

Fig. 10. Effect of individual parameters on speed-up: $LEVcri_A$ (smoother), $LEVcri_B$ (communication), and $LEVcri_C$ (smoother for coarse grid solver), T2K/Tokyo with 512 nodes, 8,192 cores, HB 16×1, "medium" size case (524,288 (=128×64×64) meshes/node)

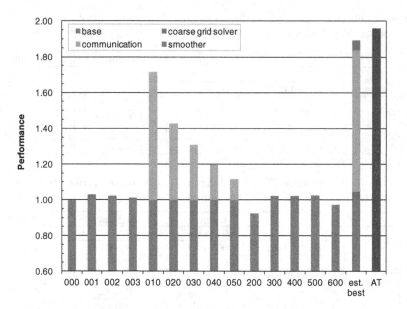

Fig. 11. Effect of individual parameters on speed-up: *LEVcri$_A$* (smoother), *LEVcri$_B$* (communication), and *LEVcri$_C$* (smoother for coarse grid solver), T2K/Tokyo with 512 nodes, 8,192 cores, HB 16×1, "small" size case (65,536 (=64×32×32) meshes/node)

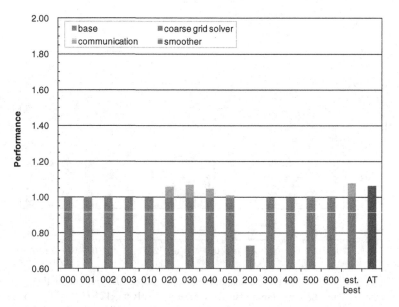

Fig. 12. Effect of individual parameters on speed-up: *LEVcri$_A$* (smoother), *LEVcri$_B$* (communication), and *LEVcri$_C$* (smoother for coarse grid solver), Fujitsu FX10 with 128 nodes, 2,048 cores, HB 16×1, "small" size case (65,536 (=64×32×32) meshes/node)

Table 6. Speed-up compared to that of the original case (policy=000) of HB 4×4/6×4 for "small" size case (65,536 (=64×32×32) meshes/node)

		T2K/Tokyo 512 nodes	Cray XE6 128 nodes	Fujitsu FX10 128 nodes
HB 4×4/6×4	Original	1.00	1.00	1.00
	AT	1.08	**1.49**	1.06
	Estimated Best	1.14	1.41	1.06
HB 8×2/12×2	Original	.963	.879	1.03
	AT	**1.35**	1.38	**1.08**
	Estimated Best	1.29	1.47	1.06
HB 16×1/24×1	Original	.572	.652	.866
	AT	1.12	1.14	.920
	Estimated Best	1.08	1.06	.932

5   Concluding Remarks

In the present work, we focus on automatic selection of single-threading or multi-threading in procedures of parallel multigrid solvers using hybrid parallel programming models. We propose a simple empirical method for AT of related parameters. The proposed method is based on the run-time tuning procedure for the optimization of communications as a subroutine of the target application and the parameter of the critical loop length for multi-threading derived from a simple off-line benchmark. The effect of the proposed method was evaluated on the T2K/Tokyo, the Cray XE6, and the Fujitsu FX10 using up to 8,192 cores. The proposed AT method is very effective, and the automatically tuned code provides twice the performance as the original code on the T2K/Tokyo and the Cray XE6 when the problem size per node is relatively small. The proposed method is very useful in strong scaling computations in these architectures. The effect is not so significant on the Fujitsu FX10, because multi-threading provides a higher efficiency even for short loops on the Fujitsu FX10 due to its hardware barrier. Because the original code is optimized for cc-NUMA architectures, such as the T2K/Tokyo and the Cray XE6, a different strategy for further optimization may be needed for the Fujitsu FX10. Generally speaking, NUMA-aware optimizations do not improve performance of the code on such architectures like Fujitsu FX10, where each core of the computing node can access the memory in a uniform manner.

In the present work, we focus on the choice of single-threading or multi-threading. A more sophisticated method that defines the optimum number of threads at each level may further contribute to optimization. For example, current choice is only 16-threads or a single thread for HB 16×1 of T2K/Tokyo and Fujitsu FX10, but 2-, 4- or 8-thread procedures at certain levels of the multigrid may provide further improvement of the performance. This type of more flexible approach is an interesting topic for future works. Because the topic of the present work covers only a small aspect of parallel multigrid methods, other directions of optimization, such as reducing communications, should also be considered in future works.

Acknowledgements. This work is supported by Core Research for Evolutional Science and Technology (CREST), Japan Science and Technology Agency (JST), Japan.

References

1. Nakajima, K.: Parallel Multigrid Solvers Using OpenMP/MPI Hybrid Programming Models on Multi-Core/Multi-Socket Clusters. In: Palma, J.M.L.M., Daydé, M., Marques, O., Lopes, J.C. (eds.) VECPAR 2010. LNCS, vol. 6449, pp. 185–199. Springer, Heidelberg (2011)
2. Nakajima, K.: New strategy for coarse grid solvers in parallel multigrid methods using OpenMP/MPI hybrid programming models. ACM Proceedings of the 2012 International Workshop on Programming Models and Applications for Multicores and Manycores, ACM Digital Library (2012), doi:10.1145/2141702.2141713
3. Information Technology Center, The University of Tokyo, http://www.cc.u-tokyo.ac.jp/
4. The T2K Open Supercomputer Alliance, http://www.open-supercomputer.org/
5. Baker, A., Gamblin, T., Schultz, M., Yang, U.: Challenge of Scaling Algebraic Multigrid across Modern Multicore Architectures. In: Proceedings of the 2011 IEEE International Parallel & Distributed Processing Symposium (IPDPS 2011), pp. 275–286 (2011)
6. Hypre Library, http://acts.nersc.gov/hypre/
7. NERSC, Lawrence Berkeley National Laboratory, http://www.nersc.gov/
8. STREAM (Sustainable Memory Bandwidth in High Performance Computers), http://www.cs.virginia.edu/stream/
9. Deutsch, C.V., Journel, A.G.: GSLIB Geostatistical Software Library and User's Guide, 2nd edn. Oxford University Press (1998)
10. Washio, T., Maruyama, K., Osoda, T., Shimizu, F., Doi, S.: Efficient implementations of block sparse matrix operations on shared memory vector machines. In: Proceedings of the 4th International Conference on Supercomputing in Nuclear Applications (SNA 2000) (2000)

A Predictive Performance Model for Stencil Codes on Multicore CPUs

Andreas Schäfer and Dietmar Fey

Friedrich-Alexander-Universität Erlangen-Nürnberg, Germany

Abstract. In this paper we present an analytical performance model which yields estimates for the performance of stencil based simulations. Unlike previous models, we do neither rely on prototype implementations, nor do we examine the computational intensity only. Our model allows for memory optimizations such as cache blocking and non-temporal stores. Multi-threading, loop-unrolling, and vectorization are covered, too. The model is built from a sequence of 1D loops. For each loop we map the different parts of the instruction stream to the corresponding CPU pipelines and estimate their throughput. The load/store streams may be affected not only by their destination (the cache level or NUMA domain they target), but also by concurrent access of other threads. Evaluation of a Jacobi solver and the Himeno benchmark shows that the model is accurate enough to capture real live kernels.

1 Introduction

Some of the largest supercomputing applications are stencil codes [1,2,10,11]. These codes are usually bandwidth limited, which means that their computational intensity according to the roofline model [15] is below the memory bandwidth of the CPU. This situation is aggravated by the increasing core numbers per chip, which outpace the growth of the memory bandwidth. Better cache utilization can mitigate the bandwidth starvation [8]. However, not all codes will benefit equally from multi-threading, cache optimizations or vectorization. Given the fact that successful implementations may require significant amounts of development time, it would be desirable to have a performance model which yields lightspeed estimates – or upper bounds – so that the most profitable optimizations can be determined.

Another use case of our performance model is to explore which changes to the *simulation model* might reduce the execution time. Often the model developer has several degrees of freedom when formulating the model, e.g. in the case of multi-phase Lattice Boltzmann methods sometimes two grid sweeps may be fused into one – at the cost of increasing the stencil radius.

Current multicore processors offer multiple levels of parallelism. The most obvious one is the threading level, which can be further divided into simultaneous multi threading (SMT), multicores, and NUMA and multi-socket systems. But even from a single instruction stream a CPU will try to distill instruction level parallelism (ILP) via dependency analysis and out-of-order execution in order to

M. Daydé, O. Marques, and K. Nakajima (Eds.): VECPAR 2012, LNCS 7851, pp. 451–466, 2013.

exploit pipelining and superscalar execution units. Within the execution units the programmer may additionally use vectorization.

Most current CPU designs are bandwidth-starved, which means that they feature a high ratio of FLOPS to memory bandwidth. A common technique to mitigate this shortcoming is cache blocking. The goal is to perform multiple time step updates within one sweep through the matrix. Cache blocking can be carried out in two different ways: spatial and temporal blocking.

Spatial blocking [2] is also referred to as tiling. The matrix is split into smaller blocks (or tiles) which fit into the caches. Boundary elements of the blocks cannot be updated since their neighbors are not available. For instance if a block of b^3 cells is to be updated x steps, then only $(b - 2 \cdot x)^3$ cells can be written back. The lost elements have to be compensated by overlapping the blocks. Choosing an optimum block size and number of in-cache update steps is not trivial: the more steps, the more data is read from cache, but also the more cells are lost. The larger the blocks, the lesser this influence is, but also the larger the cache needs to be.

Temporal blocking differs from this by continuously streaming data in the form of a wavefront from memory and performing multiple updates on consecutive slices from different time steps [14]. The advantage of this is that fewer boundary cells are lost. E.g. if the wavefront traveling through the 3D volume is a tile of $b \times b$ cells in size and x updates are performed per sweep, then $(b - 2 \cdot x)^2$ cells are written back to memory. For this roughly $2 \cdot x + 1$ tiles have to be kept in memory.

2   Loop Model

A performance model with perfect accuracy would have to reimplement not just the exact behavior of the processor, but also the whole simulation code. The roofline model considers only the arithmetic intensity and the machine balance, which represents a huge abstraction. It does not take into account the limitations of superscalar designs, e.g. when a code has to execute excessive shuffle operations to fill all vector floating point units (FPUs). We attempt to capture more effects in our model while still keeping its complexity manageable.

Let I be the instruction set and F be the set of function units in the CPU. A linear, jump-free sequence s of instructions can be described as a tuple $s = (i_1, i_2, \ldots i_{n_s})$. The scheduling function $\text{sched}(s, j)$ defines to which function unit $f \in F$ an instruction i_j within s will be mapped. In an instruction sequence we identify sub-sequences s_k which are defined by the function unit $f \in F$ to which the instructions i_j get mapped. Thanks to pipelining, most CPUs can retire one instruction per clock cycle and functional unit – at least unless the design is prone to front-end starvation. Thus we assume the execution time $t(s_k)$ (measured in clock cycles) to be simply the length of the sub-sequence:

$$\text{sched}: \quad I^n \times \mathbb{N} \to F$$
$$s_k = (i_j | f(s, j) = k)$$
$$t: \quad I^n \to \mathbb{R}$$
$$t(s_k) = |s_k|$$

For our model we consider nested loops of code sequences. Thanks to ILP a CPU may often execute multiple $i_j \in I$ in parallel. We disregard the dependency graph of the individual instructions and assume that out of order execution sufficiently overlap multiple loop iterations, so that no pipeline stalls caused by delayed computations occur. This simplification may lead to overly optimistic performance estimates, but the reorder buffer of today's CPUs may store 30 micro-ops or more, which is enough to encompass the inner loops of many kernels.

As said, most stencil codes which we have encountered are memory-bound. Cache blocking can sometimes mitigate this so that codes are add-limited, meaning that their throughput primarily depends on how many floating point additions the CPU can perform per cycle. In the case of memory-bound, or more exactly bandwidth-limited loops, we base our model on the model of Treibig et al. [12], but will now extend this model to encompass an arbitrary number of data streams:

Data may have to traverse multiple levels, depending on the caches organization. For instance on Intel CPUs the caches are inclusive, meaning that data loaded from memory will be present in all cache levels. This also means that bandwidth from all affected caches is consumed to make the data travel from the memory controller to the registers. Let A_r be the set of all arrays being read by s and likewise A_w be the set of all arrays being written. $l(a)$ yields the level in the cache hierarchy in which a resides and $b(a)$ is the number of bytes being accessed from a by executing s. We can now calculate the total read traffic $b_r(k)$ and write traffic $b_w(k)$ on each level k of the cache/memory hierarchy (0: L1 cache, 1: L2 cache, 2:L3 cache, 3: RAM).

$$b_r(k) = \sum_{a \in A_r, l(a) \leq k} b(a) + \begin{cases} 0 & \text{if } k = 0 \\ \sum_{a \in A_w, l(a) \leq k} b(a) & \text{if } k > 0 \end{cases}$$
$$b_w(k) = \sum_{a \in A_w, l(a) \leq k} b(a)$$

The definition of $b_r(k)$ includes the write streams A_w because of the additional read required for write-allocate. Streaming stores could also be modelled, but are not considered for the sake of simplicity. The number of cycles $t_{\text{mem}}(k)$ spent on memory traffic per level k can be derived from the level's bandwidth $B(k)$. The total time for memory traffic required by s is $t_{\text{mem}}(s)$.

$$t_{\mathrm{mem}}(k) = \begin{cases} \frac{b_r(k)+b_w(k)}{B(k)} & \text{if level } k \text{ is single-ported} \\ \frac{\max(b_r(k),b_w(k))}{B(k)} & \text{if level } k \text{ is dual-ported} \end{cases}$$

$$t_{\mathrm{mem}}(s) = \sum_k t_{\mathrm{mem}}(k)$$

If the code is multi-threaded, then $b_r(k)$ and $b_w(k)$ have to add up the traffic from all threads which access a common resource. On current multicores this typically means that the bandwidth of the L3 cache and memory controller will be shared. We define $b_r^i(k)$ and $b_w^i(k)$ (i.e. the memory traffic at level k, which is shared by i threads) as the single threaded traffic, scaled by the number of threads:

$$b_r'(k) = b_r(k) \cdot i$$
$$b_w'(k) = b_w(k) \cdot i$$

This is a simplification, but as our measurements below show, it is still valid for stencil codes. Most likely this is because they exhibit a very regular memory access pattern.

All sub-sequences s_k of s are assumed to be executed in parallel. The slowest sub-sequence determines the effective runtime $t(s)$ for arithmetically limited sequences. Otherwise it's the time for memory traffic. We model the runtime $t(t)$ of a loop $l = (s, m)$ which executes s exactly m times by estimating a certain overhead $t_{\mathrm{overhead}}(l)$ (e.g. for initializing counters or writing back results at the end) and adding to this the execution times for the code sequence. This overhead can be determined by analyzing the assembly code of an existing implementation of the stencil code under test, or by instrumenting and benchmarking the code. Both ways are tedious, but as we will see in the experimental evaluation below, a bad estimate for the overhead will only spoil the estimates for small arrays. For production codes we are generally interested in the performance for large arrays.

$$t(s) = \max\left(\max_{f \in F}\{t(s_f)\}, t_{\mathrm{mem}}(s)\right)$$
$$t(l) = t_{\mathrm{overhead}}(l) + m \cdot t(s)$$

3   1D Diffusion Stencil

Consider the code fragment in Fig. 1 which is a naïve implementation of a 1D diffusion stencil code. We will use this kernel to illustrate how code and model correspond. For brevity a detailed analysis is given only for the more interesting 2D code below.

If the compiler doesn't optimize away the two redundant loads then this code exhibits four data streams. Because of the preceding load stream of a[i + 1] from the previous loop iterations, the two other load streams (which read a[i - 1] and a[i + 0]) will be served from L1 cache, no matter how large the arrays are. Only the load of a[i + 1] my require accesses to either higher cache levels or main memory. A premature expectation would be that this code took two clock cycles per iteration, as each iteration requires two adds, which can be overlapped with the single multiplication by one third.

However, our model predicts that the loop will take three cycles on a pre-Sandy Bridge Intel chip, if running in L1 cache: For each loop iteration the CPU will execute three loads, two floating point additions, one multiplication and one store.[1] The longest sub-sequence in this instruction sequence are the loads, which get all mapped to the single load unit found in Intel older CPU cores (Sandy Bridge has two load ports). Each loop iteration performs three FLOPs, so according to our model we can expect the code to run at 2.8 GFLOPS on the Core 2 machine in our testbed. Measurements have confirmed this to be exact. Naturally, such scalar code is not very efficient. Fig. 2 computes the same function, but uses only two streams.

```
for (int i = 1; i < n - 1; ++i)
  b[i + 0] = (a[i - 1] + a[i + 0] + a[i + 1]) * (1.0 / 3.0);
```

Fig. 1. Naïve 1D stencil

```
double buf1 = a[0];
double buf2 = a[1];
for (int i = 1; i < n - 1; ++i) {
  double buf3 = a[i + 1];
  b[i + 0] = (buf1 + buf2 + buf3) * (1.0 / 3.0);
  buf1 = buf2; buf2 = buf3;
}
```

Fig. 2. Element-reusing 1D stencil

```
for (int y = 1; y < DIM - 1; ++y)
  for (int x = 1; x < DIM - 1; ++x)
    b[y][x] =
      (a[y-1][x] + a[y][x-1] + a[y][x] +
       a[y][x+1] + a[y+1][x]) * (1.0 / 5.0);
```

Fig. 3. Vanilla 2D diffusion stencil

---

[1] The additional jump and counter/index handling are required can for now be ignored as the relevant function units are not saturated.

4   Handling Multiple Dimensions

The previous model can capture arithmetic kernels with multiple linear data streams. We will now show how this model can be extended to cover multi-dimensional codes. Also, this section serves to bridge the gap between the analytical formulation of our model and its application to actual implementations.

For brevity we will analyze a 2D code, but 3D codes can be evaluated likewise. For the following analysis we will assume a single socket, quad-core Sandy Bridge system as detailed in Tab. 1. Consider the diffusion stencil shown in Fig. 3. A cell's new state is determined by averaging it with its four neighbors in the cardinal directions. The boundaries are constant. Per cell update four additions and one multiplication have to be performed. The inner loop of the code updates row y of the destination matrix. During this update the three lines $y - 1, y$, and $y + 1$ of the source matrix have to be read. The key to applying the loop model to such a 2D code is to determine on which level of the memory hierarchy each row will reside. As an introductory example we assume that both matrices are 320×320 elements in size. At the beginning all caches are empty and both matrices are stored in main memory. After updating row 1 the first three rows of the source grid and row 1 on the target grid have been loaded to the L1 cache. For the update of row 2 and all remaining rows in this timestep, two rows in the L1 cache can be reused and only one row has to be read from main memory. The newly updated row can be stored in L1, too. No cache evictions take place as the L1 cache is large enough to store both matrices. In the subsequent timesteps all memory accesses are served by the L1 cache.

We can now apply our loop model to this code by counting arithmetic instructions and memory accesses. A vectorized assembly version of the inner loop shown in Fig. 3 can be seen in Fig. 4.[2] This analysis could theoretically be done without the having the source code of a kernel, but just by reasoning which array elements get read and written, and which computations need to be carried out. However, we will use the assembly code since be believe that it makes the analysis much more descriptive.

Per iteration of the loop 8 cells are updated. The code issues 16 loads, 4 stores, 16 floating point additions and 4 floating point multiplications.[3] In a Sandy Bridge CPU all these instructions would be scheduled to different pipelines (loads: ports 2 and 3, stores: port 4, multiplications: port 0, additions: port 1, shuffles: port 5)[7]. Equations 1-6 list the execution times in machine cycles (c), as predicted by our model. Note that the loads get distributed among two pipelines as Sandy Bridge has two load ports[16].

---

[2] The original C++ code, which is too long to be included here, but possibly easier to read, is available at
http://svn.inf-ra.uni-jena.de/trac/libgeodecomp/
browser/src/testbed/testbed.cpp

[3] We disregard register copies and integer instructions as there are enough functional units to execute them and they do therefore play a negligible role in our analysis.

$$t(s_{\text{load}\_1}) = 8\,c \tag{1}$$
$$t(s_{\text{load}\_2}) = 8\,c \tag{2}$$
$$t(s_{\text{store}}) = 4\,c \tag{3}$$
$$t(s_{\text{mult}}) = 4\,c \tag{4}$$
$$t(s_{\text{add}}) = 16\,c \tag{5}$$
$$t(s_{\text{shuffle}}) = 4\,c \tag{6}$$

The slowest sub-sequence in this code is obviously formed by the 16 adds. Thus:

$$\max_{f \in F}\{t(s_f) = t(s_{\text{add}})$$
$$= 16\,c$$

Assuming that all data is in L1, we can now proceed to calculate t_{mem}. In total $3 \cdot 4 \cdot 2$ elements (3 arrays, 4x unrolled, 2 double precision numbers per vector register) are read, and 8 are written array (one array). L1 bandwidth ($B(0)$) is assumed to be 48 Bytes per clock cycle[7].

$$b_r(0) = 3 \cdot 4 \cdot 2 \cdot 8\text{Byte} = 192\,B$$
$$b_w(0) = 1 \cdot 4 \cdot 2 \cdot 8\text{Byte} = 64\,B$$
$$t_{\text{mem}}(s) = \frac{b_r + b_w}{48\frac{B}{c}}$$
$$= 5.333\,c$$

Now, the only number we are lacking is $t_{\text{overhead}}(l)$. Overhead is incurred before/after the inner loop, mostly by loop peeling: the vectorized code in Fig.4 operates on eight (aligned) vector elements per iteration. If the elements which are to be updated are not properly aligned and/or not a multiple of eight, then some preceding and trailing elements may have to be updated by slower scalar code. Additional overhead is caused by incrementing loop counters and updating array pointers. For our implementation of the 2D diffusion stencil this overhead is roughly 40 machine cycles.

Finally, we can calculate the total time $t(l)$ which it takes to execute the inner loop:

$$t_{\text{overhead}}(l) = 40\,c$$
$$t(s) = \max{(16\,c, 5.333\,c)}$$
$$m = \frac{320}{8} = 40$$
$$t(l) = 40\,c + 40 \cdot 16\,c$$
$$= 680\,c$$

Assuming a clock speed of 3.4 GHz and 5 FLOPs per cell update, we would estimate that the code delivers 8 GFLOPS for our 320×320 example. Our measurements in Fig. 7 have shown a peak performance of 7.8 GFLOPS, which corresponds to an error of 2.5 %.

For larger matrices the L1 caches would not be able to hold all data. E.g. for a matrix of 15024×15024 elements L1 would not even be able to hold one row of the matrix. To update our performance estimate, we only have to update $t_{\mathrm{mem}}(s)$: Per iteration we have to read and write one cache line from/to main memory. The line which gets written has to be read from memory first because to serve the write allocate. Our test machine has a memory bandwidth of 9.39 Byte per cycle.

$$b_r(2) = 64\,B$$
$$b_w(2) = 64\,B$$
$$B(2) = 9.39\frac{B}{c}$$
$$t_{\mathrm{mem}}(2) = \frac{128\,B}{9.39\frac{B}{c}}$$
$$= 13.63\,c$$

The L2 cache has a bandwidth of 32 Bytes per cycle, L1 of 48 Bytes. Two lines from L2 cache can be reused, none of L1, thus:

$$b_r(1) = 192\,B + 64\,B$$
$$b_w(1) = 64\,B$$
$$B(1) = 32\frac{B}{c}$$
$$t_{\mathrm{mem}}(1) = \frac{256\,B}{32\frac{B}{c}}$$
$$= 8\,c$$
$$b_r(0) = 192\,B$$
$$b_w(0) = 64\,B$$
$$B(0) = 48\frac{B}{c}$$
$$t_{\mathrm{mem}}(0) = \frac{192\,B}{48\frac{B}{c}}$$
$$= 5.33\,c$$
$$t_{\mathrm{mem}}(s) = 5.33\,c + 8\,c + 13.63\,c$$
$$= 26.97\,c$$

This leads to an estimate of 5.04 GFLOPS, which is close to our measurements of 5.14 GLOPS.

```
0x0403640:   vmovapd  %xmm7,%xmm9
0x0403644:   vmovapd  %xmm8,%xmm3
0x0403648:   vmovapd  0x20(%rdi),%xmm2
0x040364d:   add      $0x8,%edx
0x0403650:   vmovapd  0x30(%rdi),%xmm1
0x0403655:   vshufpd  $0x1,%xmm2,%xmm3,%xmm6
0x040365a:   vaddpd   %xmm9,%xmm3,%xmm3
0x040365f:   vmovapd  0x40(%rdi),%xmm0
0x0403664:   vshufpd  $0x1,%xmm1,%xmm2,%xmm5
0x0403669:   vshufpd  $0x1,%xmm0,%xmm1,%xmm4
0x040366e:   vmovapd  0x50(%rdi),%xmm8
0x0403673:   vaddpd   %xmm6,%xmm2,%xmm2
0x0403677:   add      $0x40,%rdi
0x040367b:   vshufpd  $0x1,%xmm8,%xmm0,%xmm7
0x0403681:   vaddpd   %xmm5,%xmm1,%xmm1
0x0403685:   vaddpd   %xmm4,%xmm0,%xmm0
0x0403689:   vaddpd   %xmm6,%xmm3,%xmm6
0x040368d:   vaddpd   %xmm5,%xmm2,%xmm5
0x0403691:   vaddpd   %xmm4,%xmm1,%xmm4
0x0403695:   vaddpd   %xmm7,%xmm0,%xmm0
0x0403699:   vaddpd   0x00(%rcx),%xmm6,%xmm6
0x040369d:   vaddpd   0x10(%rcx),%xmm5,%xmm5
0x04036a2:   vaddpd   0x20(%rcx),%xmm4,%xmm4
0x04036a7:   vaddpd   0x30(%rcx),%xmm0,%xmm0
0x04036ac:   add      $0x40,%rcx
0x04036b0:   vaddpd   0x00(%rsi),%xmm6,%xmm6
0x04036b4:   vaddpd   0x10(%rsi),%xmm5,%xmm5
0x04036b9:   vaddpd   0x20(%rsi),%xmm4,%xmm4
0x04036be:   vaddpd   0x30(%rsi),%xmm0,%xmm0
0x04036c3:   add      $0x40,%rsi
0x04036c7:   vmulpd   0x3691(%rip),%xmm6,%xmm6
0x04036cf:   cmp      %edx,%r13d
0x04036d2:   vmulpd   0x3686(%rip),%xmm5,%xmm5
0x04036da:   vmulpd   0x367c(%rip),%xmm4,%xmm4
0x04036e2:   vmulpd   0x3676(%rip),%xmm0,%xmm0
0x04036ea:   vmovapd  %xmm6,0x00(%rbx)
0x04036ee:   vmovapd  %xmm5,0x10(%rbx)
0x04036f3:   vmovapd  %xmm4,0x20(%rbx)
0x04036f8:   vmovapd  %xmm0,0x30(%rbx)
0x04036fd:   jg 0x403640
```

Fig. 4. Assembly code for the inner loop of the 2D diffusion stencil. For clarity we have highlighted instructions and memory references according to the sub-sequences of the instruction stream to which they belong. The code is 4x unrolled and vectorized via SSE. Unaligned loads (for accesses to left and right neighbors) are avoided by combining aligned loads and shuffle operations. rdi points to the current row in the source grid while rcx and rsi point to the rows above and below. The target row is at rbx. The peculiar reference accessed via rip is an operand stored within the code segment. It contains the factor one-fifth.

5   Experimental Testbed

Tab. 1 summarizes the machines we used for validating our model. All machines run Intel chips which use an inclusive cache hierarchy with 64 Byte cache lines. Stores are handled via write back/write allocate – unless streaming stores are used. Write allocate means that every cache line which is to be written is read from memory first. This effectively doubles the bandwidth required for stores. Streaming stores (e.g. MOVNTPD) may be used by the CPU as hints to avoid this additional read. To determine basic machine properties, such as the maximum cache bandwidths, we used the micro benchmarks likwid-bench that come with the Likwid suite [13]. We did not test our model on AMD chips, as their cache hierarchy is much different from Intel chips: Opterons use mostly exclusive cache storage with all caches connected via a common bus. We do not yet have enough details on their caches hierarchy to adapt our model to it.

Table 1. Hardware details of the servers we used for our benchmarks. The bandwidths $B(k)$ are shown in Bytes per clock cycle for the caches while the raw memory bandwidth $B(3)$ is listed in GB/s. The L1 cache is assumed to be dual-ported, we do thus list read and write bandwidths. All other caches are assumed single-ported.

	Core2	Nehalem	Sandy Bridge A	Sandy Bridge B
Model	Core2 Quad Q9550	Xeon X5650	Xeon E31280	Core i7-2600K
Cores	4	6	4	4
Clock	2.83 GHz	2.66 GHz	3.50 GHz	3.40 GHz
L1 Cache	32 KiB	32 KiB	32 KiB	32 KiB
L2 Cache	2x 6 MiB	6x 256 KiB	4x 256 KiB	4x 256 KiB
L3 Cache	-	12 MiB	8 MiB	8 MiB
RAM	DDR2-800	DDR3-1333	DDR3-1666	DDR3-1666
Channels	2	3	2	2
$B(0)$	16/16	16/16	32/16	32/16
$B(1)$	32	32	32	32
$B(2)$	-	32	32	32
$B(3)$	12.8	32.0	26.6	26.6

Table 2. Himeno benchmark on Sandy Bridge. Performance given in MFLOPS.

	Prediction	Benchmark	Error
Small (128 x 64 x 64)	3369.75	3365.96	0.00113
Middle (256 x 128 x 128)	3435.51	3345.45	0.02692
Large (612 x 256 x 256)	3469.05	3363.69	0.03132

6   Measurements and Validation

To verify the soundness of our model, we have tested it against a number of different stencil codes on multiple generations of CPUs. We present only the plots of our measurements versus our predictions, as a a detailed analysis for each stencil, problem size, and CPU, comparable to the one in Sec. 4, would be would be lengthy, mechanical, and thus beyond the scope of this paper.

In Fig. 5 we have evaluated an 4x loop-unrolled vectorized (SSE) version of the 1D diffusion stencil on Core 2, Nehalem, and Sandy Bridge CPUs. Despite good cache hit rates, the performance for small arrays is comparatively low, which is caused by the fixed overhead. As the arrays grow the overhead becomes negligible, but after a certain point the arrays drop from L1, L2 and, if available, L3 cache.

The predictions are most accurate for all architectures if the arrays fit into the L1 cache. For larger arrays Sandy Bridge performance is still predicted with high accuracy while Core 2 and Nehalem measurements deviate from the prediction. The discrepancy is at about 20 %, so improvements are clearly desirable, but still the model remains usable in the sense that performance critical array sizes can be detected and their relative performance can be estimated.

The benchmarks show that our model is slightly optimistic and that actual code will in most cases be slower. A weakness of our model can be seen when the arrays hit cache size boundaries. The actual measurements reveal that the performance starts to drop even if the arrays are still slightly smaller than the caches and can retain some throughput benefits even if the arrays are larger than the caches. We would need to know how cache eviction is performed on Intel's chips to fix this inaccuracy. On the other hand, this would increase the model's complexity significantly, so for now we deem this to be an acceptable tradeoff between simplicity and accuracy.

The 2D code discussed in Sec. 4 has been benchmarked in Fig. 7. This benchmark was run on our second Sandy Bridge system. Similarly to the previous measurement the effects of overhead and cache sizes can be observed and again the prediction is accurate.

Cache blocking (spatial blocking) and multi-threading are evaluated in Fig. 6. The code was parallelized via OpenMP. For small arrays we can observe a poor accuracy of our model. Apparently thread synchronization and loop scheduling cannot be represented by a fixed overhead. This is not a surprise, but for larger arrays this error shrinks, which is why we are confident, that our model may still be applied in situations where the overhead can be determined a priori.

6.1   Himeno Benchmark

So far we have only tested our model against 1D/2D variants of a diffusion kernel. Actual solvers are more complex: they use more data sets and are usually 3D. As a more realistic example we chose the Himeno benchmark[4], a well

---

[4] http://accc.riken.jp/HPC_e/himenobmt_e.html

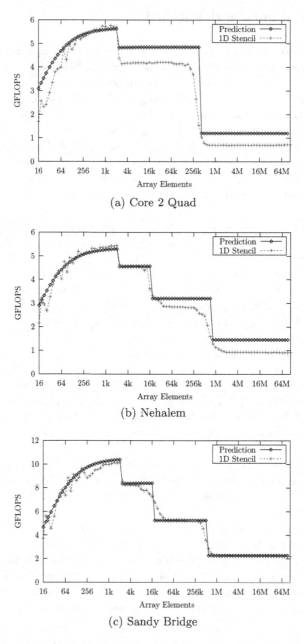

Fig. 5. 1D diffusion stencil results on three Intel architectures. The actual code is a 4x unrolled SSE version of the code seen in Fig. 2. Despite sharing many architectural traits, the Sandy Bridge system comes much closer to the theoretical limits. The convex growth on the left side is caused by the fixed initialization overhead (e.g. for setting up counters and pointers). It becomes negligible as the arrays grow and the loops make up the major part of the running time. Please mind the logarithmic scale on the x-axis.

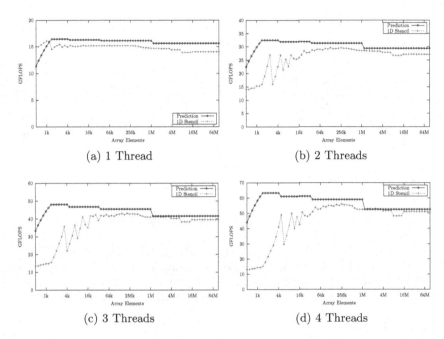

(a) 1 Thread

(b) 2 Threads

(c) 3 Threads

(d) 4 Threads

Fig. 6. Multi-threaded implementation of the 1D diffusion. The code is 4x unrolled and vectorized via AVX. To conserve memory bandwidth we implemented spatial cache blocking, which is also known as tiling. As predicted by the model, scaling is not perfect, as multiple cores compete for memory bandwidth. Yet, the speedup is still about 3.25 for 4 threads. All stores to the destination array are non-temporal stores. These bypass the caches, which reduces cache pollution. To avoid unaligned loads and costly cross-lane shuffles for accesses to left/right neighbors, we use an element-interleaving technique as shown in [3]. All measurements were run on the *Sandy Bridge A* system in our testbed. `likwid-pin` was used for pinning the threads to CPU cores.

established code which is often used to assess the performance of stencil code implementations [5,9]. Table 2 compares the benchmark's performance on the Sandy Bridge A system – measured in MFLOPS – with the performance predicted by the model. The relative error for all grid sizes is well below 4%. For each cell update the code issues 33 loads and 2 stores. 20 floating point additions and 14 multiplications have to be carried out. Because of number of coefficients required for each cell, a listing of all equations required to calculate $t(l)$ would take up several pages. In the end, it turns out that our model claims the code is bandwidth-limited. This matches our experience from analyzing the 2D diffusion code, which was bandwidth-limited as well. Even small instances of the benchmark exceed the CPU's cache capacity. This explains why the performance does hardly increase when the problem size is reduced from *Large* to *Small*.

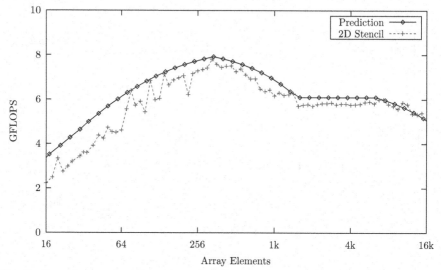

Fig. 7. Comparison of predicted and measured performance of the 2D diffusion code. The initial rise is caused by a slowly decreasing effect of the fixed overhead. The drops in performance are caused by the decreasing cache efficiency as the matrices grow.

7   Conclusion

We have presented a generalized analytical model to assess the performance of stencil code implementations on state of the art multicore CPUs. The model is applicable not just to 1D and 3D codes alike, it also allows for optimizations such as vectorization, loop unrolling, cache blocking and multi-threading.

Measurements have shown our model to be reasonably accurate on multiple generations of Intel CPUs, albeit further work is required to make it applicable to AMD or SPARC64 [6] and to improve the prediction of thread synchronization times. For CPUs whose caches are not inclusive, the sum formulas for $b_r(k)$ and $b_w(k)$ would need to be adapted: in that case the data typically does not have to travel step by step, from level to level, through the memory hierarchy. For instance current AMD chips feature a bus which connects all caches.

The advantage of having an analytical mode is that it may be used to guide application development: implementing a vectorized, multi-threaded code with cache blocking may require thousands of lines of code. Our model can give an estimate of the approximate performance before investing weeks or possibly months of work in such an implementation. It relies on basic performance data of the machine (e.g. memory bandwidth, number of functional units, etc.) and kernel (number and size of arrays accessed), all of which are either freely available or can be deduced from a naïve implementation.

A hard to determine parameter in our model is the overhead. A way to remedy this drawback would be to replace the fixed $t_{\text{overhead}}(l)$ by a function which draws its definition from a series of micro-benchmarks. But, as the multi-threading tests in Sec. 6 have shown, even a badly chosen estimate does not hugely affect predictions for larger and thus relevant problem sizes.

Unlike models based on benchmarking code [4], our model may be applied to processors of which the basic performance characteristics are known, but which are not yet available for testing.

References

1. Allen, G., Dramlitsch, T., Foster, I., Karonis, N., Ripeanu, M., Seidel, E., Toonen, B.: Supporting Efficient Execution in Heterogeneous Distributed Computing Environments with Cactus and Globus. In: SC 2001: Proceedings of the 2001 ACM/IEEE Conference on Supercomputing (2001), http://portal.acm.org/citation.cfm?coll=GUIDE&dl=GUIDE&id=582086
2. Datta, K., Murphy, M., Volkov, V., Williams, S., Carter, J., Oliker, L., Patterson, D., Shalf, J., Yelick, K.: Stencil computation optimization and auto-tuning on state-of-the-art multicore architectures. In: Proceedings of the 2008 ACM/IEEE Conference on Supercomputing, SC 2008, pp. 4:1–4:12. IEEE Press, Piscataway (2008), http://portal.acm.org/citation.cfm?id=1413370.1413375
3. Henretty, T., Stock, K., Pouchet, L.-N., Franchetti, F., Ramanujam, J., Sadayappan, P.: Data layout transformation for stencil computations on short-vector SIMD architectures. In: Knoop, J. (ed.) CC 2011. LNCS, vol. 6601, pp. 225–245. Springer, Heidelberg (2011), http://dl.acm.org/citation.cfm?id=1987237.1987255
4. Kamil, S., Husbands, P., Oliker, L., Shalf, J., Yelick, K.: Impact of modern memory subsystems on cache optimizations for stencil computations. In: Proceedings of the 2005 Workshop on Memory System Performance, MSP 2005, pp. 36–43. ACM, New York (2005), http://doi.acm.org/10.1145/1111583.1111589
5. Maruyama, N., Nomura, T., Sato, K., Matsuoka, S.: Physis: An Implicitly Parallel Programming Model for Stencil Computations on Large-Scale GPU-Accelerated Supercomputers. In: SC 2011: Proceedings of the 2011 ACM/IEEE Conference on Supercomputing, Seattle, WA (2011)
6. Maruyama, T., Yoshida, T., Kan, R., Yamazaki, I., Yamamura, S., Takahashi, N., Hondou, M., Okano, H.: Sparc64 viiifx: A new-generation octocore processor for petascale computing. IEEE Micro 30(2), 30–40 (2010), http://ieeexplore.ieee.org/xpls/abs_all.jsp?arnumber=5446249
7. Murray, M.: Sandy Bridge: Intel's Next-Generation Microarchitecture Revealed (2010), http://www.extremetech.com/computing/83848-sandy-bridge-intels-nextgeneration-microarchitecture-revealed (accessed April 09, 2012)
8. Nguyen, A., Satish, N., Chhugani, J., Kim, C., Dubey, P.: 3.5-d blocking optimization for stencil computations on modern cpus and gpus. In: Proceedings of the 2010 ACM/IEEE International Conference for High Performance Computing, Networking, Storage and Analysis, SC 2010, pp. 1–13. IEEE Computer Society, Washington, DC (2010), http://dx.doi.org/10.1109/SC.2010.2
9. Phillips, E.H., Fatica, M.: Implementing the himeno benchmark with cuda on gpu clusters. In: IPDPS, pp. 1–10. IEEE (2010)
10. Schäfer, A., Fey, D.: LibGeoDecomp: A Grid-Enabled Library for Geometric Decomposition Codes. In: Lastovetsky, A., Kechadi, T., Dongarra, J. (eds.) EuroPVM/MPI 2008. LNCS, vol. 5205, pp. 285–294. Springer, Heidelberg (2008)

11. Shimokawabe, T., Aoki, T., Takaki, T., Yamanaka, A., Nukada, A., Endo, T., Maruyama, N., Matsuoka, S.: Peta-scale Phase-Field Simulation for Dendritic Solidification on the TSUBAME 2.0 Supercomputer. In: SC 2011: Proceedings of the 2011 ACM/IEEE Conference on Supercomputing, Seattle, WA (2011)
12. Treibig, J., Hager, G.: Introducing a performance model for bandwidth-limited loop kernels. In: Wyrzykowski, R., Dongarra, J., Karczewski, K., Wasniewski, J. (eds.) PPAM 2009, Part I. LNCS, vol. 6067, pp. 615–624. Springer, Heidelberg (2010), http://dl.acm.org/citation.cfm?id=1882792.1882865
13. Treibig, J., Hager, G., Wellein, G.: Likwid: A lightweight performance-oriented tool suite for x86 multicore environments. In: Lee, W.C., Yuan, X. (eds.) ICPP Workshops, pp. 207–216. IEEE Computer Society (2010)
14. Wellein, G., Hager, G., Zeiser, T., Wittmann, M., Fehske, H.: Efficient temporal blocking for stencil computations by multicore-aware wavefront parallelization. In: Annual International Computer Software and Applications Conference, vol. 1, pp. 579–586 (2009)
15. Williams, S., Waterman, A., Patterson, D.: Roofline: an insightful visual performance model for multicore architectures. Commun. ACM 52, 65–76 (2009), http://doi.acm.org/10.1145/1498765.1498785
16. Yuffe, M., Knoll, E., Mehalel, M., Shor, J., Kurts, T.: A fully integrated multi-cpu, gpu and memory controller 32nm processor. In: 2011 IEEE International Solid-State Circuits Conference Digest of Technical Papers (ISSCC), pp. 264–266 (February 2011)

Author Index